TURBINE STEAM PATH

MAINTENANCE AND REPAIR

Volume One

TURBINE STEAM PATH

MAINTENANCE AND REPAIR

Volume One

William P. Sanders, P. Eng.

PennWell®

Library of Congress Cataloging-in-Publication Data

Sanders, William P.
Turbine Steam Path Maintenance and Repair
Volume One / William P. Sanders, P.E.
p. cm.
q.cm
Includes index
ISBN 0-87814-787-X

PennWell Corporation
1421 South Sheridan Road
Tulsa, OK 74112

800-752-9764
sales@pennwell.com
www.pennwell-store.com
www.pennwell.com

Cover and book design by Robin Brumley

Library of Congress Cataloging-in-Publication Data

Printed in the United States of America

1 2 3 4 5 05 04 03 02 01

TABLE OF CONTENTS

List of Acronyms ix

Foreword x

Preface xii

Acknowledgements xxv

Chapter 1—Considerations of a Turbine Steam Path Maintenance Strategy

Introduction 1
Considerations Relating to a Maintenance Strategy 2
The Turbine Outage 6
Establishing the Need for Unit Shutdown 7
Outage Scheduling 12
Interval Between Maintenance Outages 13
The Inspection/Maintenance Outage 17
The Available Corrective Options 19
Distinction Between Causes and Mechanisms of Failure . . 25
Component Susceptibility for Deterioration 52
Instantaneous Damage or Failure 57
Factors Contributing to Gradual Deterioration 59
Monitoring Damage and Deterioration 66
Replacement Parts Strategy and Supply 82
References 84

Chapter 2—Steam Path Component Alignment and Stage Spatial Requirements

Introduction 85
Predictable Factors Affecting Design Clearance 93
Rotor Vertical Deflection 95
Differential Expansion 108
Radial Expansion of the Steam Path Parts 145
Diaphragm Deflection at Pressure and Temperature 159
Unit "End-to-End" Lateral Alignment 166
Methods of Field Alignment 167
Unpredictable Factors Affecting Design Clearance 175
Steam Path Area Requirements 188
The Stage Operating Definition 190
Steam Path Component Arrangement (Axial/Radial Direction) . 192
Blade Vane and Cascade Geometry 229
The Effect of Vane Placement Errors 248
References 282

Chapter 3—Steam Path Damage Induced by Water

Introduction 283
Water Condensation in Expanding Steam 285
Radial Distribution of Moisture 289
Moisture Deposition 292
Measuring Moisture Distribution and Content 301
Water Removal from the Steam Path 304
Moisture-induced Damage 322
Moisture-impact Erosion 323
Blade-trailing Edge Erosion 352
Wire-drawing Erosion 370
Water-washing Erosion 379
Water Ingestion into the Steam Path 389
References 395

Chapter 4—Operational Events Giving Rise to Steam Path Damage

Introduction 397
Foreign Object Impact Damage 398
Sources of the Impacting Objects 401
Impact Damage Classification 409
Solid-particle Erosion (Abrasion) 426
Scale Formation 428
The Erosion Mechanisms 431
Material Loss Patterns Due to SPE 439
Protective Measures Against Erosion 459
SPE Influence on Stage Performance 460
Steam Path Component Rubbing 462
Fretting Corrosion 477
References 486

Chapter 5—Steam Path Damage and Deterioration from Material Property Degradation

Introduction 487
Considerations of Material Structure 489
High-temperature Creep 490
Creep Deformation 493
The Creep Mechanism 497
Creep Rate 504
Creep in Steam Path Components 507
High-cycle Fatigue 520
The High-cycle Phenomena 522
Rotating Blade Vibratory Stresses 525
Material Properties 561
Fatigue Stresses and their Representation 569
Crack Growth 576
HCF Failure Surface Appearance 579

Creep Fatigue 581
Temper Embrittlement 582
Low-cycle Thermal Fatigue 583
Thermal Transients 586
Determination of Thermal Stresses 596
Components Operating at High Temperature 600
References 604

Chapter 6
Steam Path Damage and Deterioration from the Deposition of Contaminants

Introduction 607
Source of Steam Path Impurities 610
The Composition of Deposits 619
The Removal of Chemical Deposits from the Steam Path . . 637
Steam Path Cleaning Methods 639
Deposition Patterns 644
Steam Path Efficiency Deterioration 659
Steam Path Corrosion 671
Forms of the Corrosion Process 677
References 704

LIST OF ACRONYMS

AA	arithmetic average
AISI	American Iron and Steel Institute
BHN	Brinell Hardness Number
CLA	centerline average
EDT	enthalpy drop test
EOH	equivalent operating hours
ESV	emergency stop valve
FEA	finite element analysis
HAZ	heat affected zone
HCF	high-cycle fatigue
LCF	low-cycle fatigue
LP	low pressure
NOH	normal operating hours
NPF	nozzle passing frequency
ppb	parts per billion
RMS	root mean square
SCC	stress corrosion cracking
SPE	solid-particle erosion
T-G	turbine generator
TTD	terminal temperature difference
UTS	ultimate tensile stress

FOREWORD

After more than 100 years of progressive invention and improvement, the steam turbine continues to reign as the prime mover of the world's power generation industry. Innovations such as Edison's light bulb and Tesla's induction motor resulted in a phenomenal demand for electric power driven by steam turbines, which have grown exponentially to more than 750,00 megawatts (1 billion HP) of capacity in the U.S. and almost 3,000 gigawatts worldwide. Steam turbines now generate more than 80% of the total electric power throughout the world.

Since the inception of modern day steam turbines by such designers as Parsons, Curtis, Rateau, and DeLaval in the late nineteenth century, Rankine cycle thermal efficiencies have improved from 5% to almost 42%. State-of-the-art combined cycle efficiencies employing steam and gas turbines now exceed 60%. Similarly, the power output of steam turbines has steadily increased from just a few kilowatts to more than 1,500 megawatts in a tandem compound shaft configuration.

Mr. William Sanders has followed in the tradition of such illustrious technical authors as Stodola (1905, 1927) and Salisbury (1950). Whereas these previous authors concentrated on the mechanical and thermodynamic aspects of steam turbine design, Mr. Sanders has focused his text on the equally important aspect of proper maintenance and repair, and their effect on turbine thermal efficiency and reliability. Mr. Sanders' text also includes areas not previously addressed, such as steam path solid-particle erosion.

Mr. Sanders' text is extremely well organized and contains many excellent illustrations, photos references, and case histories gathered from his now 40+ years of field experience and design background.

Mr. Sanders is to be congratulated for his dedication to the mechanical engineering profession by contributing this well written and thorough work for the benefit of the entire power generation industry. This work will provide a wealth of knowledge and practical experience for proper operation and maintenance of existing and future steam turbines well into the twenty-first century to future design engineers, maintenance managers, and plant operators.

Thomas H. McCloskey
Formerly Manager, Turbomachinery
Electric Power Research Institute
Palo Alto California USA
November 1999

PREFACE

The Turbine Steam Path, Damage, Deterioration, and Corrective Options

This book has been prepared for those technical people responsible for the operation and maintenance of steam turbines.

Steam turbines represent a complex technology for units commonly designed to operate hundreds of thousands of hours while being subjected to a severe environment and a variety of operating phenomena capable of degrading their condition. These units are required to continually operate in a reliable, safe, and cost effective manner. Under such circumstances, these units cannot maintain their original design-specified level of performance indefinitely. All units will deteriorate with age. Owners anticipate this, and designers will normally leave an adequate margin, knowing that some level of such deterioration is tolerable.

The technology of steam turbines—while mature—continues to evolve. More accurate and time-responsive diagnostic tools and techniques are becoming available to assist in predicting when a unit has deteriorated to the extent that corrective action is required. Similarly, tools are available to assist the operator in analyzing problems and determining the effective corrective action best suited to the condition causing deterioration. The improved understanding of unit condition and rates of deterioration now achieved, together with advances in materials, should allow units to be maintained in a manner that will help minimize maintenance concerns and costs.

It is the premise of this book that units "as supplied" will fulfill two basic requirements:

- It is assumed the unit "as designed" represents an optimum selection of component sizing and arrangement
- It is assumed the unit "as delivered" meets design specification within the range of tolerances provided by the design engineer, i.e., unit components have been manufactured, assembled, tested, and installed in such a way that they are in compliance with the original design specification

The implication of this second assumption is that if nonconforming situations or conditions arose during the total manufacturing process (and exist within the unit), they have been evaluated by a competent design authority in the engineering organization of the manufacturing company and have been assessed as not having an adverse impact on the potential performance of the unit.

In terms of turbine unit components, "design optimum" is a difficult term to define. The entire design process is one of compromise by the designer who wants a unit to be both efficient and reliable. These requirements often represent competing demands, forcing the designer to select from among various elements, possibly electing to downgrade one aspect of these requirements to meet the demands of the other. This is done consciously and with detailed evaluation to provide a balanced selection.

Units delivered by a manufacturer represent the supply of elements that conform to the design principles established by his or her design function, and conform with the best technology available to that supplier at the time the design specification was prepared. However, the operator must recognize that the labor and material costs involved in building a steam turbine are high, and turbine suppliers must be able to produce units at competitive levels sufficient to allow them to achieve a profit margin enabling them to sustain business as well as finance further development.

Many power systems are currently experiencing significant changes in how they operate. Pressures from deregulation, environmental concerns and legislation, and an aging fleet of power generating equipment is shifting emphasis from the installation of new capacity to the maintenance and care of the old. There is a continuing increase in demand for electric power but new capacity installation is not keeping up with it. Operators of turbine generators are therefore required to meet this demand with their existing fleets—aging units requiring greater care to reduce the possibility of forced outages. The prospect of units experiencing extended outages as damage is found at planned outages.

Historically, as units have aged they have tended to be used less frequently. They are initially placed on spinning reserve and ultimately placed in reserve, mothballed, or retired—their capacity replaced with newer, more efficient units. An advantage of this dwindling reserve is that older units have continued to operate at high load factors and therefore become less susceptible to the rigors of start-up, shut down, and the associated thermal transients. Unfortunately, there have also been fewer opportunities for plant maintenance to proceed with the maintenance outages required to maintain unit operational health.

Maintenance problems associated with keeping aging units available are only going to increase. Operators who are expected to provide power on demand are going to experience even greater future challenges of damage and deterioration. They will be expected to identify not only the damage, but also the causative effects, and then find immediate solutions that will not jeopardize system security.

This book examines the damage deterioration and failure mechanisms occurring with unfortunate consequences—on some units, with monotonous regularity—within the turbine steam path. These various forms of degradation can be the result of a number of factors related to conditions often beyond the control of operating and maintenance personnel. However, even if the steam turbine is operated precisely as intended by design, and suffers no external degrading effects for its entire operating life, the steam environment is one that can cause components to suffer various forms of distress. Under normal circumstances, the design process selects and defines individual components suitable for the design operating life of the unit (normally about 200,000 hours). At a mean load factor of about 75%, this represents a 30-year operating life.

A number of unavoidable influences affect the operating life of the various components comprising the turbine. These include the steam environment itself, the stresses induced in the components by rotation, and stresses induced in various portions of the unit by expansion of the steam through the blade passages. There are also the effects of the high- pressure steam, causing high-pressure drops across some components that must be contained by the casings.

External factors that can affect the reliability of components of the steam path and act to lower the expected operating life include the possible formation of corrosive elements at various locations within the steam cycle, or impurities gaining access from in-leakage at sub-atmospheric pressures. There can be unit trips caused by a number of circumstances, from system trip electrical faults to lightning strikes on power lines. Many of these factors, while possibly occurring in a 30-year operating life, cannot be anticipated in terms of when, where, how many, or how severe their effects might be.

The damage and deterioration that occurs within the steam path can be of several forms. It can result in a gradual material loss—the growth of a crack—or an immediate failure causing a forced outage. Gradual deterioration can (depending upon type and location) be monitored and replacement parts made available, or corrective action taken to rectify the situation before it extends to an unacceptable degree. Immediate failure is most often the consequence of either mechanical rupture or the presence in the steam path of some foreign object, either generated within or having gained access from some external source (including "drop-ins").

In writing this book, I have tried to present information that plant personnel will be able to use to make value judgments on the type and severity of any damage, suggest possible causes, and then consider the most appropriate corrective actions that are available. To aid in the recognition and classifying of operational damage and deterioration, photographs are used to illustrate unacceptable or suspect conditions.

Many of the damaging phenomena considered in these chapters do not occur in isolation. It is possible that several can and will occur simultaneously, demonstrating that components are subjected to more than one degrading influence. A condition may initiate due to one damaging mechanism introducing a condition of weakness, which then allows another mechanism to become predominant and drive a component to failure. This situation often occurs even though the driving mechanism would not have been capable of causing failure had not the weakness been introduced by the first, or initiating mechanism.

Before considering degradation and failure in any detail, it is important to define what constitutes failure and/or deterioration. An important consideration in any case of evaluation and condition assessment of a turbine is establishing what constitutes failure. The definition I find most acceptable is this: *A condition exists within the unit that while it would not prevent the unit from returning to service and continuing to develop power, it could force it from service before the next planned outage.* Various other definitions exist, and the definition of failure used in any situation—and therefore the responsibility for correction—can be controversial. This controversy is to some extent aggravated by possibilities; *e.g.*, a crack that has been determined to exist may be predicted by the methods of fracture mechanics to be growing at a rate that would not cause complete rupture, forcing the unit from service before the next planned outage.

As reserve power margins diminish, steam turbines—that currently have operating periods between major maintenance outages of three to eight years—could be forced to operate longer than intended when they were originally returned to service. Under these circumstances, it is difficult when making a prediction of a unit's future operation, to be certain there will not be some major change in its operating parameters. Parameters that can influence an acceptable definition of failure in any situation include the exact operating period, the unit load pattern, and the steam conditions the unit will experience over a number of years.

A simple and conservative solution to this definition of failure would be to change *any* suspect component showing any crack or unacceptable damage-or-deformation indication. This may appear to be an expensive option, but is considerably less expensive than a forced outage requiring weeks or months to open, repair, await replacement parts, replace those parts, close the unit, and return it to service.

Defining efficiency deterioration is somewhat easier. It is even possible to quantify such deterioration in terms of reducing steam path efficiency and unit output. What is *not* possible to determine is the extent of any mechanical deterioration that may occur and cause efficiency deterioration. This is an unknown situation not recognized until complete mechanical rupture occurs. There is normally no manner to predict such an occurrence—damage could be in the incubation phase—even when an examination of the steam path is made at maintenance outages.

During operation, certain situations and phenomena are known to occur that have the potential to initiate damage or to cause deterioration in performance. These damaging and deteriorating phenomena can be of a continuous or intermittent nature, produced as a consequence of transient operating or steam conditions. Such phenomena can also be the result of sudden mechanical failures of components that cause more extensive consequential damage. The most commonly occurring of these degrading effects are related to the formation of moisture in the steam path or solid foreign particles, possibly from the boiler or scale generated within the superheater and reheater tubes. Other sources include chemical contaminants that are introduced, or gain access to the steam path on which they are deposited, and possibly act as corrosive elements. The other principal degrading condition is the operational phenomena occurring during the operating life of the unit.

The first two chapters of this book provide general information. The first outlines what is considered necessary to define and constitute a maintenance strategy that represents management's commitment to maintaining a healthy system. This chapter also outlines means of monitoring conditions indicative of damage. The second chapter deals with the spatial arrangement within the steam path and the factors that affect it. This is important because the performance (efficiency and reliability) of a turbine is influenced considerably by the alignment of the unit and the resulting axial and radial clearances and "laps" that are achieved in the hot operating condition.

Chapters 3, 4, 5, and 6 discuss the various phenomena known to affect both the efficiency and structural integrity of the components. In the second volume, chapters 7, 8, and 9 consider repair and refurbishment options currently available. Fortunately, there are ever-present advances in these technologies, and as experience is gained, newer and improved methods develop to be applied to older units so they can continue to operate with high levels of availability—often with improved efficiency. Chapter 10 considers seal systems and gland rings, and provides means of estimating the financial penalties associated with excessive leakage. Seals are one area where operators and maintenance personnel can influence the cost of power generation, and help reduce the cost of power to their customers.

The final two chapters, 11 and 12, relate to quality and the inspection of elements being manufactured to replace damaged components. This is an area where many engineers feel the cost of undertaking such inspections is difficult to justify. However, what happens when components—manufactured when they are required in an emergency to return a unit to service—have any form of fault and force the unit from service prematurely? In such a case, the cost of inspection—ensuring that a supplier's quality program is prepared and operating properly—is well justified. It is often said, "There isn't time and money to do it right, but there is always time and money to correct it." This statement is well applied to the manufacture or repair of components in an emergency, because the cost of a second outage is just as high as the first, and far more embarrassing.

Because the steam turbine is a thermal machine designed to convert thermal energy to rotation kinetic energy, I have included an appendix that provides the basic thermal relationships required to understand the turbine and its operation.

Situation evaluation

The more susceptible areas in any turbine unit are a function of many complex factors—individual stress levels, stress concentration, mode of operation, and the operating environment. Individual components are also greatly influenced by the expertise with which the parts were designed, manufactured, and assembled, and the operating transients to which they have been subjected. The diversity of the factors that can contribute to damage precludes any generalization of cause or value. Steam path components are subjected to high stress, both direct and alternating. Many parts operate at high temperatures and are of complex forms interacting with one another in unpredictable ways. These factors, when combined with load and temperature transients that occur during operation, combine to make the steam path highly sensitive and a major source of concern to the designer and operator.

While some concerns are common to most operators, the type of deterioration or damage to which any component or area is subjected normally varies from unit to unit. This accounts for the variety of concerns expressed by maintenance staff, and the different dispositions of the various nonconforming conditions that will be developed in any situation.

In many instances when corrective action is required, there is no optimum solution that can be followed without deviation. Operation and load demands will often negate the optimum. At other times, costs, special tools, skills, and the availability of replacement parts could require some form of compromise. These compromise solutions may have to be adopted from necessity, but the final disposition should provide the best balance between cost, risk, and the immediacy of returning the unit to service.

The logical approach to maintenance and repair dispositions is:

- Consider the available alternatives in terms of the original design requirements of the affected components
- Evaluate possible solutions in terms of departure from the design specified requirements

Many "repair" or "accept-as-is" dispositions will have only a limited effect on unit performance, and can be readily accepted. Other repairs can be proposed and accepted, representing a compromised condition. Such options should only be accepted on the basis that the unit will be operated with this compromised solution for as short a period as possible, and that the selected option does not represent a significant level of risk in the short term. If this is possible, plans should be put into effect immediately to develop an acceptable solution that can be undertaken within a reasonable time.

The maintenance options

The satisfactory performance of a steam turbine is influenced considerably by the manner and expertise with which it is maintained, and the load patterns it follows. While the plant operating engineer can control, to a large degree, the maintenance of the units for which he is responsible, he is unfortunately unable to exercise little influence on operating patterns. This is a responsibility of dispatchers who have a mandate to serve the demands of their clients rather than the turbine generators of their system.

For maintenance to be cost-effective, it must be planned. When signs of distress, excessive wear, misalignment, or component deterioration are detected, the need for corrective action must be considered. These corrective actions should help ensure the situation does not deteriorate further, to the extent the unit is placed on a forced outage status, severely load limited, or suffers an unacceptably high degree of deterioration in efficiency.

There are general maintenance requirements for any unit. Guidance for these is provided by the designer and should be followed for all routine matters. The designer will also provide recommendations for the operating time between opening sections of the unit for periodic maintenance and examination. During these maintenance outages, any findings that could affect unit performance must be reviewed in relation to their possible long-term effects.

Maintenance actions

Opening a unit for maintenance provides the opportunity to make repairs or to install replacement parts when the necessary skills and special purpose tools are available. Such an opening also allows replacement parts to be ordered, which can be placed in the unit at the current or later outage, depending upon the delivery and required period of the outage. Replacement is made when an evaluation of any found operational nonconformance is judged to be placing the unit at risk if returned to service without correction. A detailed evaluation of each nonconformance should be made and it should indicate if, and what actions are required.

The principal purpose of a steam turbine maintenance inspection is to detect potential problems at an early stage. If this is not done, relatively minor situations could progress to the extent a forced outage or excessive loss in unit output and efficiency could occur. During such a maintenance inspection outage, parts can be examined visually for indications of failure, wear, or distortion. Also, nondestructive tests can be applied to critical components to determine if their ability to continue to perform satisfactorily has deteriorated and if so, what remedial action should be taken, or planned.

A nonconformance in any part of the steam turbine unit is considered to have occurred when there are signs of mechanical failure, excessive wear, or any form of deterioration that has the potential to adversely affect the performance of the unit. Such nonconformances must be reviewed for its short- and long-term effects.

As soon as unit inspection indicates a nonconforming condition has been found, it must be evaluated. The logic process of evaluation for both availability and efficiency is considered in chapter 1. This chapter outlines avenues the maintenance engineer should explore in deciding what corrective action needs to be taken. There are four decisions that can be reached. In some circumstances the decision is relatively simple, and is in fact obvious. In other situations, a decision is made based on the probability of failure, the possible cost of repair, and ultimately, the reparation of consequential damages that are the result of not taking corrective action. These four options can be considered:

- scrap and replace
- repair
- rework
- accept-as-is

Of these decisions, possibly the most difficult and potentially most controversial is the latter—accept as is—a disposition that allows a component to return to service with no effort made to correct the nonconforming condition. There are two reasons for reaching and deciding upon this course of action:

- *There is little need to make any corrections.* To make them will add no or marginal improvement to unit performance and the condition will not place the unit at risk

- *The cost of replacing, repairing, or reworking cannot be justified.* This is often a judgment call on the part of the engineer and can only be made if he or she is aware of any risks involved

Such a decision should not be made as a desperation measure. The risks, if any, should be fully evaluated. The options and the probability of failure—from an extended outage to operation—must be fully considered.

Therefore, the evaluation process can be a complex one. Occasionally, the solution is self-evident—such as when partial failure has occurred, or when excessive damage exists. The most difficult decisions are those related to suspected damage or deterioration, and those for which it is difficult to determine the cause. In these instances of uncertainty, mature judgment is required, together with knowledge of the operating and maintenance history of the unit. This knowledge should help in the evaluation. The information in this book can also provide confidence in the selection of the final disposition.

The availability of replacement parts, special skills, and tools will often influence which decision is reached. Care must be exercised to ensure that availability or non-availability of replacement parts does not force the owner/operator into a decision ultimately causing more expense and increasing the overall risk level to an unacceptable degree.

Often, alternatives to these potential solutions are available. Some may degrade a unit's rating or impose other restrictions in terms of maximum output, or the time for which a unit can be operated. The compromise correction is ultimately more acceptable over the short-term, while the owner/operator arranges for a more palatable long-term solution.

William P. Sanders
Richmond Hill, Ontario, Canada
August, 1999

ACKNOWLEDGEMENTS

In preparing this book, I would like to acknowledge the assistance given me by many engineers employed in the power generation industry for their efforts to read and comment on the contents. While these people have reviewed the manuscript and made many valuable comments, the responsibility for the material contained in this work is entirely my own. Each of these engineers have helped me to avoid ambiguity and error. Specialists in their field, these individuals represent many hundreds of years of design and practical experience. Their reviews and recommendations for inclusion, deletion, and modification have added significantly to this volume's value for the operating engineer.

Various companies have made drawings and photographs available to me, and I have used many more from technical papers presented at power generation industry conferences. These are acknowledged under the illustrations. I wish to thank these organizations for this information, which has helped in many instances to make my meaning clearer.

My special thanks to J. Scott Shearer, my friend of more than 40 years, who has commented extensively on the contents and arrangement, and with whom I spent years designing turbines—when we still used slide rules and Mollier diagrams. The value of his assistance cannot be underestimated.

Also I would like to acknowledge the assistance of Dr. Graham Bellamy and Colin Moore of Ontario Hydro, John A. Cizek of Nebraska Public Power District, Michael J. Fraser of Turbine Blading Group, (U.K.) Ltd., Thomas J. Hix, Dan J. Storey, and Cliff. L Baldwin, of Omaha Public Power District, Janusz A. Kubiak of Instituto de Investigaciones Electricas, Dr. Neville F. Rieger of STI

Technologies, David Rasmussen of Turbine Consultants, Anthony F. Mitola of Turbo Parts, Ingo Schulz of Steam-Path Support, Charles G. Turner of Charles G. Turner & Associates, Doug Whitney of TAPPS, Elmar Locker of H&E Machinery, and Ralph J. Ortolano of Turbine Rescue. Their help in various sections and on topics in which they specialize has been of considerable value.

WPS

Chapter

1

Considerations of a Turbine Steam Path Maintenance Strategy

INTRODUCTION

Although most parts of the steam turbine are capable of suffering mechanical damage—and do sustain it—some areas or components suffer greater levels of deterioration than others. Why some areas in any unit are more susceptible than others is a function of many complex factors—individual stress levels, stress concentration, the mode of operation, operating environment, and the manner in which the unit is maintained. Other critical factors involve the operating transients to which components are subjected. This diversity of factors that can influence the potential for damage precludes any ability to state a generalization of causes or of value. Despite this fact, the area of the unit having a considerable potential to affect performance—and of raising the concern of the operating and maintenance engineer—is the steam path.

Components of the steam path are subjected to high stresses, direct and alternating. Many parts operate at high temperatures. Often they are complex. These elements will also interact with others in unpredictable ways. These factors, when combined with load and steam temperature transients that commonly occur during operation, make the steam path highly sensitive and a prime candidate for deterioration. If populations of steam turbines were to be reviewed, there is a high probability that failures or problems within the steam path would be a significant cause of deteriorating performance for many of them. The rotating blades would pose a major cause within the steam path itself.

This chapter discusses strategies available to owners when they determine their units have suffered some form of structural deterioration within the steam path (to the extent corrective action is considered necessary, or at least considered prudent to evaluate the condition before the unit is returned to service).

When it is determined that a nonconforming situation (damage) exists within the steam path, the user must consider available options. In these circumstances, it is necessary to evaluate the situation and decide which action will properly utilize the technical capabilities and skills available to the operator. Decisions relating to the condition and the evaluation include considerations of both cost and time, and the potential costs of electing not to take corrective action during the current outage.

CONSIDERATIONS RELATING TO A MAINTENANCE STRATEGY

In order to operate turbine units in the most cost-effective manner, each operating organization must define a specific policy governing the maintenance of T-G units. It is normally inefficient and very costly to operate units until they are forced from service and maintenance undertaken on an "emergency" basis. An effective policy must recognize that some spare capacity must exist within the system, and that power must be able to be purchased or exchanged between neighboring and interconnected systems so units can be shut down for periods sufficient to allow them to be opened, examined, and where necessary, corrective action taken.

An effective maintenance strategy for any population of steam turbines owned and operated by a single authority carries certain requirements:

- The strategy and the corrective actions that an evaluation for any condition selects must be cost effective
- The selected corrective actions (if actions are required) must be able to be performed within a reasonable time frame
- The selected corrective actions must not create a condition that will in any way compromise the availability of the unit
- The selected corrective actions must not modify the physical characteristics of the steam path components to the extent they cause a deterioration of unit efficiency beyond what is determined to be acceptable
- The selected action must be able to be performed by maintenance personnel available from within the plant or specialized outside workers. Any required, specialized tools must be available as needed

Considerations of three separate outages influence (or are influenced by) the work scope and corrective actions at each maintenance outage, as shown in Figure 1.2.1, *i.e.*, the preceding outage, the one being entered, and the next one planned (presumably in four or more years time) have to be considered. The major objectives of the planned outage and subsequent inspection can therefore be considered:

- To undertake normal maintenance (cleaning and replacement of consumable parts of the unit, such as seal strips, which upgrade or restore the efficiency level)
- To determine the unit's condition and the possible need for corrective action at some future outage
- To take corrective action on conditions noted at the previous outage. These previous conditions may have been judged at that time to be acceptable and the unit returned to service with a recommendation for deferred action
- To record measurements and other conditions being monitored to determine the rate of deterioration of any condition that could at some future time require corrective action
- To take emergency corrective action on conditions found since the last inspection and discovered upon opening the unit. These are conditions that could prevent the unit being returned to service with a probability of performing at an acceptable level

If the condition(s) discovered could prevent the unit from being returned to service, there are three possible courses of action available. Each must be separately evaluated:

- To replace the damaged or deteriorated parts. To do this, replacement parts must either be available in inventory or on "just in time" delivery. The owner must evaluate the cost of extending the outage for hard-to-get parts
- To refurbish components as needed. This action is only possible to the extent that components are capable of refurbishment. There are circumstances that would allow permanent

repairs and others that would allow refurbishment sufficient for the unit to be returned to service until the next outage, when replacement parts could be installed

- To remove affected parts. This can sometimes allow the unit to operate at a de-rated condition. However, even to do this often requires the installation of components such as pressure reducing plates (to correct pressure distribution throughout the unit to meet more closely the design conditions), which will allow the unit to operate without further unacceptable deterioration

The action the owner selects depends upon economic considerations of the total situation. This requires a careful evaluation of the options available under the actions items listed above. This evaluation should determine the most economical solution, consistent with returning the unit to a satisfactory mechanical condition.

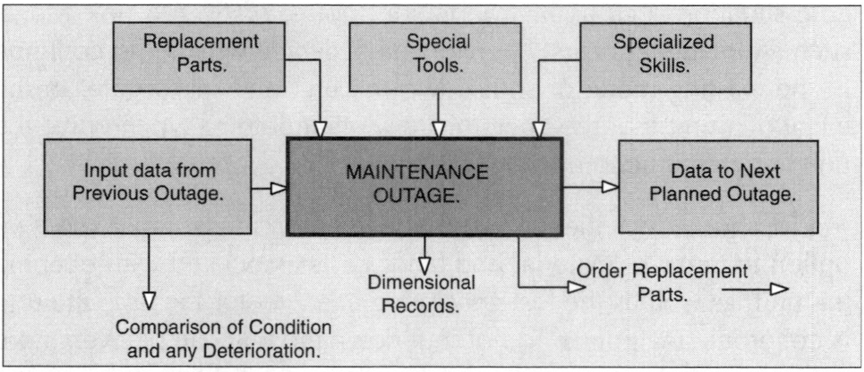

Fig. 1.2.1—The results of three openings impact on each outage, the previous in terms of the preparation for the present, the present for corrective actions identified for the future, and also to develop plans and work scopes for repair or refurbishment at the next.

THE TURBINE OUTAGE

Maintenance or planned outage

This outage is scheduled. It is part of the established plan for maintaining the turbines in an acceptable condition, and may require special skills and tools be made available to undertake the required work. These outages may also be an opportunity to replace parts that were damaged or badly worn and noted as damaged or deteriorated at a previous outage. Replacement parts would have been specially ordered for this outage, and its occurrence could have been time dependent upon their delivery.

Forced outage

This is an unplanned outage that occurs as a consequence of some unexpected failure or damage occurring within the turbine unit. The damage may result in a condition indicating an unacceptable situation such as high vibration, noise, or even a unit trip. If such an incident occurs, operators must decide whether to continue in the existing mode or shut down the unit and investigate. In the event of a unit trip, the operators may attempt to re-synchronize the unit and continue to operate.

To shut down the unit and investigate can be an expensive option in terms of material and labor costs associated with opening the unit, as well as the lost generating capacity for the time the unit is not producing power. To *not* shut down the unit can be even more expensive, however, if such action neglects a condition that is in the early stages of development, and which has caused the early system interruption in the first place. An uncorrected condition could engender more serious damage to other components within the unit, and could ultimately cause extensive damage or force major components to be scrapped. Safety has to be considered in any decision not to shut down a unit. A condition may exist that indicates a major and possibly catastrophic failure may occur.

Extended outage

An extended outage is the result of finding a situation during a planned outage that causes the planned time frame for the unit inspection, cleaning, and refurbishment to be extended. Normally this occurs when damage or deterioration is found requiring the use of nonstandard tools or skills for evaluation, or the necessary replacement parts being unavailable.

ESTABLISHING THE NEED FOR UNIT SHUTDOWN

Outside of the normal period during which a unit is returned to service and expected to operate at an adequate level of performance, there are certain indications from a unit denoting a need to consider the situation, investigate it, and possibly shut down for an unscheduled outage.

The signals indicating damage or deterioration

Station operating procedures include the need to monitor conditions of certain operating parameters and, when they vary outside expected norms, can indicate the need for emergency corrective action. The more common of these are shown in Figure 1.4.1, and include:

Vibration levels. An increase in the vibration levels of the unit rotors is an indication problems exist. There are various possible causes for this amplitude increase. Among the more common are:

- A change in the unit alignment. This loss of alignment may be caused by various phenomena, and the possible causes will need to be investigated separately. A vertical or horizontal shift of the bearings position can cause such a situation. If this has occurred it will be necessary to realign the shaft. The extent to which the unit must be opened and rotors and other components dismantled will depend upon its configuration and the access that can be made to the bearings

- The mechanical rupture of a component that has disrupted the dynamic balance of the unit. Such mechanical failure will not necessarily cause an imbalance. Shorter blades or whole portions of a coverband may detach without significantly affecting the balance and therefore the vibration level

- Heavy rubs that have caused a shaft bend. If there has been a heavy rub caused by some transient or other condition, it is quite possible the shaft will bend to the extent balance is upset. If this has occurred it becomes necessary to open the unit, remove the affected rotor, and undertake major restoration

- Starvation of oil to the bearings (axial or thrust). If this situation occurs it could cause significant, and possibly irreparable damage. This situation should be rectified as quickly as possible

Noise. One of the most significant and useful indications of damage or pending failure is noise from the unit. Because of the high rotations speeds, even the lightest rubs can be indicative of pending problems. Often these noises appear and then disappear, so it should be noted if the noises are associated with transient conditions, *e.g.*, changes in steam conditions or load. In any event:

- If possible, determine if the noise is sensitive to transient conditions, speed, and temperature
- Identify the location of the rub as accurately as possible
- If the noise is continuous, try to determine if it is at a constant level and frequency. This is subjective judgment, but careful observation can assist

When the unit is opened, examine the region where the noise appeared to be centered and look for rubbing damage in the form of grooves, lifted coverbands, and changes in material hardness. There is often a change in hardness associated with rubs that could cause later failure—the material structure on the steel will have changed, making it more brittle.

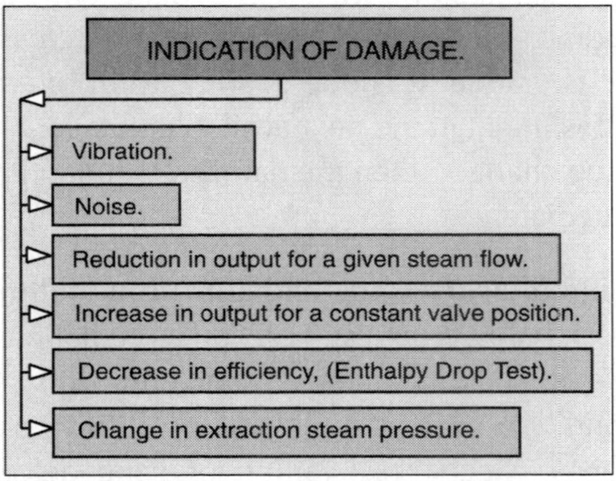

Fig. 1.4.1—Indications of a distress or a damage condition.

A reduction in unit output for a given steam input. If there has been a step change in the output for a given steam input, this can be indicative of mechanical damage in the blade system—either a blockage early in the steam path, or some broken elements that are reducing the efficiency of energy conversion. Often the region where this damage is occurring can be identified by diagnostic means and a determination made of the need to shut down and correct.

It is necessary to differentiate between a gradual reduction in output and the gradual wearing of seal strips. This is particularly true after returning to service from maintenance, and is a normal deterioration caused by deposits on the blade elements. Step changes are difficult to identify, as damage will often occur during transients and the change may not be noted when the same conditions are reestablished. Operating staff, particularly those with on-line monitoring capabilities, best do this type of monitoring.

Increase in unit output at a constant valve position. Such a condition normally occurs because of damage in the control stage nozzle block. This causes a change in wheel case and stage pressure throughout the steam path. The nozzle plate will have sustained damage and admits more steam because of an increased flow area. If this increase is gradual it is due to some form of erosive damage requiring refurbishment at the next available outage. However, if the change is a "step change," then the damage is sudden and should be examined immediately.

Decrease in efficiency using enthalpy drop methods. If a utility performing an enthalpy drop test (EDT) on a regular basis finds a significant change in section state line efficiency, this normally indicates some condition causing a disruption of flow and energy conversion efficiency. Mechanical failure or vane deformation commonly causes the change of efficiency and should be considered an indicator of existing problems.

Again, only step changes are significant. As EDTs may be run only on an annual or six-month basis, a long time may pass before such a condition is determined to be present by testing methods. Station instrumentation used to monitor state line efficiency is usually not sensitive enough to note changes with sufficient reliability so a unit could be opened on the basis of such results. Other forms of analysis would be required to confirm or deny the possibility of damage.

Changes in steam extraction pressures. Blade system damage will cause a redistribution of steam pressure throughout the steam path. These changes may be minor, but if the unit is returned to stable and known operating conditions the changes may be determined to exist.

The indicators discussed above do not represent an exhaustive list, and in many units, operators and maintenance staff become sufficiently familiar with individual units on their system to be aware of unit peculiarities. Under these conditions, operating staff can often act in an anticipatory manner and recognize characteristic changes in the units. This ability represents a valuable skill to the owners, and is only obtained from considerable experience with particular units.

System requirements

Forced outages mandate that owners arrange to meet system requirements by obtaining power from alternative units. This alternative power can come from several sources. These include:

- Line units on spinning reserve or stand-by. These units may be at or close to the facility but the fact that they are spinning reserve indicates their operation is less cost effective than the units forced from service
- Older, less-efficient units on the system. This may cause a time delay but will allow system requirements to be met
- Delaying any planned or scheduled outage on other units on the system
- Purchasing power from another utility connected to the system. This can be expensive, but may be necessary in certain

circumstances to meet requirements. Industrial installations normally have ties to the local utilities, but this can represent a significant economic penalty, particularly if boilers have to be kept running to meet process steam demands

- Adjustment of power factors and distribution voltage. There is a limit as to how much can be achieved by this method. On a large system this can make a difference but there is a possibility of damage to certain electrical appliances, particularly from the lowering of voltage

In the event of a forced outage the owner must decide which option offers the best alternative for meeting system demands. An initial decision may be made in terms of returning service to customers, but a long-term solution may require a change of supply to meet the most cost-effective means of meeting demands over the period of the forced outage.

OUTAGE SCHEDULING

A unit in service requires regular inspection or preventative maintenance outages so it can be examined and corrective action taken to allow it to remain in service in an acceptable condition. There are two basic approaches to scheduling outages that provide sufficient time to allow examination, the correction of known defects, and the detection of any deterioration that will require attention at the next available outage:

- *Shut down the unit and expose sections and valves to make an examination and determine required repairs.* This is probably the most common approach, and is used by the majority of large systems. This requires the entire unit to be made unavailable to the system for a period so all sections can be worked on at the same time

- *Shut down sections selectively, opening one or possibly two during a regular outage.* The advantage here is that the "unavailable time" is somewhat shorter and skills are concentrated on the opened sections. It also means that sections prone to damage or significant deterioration can be examined more frequently

This dual approach has been found to be effective on systems that have both summer and winter peak load demands combined with relatively short periods in between, when the unit can be made available for maintenance. It has also been used on large nuclear units that must be shut down for six to seven weeks for refueling. At that time one section of the unit is opened and made available for replacements or refurbishment of damage.

INTERVAL BETWEEN MAINTENANCE OUTAGES

Operators of steam turbine generator units must determine the optimum, or acceptable interval between maintenance outages and inspections. This interval between outages should be determined for each unit within the system and must reflect not only operating and repair costs, but also the consequences of not making such inspections and detecting faults in their early stages of development. One manufacturer has determined that under certain conditions, and with recognition of limitations, their unit rotors may operate for up to 10 years between major re-inspections.

Apart from the technical factors, the following issues should be considered in establishing the period between inspection outages:

- The fuel costs of the unit being removed from the system for inspection and the unit(s) started to meet load requirements, or the cost of the replacement power that will be purchased
- The anticipated load factor of the unit during the outage period and its output rating
- The differential heat rates of the unit being inspected and the units being started to meet load requirements
- The generic or historic forced outage rate of the unit
- The anticipated maintenance period
- Known labor costs and the anticipated cost escalation of replacement parts
- The anticipated improvement in unit heat rate that occurs as a consequence of the outage
- The reserve capacity on the system and the level of risk the owner is prepared to accept in continuing to operate if the outage is foregone
- Problems discovered on similar units owned and operated by the owning and other utilities

These factors influence the cost of an outage and the period for which a unit should operate between major inspections.

The ideal situation for scheduling outages would be for a unit to be open very infrequently; for no damage (or a minimum amount of damage and deterioration) to be found at the outage, and then for the unit to be successfully returned to service. In fact, units are opened periodically on a "life cycle basis"—usually between four to eight years—with the possibility of this operating period being extended as system reserve declines.

The actual opening time between major outages is dependent upon various factors including:

- *the manufacturers' recommendations.* These are most often based upon observations of a large population of similar units and the experience established with these units after different periods of operation

- *known problems within the unit.* These can be related to both supplier and user-induced situations. Design weaknesses require opening the units more frequently until a permanent solution to a particular problem is fully identified and solved

- *the manner in which the unit has been operated since the last outage.* This is a factor over which the supplier has no control and, to a degree, the operator has only limited influence

Factors that influence the determination of when the unit should be opened include:

- the number and type of starts since the last outage (including very hot, hot, warm, and cold starts)

- any temperature transients experienced since the last outage

- any excessive overspeed transients, and their duration since the last outage

A new unit represents the latest technology available to the system, and it is normally operated at maximum capacity. Unfortunately, as the unit ages, it becomes less important to the system, and component material properties will tend to degrade. The unit is then normally subjected to more cyclic operation, which consumes component life much faster.

There is at present no absolute method of defining when a unit should be removed from service for maintenance. On those occasions when operators know it should be shut down, some operators cannot because of system or process demands. Under these circumstances, the unit is operating at risk. If the operation cannot be avoided entirely, it should be minimized and the transients to which it is exposed limited.

Other factors and considerations influencing the period of operation before a unit is removed from service for a maintenance outage are discussed later in this chapter.

Equivalent operating hours

A suitable method of assisting in "opening decisions" is to accumulate the equivalent operating hours (EOH) for any service period after the unit is returned to service from a major outage. In accounting for this period of operation, the EOH factors effect start ups and can be made to include the effects of some transient operation. A formulation developed by one major manufacturer provides the following equation that is applicable to their units, but with minor modification can be made to suit any unit:

$$EOH's = (Nc \times TC) + (NW \times TW) + (NH \times TH) + (NV \times TV)$$

where:

EOH's = Equivalent operating hours due to starts
Nc = The number of cold starts
Nw = The number of warm starts
Nh = The number of hot starts
Nv = The number of very hot starts
Tc = The weighing factor for cold starts
Tw = The weighing factor for warm starts
Th = The weighing factor for hot starts
Tv = The weighing factor for very hot starts

These "time equivalent" notations for starting the unit are added to the normal operating hours (NOH), which are included irrespective of the load generated by the unit. When the total of normal operation plus start up reaches some predetermined value, the unit is due for an inspection/maintenance outage. Therefore:

$$EOH = \Sigma NOH + EOH's$$

These equations can be modified to include terms accounting for both overspeed and temperature transients. A more difficult determination is the number of EOH that should form the basis of the times at which the unit should be shut down for inspection. It is suggested that an EOH of 30,000 to 35,000 be considered between major outages. The initial and reheat temperature and the operators' experiences should modify this value with the unit.

THE INSPECTION/ MAINTENANCE OUTAGE

A principal purpose of the steam path maintenance inspection is to detect potential problems at an early stage. If this is not done, minor deterioration could progress to the extent that a forced outage or excessive loss in unit efficiency could occur. During such an inspection, parts can be examined visually for indications of failure, wear, or distortion. Non-destructive tests can be applied to critical components determining if their ability to continue performing satisfactorily has deteriorated.

The satisfactory operation of a steam turbine within a utility is influenced to a large extent by the manner and expertise with which it is operated and maintained. For maintenance to be cost effective it must be planned. When signs of distress, excessive wear, misalignment, or component deterioration are detected, the need for corrective action must be considered. This corrective action should ensure the situation does not deteriorate further, or to the extent the unit is placed into a forced outage status, is severely load limited, or suffers an unacceptably high degree of degradation in efficiency.

The cost of forced outages can be extremely high. In the case of large utility units this often requires that older, less efficient units be started to meet system demands or that units employing an alternate, more expensive fuel be employed. The most severe situation for utility systems is when reserve capacity does not exist and replacement power cannot be purchased. At that time a blackout or brownout situation occurs.

In addition to routine maintenance, monitoring, and care activities, there are surveillance actions associated with a planned maintenance outage. In such instances, upper half covers are removed to expose the steam path for examination. This opening of the unit allows repairs to be made or replacement parts to be installed. At this time, necessary skills and any special purpose tools that are required can be made available.

Such a maintenance opening also allows replacement parts to be ordered. This is done when an evaluation of any "found operational nonconformance" is judged to place the unit at risk. An evaluation of the nonconformance will normally indicate whether a unit can safely be returned to service, or if some temporary remedial action is required so the situation can be corrected before restarting.

THE AVAILABLE CORRECTIVE OPTIONS

The type of deterioration or damage to which a component or area is subjected varies from unit to unit. It depends upon a variety of causes that exist, or are present as a function of the component's design details and unit operating mode. These various causes can normally account for the degree of concern the maintenance staff expresses and they influence the different dispositions that will be developed regarding the nonconformances.

In many instances of damage or failure there is rarely any "best" or "perfect" solution. Instead, compromises will have to be made based on the availability of replacement parts, overall costs, the requirements for special tools, and the skills and time available to take corrective action. The selected disposition will also be influenced by the level of risk associated with each possible solution in any situation. The selected action should provide the best balance among cost, risk, and how quickly the unit can be returned to service.

In order to rationalize the evaluation process, conditions and maintenance decisions have to be defined. The most important follow:

A field found nonconformance

A nonconforming situation is considered to exist when the components of the unit have changed or deteriorated to the extent the design requirements are no longer present. A nonconformance in any part of the steam path is considered to be present when there are signs of structural failure, excessive wear, or any form of deterioration having the potential to adversely affect the performance of the unit. When such a nonconforming condition is confirmed, it must be viewed for both its short- and long-term effects (see chapter 11 for manufacturing nonconformances).

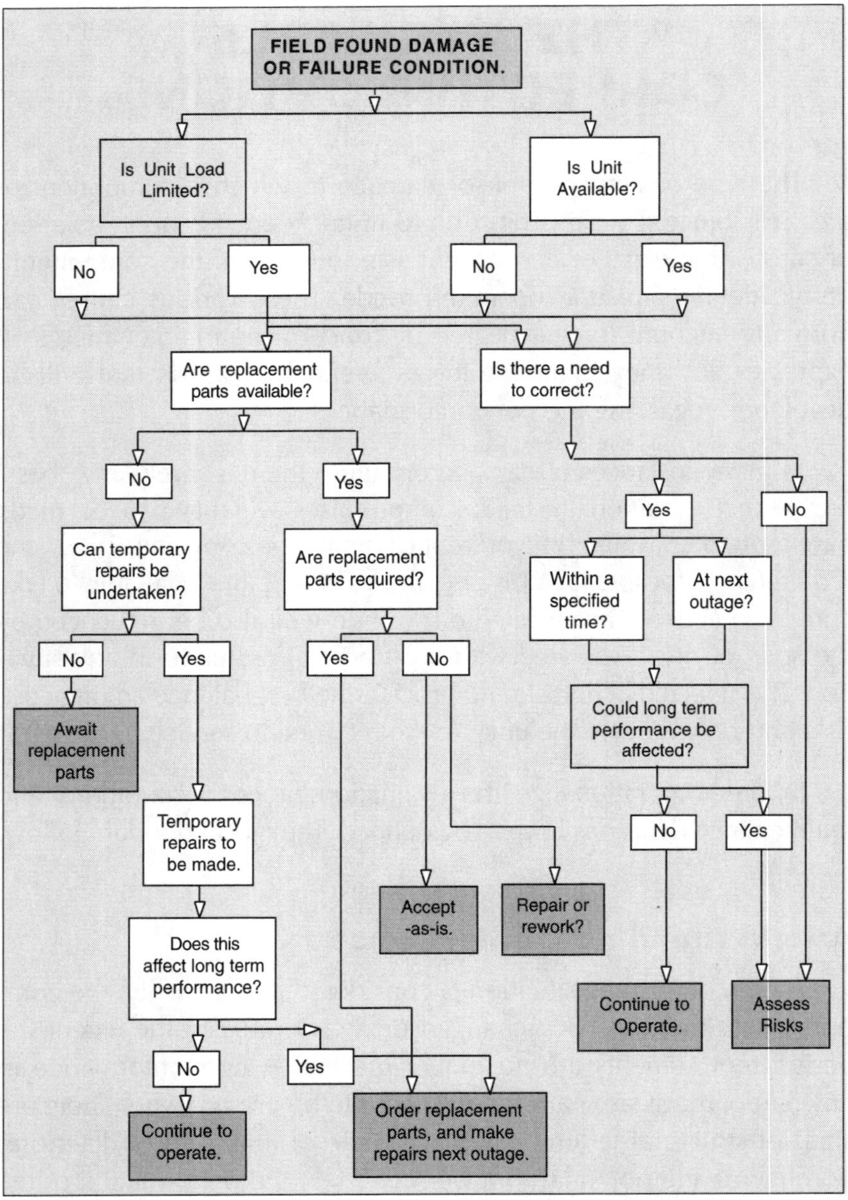

Fig. 1.8.1—The "Logic Review Process" when a nonconforming condition is found in the unit at maintenance inspection. The final decision of corrective action is dependent upon many factors including the availability and delivery of replacement parts. The maintenance engineer must evaluate the options and make a decision of the best long term solution.

When an inspection indicates a nonconforming condition, it must be evaluated. The logic process of evaluation for performance potential is shown in Figure 1.8.1. This figure outlines the options the operator or maintenance engineer explores in deciding what corrective action is most appropriate. An evaluation leads to one of four alternate decisions. In certain circumstances, the decision is relatively simple to make and in fact, is obvious. In others, options are available, and a decision is made based on the probability of failure, the possible cost of repair, and the ultimate consequences, including the correction of consequential damage that results from not taking corrective action.

The four basic decisions that can be reached are detailed below:

Scrap and replace. Such a decision is made when the situation has deteriorated to the extent the component must be replaced, either because it has failed, or returning it to service will jeopardize structural integrity and safety of the unit. Often this is a self-evident decision with little need for evaluation. At other times, this decision is reached only after extensive review of options and possible consequences. In this latter situation, it is judged that the risk associated with continued use is too great, and the part must be scrapped and replaced, even if this requires waiting for the delivery of replacement parts.

Under circumstances described as "uncertain," a unit can be operated at part load and/or with reduced steam conditions, if the component suffering the nonconformance does not affect other parts of the unit. Such a possibility must be evaluated for each nonconforming condition.

Repair. A repair corrects a nonconforming condition, but does not re-establish original design characteristics within the element or unit. It is often possible to make repairs to components, sufficient to allow them to be returned to service. Depending upon the nature of the nonconformance and of the repair, the affected component may or may not ultimately require replacement.

During the past decade there have been significant advances in many repair and refurbishment techniques. This is particularly true in cases involving welding, where new technology has made available materials and techniques capable of extending the useful life of many apparently failed or badly deteriorated components.

The technical requirements for performing such a repair are normally stringent. However, if they allow a unit to be returned to service within a short period (rather than require an extended outage), or until replacement parts become available, then the costs and minimal change in risk levels associated with such repairs can often be justified.

Again, the repair decision is normally made after a review of the nonconformance, an evaluation of the possible repair procedures, and the level of risk involved.

Rework (refurbishment). A refurbished component is considered to be returned to its design condition (or better). The decision to refurbish a nonconforming condition is, in some respects, similar to that of repair but implies complete conformance with the original design conditions.

In a number of situations, reworking involves reforming the existing material and may or may not require the addition of heat. In many instances this is an easy decision to reach, particularly when applied to stationary parts of the steam path. There can be an extended time associated with major rework decisions, but many techniques can now be undertaken without extending an outage.

Recent improvements in design and manufacturing technology mean that for certain damage-condition refurbishments, components that have the potential to perform at improved levels compare to the original design. Also, there is a growing tendency among owners and operators to repair or refurbish components once they have been removed from the unit and replaced. The repaired/refurbished components can then be carried as inventory spares. If the component is one that deteriorates during operation, the two elements can be alternated at each outage. This is an attractive proposition, particularly if there are a number of units in a plant or system utilizing these same components.

Accept-as-is. This decision permits a component to be returned to service with no effort made to correct the nonconformance. There are three criteria involved in deciding upon this course of action:

- *There is little need to make any corrections.* To do so will add no more than marginal improvements to unit performance
- *The cost of replacing, repairing, or refurbishing cannot be justified.* This extends to the unit and the degree of deterioration present as rework or repair could often increase risk to performance level
- *There is insufficient time to take corrective action.* Correction, repair, or refurbishment would extend the outage period beyond an acceptable time frame

This decision is often a judgment call based on the experience of the engineer, and can only be made with awareness of any risks involved. Such a decision should not be made as a desperation measure. The risks, if any, should be fully evaluated and the options considered, from an extended outage to operation and the probability of failure.

An accept-as-is decision can often be made (again, being aware of the risks) while replacement parts are obtained.

The decision or evaluation process can be a complex one. Occasionally the solution is self-evident, *i.e.*, when partial failure has occurred or excessive damage exists. The most difficult decisions are those related to "suspected" damage or deterioration, and those for which it is difficult to determine the cause. In instances of uncertainty, mature judgment is required, together with knowledge of the operating history of the unit. This knowledge can help in evaluating an existing situation and can also provide more confidence in the selection of the final disposition.

The availability of replacement parts, special skills, and tools will often influence which decision is reached. Care must be exercised to ensure that availability or non-availability of replacement parts does not force the owner/operator into a decision that will ultimately be more expensive and increase the overall risk to an unacceptable level. Often alternate solutions are available that may degrade a unit's rating or impose other restrictions, but are ultimately more acceptable short-term, while the owner/operator arranges for a more palatable long-term solution.

A common, logical, and recommended approach to maintenance decisions and repair dispositions is to:

- consider the alternate actions in terms of the design requirements of the affected components
- evaluate each possible solution in terms of the departures from the design-expressed requirements they represent

Many repair or accept-as-is dispositions will have only a minor effect on unit performance and can be readily accepted. Other repairs can be accepted if the unit can be operated in a compromised condition, such as reduced steam conditions, or limited load swings.

DISTINCTION BETWEEN CAUSES AND MECHANISMS OF FAILURE

When a unit is opened—either in a planned outage or a forced outage situation—there is the possibility that a failed, damaged, or deteriorated condition will be found. It is necessary then for the owner and the maintenance and operating staff to analyze the situation and select a corrective action plan.

A first step in discovering the existence of such a situation is to identify the cause. This is of particular importance when a new or relatively new unit is examined, as this condition can be symptomatic of a design fault that must be corrected. A second step is to identify the mechanism of failure.

It is important to recognize at the beginning of such an investigation, that the "cause" and "mechanism" causing the deterioration are not the same things. The failure mechanism is the form of material deterioration that has occurred and ultimately consumes the life of the material or component, often resulting in material rupture or severe deformation. The cause is that feature of the operation or structure of the unit that has introduced the environment within which the mechanism can initiate and develop.

The causes of failure

When a mechanical component fails, there is obviously a cause. Failures do not occur if they are not initiated by some condition existing within the unit. The possible causes are shown in Figure 1.9.1 and examined here.

Poor or inadequate design. Some possible causes of design-initiated failures are shown in Figure 1.9.2.

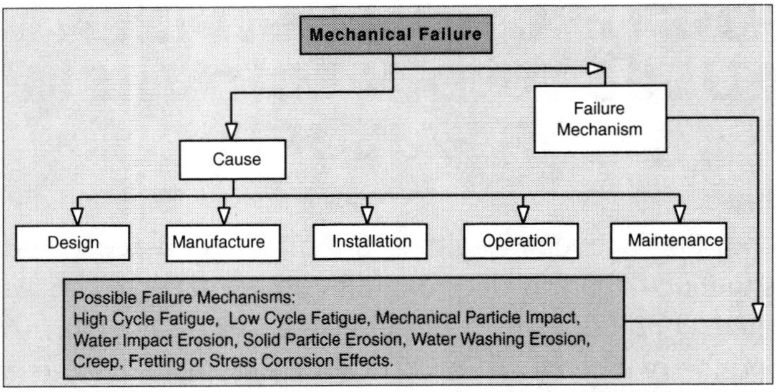

Fig. 1.9.1—The major causes of failure in the steam path, together with the principle failure mechanisms.

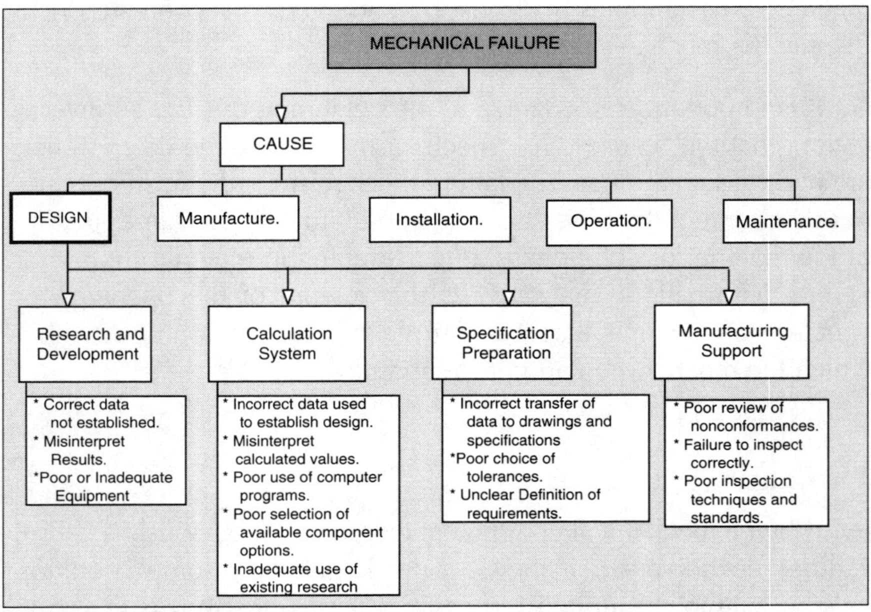

Fig. 1.9.2—The most likely influences from design error which could contribute to component failure.

The design process produces a manufacturing specification that identifies dimensional requirements, materials, and process requirements sufficient to make the unit suitable for its intended application. To achieve this specification the design process employs the results of research, development, and experimentation, all of which are initiated by the need to identify and solve potential problems that could arise with the performance of the unit.

In all large design organizations, qualified procedures exist for both the thermodynamic and mechanical design procedures. These are well understood by the industry, and applied by the individual engineers undertaking the design tasks. There are situations, however, in which the extrapolation of existing designs, errors in calculations, or the incorrect specification of materials will cause deterioration in a unit sooner than normally expected.

Errors caused by the design process can correctly be attributed to faulty design specification or technology application and implementation. With the use of computerized design techniques, many of these errors, and the opportunity for such errors to occur, have been lowered.

A basic premise of the design process is to define a component that will operate without failure for a minimum of 200,000 hours (about 30 years of normal operation). This philosophy is applied to major components. It is recognized that other components are considered to be "consumable" and will be replaced on a regular basis to maintain both efficiency and availability at acceptable levels. For those components designed to achieve 200,000 hours of operating life, it is assumed their units are operated as intended and that operating parameters, as defined by the designer, are maintained. In defining these operating limitations, some or all of the following restrictions may be addressed:

- Initial steam pressure rate of variation and total hours of overpressure operation

- Initial and reheat steam temperature and hours of over-temperature operation
- Initial and reheat steam temperature; rate of variation
- Condenser backpressure
- Unit speed and speed swing limitations
- Overspeed testing recommendations
- The removal of feed water heaters
- The rates at which the unit should be started, loaded, and shut down

Normally these limitations are addressed in terms of the number of hours of operation under "off design" conditions—the number, frequency, and severity with which the unit could experience various transient conditions (recognizing many of these conditions are difficult or impossible to control). These limitations vary from manufacturer to manufacturer and reflect to some degree each manufacturer's operating experience under such conditions. They may also reflect the basic design philosophy and, to some extent, the level of conservativeness in both the design and operating philosophy.

The design engineering function is responsible for unit and component quality, as well as the definition of the tolerances to which the unit and its components will be manufactured and tested to achieve the level of required performance. Therefore, when nonconforming conditions result within a manufacturing department, the design engineer must evaluate them, rule on their acceptability, and decree the corrective action required.

In the event the design engineer accepts a situation or condition not within his defined tolerances, and the unit is forced from service or a planned outage has to be extended to correct such a situation, this is considered a design initiated failure. The root cause is poor design.

Poor quality manufacture. It is the responsibility of the manufacturing department to convert an engineering specification to a finished product. The manufacturing department has no responsibility for establishing the quality level of the individual components—only to ensure that quality is met (as defined by the design department). "Quality" is defined in a series of documents from drawings to process and material specifications. Normally each of these documents are supplied with production and application tolerances clearly defined, either as limitations placed on the individual documents or as standards established and published within the manufacturing department.

It is obvious, therefore, that there exists considerable opportunity for manufacturing errors to occur. To help prevent or minimize them, the engineering department employs quality control and quality assurance methodology. (The industry is fortunate that there are not a greater number of faults and failures occurring due to inadequate manufacture.) These quality programs are a management-imposed discipline. However, manufacture is still a human operation and there are still a number of hand operations and judgmental situations associated with manufacture. Faults can and do occur and are incorporated into a unit.

The major areas of manufacturing activity where faults can occur are shown in Figure 1.9.3.

Component assembly is an integral part of the manufacturing process. It carries with it the possibility of "residual stresses" which, compounded by operating stresses, can cause premature failure of components. Unfortunately, such failures are difficult to trace and cannot be anticipated, since the deformed or over-stressed element, or the lack of inspection, is often hidden by the assembled components.

In turbine manufacture, many operations are termed "special" by quality definition. The term indicates that at the completion of the process, there is little or no opportunity to check conformance without destructive testing or significant degradation of the final product. Such processes are often associated with heat fusion or assembly. In these instances, quality assurance involves calibrating the process to ensure conformance with the process. This is done by the detailed examination of samples to achieve the correct product, including destructive testing and metallurgical examination. However, during the application of such a process it is possible for calibration to slip, and so these processes often represent manufacturing areas where nonconforming products can be incorporated into the unit if careful control is not exercised.

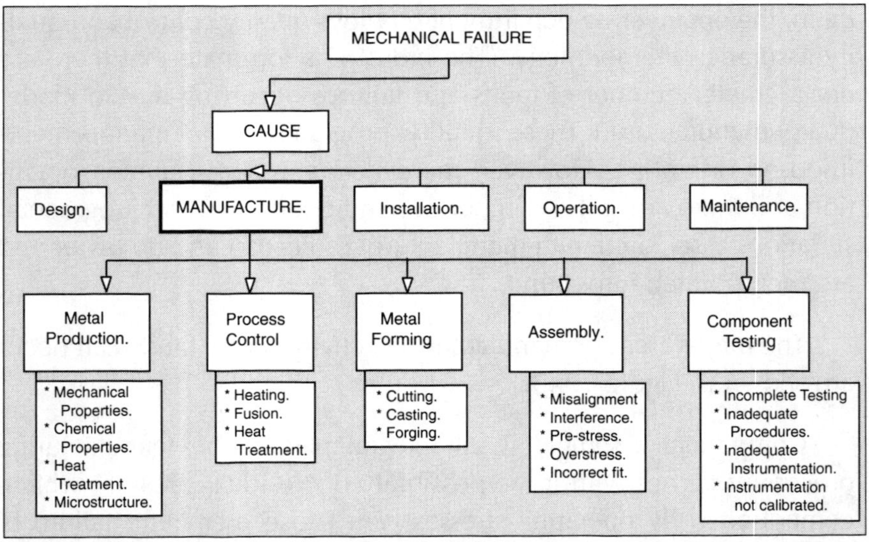

Fig. 1.9.3—The most likely influences from manufacturing error which could contribute to component failure.

Poor quality installation. Assembly and installation on-site is a complex operation that includes the installation of individual components and sub-assemblies. This should also be considered to include commissioning, which is normally an integral part of the turbine supply process. If these activities are not performed in a correct, controlled manner they will often result in poor operating characteristics and the imposition of undue stresses on the component parts of the unit. These additional strains can include residual stresses leading to premature failure within the components of the unit if not detected and corrected. Some possible causes of installation- and total assembly requirement-initiated failures are shown in Figure 1.9.4.

Site assembly requires that multiple section units be aligned in vertical and horizontal locations and all clearances are at design specified values to achieve the correct relationship between the

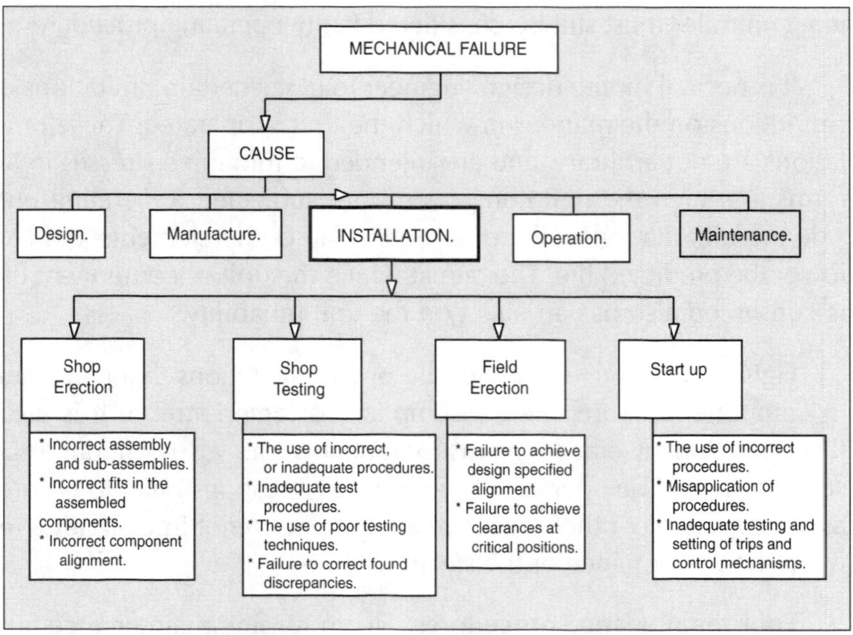

Fig. 1.9.4—The most likely influences from poor installation which could contribute to component failure.

rotating and stationary components during operation. Failure to achieve the correct alignment will result in higher than expected levels of vibration and possibly the imposition of alternating stresses on the normal operating levels predicted by design. Poor control of alignment can also result in operating "rubs" that have the potential to degrade efficiency and reliability.

Control gear and instrumentation must also be installed and calibrated to ensure required levels of control and protection.

Poor or incorrect operating practices. A common cause of component deterioration or failure is the application of incorrect operating procedures, or violation of the operating parameters. This is particularly relevant in temperature variations. In many instances the unit operator may have little control over such variations and may have to respond to the requirements of the system controller so customer demands can be met. However, such conditions required by the system controller must still be considered faulty operating procedures.

It is normal for the design engineer to place certain limitations or restrictions on the manner in which the unit is operated. These limitations are not arbitrary, and are intended to minimize stresses in all forms to which the unit components are subjected. Operating outside these limitations will consume the life of the elements and can lower the predicted life. The rate at which the unit or component life is consumed also has an effect on the unit reliability.

Figure 1.9.5 shows some of the operating actions that can result in component failure, life consumption, or deterioration. It is clear the operation of other system components can also influence the steam turbine. The operator, therefore, has a responsibility to ensure, as far as possible, other pieces of equipment do not impact adversely on the performance of the steam turbine.

Poor maintenance procedures. Steam turbine maintenance falls into two categories:

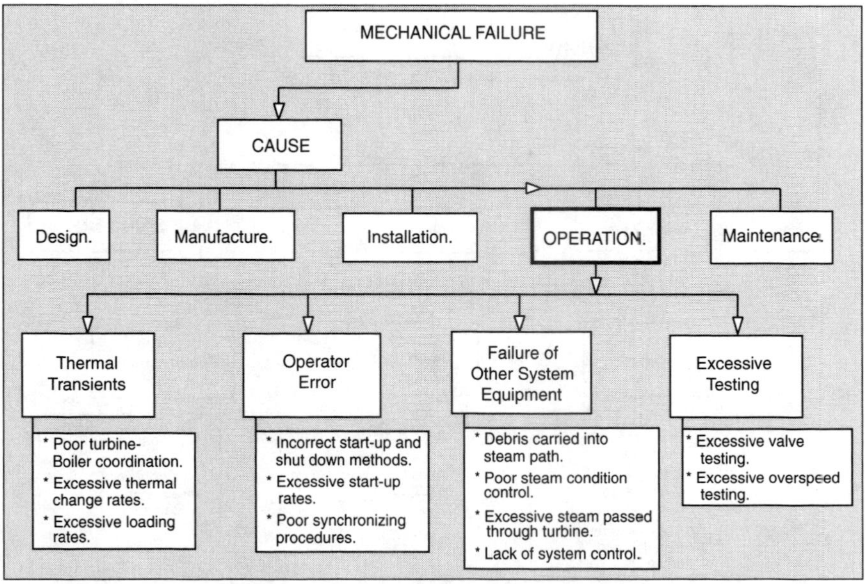

Fig. 1.9.5—The most likely influences from poor operating practice which could contribute to component failure.

- Routine maintenance activities, *i.e.*, valve testing, oil level maintenance, greasing slides and supports, etc.
- Maintenance performed to upgrade or maintain an acceptable condition when a unit is opened up

In terms of unit performance, this second aspect of maintenance is related to component evaluation and the correction of nonconforming conditions. This tends to be more critical in terms of discovered existing and potential failure, but the routine maintenance, if neglected, can also have serious and long consequences for the unit.

Figure 1.9.6 lists some maintenance activities that can influence the performance of the steam turbine.

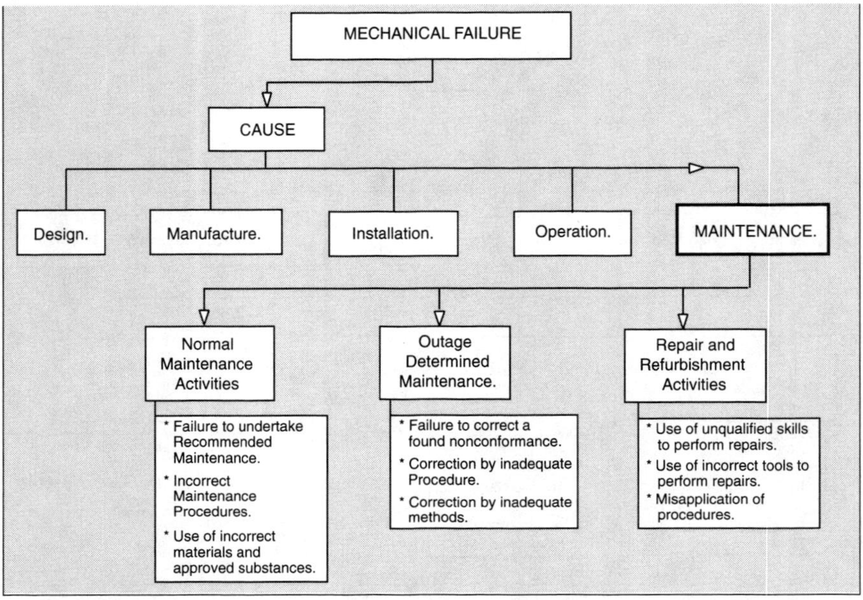

Fig. 1.9.6—The most likely influences from poor maintenance procedures which could contribute to component failure.

The examination of a failure

Considering the causes listed above, when a failure or deteriorating condition is found to exist, it is important to identify the actual cause and, to the greatest extent possible, establish if it was the result of design, manufacture, installation, operation, or maintenance. This can become a difficult and often impossible task. There are no rules or guidelines that can be applied. It often takes considerable investigation to identify both the cause and the initiating condition.

Often, if the initiating condition can be identified (*i.e.*, inferior or inadequate material, poor process control, overheating of the components, etc.), it becomes considerably easier to establish which of the five causes discussed above is the initiating condition.

The cause of a nonconforming situation within a unit—even when the mechanism is well defined and agreed upon—is often a contentious matter among the parties involved in correcting it.

In fact, in some instances, it becomes impossible to isolate a single cause beyond a reasonable doubt. It should be remembered that the primary reason for identifying the cause is not to apportion blame (except in the case of new units or components where warranty considerations exist), but to allow the manufacturer and plant staff to initiate corrective actions that will prevent or minimize the possibility of its reoccurrence.

Figure 1.9.7 is a logic diagram for the investigation and identification of causes and mechanisms for a damaged or deteriorated condition.

This investigative process is shown to comprise two phases. The first might be termed a "field phase"—while all material and information is at the site, work is undertaken to allow the condition to be corrected and the unit returned to service. There are many occasions when this will suffice and no further investigation is required.

However, if this field phase does not identify the cause and/or mechanism, then the second is initiated—normally after the unit has been returned to service. There are instances when a condition cannot be corrected and restorative action taken until this "laboratory phase" is complete—possibly after some redesign and/or corrective manufacture has been undertaken. This need is determined by the extent and severity of the failure.

Initiating and driving mechanisms

When attempting to analyze a failure, it is necessary to identify the cause as far as possible. It is also necessary to identify the mechanisms involved in the failure. This determination can be difficult. For a well-designed and well-operated unit, the stress levels and

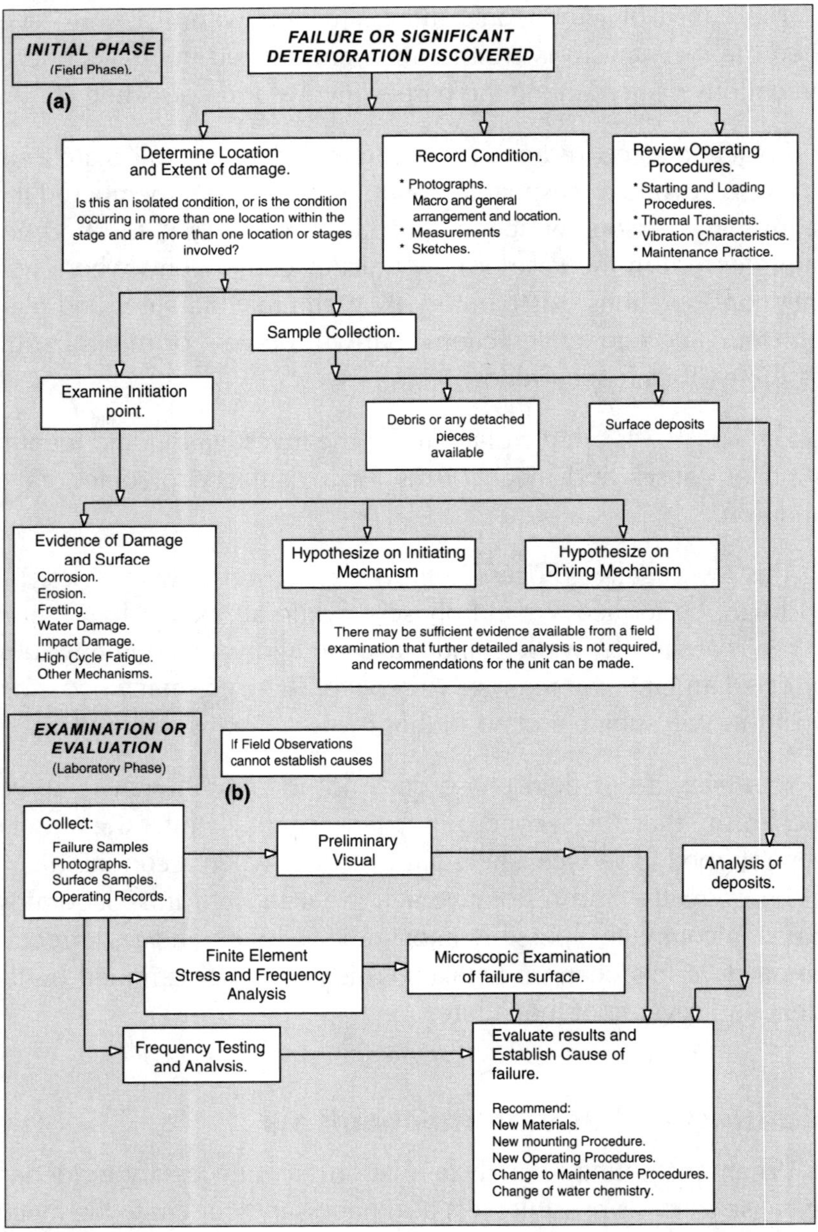

Fig. 1.9.7—Two phases of the damage evaluation process for identifying and quantifying a field found failure.

operating procedures should guard the various components against premature failure. Therefore, when a failure or deteriorated condition is found, it is important to identify the mechanisms producing this condition.

In making this determination it is normally necessary to identify two separate mechanisms. The first is that which initiated the condition—which can be any of a number, as discussed below. The second is the mechanism that is driving the condition, often to failure. In the majority of situations, this mechanism is high-cycle fatigue, ultimately producing material rupture at stress levels well below those able to produce failure if an initiating condition or fault had not been induced by some other stress-concentrating condition (*i.e.*, the initiating mechanism).

The mechanisms of damage and deterioration

The mechanisms of failure are the normal deteriorating processes such as fatigue and corrosion, which can occur in the majority of mechanical components if the conditions are right. Many of these mechanisms are relatively easy to recognize. They are known to occur as a result of the environment in which components of the unit operate and the loads to which they are subjected. In terms of the steam turbine the major mechanisms are:

High-cycle fatigue (HCF). Possibly the most damaging effects present in the turbine steam path are those associated with HCF. *HCF* is a mechanism that is a primary or initiating cause of failure in many components, both rotating and stationary. It is also the major damaging mechanism that will drive damage initiated by some other mechanism, forcing the situation to propagate to destructive levels.

Cyclic effects (or impulses) are generated in the flowing steam due to a number of effects (discussed in chapter 5). The frequency of the impulse is dependent upon the nature and source of the excitation, and the magnitude is dependent upon the flow characteristics at that location in the flow path. (Table 5.10.1 of chapter 5 lists various exciting sources within the steam path.) It is not necessary for the impulse magnitude to be high for failure to occur. Also, if the frequency of the impulse is near one of the natural frequencies of the damaged component, it requires only a relatively low magnitude impulse to cause failure.

A typical surface showing the effects of HCF is shown in Figure 1.9.8. The "beach" marks characteristic of HCF can be clearly seen. In fact, the true HCF surface shows a fine, uninterrupted crystalline structure, and the final rupture is in the form of a tensile or torsional overload failure, depending upon the form and type of stress causing final separation. The beach marks are normally present and represent transient conditions causing a fracture arrest or change in the crack propagation rate. This can be a massive progression of the crack at the tip of its current position.

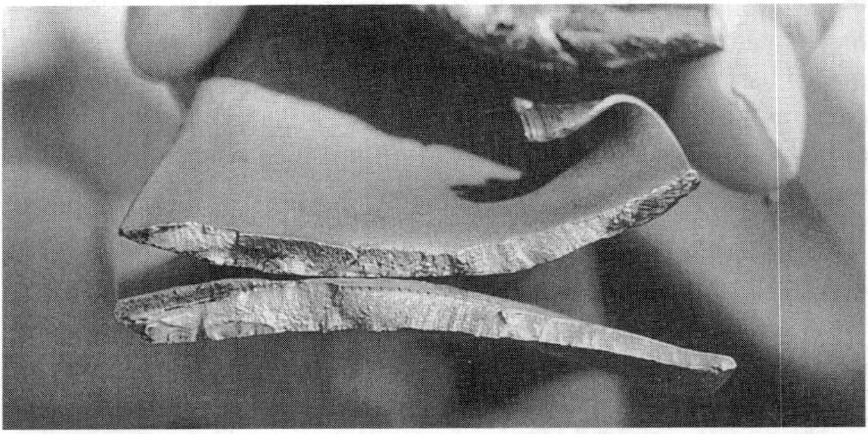

TAPPS

Fig. 1.9.8—The striations or beach marks on the vane of an L-1 blade, and the mating tip section recovered from the internals of the turbine.

Section	Harmonic	Typical Sources
High Pressure Section	• High per rev (40x)	• Nozzle tolerance limits • Upstream wake degeneration • Structural turbulence
Intermediate Pressure Section	• Nozzle passing frequency (NPF)	• Nozzle wakes
	• 2 x NPF	• Diaphragm harmonics
	• 3 x NPF	• Diaphragm harmonics
Low Pressure Section	• One per rev	• Relative displacement nozzles to blades
	• 2 per rev	• Diaphragm joints
	• Multiple/rev	• Structural supports in flow path
	• Medium/rev	• Diaphragm harmonics Aeroelastic disturbances
	• High/rev	• Nozzle turbulence harmonics • Upstream wake degeneration Structural turbulence

Table 1.9.1—Steam Path Exciting Forces

The true HCF surface has no indication of mechanical deformation, and if a surface were to fail entirely by HCF, the mating surfaces would "fit" together perfectly.

Low-cycle fatigue (LCF). Low-cycle fatigue is distinct from the HCF phenomena in several ways. These are best summarized as:

- the number of cycles required to cause failure are significantly less than for HCF—the time to failure being a function of the magnitude of the stress levels

- the number of stress cycles to cause failure are significantly less than for HCF. An arbitrary demarcation is often taken to be 100,000 cycles. However, this is simply a convenient means of definition. Surface appearance and other characteristics provide a better delineation between the two

- the individual stress levels developed at each cycle are normally considerably higher with the low-cycle effects
- there is mechanical deformation at the failure surface, and failure marks show individual evidence of rupture, *i.e.*, the stresses are large enough to cause plastic deformation at the leading edge of the crack

The most common forms of LCF failure are those attributed to thermal cycling, when large temperature changes occur causing localized thermal expansion or contraction of the material surface. These thermal movements introduce large internal constraining forces within the material introducing stresses of a magnitude sufficient to initiate and drive to rupture.

Figure 1.9.9 shows a normal LCF crack that originated at a sharp corner within a high temperature casing. Such cracks can be repaired normally by welding or by stitching.

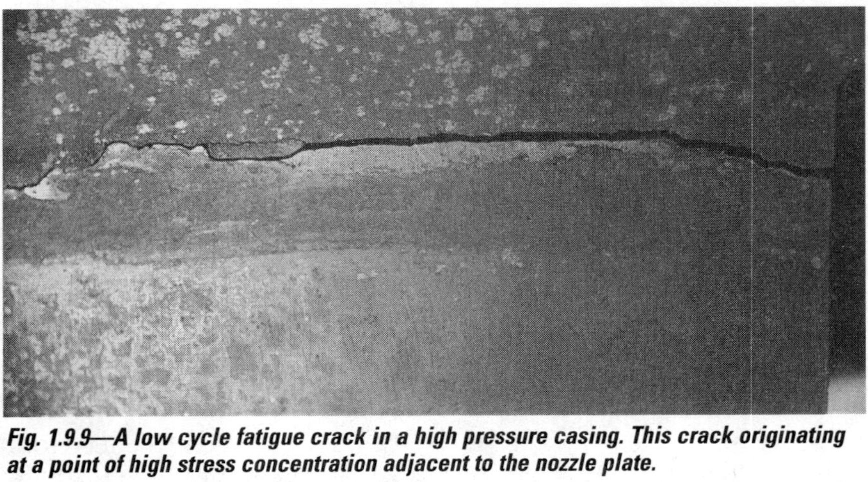

Fig. 1.9.9—A low cycle fatigue crack in a high pressure casing. This crack originating at a point of high stress concentration adjacent to the nozzle plate.

Creep deformation. Steels that are subjected to elevated temperatures for extensive periods exhibit changes in their microstructure. This has a marked effect on the material creep characteristics. Under conditions of high tensile stress, failures are normally trans-granular. However, at elevated temperatures and with the application of tensile stress, it is common for inter-granular failure to occur due to the effects of creep. *Creep* is a granular "slip" mechanism within the material that occurs at dislocations of the grain boundaries. This slip causes a plastic distortion of the components that is not removed when the stress is removed from the component.

On high temperature stages there are instances in which creep deformation causes plastic deformation of the steam path components. This is not a common type of mechanism, and occurs in only the highest temperature elements. It is often difficult to detect without the aid of detailed and repeated measurements. However, in those instances where this phenomenon is present, it contributes to the general degradation of a stage or component.

It is uncommon for elements to operate for periods of time after they go into the tertiary region of creep—and fail—as the damage is normally obvious before this level of damage is reached. Figure 1.9.10 shows the notch block in a high-temperature stage in which the centrifugal load of the blade is causing creep deformation of the pin and wheel material. Such damage presents an increase of the gap between the closing block and wheel. This type of gap is common but does not necessarily indicate that failure will occur. The width of the gap should be monitored until a maximum stipulated gap has been formed.

Solid-particle erosion (SPE). SPE is a damaging mechanism found in many front-end stages of high-pressure and reheat sections. SPE removes material from surfaces through a combination of gouging and the impact of solid particles of hard scale exfoliated from boiler tubes and carried into the steam path.

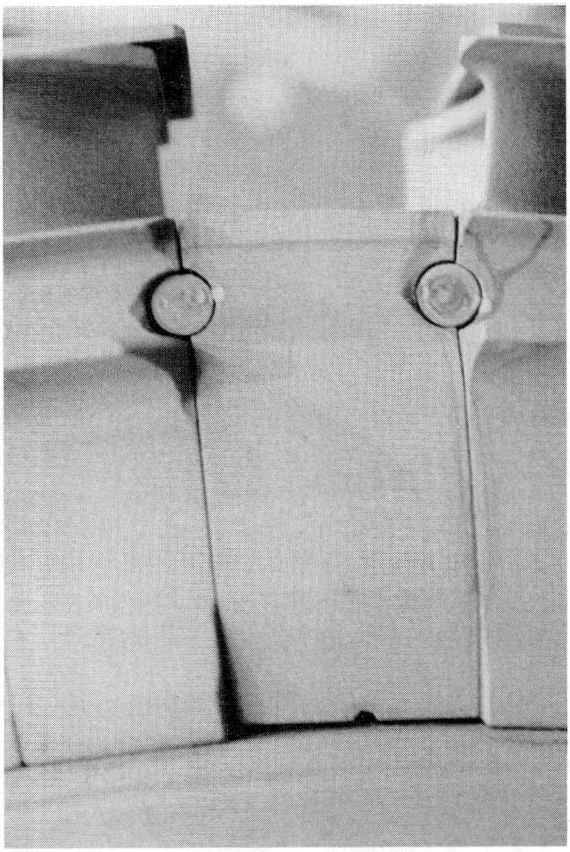

Fig. 1.9.10—The notch block from a high temperature stage which has crept out radially under the influence of centrifugal loading.

As scale enters the steam path, it impacts on and removes material from the stationary and rotating blade elements, the sidewalls, and the coverbands. It is also capable of removing coverband tenon material, thereby weakening the attachment of the coverband to the blades.

On stationary blade rows, the major effect is to remove material from the vane discharge edge, causing thinning and an increase in

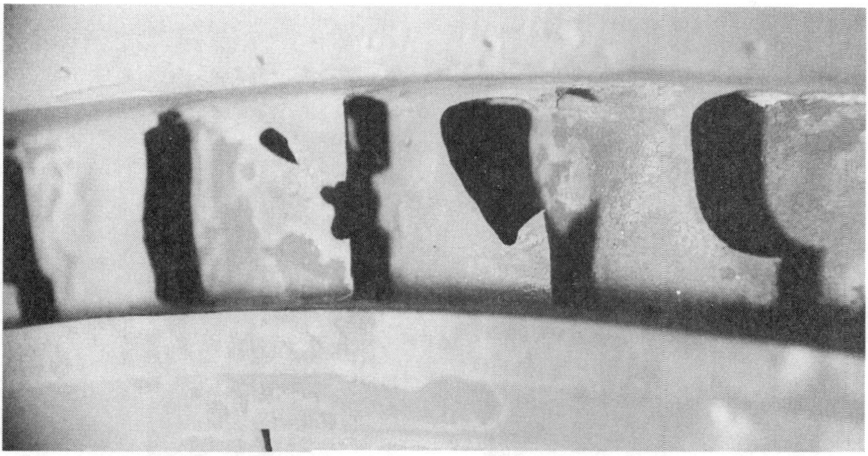

Fig. 1.9.11—Solid particle erosion showing material loss from a fixed blade discharge edge. In this stage the material loss is from previously weld repaired elements, and the loss has been severe. This material loss modifies the total discharge area from the stage.

the stage discharge area and angle (Fig. 1.9.11). The sidewalls can also suffer material removal, normally at the outer sidewall. The inlet to the stationary blades can also be affected by the gouging action of the particles (Fig. 1.9.12).

On the rotating blades, material is removed from the inlet edge, modifying the vane section and inlet angle (Fig. 1.9.13) and can remove material from the discharge edges (Fig. 1.9.14). From a structural consideration, a more insidious material loss comes from the pressure or concave face of the profile (Fig. 1.9.15). Such damage should be looked for when damage of the form shown in Figure 1.9.14 is evident. This is because many stages subject to this erosive phenomenon have an integral coverband. This material loss represents an undercutting of the integral coverband that will weaken the outer section of the blade and modify its vibratory characteristics. For stages without integral coverbands, there can be an undercutting of the tenons.

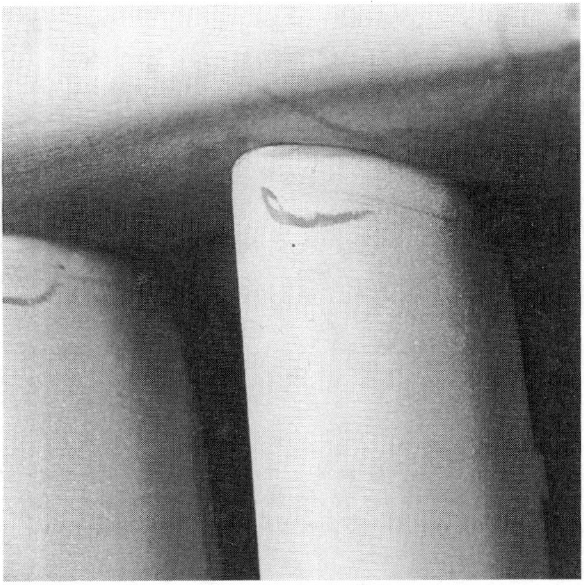

Fig. 1.9.12—Material removed from the inlet nose of a stationary blade row by solid particle erosion.

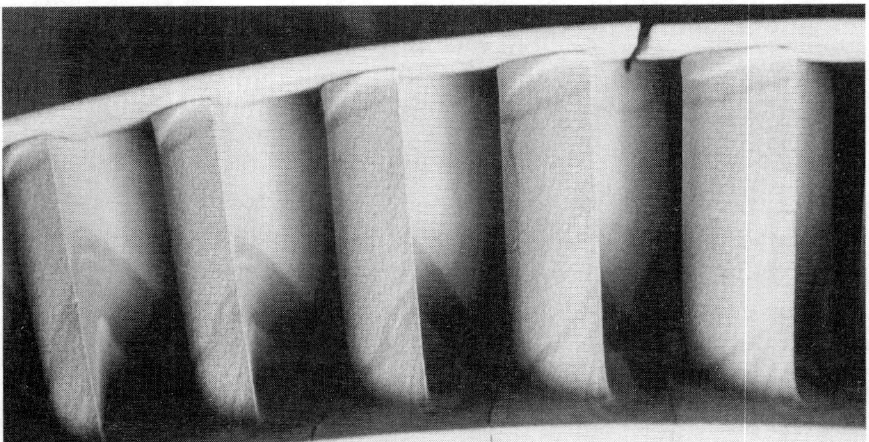

Fig. 1.9.13—Material removed from the inlet edge of a rotating blade row by solid particle erosion.

Fig. 1.9.14—Material loss from the discharge edge of a rotating blade row by solid particle erosion.

Corrosive effects. The ingress of corrosive ions into the steam path is always potentially possible. These ions have the potential to cause dramatic and expensive outages within the unit, and many stages of the turbine can be affected by this condition. However, the most seriously affected stages are usually those in which moisture forms after expansion of the working fluid into the moisture region. As this occurs, many of the corrosive ions precipitated from the steam are deposited on the steam path surfaces and the deposited moisture carries them into "hideouts" where they concentrate. At these locations they become aggressive when the correct environmental conditions occur.

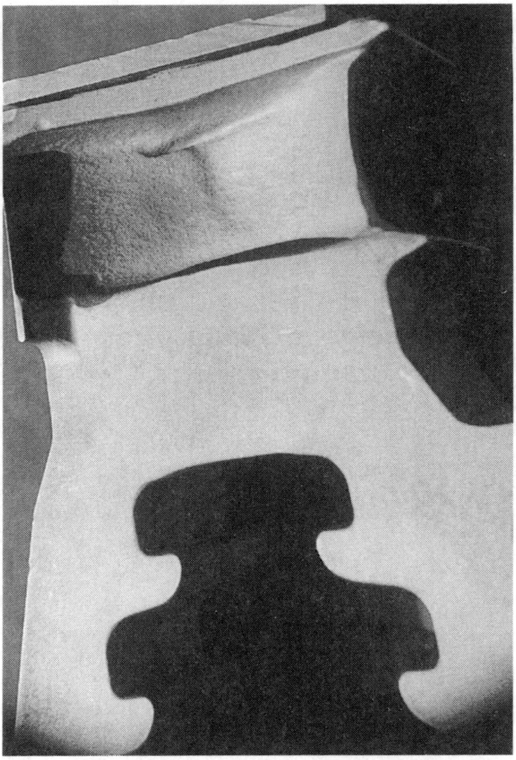

Fig. 1.9.15—The effect of "caving" on the pressure surface of a rotating blade.

There are several corrosive conditions having the potential to cause damage or deterioration:

- *Stress corrosion cracking (SCC)*—For SCC to occur, three conditions must be present in the elements: the presence of aggressive ions, a tensile stress in the component, and the material must be susceptible. The rate at which such cracks initiate and propagate depends upon the environmental temperature present to support the chemical reaction. In many portions of the steam path, residual tensile stress exists—both by design and also accidentally—as a consequence of manufacture. At these locations SCC can readily occur. The rate at which corrosive cracks are generated and then propagate depends upon the concentration of corrosive products and the magnitude of the stress and local temperature

- *Corrosive pitting*—Figure 1.9.16 shows a rotor portion where "High-Lo" seal castellation occurred. An environment of geothermal steam with a high corrosive content has corroded these

- *Corrosion fatigue*—Many components are subject to alternating stress of a magnitude that ultimately, fatigue contributes to its deterioration. If a corrosive element further contaminates the component, then the component's ability to resist failure is further impacted. Under these circumstances, failure will occur at either lower stress levels or in a shorter period of time at the same stress. Figure 1.9.17 shows a crack emanating from a tie-wire hole, where the initiating mechanism of failure was found to be corrosion fatigue

Fig. 1.9.16—Corrosive pitting on a rotor surface.

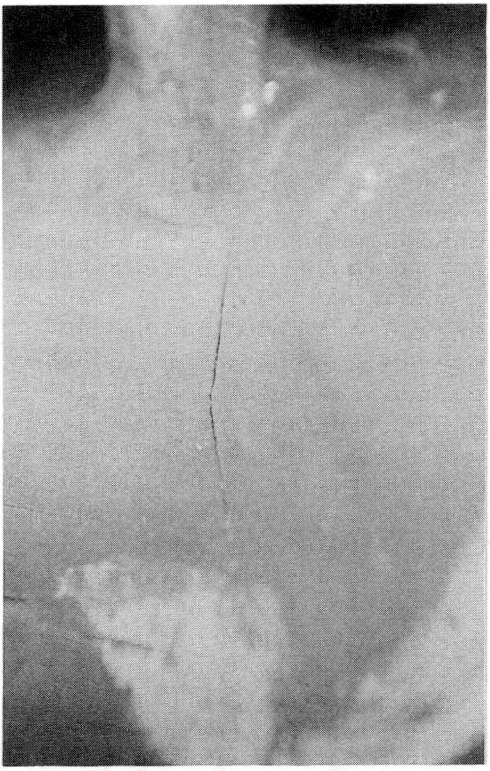

Fig. 1.9.17—A crack initiating at the tie wire hole.

This latter form of failure is distinct from SCC in that the applied stress is cyclic and the failure is a fatigue-type mechanism. Such failures tend to initiate at those points where corrosive products can concentrate and where a level of stress concentration causes high local tensile effects. Such cracks can be either trans-granular or inter-granular.

Moisture effects. Moisture forms in the working fluid due to condensation from the expanding steam (see chapter 3). Water droplets that form are transported through the steam path by the parent steam. Some of these droplets are ultimately deposited on the component surfaces where they have the ability to flow across them and accumulate into larger droplets. These larger droplets can re-enter the main steam flow and ultimately impact with the metal surfaces of the steam path components, causing erosion. However, more than one form of water damage can occur:

- *Moisture-impact erosion*—Once water droplets have formed, accumulate, and re-enter the steam path, they have the potential to erode the blade material if they impact at a value above a threshold velocity. This form of damage is common on the majority of large last-stage rotating blades (Fig. 1.9.18)
- *Trailing edge erosion*—This damage is caused on the discharge edge of rotating blades by water introduced into the steam path to cool the exhaust (Fig. 1.9.19). It occurs in high stress areas and is capable of causing ruptures that initiate at one of the erosion/cavitation grooves at the edge
- *Washing erosion*—When water flows at high velocities across metallic surfaces there is always the possibility of surface material loss. As this occurs, material grains break away and cause a roughening of the surface. Figure 1.9.20 shows the outer ring of a cast-iron diaphragm where material has been removed by washing, causing grooves to be produced emanating from the suction surface of the vanes
- *Wire drawing erosion*—When water exists at high-pressure differentials across a joint, there is always the possibility this moisture will be forced across the joint, and will remove surface material. This loss is also sometimes called "worming" (Fig. 1.9.21)
- *Water ingestion*—If water is returned to the turbine steam path as a "slug" this can cause various forms of damage from blade rupture to rotor bending

Fig. 1.9.18—Moisture impact erosion.

Transient operating conditions. Severe temperature ramps are a common damage phenomenon associated with transient operation. Other considerations include overspeed transient, which occur on a sudden load rejection or when testing the valve mechanisms. Each of these mechanisms or situations has the potential to cause a deterioration in the operating capability of the unit.

Fretting corrosion. When there is relative motion between surfaces that are in nominally "tight" contact, heat generation can result from this movement. The movement may be of the "sliding" or "impact" type. In either case there is a generation of excessive heat that removes surface material by causing localized oxidation. The material removed has the characteristic red "rust" appearance.

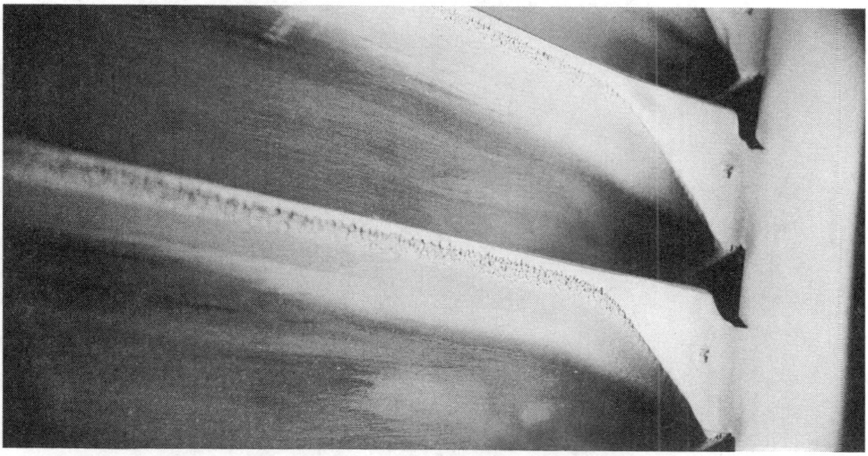

Fig. 1.9.19—Trailing edge erosion occurring on the last stage blades.

Fig. 1.9.20—Washing erosion on the outer ring of a cast iron diaphragm.

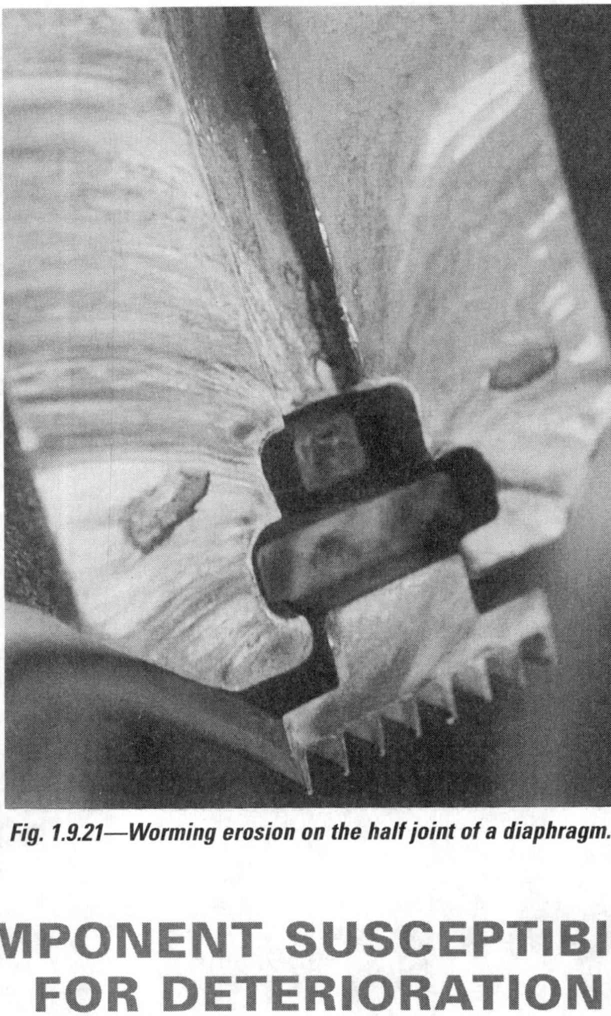

Fig. 1.9.21—Worming erosion on the half joint of a diaphragm.

COMPONENT SUSCEPTIBILITY FOR DETERIORATION

Should either the critical fixed or rotating components within the steam turbine sustain damage, it could be sufficient to force the unit from service. Such damage can be caused by various mechanisms associated with operating phenomena resulting from the characteristics of the working fluid, its pressure, temperature, moisture content, and possible rate of change during operation (these were discussed in the previous section). Deterioration can also be related to the stress levels within the components, or the result of matter carried into the unit from the boiler or other parts of the system. Problems of deterioration can also be the consequence of defects in the manufacture of the component.

In terms of the level of damage or deterioration with which the maintenance staff should be concerned—and possibly be monitoring—the following categories of "deterioration potential or susceptibility" can be ascribed to the various locations and components within the steam path. While not exhaustive, they will provide some general indication.

High susceptibility

Typically, these components include the last (L-0) stage rotating blades and many L-1 or L-2 stages. These longer blades are tuned, and the last stage blades are subject to variable pressure ratios and damage as a consequence of their interaction with the moisture present in the steam. This category would also include the high temperature stages of the high pressure and reheat sections that operate on steam returned from the boiler. In these stages, exfoliated scale may be carried in with the steam, causing stationary and rotating blade vane material loss. If the first stage of the high-pressure section is also subject to the effects of partial arc admission, that can also cause high impulse loads on the stationary and rotating blades, possibly leading to high-cycle fatigue failures.

Wheels shrunk onto a central spindle can have a high potential for damage due to corrosive action. This—in terms of serious damage—can be avoided to some extent by the stringent control of water quality and careful examination at outages.

Intermediate susceptibility

Typically, these components include those subject to corrosive attack, such as the last minus one (L-1) stage in the low-pressure section. Depending upon the pressure distribution, it could also include the last minus two (L-2) stage. This group also includes stages with a drilled hole in the vane for the admission of tie wires. The second and third stages after admission or return of hot steam can also be included because of degraded material mechanical properties at high temperatures. Because pressure loads on the various stage components can be high, creep deformation is a consideration. The possibility still exists with these front-end stages that hard scale from the boiler will cause an erosive material loss.

The diaphragms of the high temperature stages can be subject to creep, and rubbing should be considered as a potential. Nozzle boxes in the high pressure section can be a cause for concern due to solid-particle erosion and possibly other damage resulting from temperature transients and debris carryover from the boiler, particularly after boiler repairs.

Low susceptibility

This category includes all other components and stages in the unit. The blade rows operate at substantially constant pressure ratios. Operating temperatures are normally below the threshold temperature where material properties are reduced significantly, so components are not normally subject to creep deformation.

The rotor wheels, in many stages—particularly if there are pressure balance holes—have a potential to cause problems, and fillet radii can eventually be the initiation point of low-cycle fatigue cracks. In wet stages, these wheels can be subject to moisture-impact erosion under certain operating conditions, although the threshold velocity at which such erosion normally occurs has not been achieved.

Categorization

In categorizing components, it must be remembered that many units appear to have characteristics which make certain stages and components within those stages more susceptible to damage and deterioration. Therefore, categorization as given above should be considered as only a guide, and should not be interpreted to mean the "low susceptibility" elements will not suffer damage until after the 'high' and 'intermediate susceptibility' elements have been affected. Most owners will be aware of "rogue" stages and components that appear to fail with monotonous regularity in an otherwise acceptable unit.

While some elements may have a high susceptibility to failure or deterioration, it is possible that in the event of failure, these can be removed and the unit will continue to operate. Conversely, it is possible that low susceptibility elements must be replaced or refurbished when they do deteriorate, before the unit can be returned to service. The susceptibility level does not designate or indicate the ability of the component to force the unit from service for extended periods; rather, it reflects only the component's propensity to damage and/or deterioration.

Owners will often identify potential problem areas in their units at the first (warranty) inspection. At that time, the manufacturer will make a detailed examination of the unit and be able to identify to the operators any area or areas that have shown deterioration levels above the "norm" for the units. The owner should then decide to monitor these components, make a record of the existence of any damage present, and define or record the extent in some manner. The owner may also elect to carry replacement parts as inventory spares in the event they are required.

Monitoring efforts, when undertaken, should be concentrated on the most susceptible locations and components. A monitoring system can be established in conjunction with the unit manufacturer, who is normally interested in such information for future designs.

Steam seals are one series of components of the steam path that do not have a significant susceptibility to cause a forced outage—though their performance can influence the operation of the unit considerably. High leakage rates in seals will cause extensive losses to an owner in terms of higher fuel costs. It is generally prudent to have replacement seals or gland rings available at each outage, and the owner should know the limit at which deterioration (increased leakage area) can no longer be accepted. The owner should be prepared to replace worn seals whenever a unit is available.

Many seals are produced as an integral part of another component. These cannot be easily or economically replaced. However, refurbishment actions can often be taken to improve their effectiveness without major replacement. Recent advances in repair and refurbishment technology have added considerably to the ability of owners to refurbish existing components. This is particularly true for rotating blades (see chapter 8). Such refurbishment techniques can often reduce maintenance costs significantly, and return the elements to a fully acceptable mechanical condition within periods of time sufficient to support a normal maintenance outage.

While these refurbishment techniques are often able to return steam path elements to an acceptable condition, it may be necessary to plan that such refurbishment is undertaken during periods of low-load demand (when unit output is not required or can be replaced from other sources and the unit can be made available for this work). If the refurbishment process is protracted, the resulting extended outage, and the cost of purchasing replacement energy may more than offset any potential savings from refurbishment. Some refurbishment techniques may be undertaken on-site and in place; others may require the components be removed from the unit.

INSTANTANEOUS DAMAGE OR FAILURE

An instantaneous failure is one that occurs when the owner has no prior or minimal warning from a previous outage or observations, to the extent preventative action could be taken. In such cases, a significant failure is suspected from a change in operating characteristics and a unit is opened for inspection—and the failure is confirmed. At this point, the owner must make some determination regarding the actions required to return the unit to an acceptable condition, and then to service.

The instantaneous failure can be severe enough that refurbishment actions required are self-evident. Other failures or damage leave a degree of options available. In some instances the unit can be returned to service for an acceptable period with partial repair undertaken. A final decision concerning required corrective actions may be influenced by diverse factors such as the amount of reserve generating capacity on the system and the cost and availability of replacement or incremental power from other sources. The owner is also influenced by the extent to which replacement parts are available.

The ability to refurbish the damage within an acceptable time must be considered, as must the cost and potential of success likely to be achieved. The ultimate selection of corrective action will most probably involve at least the following factors:

Parts. Are replacement parts available from inventory? How long would it take for such parts to be manufactured and installed? Can the existing components be repaired or refurbished?

Power. Can the unit be operated with the failed elements removed? What is the level of power lost, and what effect will this have on unit heat rate and availability? Will loading limitations be required to protect the remainder of the unit? If the unit is operated without the failed components, will the steam conditions to the unit need to be reduced, or will the unit be load limited?

Time. Will such refurbishment be less expensive than replacement, in terms of both initial cost, and the time to complete the work? What are the possible effects on unit availability?

The effect of a major component failure or deterioration concerns any owner. The owner will normally suffer a financial penalty associated with such damage, through the cost of rectifying the situation and/or additional generating costs during the unit's unavailability. It is therefore necessary, in such circumstances, to determine the most cost effective method of correcting the situation and preventing reoccurrences.

Often when new parts are installed after some form of failure, possibly the removed parts can be refurbished and carried as inventory spares. This would allow a similar situation in the future to be dealt with in a more rational and expeditious manner. This is particularly so if the components are prone to failure, or the unit is one of a multi-unit installation. The owner of a unit that has suffered some form of damage or failure should always consider this approach.

It is not possible to formulate any rules to assist in decision-making related to instantaneous or sudden deterioration. This damage is only detectable from a change in the mode of operation of the unit, or its presence is determined at maintenance inspections when the steam path is exposed. At the time damage is found, the owners must determine the most expedient course of action. Such a determination may be to replace the elements or to undertake refurbishment, should this be possible. The logic for such an evaluation indicates the analysis and decisions required to establish a repair procedure (Fig. 1.11.1).

Often circumstances of instantaneous failure or damage mean insufficient time to make extensive plans without extending the outage,

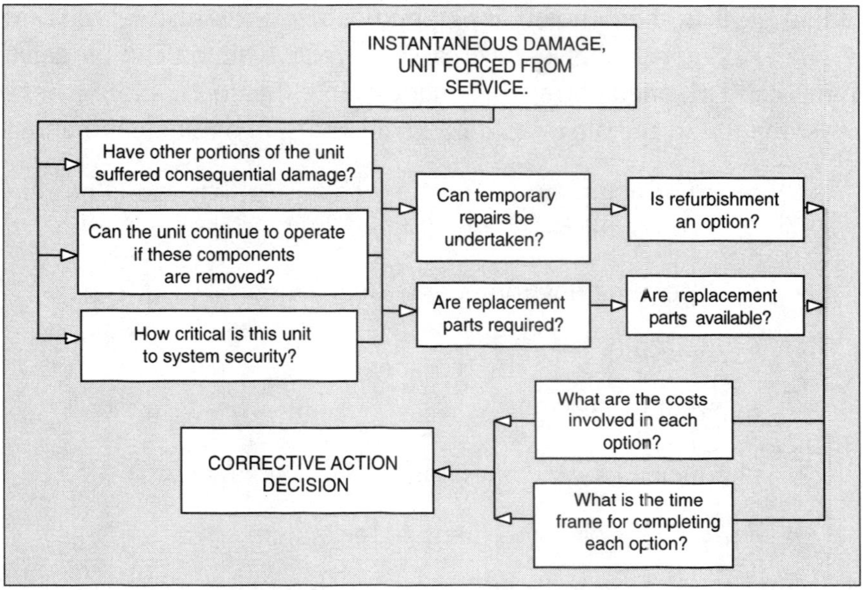

Fig. 1.11.1—The evaluation of a failure and the considerations required to arrive at an acceptable solution.

and the owner is forced into making a decision based on the most expeditious course at that time. In the long term this may not be the most suitable for rectifying the condition of the unit, and if doubt exists, a decision and action plan should be delayed until a more favorable corrective action is possible.

FACTORS CONTRIBUTING TO GRADUAL DETERIORATION

A gradual deterioration is one that occurs due to some mechanism slowly lowering the performance potential of the unit. Such a situation can be monitored from station instrumentation or is a condition measured and noted at planned outages. It is also a condition that allows monitoring so acceptable corrective action can be determined and planned corrections made before the condition reaches a level the unit is unable to operate in either a safe or efficient manner.

There are principal and various mechanisms that can contribute to gradual deterioration—all possibly leading to failure:

- the operating environment within the unit
- stress levels
- material properties
- chemical contamination within the working fluid
- carry-over from other parts of the system

Here a brief survey of these factors will be considered.

Operating environment considerations

The steam path is designed to operate and achieve its output under stipulated steam conditions. The flow quantity required for a specific output—the initial, reheat, and exhaust steam conditions—dictate stage geometry, construction materials, and design details. System economics tend to dictate unit and cycle configuration.

While steam temperatures and pressures at operational speed may introduce high levels of stress on components under normal operating conditions, modern design technology can help ensure these will not induce failure if the components are correctly designed. Under normal circumstances these are not factors that would cause elements to deteriorate or fail.

There are, however, instances where transient conditions (particularly those of initial and reheat temperature) will induce high thermal stresses in the components, initiating failure. The components most affected by these transient temperatures are the rotors, rotating blades, and casings. Manufacturers normally provide guidelines to operators defining a means of assessing potential damage accumulation as a function of the rate of temperature change. A typical curve of "life expenditure indices" for a high-pressure rotor is shown in Figure 1.12.1.

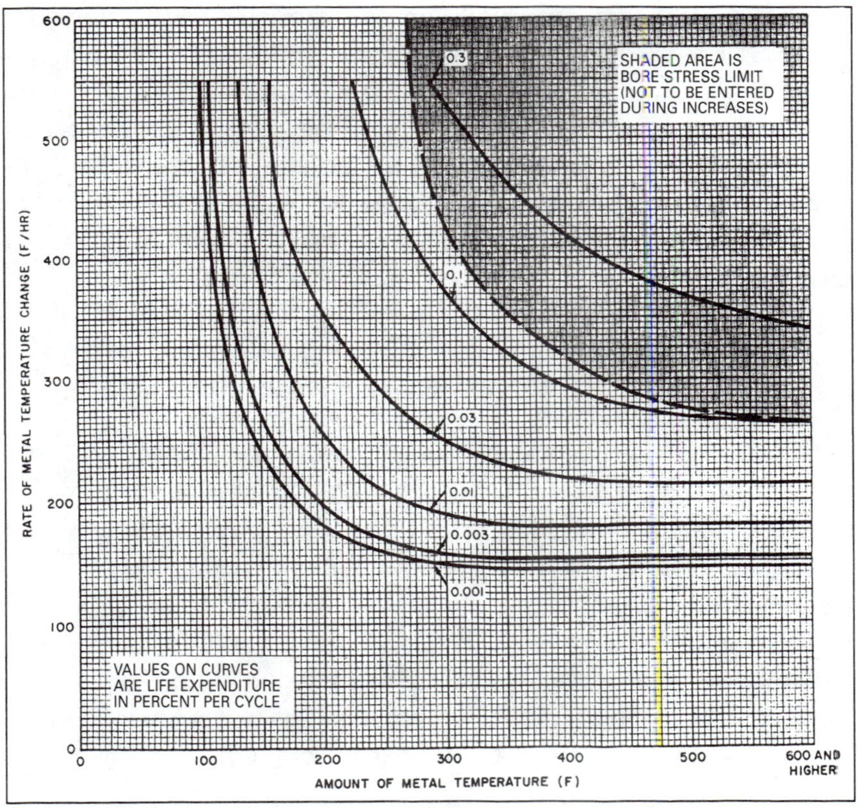

Fig. 1.12.1—Life expenditure curve for a high temperature, high pressure rotor.

Initial steam pressure does not normally cause concern in terms of placing the unit at risk. However, most manufacturers will place a limit on the amount of increase for which their units are suited, and the number of hours of operation at these higher pressures. A pressure limit is normally stated as a percentage of the initial design value. Such a condition is generally of more concern in industrial installations where other sources of steam may become available. Under these conditions if applied without a full evaluation of the unit (and possible steam path modification), it could induce failure due to one of several mechanisms.

For condensing units, minimum pressure must be maintained in the condenser. The last stage blades of large utility units are a variable pressure ratio element (this ratio also varies with radial height). If the exhaust pressure is raised too high (in excess of 6″ Hg, or inches of mercury), there is a possibility of inducing "flutter" type vibrations within the L-0 blades. There is also a possibility of deterioration caused by re-circulation through the lower vane sections of the discharging steam. This could occur because of the reduction in volumetric flow. Frictional heating of these last stage blades will cause a change in their operating temperature, and therefore could modify their vibration characteristics, which are temperature sensitive.

The one environmental effect that causes some level of deterioration in most condensing units is moisture-impact erosion. However, in a well-designed unit, this type of damage should not be an impetus to change blades. In fact, few units suffer material loss and deterioration to the extent blades cannot operate for 30 years. It is possible that towards the end of this period, there could be a significant deterioration in stage efficiency, but this is not a consideration of forced outage. At this time a unit normally operates at lower load factors and is not as important to the security of the power system.

More problems are induced by the methods of shield attachment, which is a process involving the use of heat. It can expose the blade material to the effects of heating cycles.

Stress level considerations

Stresses are induced in the unit during normal operation due to the steam environment, the effects of centrifugal loading, and the bending loads induced by the expanding steam.

As force develops on the blade rows, it is transmitted through the rotor to the generator or mechanical device that the turbine drives, causing torque in the shaft. Stresses in the stationary portions are of considerably less magnitude than those in the rotating components. In general, stationary components do not fail due to high-level stress, but there are instances in which stationary blades can suffer high vibratory loads and fail. Also, parts affected by creep will occasionally deflect axially to an unacceptable level.

The dangerous levels of stress induced in the rotating portions of the unit are most often those caused by stress concentration—a function of the geometry of the components. During the design phase, it is normal to evaluate the effect of stress, and modern design methods are fully capable of determining even the most complex forms. Unfortunately, design conditions may not be maintained for long periods. There are many regions of the rotating component where stress concentration can be the result of mechanical damage, where impact or other types of deformations are produced as the result of operation. Such deformations should be looked for and some evaluation made of their possible consequences. Non-destructive examination should be undertaken of such discontinuities when they are found.

Pressure impulses generated in the steam path have the potential to amplify the dynamic loading applied to the various components. The design process attempts to recognize, predict, and quantify these potential effects, minimizing their damage-causing potential. However, they continue to occur and units are often forced from service as a consequence of their presence.

Material property considerations

For each component comprising the steam turbine, materials are selected and their properties evaluated for suitability. Materials are selected to help ensure the unit will operate for its entire life with the minimum probability of failure or need for replacement, if operated to design specification. There are, however, some components considered consumable, and it is the intention of design that these will be replaced as required by the user.

The designers of steam turbines evaluate materials and establish the variation of long life properties at various required temperatures. During the design phase the components are selected and dimensioned to achieve an acceptable "factor of safety" at the predicted stress levels and local environmental temperature.

To ensure that materials meet their design specified requirements, the manufacturer stipulates chemical composition, material properties, method of manufacture, heat treatment, grain structure, and level of non-metallic inclusions—all of which could affect life and performance. When the materials are received from the supplier, and before use, the manufacturer will evaluate them. The manufacturer should not knowingly use any that do not meet the stipulated requirements.

All high temperature components (above 900°F) and many operating at high stress levels are particularly susceptible to variations of material properties. These components in particular should be evaluated in detail and not produced from any material that does not meet the highest standards as specified by design.

One aspect of material suitability that should be considered is many material properties, especially at high temperatures, are normally determined from short time test data. If the material is of a new design, there could be some small margin of error in the stated values. A manufacturer will normally allow for this possibility as a factor of safety in the design.

A designer's additional concern for components designated for high temperature operation is that material properties in these environments—operating at high stress levels—can deteriorate with time. This is an acknowledged fact and many components must have either an adequate factor of safety to prevent significant deterioration (blades), or be classified as a consumable (studs and nuts).

Considerations of chemical contamination

There are two aspects of chemical contamination that need to be considered:

- Possible deposits of compounds on steam path components that will lower the efficiency of expansion and therefore introduce extra aerodynamic losses within the unit. This is undesirable in terms of operating cost, but is not an effect that will lead to mechanical deterioration. It should be a part of each normal maintenance outage to remove these deposits by some suitable cleaning procedure

- Contaminants could gain access within the total thermal cycle and form corrosive compounds. These compounds can be carried over from the boiler and other portions of the cycle, and will eventually be deposited on some steam path components. Once deposited in the steam path, they will, at some locations, be the constituents of failure in the unit—an aggressive chemical compound, tensile stress, and an elevated temperature. These contaminants can react with construction materials and could eventually lead to mechanical failure, forcing the unit from service

Considerations of "carry-over" from other portions of the system

There are two types of carry-over into the turbine that need to be considered:

- The possibility of metallic or other solid debris carry-over from other parts of the system. This is concerned principally with the flow of solid particles, such as a weld bead from the boiler. This is normally most severe during initial start up, commissioning, or during early stages of operation. During these periods it is common to use a fine mesh screen over the main and intermediate valves. Also, after some years of operation, hard oxide scale exfoliated from boiler tubes will enter the unit and cause solid-particle erosion (Figure 4.3.3 in chapter 4 shows a fine mesh screen with contained weld bead)
- There is also the possibility at any time during operation for water to collect in a pipe above a choked drain. Under certain circumstances, this water can be carried over into the unit. The extent to which this will cause damage depends upon the quantity of water in the line, the manner of its entry to the steam path, and the extent and duration of the ingestion

MONITORING DAMAGE AND DETERIORATION

There are certain types of damage or deterioration that, when they occur, may not force a unit from service immediately but could have a long-term effect on availability. These are recognized as gradual deteriorating mechanisms. When such gradual deterioration is detected, it is often necessary and expedient to return the unit to service. This represents a situation under which the operator can, and often should, initiate some form of monitoring program designed to establish the deterioration rate. If necessary, corrective action can be planned for a future outage. There are some forms of deterioration that can be accepted indefinitely, but this is usually done at some loss in unit efficiency.

One of the most complex and potentially controversial decisions the maintenance staff will be required to make when responsible for steam turbine operation and performance is: *"At what stage in their operating life has the condition of the components of the turbine steam path deteriorated to the extent corrective action must be taken—either to restore the efficiency to an acceptable level or to prevent failure which would force the unit from service"?*

The factors that must be considered when damage is found, and the decisions—possibly backed by financial evaluations—include the following:

- Does the situation require immediate action to correct this deterioration, or can the unit continue to operate at acceptable levels of performance (efficiency and reliability) with the damage it has sustained?
- If corrective action is considered necessary, is it appropriate to replace the components with new, or can existing elements be refurbished? If refurbishment is considered adequate, can the work be undertaken on-site and in place, or must the components be removed from the unit?

Since the mid '60s, vibration signature analysis has provided an indication of the overall condition of a unit, and regular monitoring has allowed owners to detect and possibly identify pending problems. However, signature analysis provides evidence of change in operating characteristics once they have deteriorated to a point where possibly significant damage has occurred. Decisions related to the timing of corrective action and outages are made easier if the condition of specific components, with a known potential to deteriorate, are monitored at regular inspections and their deterioration rate established.

A suitable method of monitoring, and one that can be conducted by plant operators without the use of any expensive equipment, is to examine the unit when it becomes available at regular maintenance inspections. At that time, the unit should be examined for evidence of abnormal or excessive deterioration.

The condition of the components should be quantified and recorded through measurement, castings, and photographs. It is unfortunate, but with few exceptions it is not possible to make an examination at other than periodic inspection outages. This means the steam path may only be available for detailed examination on a normal outage cycle. Because of this, it becomes necessary to examine components in detail at these outages, note the extent of all damage or deterioration, refurbish to the extent possible, and then plan to have replacement parts available for installation at subsequent outages.

Any component of the steam path that deteriorates can be subject to two levels (or rates) of deterioration. First is that which occurs suddenly, possibly instantaneously, as the result of some transient condition or other phenomena of operation. These phenomena can include the failure of an upstream component that causes consequential damage, or the ingestion of large quantities of water. Such damage is normally severe and may give little or no warning of its pending occurrence. Therefore, little or no indication is offered that the unit should be shut down prior to the damage. After the damage has occurred (and depending upon its severity), there may be a small increase in the vibration level or minor change in other operating characteristics to provide evidence of its occurrence. Therefore, the damage may not be detected for some period, during which time the condition could deteriorate further.

The second level of deterioration is that which occurs as a gradual effect, such as the growth of a crack or component material loss due to some form of erosion. This second level of deterioration—the gradual worsening of the component condition—will often allow the owner to select and plan his remedial actions, and to make replacement parts and services available to meet an outage schedule. Such a refurbishment outage may require, however, that a normal maintenance period be extended to allow satisfactory reconditioning.

The following sections consider various causes of gradual deterioration and, where possible, provide some guidelines that will assist operators in determining the extent to which components need to be replaced or refurbished. Unfortunately, there are no general rules that apply to any set of damage conditions. Obviously, the most satisfactory method is to replace or refurbish components when damage is first noted. However, this is not only expensive in terms of the replacement cost, but could require an extended waiting period for such parts to be delivered or reconditioned. Normally, a plant's outage plans do not support such a delayed decision. There are instances when damage can be accepted and monitored but this possibility must be evaluated in each case.

The mechanism of gradual deterioration

Those principal and various mechanisms that can contribute to gradual deterioration (possibly leading to failure) are considered in the previous section *Factors Contributing to Gradual Deterioration.* When damage due to these forms of deterioration is found, it is recommended that the condition be monitored so corrective action can be taken before the unit is forced from service.

Monitoring techniques for gradual deterioration

There are certain types of damage or deterioration which, when they occur, may not force a unit from service immediately—but could have a long-term effect on unit performance. When gradual deterioration is initially detected, it is normally necessary and expedient to return the unit to service. This represents a situation by which the owner can (and often should) initiate some form of monitoring program to determine the deterioration rate. If necessary, corrective action can be planned for some future outage. Some forms of deterioration can be accepted indefinitely, but this offers the possibility of a loss in unit efficiency.

This section uses examples to discuss typical monitoring techniques that can be applied to portions of the steam path components. They provide guidance of methods suitable for operators to use in establishing the rate of deterioration within the unit. They also provide, where possible, some guidance when conditions are deteriorating to an unacceptable level and corrective action should be considered.

In providing this guidance it must be remembered each unit is unique and represents a different set of parameters. Judgment is very much dependent upon the mode in which a unit is operated. Also, the owners' previous experience is often a good indication as to the danger posed by any situation. With these provisos, then, the following observations and situations should be considered as a guide, subject to calibration from system, unit, and manufacturer experience.

Moisture-impact erosion. In general, moisture-impact erosion will not by itself cause sufficient material loss that a blade will need to be replaced. Damage and material loss that occur in the outer portions of the blade vane should not impact upon element reliability. This is true to the extent the erosion occurs on the inlet edge suction face.

There are, however, certain aspects of erosive material loss that can cause concern, and possibly require some form of corrective action:

- Excessive erosion below the erosion protective shield
- Erosive undercutting of tenons attaching a coverband
- Severe erosion beyond, or behind, the erosion shield
- Severe erosion endangering the attachment of the erosion shield
- Severe erosion endangering the attachment of a vibration-tuning device

Moisture-impact erosion is a time dependent phenomena, though the damage or material loss does not vary linearly with time. The curve (Fig. 1.13.1) indicates erosion penetration ('ε') as a function of time in hours and shows the approximate time in years. The unit load factor is 70%. This curve can be used to predict future penetration as a function of time as the value of 'ε' changes. This will

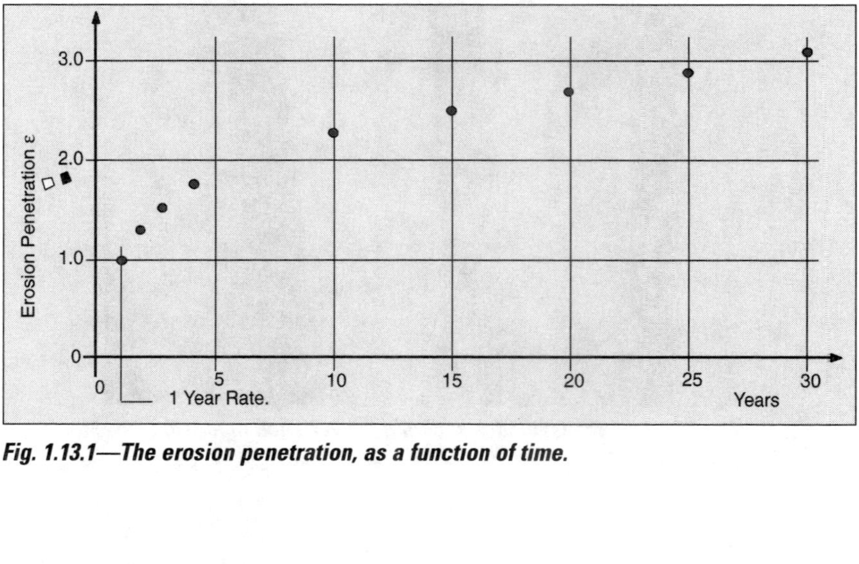

Fig. 1.13.1—The erosion penetration, as a function of time.

allow some estimation of further damage that could occur. (This relationship between penetration 'ε' and time is considered in greater detail in chapter 3.)

Localized or secondary erosion can be examined and its possible effects estimated from the dimensional changes this produces on the blade. Blades subject to moisture-impact erosion are also elements that are tuned. These elements normally have high stress levels and their frequencies can be sensitive to changes in blade section. Figure 1.13.2 shows a blade row that has suffered some small level of local secondary erosion that removed a greater amount of material at one specific location.

Fig. 1.13.2—Exhaust blades showing the variation of erosion penetration in the outer flow sections.

An example of monitoring the last-stage blades of a unit with predicted heavy erosion is shown in Figure 1.13.3. Here the mean values of erosion in the outer section is shown for the turbine end 'A' and the generator end 'B.' In addition, the worst blade is shown. These blades were eventually replaced/refurbished after 217,000 hours of operation as part of a general refurbishment program. The condition of the blades was monitored using casts and measurements.

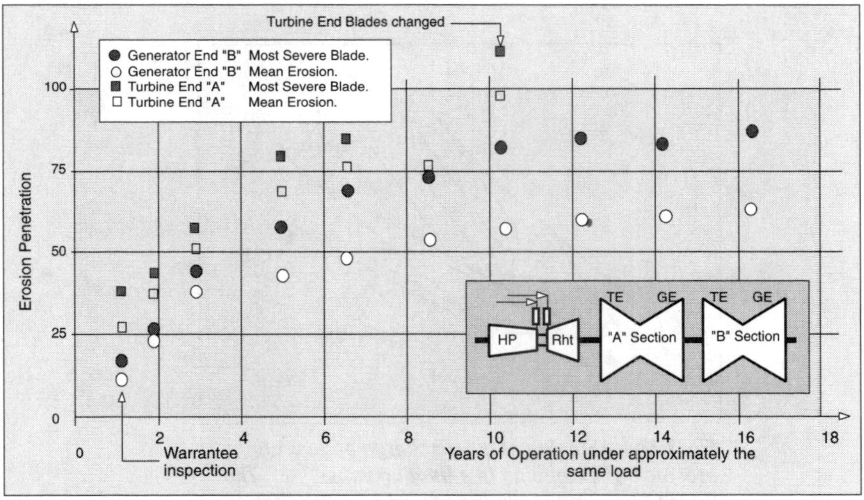

Fig. 1.13.3—Erosion penetration history of a turbine back end with heavy erosion.

Trailing edge effects of erosion. Erosion of the exhaust stage blade trailing edge occurs due to moisture particle re-circulation through the root section. This material loss occurs when cooling sprays are used to cool the exhaust hood. Figure 1.13.4 shows a typical cross section of an exhaust blade near the discharge edge of the root profile. The damage from this type of erosion removes material at a location where vane stress levels are high. Therefore, only relatively small levels of material loss can be tolerated.

No general rules exist for estimating failure potential. To determine such detail, an in-depth stress analysis (probably by finite element methods) is required. However, the operator should make a true evaluation of the stress levels and concentration potential once the penetration 'd' has extended to 0.050" or half the vane thickness 't' at discharge—whichever is less. Such an evaluation should be made prior to reaching a dangerous level of penetration. If analysis proves the material loss to be dangerous, efforts should be made to limit operating conditions and cooling water flow quantities to the extent the damage is limited. Such blades will eventually require replacement.

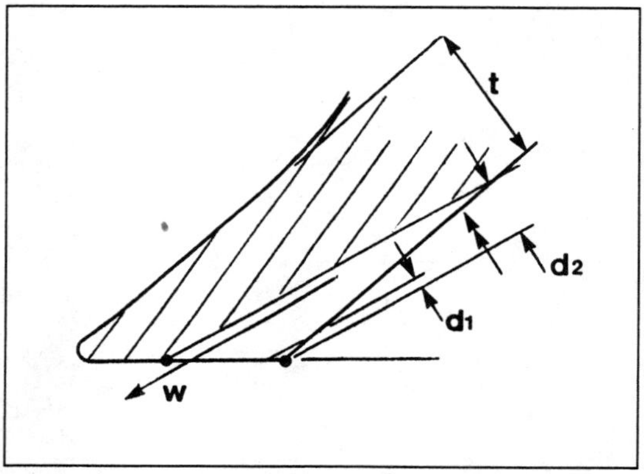

Fig. 1.13.4—Erosion of the discharge edge with erosion "d" extending to a limiting value "d2". The water discharges from the groove in a direction "w".

When such damage is initially noted, it can be monitored by means of a plot of penetration 'd' as a function of hours of operation with the cooling sprays operating. If the material loss becomes severe, the owner should consider plans to replace the blades.

Fortunately, the last stage blades (and particularly the discharge edge) can be inspected easily. When this type of damage reaches a critical level, such examination should be undertaken regularly, including the application of non-destructive methods.

Solid-particle erosion. When oxide scale—particularly hard and abrasive—is carried into the turbine steam path from the steam generator or reheater, it has the potential to remove material and affect both the mechanical integrity and thermodynamic efficiency of the steam path. Such scale can remove material from the stationary and rotating blades and, after being centrifuged to the casing, can rebound and remove material from the tenons attaching the coverband.

While the *stationary blades* may suffer little adverse influence from stress effects, they have the potential to significantly influence the efficiency and mechanical integrity of the remainder of the expansion. The following rationale applies to material loss from stationary blades:

- The quantity of steam flowing through any stage, and the discharge pressure from a blade row, is a function of the discharge area from that row. When the affected stage is of a two-flow arrangement in any pressure section, there could be an unequal sharing of the flow between the two halves. This occurs when erosion is not equal on the two halves of the flow
- The effective discharge angle from any throat formed by a blade pair is a function of the ratio "throat opening" to "pitch." Figure 1.13.5 illustrates the variation in discharge angle (or ratio) "O/P" as material is removed from the discharge edge due to solid-particle erosion

Wear is normally not uniform along the radial height of the vane. The effects of uneven wear will be to cause a significant variation of discharge angle on the stationary blade and the magnitude of the steam impulse from it.

While the *rotating blades* are subject to the same considerations as apply to the stationary blades (in terms of discharge area and angle), they are also subject to centrifugal loading. This means considerations of stress are often more significant. Fortunately, the scale that causes erosion (because it experiences the centrifugal actions imposed by the rotational effect) will tend to attack the outer regions of the blade, where centrifugal stresses are lower. However, if the affected blade elements carry a coverband, the bending stresses imposed on the outer portion of the blade vane can be severe. It is possible for these stages to suffer damage.

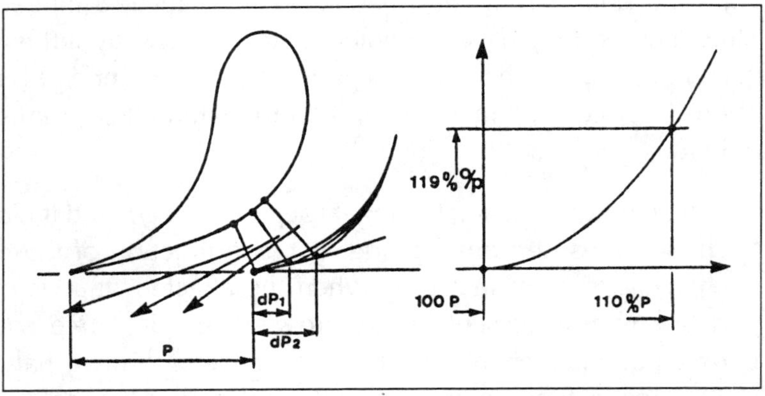

Fig. 1.13.5—The variation in the ratio "O/P" as material is lost from the discharge edge due to SPE.

An insidious aspect of solid-particle erosion (SPE) is the erosion of tenons between the vane tip and the underside of the coverband. This erosion can weaken the coverband attachment and possibly lead to coverband detachment.

SPE is time dependent—effects are dependent upon operating hours. However, units with a large number of start ups are more susceptible to this type of damage because that is when scale is exfoliated from the boiler tubes. A suggested method of monitoring penetration 'd' is shown in Figure 1.13.6. It shows that monitoring a simple calculation of the effects on stress levels in both the blade vane and shroud band will provide some indication of when corrective action needs to be taken, usually in the form of new blades.

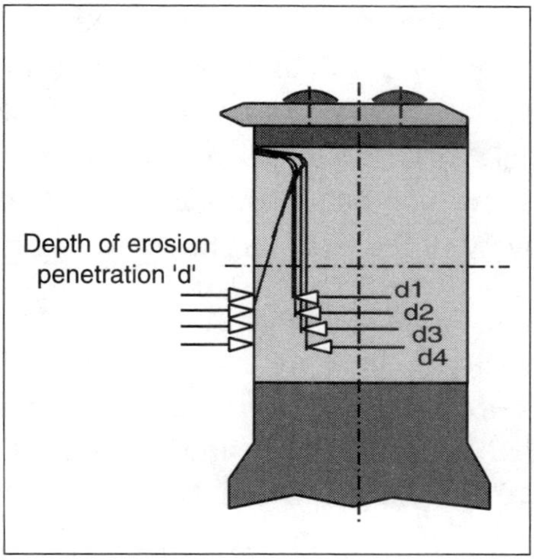

Fig. 1.13.6—Showing the measurement of "SPE" penetration at the vane inlet edge.

Crack indications in the blade vane. The appearance of a crack in the rotating blade vane or transition from the vane to root platform is sufficient warning that the blades may have reached the end of their useful life. Fracture mechanics methods exist and can be applied to indicate the rate of crack growth and eventual fracture. However, it is assumed that any crack has sufficient potential for catastrophic failure in a steam turbine and the element should not be returned to service.

One "necessary" exception to this was the case of last stage blade elements in which cracks appeared at the tie wire hole. They were evident below the braze material. These cracks were initially found in three elements. This was a utility unit, whose continued operation was essential to meet demands for power and heating steam. After the utility ordered replacement blades, it returned the unit to service and conducted monthly inspections and measurements of the cracked elements. These condition reviews were at weekend outages, when the unit could be made available for inspection from within the exhaust hood. After some months, a weld repair of the cracks was made on-site and in place. The blades continued to operate without problems. A year after the weld repair, the blades were replaced. At that time the cracks had not reappeared, but the utility elected to replace the blade row with new elements. The history of these blades from initial discovery of the crack is shown in Figure 1.13.7.

On many units, a crack appears at the fillet radius between the vane and root platform. In general, this type of crack cannot be accepted. However, there have been instances in which such cracks have been removed by grinding and/or polishing and the unit returned to service. The extent to which this is possible is a function of the depth of the crack. Such a practice is not recommended as a permanent "fix." Instead, the blades should be removed and replaced, as the risk level is high.

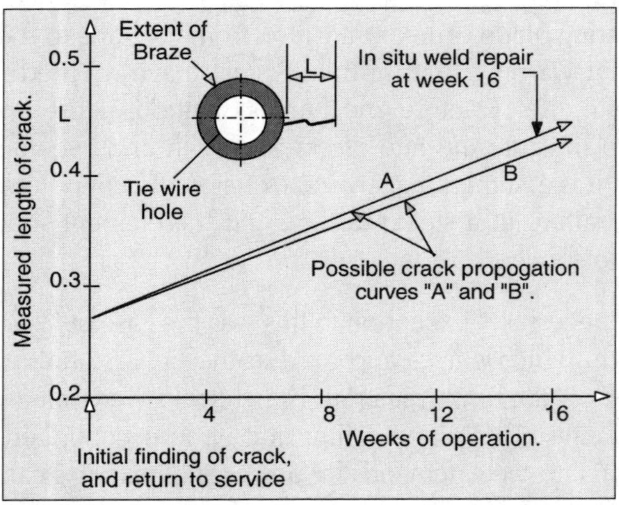

Fig. 1.13.7—Measured crack growth in a last stage blade. The crack initiating at the tie wire hole. See figure 1.8.17.

Diaphragm vane cracks. Diaphragm vanes can crack because of a variety of phenomena that do not normally represent a serious situation. However, a crack initiating at the discharge edge can grow. The edge can become unsupported and sufficient flutter can be introduced to cause the discharge to break away. Therefore, such cracks should be repaired. Depending upon the selected weld filler material, such a repair can be undertaken on-site and in place using a weld rod selected for the temperature of the stage. Low-pressure units—particularly those manufactured by welding techniques—will often develop cracks running across the vane. The growth of such cracks should be monitored and the element repaired at some suitable outage. Because it is relatively easy to make such a repair, it is not necessary to monitor these cracks. Under normal circumstances, the diaphragms should be repaired and returned to service.

Seal system wear. Seals within the steam path provide a constriction between the stationary and rotating parts. Steam that leaks (or expands) past these constrictions represents wasteful expansion and a reduction in the energy generated. Seal wear should be monitored, and as such wear becomes excessive, seals should be replaced when and where possible.

The "cost" that excess clearance represents can be calculated in terms of geometry of the stage, thermodynamic conditions at the seal point, and fuel cost. Owners must understand the cost of incremental power as a function of seal clearance, and be prepared to upgrade these seals as wear increases. It is normal practice within the steam turbine industry to monitor clearances throughout the unit each time it becomes available for such measurements. This information is valuable in a variety of ways. One of the most important is expensive fuel, which can be a predictor of when seals should be changed. Losses associated with seal wear are considered in chapter 10.

Corrosive damage. Corrosion is the result of chemical action in the steam path caused by corrodents introduced into the cycle by external sources. Such corrosion can occur on visible surfaces during normal inspection outages. Surface-type damage is not normally significant in terms of the reliability of the unit. The more serious aspect of corrosion occurs at points within the steam path where "hideouts" occur, such as at the shrink fit between the wheels and shaft, and the tie wires and tenons that attach the coverband. Units should be inspected (as far as practical) for this type of damage—when damage is discovered, components should be replaced. Unfortunately, much of this type of damage is usually invisible and can neither be seen nor adequately monitored.

Surface corrosion is less severe in terms of unit performance, but can indicate hidden corrosion. Therefore, when surface damage is found, it is a good practice to use photographs to monitor its extent to compare deterioration and consider future damage at hideouts. Unfortunately, the corrodents causing surface damage may not be the same as those that concentrate in hideouts and lead to component cracking.

Checking for crack initiation in hideout regions is complex and often expensive. Therefore, when such damage is suspected, an evaluation should be undertaken. If visual inspection confirms the possibility, an ultrasonic examination can be undertaken. The shrink fits between wheels, and central spindles are particularly difficult to access and inspect—it can represent one of the more complex evaluations expected of the operator engineer.

Fretting damage. Relative movement between components causes fretting corrosion as pressure develops between their contacting surfaces. By definition these surfaces are rarely visible and the damage is hard to detect. However, when such damage is found, its extent should be monitored. It has the potential to loosen fits, which if design does not allow, can cause a change in the mechanical dimensions of the component and aggravate what can be an already serious situation.

There are situations in which relative movement between surfaces is required. This means that conditions can occur in which the oxide products of fretting can cause a "designed" sliding surface to become "solid" and defeat the design requirement.

When this type of damage occurs, it is best recorded by means of both photographs and casts of the surface, if accessible, from which depth and affected area can be determined in extreme cases.

Water washing erosion. High-pressure water (or water at high velocity) has the potential to cause significant damage within the steam path. It can also compromise the integrity of the main structural components of the unit.

Washing or wire drawing type damage on a seal surface can reduce the stage efficiency. Also, the leakage flow can interfere with the free flow of the expanding steam, possibly causing circulation losses to be introduced into the flow path. When this type of damage is discovered, it is normal to assess the ability of the component to continue to form effective seals, and if necessary, make repairs.

When washing damage is not sufficient to require immediate repair action, a monitoring program (as a function of time) should be introduced. This is a type of damage best recorded by photographs.

The steam path requires exact dimensions of the passage through which the steam is expanding. Washing damage that modifies the shape of the flow passage must be monitored, then repaired or replacements parts used as soon as the steam path integrity is compromised.

The other type of material loss mechanism that requires some monitoring (particularly in older units) is the combined "washing/impact" material loss (Fig. 1.13.8). Here, material is removed from a structural component to the extent wall thickness is reduced and the vessel possibly weakened. Damage of this type is not common, but when it does occur, the results can be severe.

Fig. 1.13.8—Combined "washing" and "impact" erosion on the cast outer sidewall of a diaphragm.

REPLACEMENT PARTS STRATEGY AND SUPPLY

Alternate suppliers of steam path components have evolved, and have had a dramatic effect on the manner in which many owners plan and undertake their maintenance programs. The most significant of these changes include:

- the owners' ability to carry a smaller inventory of consumable parts. This minimizes the cost of maintenance by reducing the funds tied up in components not required for years

- the ability to order known, required parts weeks, rather than months, before an outage. This reduces costs and concerns regarding on-time delivery

- the owners' ability to make replace-or-repair decisions for unanticipated damage and deterioration when the unit is opened and its condition determined at the outage

Despite these advantages, the owner will not be able to order all required material for each outage and every contingency after the unit is opened. However, many of the emergency parts previously unavailable from inventory—parts not generally carried as stock—can be made available from alternate suppliers when inspections establish their need.

REFERENCES

1. Dewey, R.P., and N.F. Rieger. *Survey of Steam Turbine Blade Failures,* Research Project TIE, Final Report, March 1985

2. *Open letter from General Electric: Extending Re-inspection Intervals-GE Turbine Rotors,* Made available at the EPRI Steam Turbine-Generator Workshop, Milwaukee, Wisconsin, August 1995

3. Sanders, W.P. *The Review, Assessment, Refurbishment and Efficiency Upgrading of Steam Turbine Units,* Canadian Electrical Association, Halifax, Nova Scotia, 1985

4. ABB Power Generation Ltd, Baden Switzerland: *Overhaul Plans for Large Steam Turbines-HTGD 690 218 E and HTGD 90 700E*

5. *Recommendations for the Inspection of Steam Turbines: - VGB - R115 Me.* VGB Essen, Germany

6. *Sanders, W.P. Potential for Manufacturing Problems to Cause Steam Turbine Blade System Failure,* Turbomachinery International, Vol. 27, No 7, September 1986

7. Sanders, W.P., and W.R. Southall. *A Guide to Mechanical Condition Assessment in the Turbine Steam Path,* EPRI Steam and Combustion Turbine Blading Conference, Orlando, Florida, January 1992

8. Timo, D.P. *Design Philosophy and Thermal Stress Considerations of Large Fossil Steam Turbines*

9. Rieger, N.F. *Blade and Rotor Reliability Improvement,* STI/ERPI Seminar, Rochester, New York, June 1991

10. Greco, S. Private correspondence

Chapter

2

Steam Path Component Alignment and Stage Spatial Requirements

INTRODUCTION

In order to achieve an acceptable level of performance from the steam turbine, two things are essential:

- individual components comprising the steam path are manufactured in accordance with the design specified requirements
- components are arranged or assembled within the unit so their spatial relationship, relative to the other components with which they will interact, are correct

This compliance with design requirements will optimize the energy conversion process and help ensure the structural reliab lity of the system is at an optimum level. The spatial requirement within the steam path will be considered in this chapter, along with factors influencing the relative positions of the various components one to the other.

Note: For the purpose of this work the steam path is considered to comprise the stationary and rotating blade rows, their means of attachment, the rotors, and the casings. To complete the arrangement of the components and help ensure a satisfactory design, it is necessary to consider the arrangement made to minimize internal leakage and the leakage occurring from within the casing to other portions of the steam power cycle. In the case of low-pressure sections, leakages into the system must also be considered.

Before considering details and requirements of the individual components comprising the steam path, we must review their positional needs relative to each other, and those factors influencing them when the unit is in operation. Unfortunately, the steam path is available for measurement, adjustment, and gauging only in its cold stationary position—conditions under which components must be erected and aligned. When the upper half casing is assembled to the lower and steam is admitted to the unit, the component parts change relative to one other. These changes are due to various factors influenced by steam temperature and pressure and by the mass and rotational effects of the rotor system.

During initial manufacture (or repair or replacement of various path components) certain spatial relationships in the axial, tangential, and radial directions must be met. Other relationships, by preference, should be met to assist in optimizing performance. The turbine supplier will manufacture and assemble the component parts of the steam path to ensure design requirements are achieved within specified tolerance bands. The builder monitors this process. In addition, within the organizations of some purchasers, groups have a responsibility for monitoring these operations.

It is not unreasonable to assume a unit, as originally supplied and erected within the operator's plant, will conform to design requirements. Discrepancies with design specification must be recorded and information supplied to the owner, so corrective action can be taken as the opportunity arises.

A unit removed from service for inspection and maintenance often requires some level of remedial work. Remediation aims to return the components and their overall arrangement to conditions as close to original as possible, consistent with preserving the unit's performance potential. It is therefore valuable for plant maintenance staff to have sufficient knowledge of the component parts and their assembled requirements to determine the most appropriate course of action in any situation where adjustment or corrective action is required.

In terms of manufacturing requirements, the turbine designer must provide exact definitions. These must be made available to the manufacturing portion of the supplier organization to ensure design requirements are first understood, and then met within specified tolerances. This information must be provided in such a form that it allows design needs to be understood and followed by the manufacturing department without ambiguity. Design requirements will also identify boundaries or engineering tolerances within which the design requirements must be met. Tolerances specified by the designer should reflect the limits within which components can be manufactured and still achieve a level of performance consistent with design predicted values. These tolerances must also reflect achievable values for the processes and component being considered.

The operating engineers and technicians within the purchaser's plant normally do not have access to detailed design information—this is normally considered proprietary. They should, however, have sufficient opportunity during the manufacturing and erection phase to ensure that components meet design approval. They will normally have sufficient information on clearances, unit alignment, and the initial design conditions, and can reestablish conditions to ensure correct spacing between the parts is maintained (at least in axial and radial directions).

Engineering tolerances are selected to ensure that the design expectations of performance levels are met, and where appropriate, components can be disassembled for repair or replacement. Tolerances should also help ensure components can be interchanged within similar units in and between stations. Unfortunately, there are times when the requirements of design are not sufficiently explicit, or the unit operator has been provided insufficient information. In such cases, maintenance staff is unable to evaluate all possible courses of action and achieve the most appropriate corrective solutions. When mechanical deterioration is found at maintenance outages, it is necessary to evaluate the situation and then select corrective actions that will restore the unit to a condition enabling it to be returned to service and operate in a manner acceptable to the owner. This condition may not represent a *final* repair condition, but will be sufficient to make the unit "safe" to the extent it is able to generate power at an acceptable level of output and efficiency.

The utility engineer cannot be responsible for ensuring the adequacy of the design definition provided within the manufacturer's plant. However, the purchaser has an implied responsibility to ensure this definition exists and that it is applied without compromise during manufacture. Such definition should provide sufficient detail that critical components and specific areas of risk to individual elements are addressed within the utility's experience. It is also realistic to expect that sufficient information is made available to the owner allowing minor repairs and adjustments to be made on-site. The operating engineer should be able to question design detail.

When individual unit components have been manufactured and assembled, they can be erected to form the complete machine. For smaller units, such assembly may take place in the manufacturer's plant and the components shipped pre-assembled, or the unit broken down to major components for shipment. With larger multi-section designs, shop assembly is normally impractical (except in the case of prototype units) and complete assembly would add considerably to the total costs of the unit.

The process of site erection should result in a unit that is aligned to maximize the unit's ability to generate power at an acceptable level of efficiency—immediately—and to continue to operate reliably over the designed plant life. Irrespective of the adequacy and quality of design, if a unit is not installed correctly, it will introduce operational problems reducing the overall utility of the unit to the owner.

It is also necessary to interface the turbine generator with other major pieces of equipment comprising the thermal cycle. These important interfaces must be made so that they do not prejudice the operation of either the turbine generator or other equipment.

Control of various areas—and relationships within them—is a major characteristic in establishing the quality of the turbine steam path. These areas are associated with the expansion and flow direction of working fluid throughout the steam path. Some of this area control is achieved from a correct alignment of the steam path components as they can affect area relationships to a degree. Also, while alignment may not control the effective area in all cases, it can have a considerable influence on the efficiency of energy conversion by ensuring the steam is directed from blade row to blade row and "shock" and "incidence" losses are minimized.

Areas and area control requirements are considered in some detail later in this chapter. It is necessary to consider those phenomena that influence alignment and the "cold setting condition" of the components—phenomena that are predictable and determined by the design function for each unit. There are also phenomena that affect alignment, are unpredictable, and occur normally as a consequence of operating conditions. These have the potential to modify both material properties and physical characteristics of the individual components of the steam path, and ultimately affect alignment.

Stage terminology

A number of terms and phrases describe and define steam path components. Problems in discussing blades are compounded by manufacturers using terms not consistent one from another. In this chapter (and this book), the terms used are defined in Figure 2.1.1, and include the following:

- *Rotating blade*—the total component, comprising the vane, the root, and all other component attachments machined as part of the total.
- *Blade vane*—the airfoil portion, extending from the root block to the tip. For rotating blades, this tip is the largest diameter of the airfoil; for stationary elements, it is the smallest.
- *Profile*—the form (or cross section) of the airfoil. This form may be constant along the vane radial height (cylindrical) or can vary from section to section (vortex) to accommodate the changing steam parameters.
- *Root or fastening*—that portion below the vane and the load transfer point from the vane to the rotor.
- *Root platform*—that portion of the root block between the vane and the root ligaments that carries the load-bearing surfaces.
- *Tenons*—an integral part of the blade located at the tip diameter. They are used to attach an inner or outer coverband.
- *Tie wire hole*—those stages that have a continuous tie wire require an access hole in the vane. This is considered a portion or characteristic of the blade.

- *Stationary blades*—the stationary blade elements mounted directly into the casing or blade carrier.
- *Diaphragm*—comprises three major elements that locate a stationary blade row in the casing (inner or outer) and are normally a welded or cast assembly.
- *Outer ring*—locates the stationary blade elements in the casing, providing axial position and a steam seal from the high to low-pressure side of the diaphragm.
- *Stationary vanes*—perform a similar function to the rotating vanes and have the same stringent requirements for the profile.
- *Inner web*—the material forming the expansion passage inner surface and a steam barrier minimizing steam leakage from the high- to the low-pressure side of the stationary blade vane.

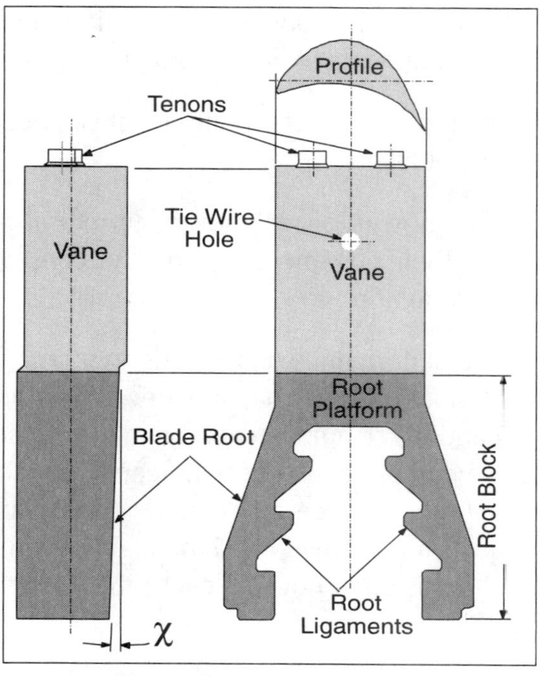

Fig. 2.1.1—Definitions of the blade components.

The turbine unit, as installed, should present to the operator/owner the optimum setting of the components, so when the unit goes into service it will provide a maximum return on the money invested in making the purchase and justify the selection among competing bids. While many design-selected options of components and arrangements within the steam path are a compromise between competing requirements, produced within design specified tolerances, the final condition should be one allowing the unit to operate and achieve acceptable levels of performance. Over time, this condition will modify, efficiency will deteriorate, and the possibility of mechanical damage and deterioration will arise, possibly affecting unit availability.

Such deterioration means owners should establish a clear maintenance strategy (explained in chapter 1). This allows them first to identify, and then correct, any nonconforming conditions as expeditiously as possible. To do this the maintenance staff should be aware of:

- the signs and indications of deterioration, and how to make objective judgments of when corrective action is required
- the potential or total deterioration that can occur if corrective action is not taken
- sufficient knowledge of the repair/refurbishment options available, their possible cost, and the consequences of not taking such action

Steam turbine performance depends upon some very demanding thermodynamic relationships that are developed and maintained within the expanding steam. Because relatively small changes in component geometry or alignment significantly impact the total performance of the unit, it is necessary that the critical characteristics of these requirements are achieved and maintained within the individual elements. What is often not so clear to the operating engineer is what is "critical," and what can be ignored to a degree and accepted before it becomes critical and requires corrective action.

Spatial requirements exist for steam path components in the radial, axial, and tangential directions. Requirements of radial and axial clearances should be observed to help ensure that the performance of the unit is maintained at an acceptable level. This chapter discusses these and provides some guidance of these different requirements and the possible consequences of their not being achieved.

PREDICTABLE FACTORS AFFECTING DESIGN CLEARANCE

Four basic phenomena (or characteristics of operation) are predictable, and influence the radial and axial clearances throughout the steam path. These phenomena will affect the total alignment within the stages, and at any location where stationary and rotating surfaces are close. The extent to which these phenomena influence the steam path design can be established (or predicted) by calculation and taken into consideration during the design phase when selecting and arranging "cold stationary clearances" and bearing elevations for the unit at initial assembly and subsequent alignment.

While the designer can calculate the "hot" running relationships between the stationary and rotating parts, he or she can only meaningfully define to the manufacturing and installation departments the spatial relationships, clearances, and setting requirements as they can be measured and set in the cold stationary condition. Therefore, it is necessary first to consider the types of modifying effects that need be evaluated, and then the degree to which they will influence the cold stationary arrangement of the unit. There are four predictable (and always present) phenomena, whose evaluation remains part of the mechanical design, and therefore always considered in determining the initial settings.

The influence of these four phenomena must be predicted by design, and their total effects compared on some form of clearance chart. These values must then be used to select the optimum cold setting at the various locations throughout the steam path with the final cold settings selected to help ensure "rubs" do not occur, or reduce the possibility of their occurrence during normal predictable operation. These four phenomena or areas requiring consideration include:

- the extent and shape to which the rotor will deflect vertically between the bearings due to its own weight, the extent this is influenced by the temperature of the rotors, and the effect of temperature on material properties
- the differential axial movement that occurs between the rotating and stationary portions of the unit due to thermal expansion of these stationary and rotating parts
- the radial growth of the steam path parts during operation due to temperature and stress effects
- the axial pressure deflection and thermal creep deformation (in the higher temperature stages) at the diaphragm inner web

These conditions must be reviewed, as they occur in the steady state condition (at all loads and maximum steam conditions), and the rate at which they change during transient operation.

ROTOR VERTICAL DEFLECTION

Figure 2.3.1 demonstrates the basic principle of rotor deflection. It represents a simple constant diameter spindle having four integral wheels, each carrying at least one blade row. A single span rotor of length (L) is supported on two bearings (1 and 2), one close to each end of the span. Under the action of its own weight, the rotor will deflect (or "sag") as shown in Figure 2.3.1(b). If the rotor is of uniform section along its length, the form of the deflection will be a true catenary. There are, however, factors that will influence the actual form of the deflection:

- The rotor central portion will not normally be of constant diameter. There will be step changes at many locations, particularly if the rotor has stepped segments at sealing locations. The stiffness of the rotor depends upon the rotor diameter cubed, and the deflection (δ) is a function of this stiffness
- If the rotor carries shrunk on discs (as shown in Fig. 2.3.2) these discs will normally be of different shapes—and therefore, of different mass—so the mass distribution along the length of the span will not be constant. Also, the blades will be of varying radial length and of increasing mass towards the exhaust end. These wheel-and-blade masses add weight, depending upon their total volume, but add no stiffness to the rotor section modulus. Deflection is a function of the central spindle diameter and mass distribution, but will be essentially the same as shown in Figure 2.3.1(b)

Figure 2.3.1 shows a total center deflection of δg. If the wheel and blade masses shown in Figure 2.3.2 are the same, the maximum deflection δs will be equal to δg. The deflection is essentially a function of the diameter (d) of the central portion of the rotor.

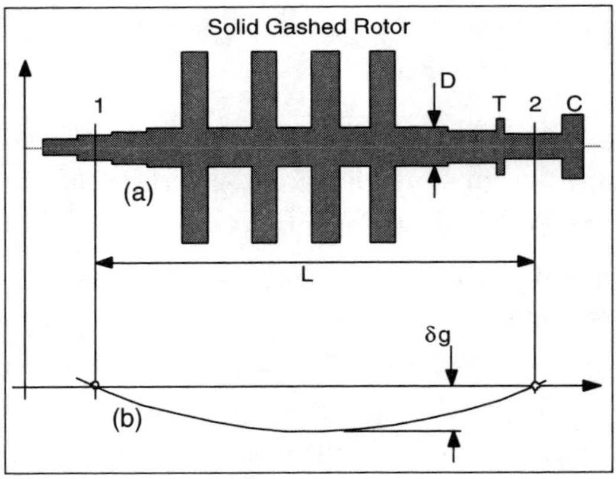

Fig. 2.3.1—Deflected form of a simple shaft and wheel construction.

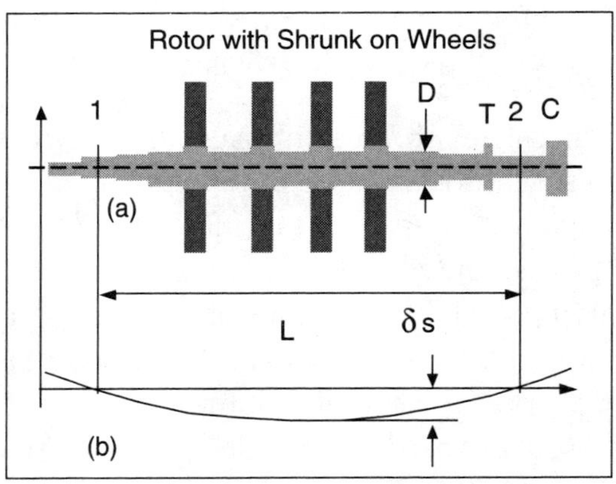

Fig. 2.3.2—The deflected form of a shaft with shrunk on wheels.

- The rotor material temperature will vary along its axial length because there is a temperature reduction in the stages; therefore a distribution along the axial length reducing towards the exhaust end. The rate of temperature reduction is dependent upon the energy dissipated within the steam path stages. With temperature variation, the mechanical properties of the rotor steel will vary along its axial length, deflecting by greater amounts at the higher temperatures

Many rotors are of the barrel form, as shown in Figure 2.3.3(a). In this figure, the basic diameter (D) has been retained at the shaft ends, but over the central span, the effective rotor diameter has been increased from D (as shown in Figs. 2.3.1 and 2.3.2) to Do. Therefore, the rotor diameter is increased, increasing the stiffness and increasing its resistance to deflection. The deflection will be reduced. As shown in Figure 2.3.3(b), the total maximum deflection is δb, which tends to be less than δg.

On each of the rotors shown in Figures 2.3.1, 2.3.2, and 2.3.3 there is a coupling flange (c) located at one end, outboard of bearing 2 and thrust collar "T." At this coupling, the turbine rotor is to be connected to the driven machine (normally a generator). These features—bearing 2 and thrust collar T of the rotor—provide no increase in the section modulus, and have only little impact on the deflected shape. However, the bearing is normally given spherical seating within its support structure so it is able to line up with the

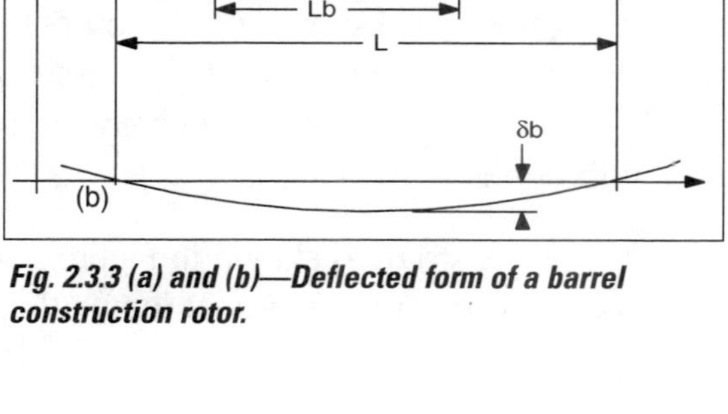

Fig. 2.3.3 (a) and (b)—Deflected form of a barrel construction rotor.

deflected angle of the rotor at this location. In Figure 2.3.3(c) the rotor centerline is inclined to the horizontal at an angle (α). To accommodate this inclination, the bearing is turned through the same angle (α) in the spherical seating of diameter (Dsp). This bearing has a white metal thickness "t" and is designed for a radial clearance to the journal of "Cl."

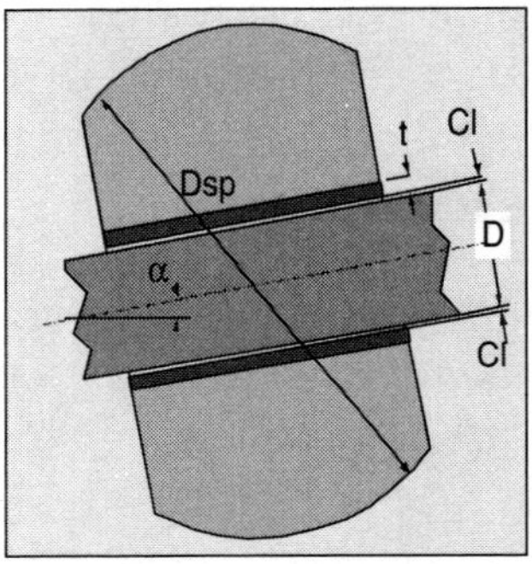

Fig. 2.3.3(c)—A bearing showing the "tilt" or "α" in the spherical seating.

The turbine rotors must be coupled, or connected to the machinery they drive. The majority of larger output units comprise more than one turbine section. When multi-section rotors are coupled, they can be connected by a flexible element or use a system connecting individual sections together in such a manner they perform as (and have the characteristics of) a single integral shaft—they are solidly coupled! All modern large rotor designs use the solid coupling system. In the following discussions of vertical deflection of multi-segment rotors, a solidly coupled system is assumed, and coupling flanges are required on both ends of such shafts. The exception is rotors located at the ends of the catenary.

Consider the three-rotor string as shown in Figure 2.3.4. There are two turbine rotors—a high-pressure section of length (Lhp) and a double flow low-pressure section of length (Llp). These are coupled to drive a generator of length (Lgen). Each rotor is individually supported between a pair of bearings (1 and 2) for the high-pressure section, (3 and 4) for the low-pressure section, and (5 and 6) for the generator.

These three rotors are shown in Figure 2.3.4 as having small axial gaps (G1 and G2) between their coupling faces, and each rotor has a natural deflection, shown as calculated for the hot running conditions. If, as shown in the lower portions of the figures, the deflection is drawn with the bearings adjusted to a true horizontal position (H-H) because of the deflected form of the rotors [with inclination of the type shown as "α" in Figure 2.3.3(c)], the gaps between the coupling faces would not be parallel. Consider the gap G-1 between the high and low-pressure elements shown in Figure 2.3.5. In this case, the gap G-1 is the mean gap as it exists at the center position of the rotor—which, due to their deflection, are at angles α1 and α2 to the horizontal as shown in Figure 2.3.6. With these inclinations, the rotors cannot be connected and the gap closed at their flanges. Even if the coupling studs could pull the faces together, this would induce a high level of bending stress into the outer fibers on the rotor material and require considerable distortion to allow the gap to become parallel and close.

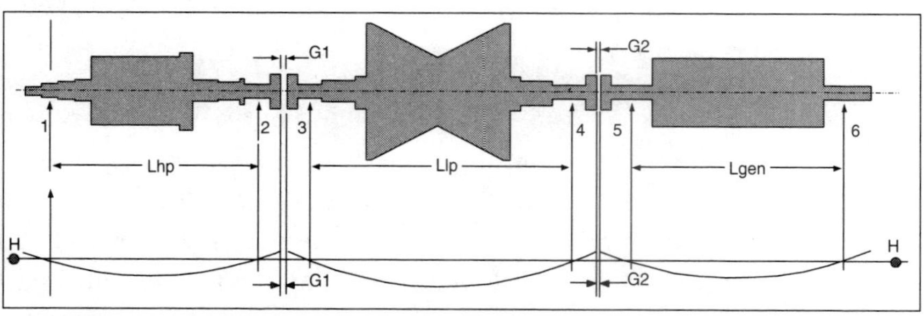

Fig. 2.3.4—A three section arrangement showing the deflections in the original setting before bearing adjustment.

Note: In fact, the gap does not close. A spacer plate between the coupling faces is adjusted in thickness to allow fine adjustment of the axial distance from the active thrust face to achieve the desired clearances in the individual sections. However, before using the spacer plate, the rotors must be adjusted so the axial gap is G-1 at all circumferential locations.

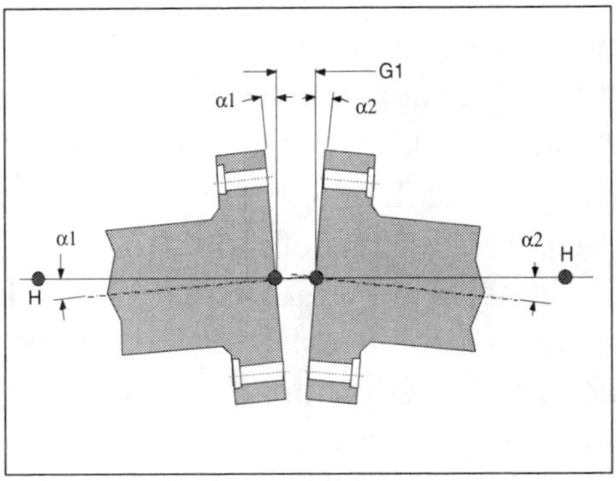

Fig. 2.3.5—Coupling gap at initial set-up.

The method used to adjust the coupling axial gaps so they are spaced equally at top and bottom positions is as follows:

- Select one section of the total string and arrange for its bearings to be horizontal
- Raise bearings on one adjacent section until the gap between its coupling flange and the horizontal section are equalized. Figure 2.3.6 is the adjusted position on the rotors, for which the high-pressure section bearing 1 and 2 have been raised by amounts Δ1 and Δ2 to achieve a gap (G-1) at top and bottom positions. Note the centerline of the horizontal set rotor is not horizontal, but inclined at an angle (α2) due to its own deflected form. Therefore, bearings 1 and 2 must be raised to turn the high-pressure rotor through an angle (α1 + 2)

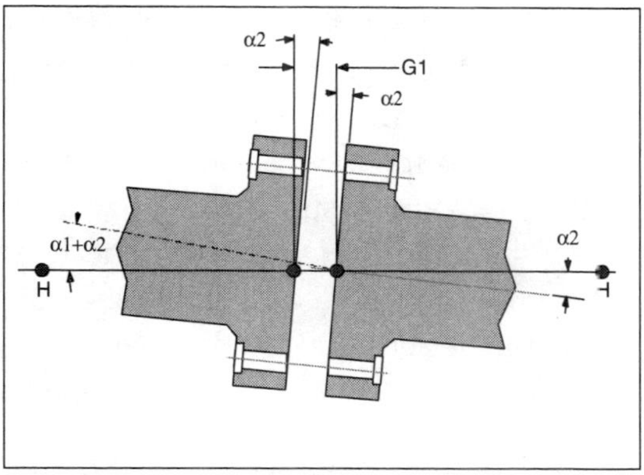

Fig. 2.3.6—The coupling faces after adjustment to achieve a constant gap "G1" at top and bottom.

- Similarly, at the other gap (G-2 between the low-pressure section and the generator), the generator rotor must be adjusted by raising bearings 5 and 6 by amounts Δ5 and Δ6. The relative coupling positions are then as shown in Figure 2.3.7, where the high-pressure rotor has been adjusted so the centerline tilt angle has been modified from α1 negative to α2 positive. The coupling face of the high-pressure rotor is now at the same angle as the low-pressure rotor. Similar adjustments have been made between the low-pressure and generator rotors

The rotors of modern turbine generators are solidly coupled and can be considered as a single shaft of varying sections with loads carried at different positions along the axial length. Consider the rotor

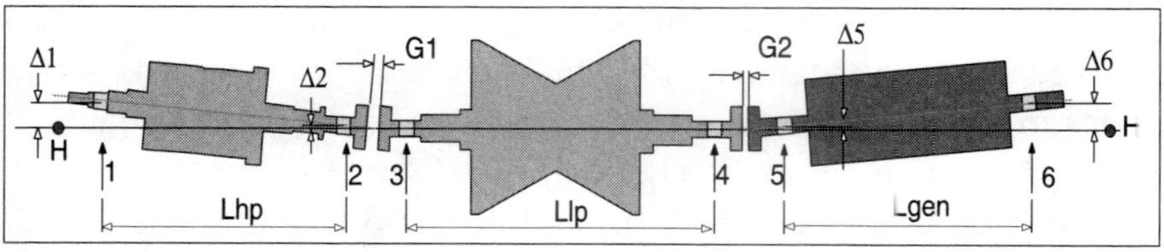

Fig. 2.3.7—The three rotor unit of figure 2.3.4 after bearing vertical height adjustment.

shown in Figure 2.3.8(a). It comprises a single-flow high-pressure section, a double-flow intermediate-pressure section, and two double-flow low-pressure sections. The rotors in these sections are coupled to a generator rotor at the low-pressure end. In the case shown, each individual turbine rotor is supported by two bearings, which between them carry the entire weight of the rotating portion (including the overhung sections from the bearings to the coupling faces). It is also possible in some designs that bearings will be shared between adjacent sections. Here, two rotors share a common bearing.

If the rotor had a uniform cross section and a uniform mass distribution along its axial length, and was at constant temperature, its deflected form would be a pure catenary. However, due to the uneven distribution of mass and section modulus and the effects of temperature variations, this pure form catenary is modified. Figure 2.3.8(b) shows a typical deflection curve for the solidly coupled rotor from the horizontal (H-H). As drawn, the vertical scale has been considerably enlarged.

In Figure 2.3.8, the five rotor sections will require 10 bearings. Because the rotor will operate in this hot, deflected form, it is necessary to adjust the bearing elevation so each will assume an inclination to permit flange face gap parallelism and an elevation such as $\Delta 1$-$\Delta 10$ in Figure 2.3.8(b).

During site installation, the manufacturer may adjust the relative bearing elevations to any convenient level. A common manufacturer's practice is to set the low-pressure sections (where the turbine has an interface with the condenser) at the same elevation. For multiple low-pressure sections, the low-pressure sections are set at the center of their span. The high and intermediate pressure and generator sections must then be set to accommodate this. Such an adjustment of the unit rotors is shown in Figure 2.3.8(c); the elevations are adjusted from the horizontal (H-H) by amounts $\delta 1$-$\delta 10$.

Using bearing 7 as an example, the initial determination of this bearing is set at a distance Δ7 below the horizontal. In the final adjustment, this bearing is raised by an amount δ7 to allow bearings 6 and 7 to be set at a common height at the center of the low-pressure sections' span. The total downward adjustment of bearing 7 is as -Δ7+ δ7 as shown.

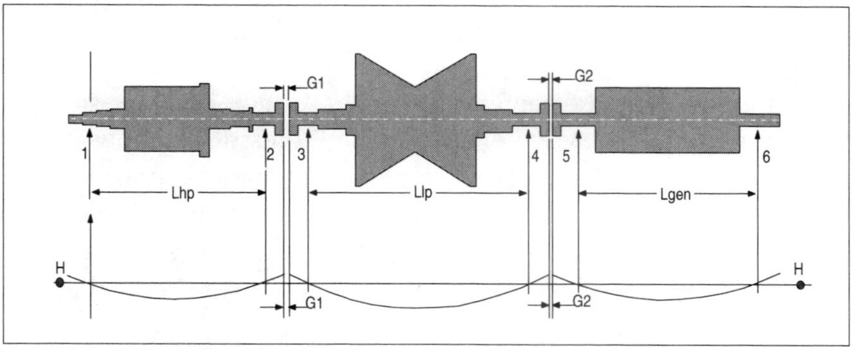

Fig. 2.3.8—The vertical alignment of a five section, ten bearing turbine-generator, with bearings "6" and "7", which are attached to the condenser set horizontal.

Factors contributing to vertical movement

The final, cold vertical bearings settings are influenced by considerations other than the rotor-deflected shape. These must be considered and accounted for in making the cold stationary setting. Principal issues among these second-order considerations are following:

Bearing oil rise. The oil film thickness causes the bearing to float the rotor vertically above the white metal surface. This bearing rise is small but should be considered in setting the nominal bearing height. This is discussed below.

The vacuum deflection. During operation, a vacuum exists at the turbine/condenser interface. If the condenser is mounted directly beneath the low-pressure sections, there will be a large net downward force on the low-pressure casing, caused by the difference between atmospheric pressure and the vacuum produced in the low-pressure hood. Often the bearings are formed as part of the low-pressure hood and carried within it. If the bearing housings are an integral part of the low-pressure structure, then these bearings will be subject to the same vertical forces causing a deflection of the hood, *i.e.*, these bearings will rise and fall with the casing. In establishing the low-pressure section bearing vertical setting, it is necessary to make allowance for this movement and consider the effects of part load and back pressure variations on total elevation.

Some low-pressure section designs have their bearings supported from pedestals located between the low-pressure sections. In this case, the casing will be subject to elevation changes introduced by the vacuum force but the rotor will be unaffected. In such designs it may be necessary to arrange for the unit to have larger radial clearances at the shaft-end positions to accommodate the vertical adjustment of the casing relative to the rotor.

Hot well water. During operation, the condenser hot well will fill with condensate awaiting removal by the condensate and boiler feed pumps. This hot well does, to a degree, act as a reservoir; therefore the level of water contained in the well will change subject to load, boiler demands, and other factors. It would be expected as water level changes, the static load produced by the weight of water in the hot well carried by the condenser will change, affecting the height of the condenser and the load transferred to the low-pressure hoods.

There are various methods of supporting the condenser. Some are suspended from the turbine and given some support from below. Others are carried on spring supports and connected to the turbine by flexible joints. Each will have a different influence on the hood deflection and needs to be factored into the total vertical setting of the bearings.

Bearing temperature effect. The bearing and its support structure are subject to temperatures above atmospheric and temperature variations. These temperatures can be marginally different from bearing to bearing and are dependent upon the location within the rotor string, the amount of heat conducted along the rotor, and the effectiveness of the atmospheric cooling. Because of this effect, the bearing supports will expand vertically by different amounts.

Flexibility of bearing supports. Because casings expand and contract due to temperature effects, they can impose varying loads on the bearing supports and can therefore modify their elevation. This effect is particularly noticeable with separate bearing pedestals that must transmit axial thrusts and at the same time maintain alignment.

Centerline support of the casings. The method of supporting the casings above the foundation, and the height at which the centerline of the casing is located, can have an effect on the height at which the centerline exists under varying temperature conditions. For this reason the design is supported from a point at the centerline, to the greatest extent possible. The upper half expands upwards and the lower half expands downwards, maintaining an equilibrium condition at the centerline.

Bearing loading. The designer can deliberately raise or lower a bearing relative to its modified catenary position obtaining a proper bearing loading. This is a practice used when there is a possibility of inducing "oil whip" into the bearings. It is known by increasing bearing loading the onset of this phenomenon will be delayed.

Bearing spherical seating. The bearings must be able to accommodate the angle of the journals' inclination so a clearance "Cl" is maintained between the bearing centerlines and the journals. If for some reason the spherical seating of the bearing does not allow adjustment to the correct angle ["α," as shown in Fig. 2.3.3(c)], then the clearance will not be correct at all axial positions of the journal. This can have an effect on the elevation at that point.

It is clear from these considerations that final bearing elevations "δ1-δ10" are selected with care after an evaluation of the various factors that can influence their operating elevation. This elevation must be achieved with considerable accuracy during site erection and subsequent adjustment. Some of these factors have minimal effect on the final design elevation—however, they do have an effect. It is necessary for the designer to consider these, if not in each specific case, at least during the development phase of the design rules formulated to ensure the final specified elevation will help ensure vertical alignment is maintained at all operating conditions.

Sideways shift

In addition to the requirements of vertical adjustment needed to achieve axial gaps of "G1" and "G2" at top and bottom of the rotors, these elements must be aligned side to side along their axial length so the rotor line is true and the coupling gaps "G1" and "G2" (Figs. 2.3.4 and 2.3.5) are identical side to side as well as top to bottom. It must be recognized, however, that the true operating shape of the rotor does not run on this "true" side-to-side line set in the cold stationary condition. There are one or two minor influences that will cause some adjustment but they are of a very low level and can be ignored for all practical purposes.

The facts of a bearing oil rise were discussed previously. If the hot running position of a bearing is considered, it will be seen that from the cold stationary position—with the rotor journal center located directly above the vertical centerline of the bearing [Fig. 2.3.9(a)]—there will be a positional shift as the unit goes into operation. In operation, an oil wedge is formed between the bearing liner's metal surface and rotor journal, so the rotor will in effect float on this oil film. This film is not of uniform thickness and will be established as a wedge [Fig. 2.3.9(b)].

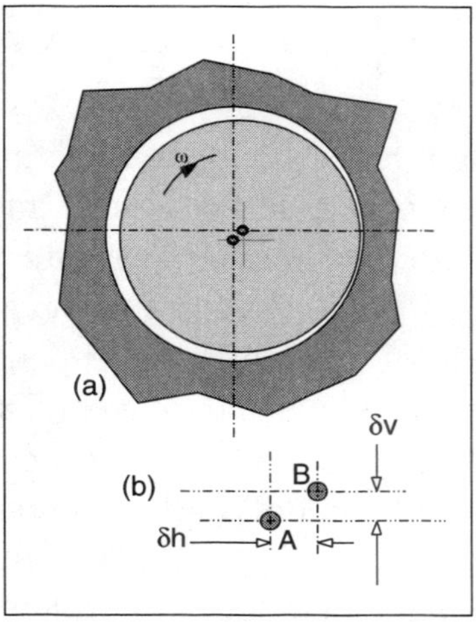

Fig. 2.3.9—The center line shift due to the oil wedge.

In the hot running position, the rotor has a "center shift" from A to B. This shift can be considered to comprise two components—"δv" in the vertical direction (discussed earlier), and a shift of "δh" in the horizontal direction. The vertical rise is considered in setting bearing elevation. Normally, this vertical rise is greater than the horizontal side shift. This horizontal shift must also be reviewed in terms of adequacy of the unit's total rotational stability. The "δh" deflection is a relatively small amount, but where bearings of considerably different diameters are adjacent to each other, this can have an effect on the oil wedge shape of both, and must be considered by design in establishing cold clearances.

DIFFERENTIAL EXPANSION

During operation, the stationary and rotating components of the turbine absorb and reject heat energy from the steam and expand and contract with the temperature changes this heat introduces. The rotating and stationary portions accept and reject this heat, both in different amounts and at different rates, depending upon their total mass and the extent to which they can radiate this heat to other media. Because these stationary and rotating components expand and contract by different amounts and at different rates, there will be a relative positional shift between them from the cold stationary position during normal operation. This difference becomes even more pronounced during transient conditions, when dramatic temperature changes and (in emergency conditions) heat generation can occur within the steam path due to frictional heating caused by rotor rotation in a partially evacuated or stagnant steam atmosphere.

Because they are anchored to the turbine foundation block at various locations, the casings—normally one for each section—will expand in an axial direction away from these anchor points. The extent of this expansion in any one casing is influenced only by the amount of heat absorbed, rejected, and retained by individual casings. The extent of expansion in any one casing is unaffected by other casings and components of the unit.

The rotating portions are solidly coupled and have a total movement from their one anchor point. The anchor point is the thrust bearing and the contact point at the active thrust face. The position of the thrust bearing is chosen to minimize the differential expansion in the higher condition stages, where the effects of differential expansion have, or can be made to have, less effect on the performance of the unit.

Figure 2.4.1 is a simple, single-section, eight-stage turbine. This unit has a thrust bearing axial location "T," as indicated. At this position there is contact between the rotating (thrust collar) and stationary (thrust bearing plates) components. This position represents the only point within the unit where there will be no relative axial movement between the stationary and rotating components as a consequence of temperature changes. In this unit, steam is admitted through a control stage "C1." This position within the unit represents the highest temperature the metal will experience. The steam will expand within the steam path through the remaining seven stages—steam will flow to the right (indicated as +"A") and will also leak through the sealing system installed at the shaft end (indicated as -"A").

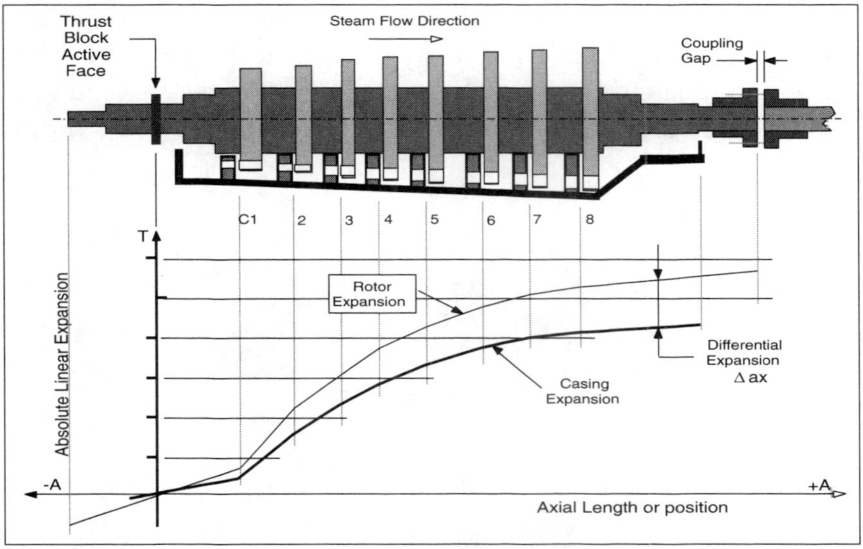

Fig. 2.4.1—The differential expansion in a single section containing eight stages.

Note that expansion to the right of the thrust bearing has been designated as positive movement and expansion to the left as negative movement. This is a convention used only to explain the effects of differential axial movement.

Consider the effects of the steam temperatures on the individual stages of the steam path. The control stage (accepting the inflowing steam) is the highest temperature region that will expand (by an amount in the positive direction). Figure 2.4.1 shows the slope of the control stage expansion curve will be the steepest of the individual stages. In the stages to the right of the control stages, temperatures will be lower and therefore the rate of expansion will be less (but still in a positive direction). The extent to which high temperature steam will expand through the rotor end seals will influence the temperature in this region. To the left of the thrust bearing, heat will be conducted down the rotor material causing a negative movement of the rotor.

Figure 2.4.2 shows the control stage, where steam is admitted to the steam path. It enters the rotating blade at a temperature "Ti," expands through the rotating blades to a temperature "To," and gives a mean blade temperature "Tb," which is a function of both "Ti" and "To."

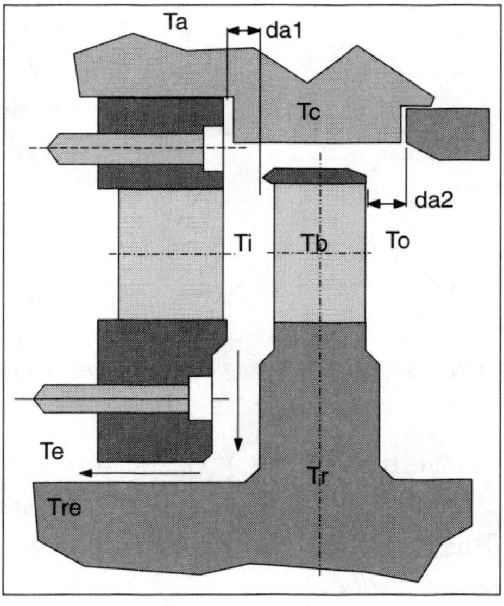

Fig. 2.4.2—The temperature variations in a stage with some level of reaction.

The rotor wheel temperatures are similar to the blade vane on its inlet and discharge faces. These temperatures will influence the rotor body, together with any heating effect of the steam expanding past the nozzle box. The temperature will range from "Ti" to "Te." The total result of this heating is for the rotor to expand away from the thrust bearing by an amount depending upon the mean temperature and the axial span of the stage.

The casing inner surface will also heat to about the same temperature as "Tb." The casing, however, has an outer surface; it is insulated but it will lose heat to the atmosphere surrounding it, and will then be at a somewhat lower temperature ("Ta"). This means that the mean temperature in the casing will be lower than the mean effective temperature of the rotating portions (considering the cooling that takes place at the outer wall). The casing temperature ("Tc") is a function of both "Tb" and other temperatures achieved by the rotor material.

The nozzle plate will be exposed to temperatures higher than "Ti"—temperatures equal to the inlet temperature of the steam. Any steam chamber within the casing will also achieve inlet temperature, which will tend to raise the temperatures of the stationary portions by some small amount in this region. They will still suffer the effect of heat loss from the outer surface. In many designs—where a double casing design is used, for instance—the outer wall temperature of an inner casing (which locates the stationary blade rows) is the temperature of the steam surrounding that casing.

With these changes in the relative axial position of the stationary and rotating components, the axial clearances (set in the cold axial position) are modified by this differential expansion. Two hot running axial clearances are of importance in determining the initial settings of the individual stages in the unit—"da1" and "da2," shown in Figure 2.4.2. These will be discussed in greater detail later.

Consider again the single casing unit in Figure 2.4.1. This unit comprises a central spindle with a control and seven impulse stages. At normal operating conditions the rotor and casing will expand from their anchor points in the rotor in the positive direction from the thrust bearing active face. The rotor has its highest temperatures at the front end, and so the growth in axial length will occur at the greatest rate at this location with the rate of growth decreasing as the exhaust end is approached. In this unit, the casing is also anchored at the thrust-block location and will also expand in the positive direction; however, because it is slightly cooler (losing some heat through its outer surface), it will not expand to the same extent.

The actual location of the thrust bearing at the front end of the unit is also important. Often this will be contained in a pedestal supporting the bearings and possibly the control mechanisms. The casing may or may not be keyed to and supported from this pedestal; it depends upon the axial location of the anchor, and is a function of the arrangement selected by design.

The curve of total rotor expansion will be of the form shown. Similarly, the casing will expand in the same direction, but to a lesser amount because of its lower mean temperatures. The difference in expansion occurring between these two curves is termed the differential axial expansion.

There is also some expansion from the thrust bearing active face in the negative direction. The temperature of this portion of the rotor (to the left of the thrust block) is established by heat conducted along the shaft from the central portion of the unit. Similarly, the casing will expand in a negative direction, again at a lower rate than occurs in the rotating portions of the unit.

Figure 2.4.3 shows the five-section unit previously shown as Figure 2.3.8—in this case, the thrust bearing axial location ("T") has been added. In this design, the thrust bearing is located at the bearing pedestal between the high and intermediate pressure (reheat) sections. Also shown in the lower portion of this figure is the axial movement from the thrust bearing of the solidly coupled rotor and the stationary portion from their anchor points to the foundation. At the thrust bearing axial position, the rotor has zero axial movement relative to the stationary elements. The casings for the high and intermediate pressure sections are normally located from the pedestal/bearing support. They often have a support platform built onto them to locate the stationary portions on either side of them.

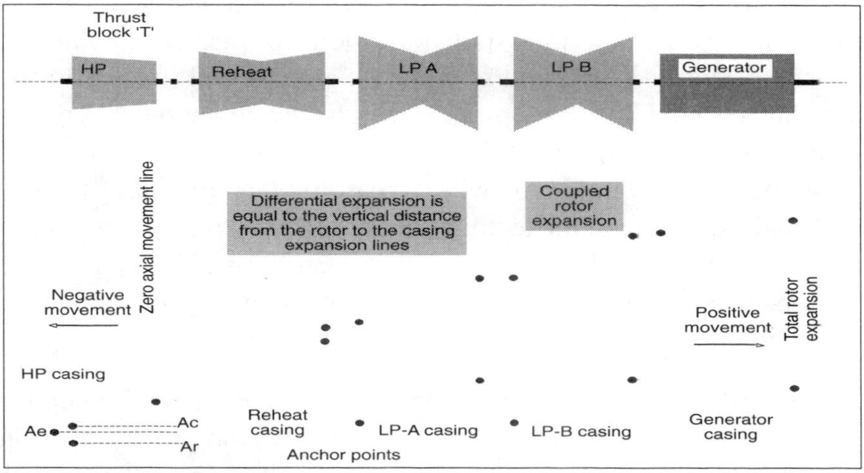

Fig. 2.4.3—Indicating the differential expansion in a five section turbine-generator, showing the rotors expanding from the thrust bearing, and the casings from their anchor points.

Using the same nomenclature in this figure as in Figure 2.4.1, it can be seen that movement (or expansion) to the left (away from the thrust bearing) is termed "negative" (for convenience), while movement to the right is termed "positive." Because of their locations, both the high-pressure rotor and casing expand away from the thrust face to the left of the thrust bearing active face. Both are therefore considered to have a negative movement. Figure 2.4.3 shows the total movement of the rotor begins at a thrust face (which is the zero point), whereas the casing initial (zero) point is located at some small distance to the left—possibly considered the support point, since it is at the outer surface of the bearing pedestal. In fact, these small distances have little influence on the differential movement, but it is wise to be aware of their existence.

An examination of the high-pressure section expansion characteristics makes it clear that modified axial clearances exist between the hot rotating and cold stationary settings during normal operation. If a stage in the high-pressure section is examined (Fig. 2.4.4), the axial clearances will modify. In this figure, clearances from the stationary to rotating blade at the steam path are shown as "Cao" and "Cai" at the outer and inner radial locations, respectively. As heat is admitted, both rotating and stationary elements will expand to the left. Because the stationary portions will have a lower mean temperature than the rotor, they will move by a smaller amount, and the clearances "Cao" and "Cai" will increase by the amount of the differential expansion at the location of transition from the stationary to rotating rows. The clearance "Ci" (between the stationary portions and the preceding rotating portions) will decrease by the amount of the differential expansion at that axial location.

Another factor to be considered is the minimum clearance between the diaphragm and adjacent wheels along their common face ("Cxi" and "Cxo"). This clearance normally occurs at the inner diameter position and will be considered in greater detail later.

If labyrinth seals are located at the inner diameter of the diaphragm and are of the "hi-lo" type, their axial position detail (shown by the dimension "g" and "h" in Fig. 2.4.4) must be monitored. Ultimately, there could be rubs of the seal strips on the rotor castellations that will normally destroy the seal strips. This will degrade efficiency and stage output. These considerations also apply to the reheat section, except the expansion of the rotor and casing are in the positive direction.

It is clear from a consideration of the low-pressure rotors (Fig. 2.4.3) that the accumulated axial movement of a solidly coupled

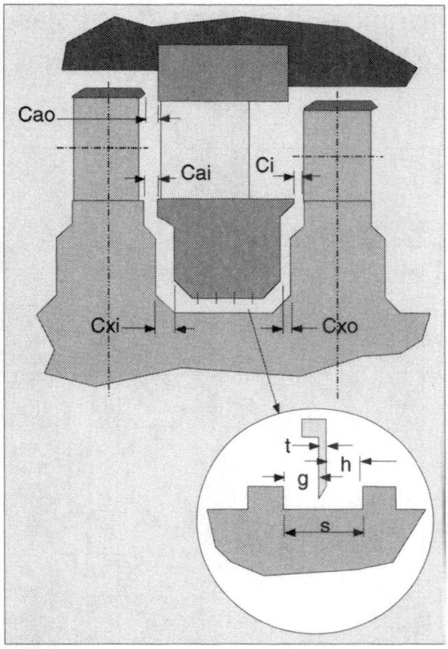

Fig. 2.4.4—Showing the axial clearances which should be considered in setting the cold stationary positions.

rotor system represents considerable movement in the positive direction. The outer casings of these sections connect to external portions of the foundation at some suitable point. Figure 2.4.3 shows how this occurs for both the "A" and "B" sections at their casing centerlines (position of steam admission). This is a convenient location, as it imposes the minimum adjustment requirement on any crossover/around piping that transports steam from the reheat section exhaust to the low-pressure sections inlet. However, there are designs that anchor the low-pressure outer casing at the point nearest the thrust bearing.

As an example of low-pressure section clearance requirements, consider the first stages of the double flow low-pressure section "LPA" shown diagrammatically in Figure 2.4.5. In the cold stationary condition axial clearances "Cat" and "Cag" stand between the stationary and rotating blade rows at the turbine end and generator [Fig. 2.4.5(a)].

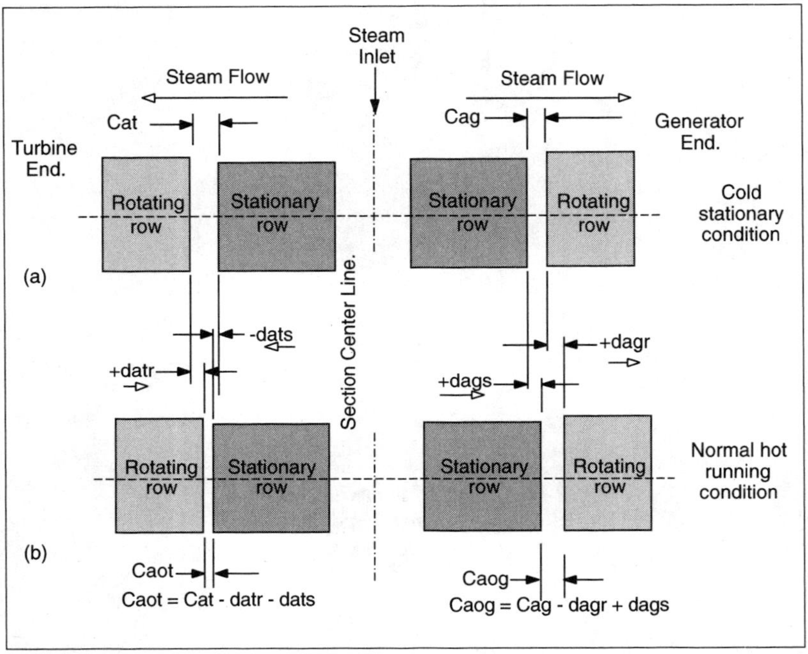

Fig. 2.4.5—Showing the change in axial position from the cold stationary position to the hot rotating position.

When the unit reaches steady operating temperatures and has adjusted to its normal axial operating position [Fig. 2.4.5(b)], positive movement occurs in the rotating components in both flows. The total axial movement at this point is the sum of the movement in the reheat section from the thrust bearing to the "LPA" coupling, plus any movement from the "reheat/LPA" coupling to the stage point being considered. This can be defined as an amount "+datr" at the turbine end flow, and "dagr" at the generator end flow. At the turbine end, the stationary blade row will have a negative movement by an amount, "-dats" towards the thrust bearing (the casing being anchored at its centerline). These two operational movements will cause a modification of the turbine end clearances. Therefore, the normal hot running clearance at the turbine end "Caot" is given by:

$$\text{Caot} = \text{Cat} - \text{datr} - \text{dats}$$

Similarly, at the generator end, there is a positive movement ("dagr") of the rotating portions of the unit, and another positive movement of the stationary components ("dags"). This produces a hot running clearance ("Caog"), which is given by:

$$\text{Caog} = \text{Cag} + \text{dagr} - \text{dags}$$

In the second low-pressure section ("LPB"), there is an even greater axial movement of the rotor since it includes the total axial expansion of the "LPA" rotor, but the reasoning in establishing these cold axial clearances is identical to that discussed for section "LPA."

In the generator, the cold setting must be adjusted to recognize this total rotor expansion. The axial clearances in the generator are not quite as critical—attention must be given to the hydrogen seals—and the effects of differential expansion on performance are minimal. There is a magnetic center to the flux field but its axial position is not critical, and can be accounted for in the cold setting.

The thrust bearing can be mounted at various axial locations. The most appropriate for any design is dependent upon the steam path or section configuration. A common location for the thrust bearing in multi-section units is between the high and reheat (or the intermediate pressure) sections. This gives both sections the advantage of minimal differential expansion and therefore the ability to maintain close axial clearances in both sections. This is done to minimize axial movement in these higher-pressure stages, regions where leakage would have the more significant degrading effect on efficiency. The cold stationary axial clearances at any position must be set according to the differential axial movement predicted at the location being considered.

The casings of the high and reheat sections are not connected, so both must have independent anchor positions. A similar logic applies to low-pressure casings that are anchored to minimize axial thrusts on crossover pipes, or in the case of nuclear units, on the valve system at the low-pressure inlet.

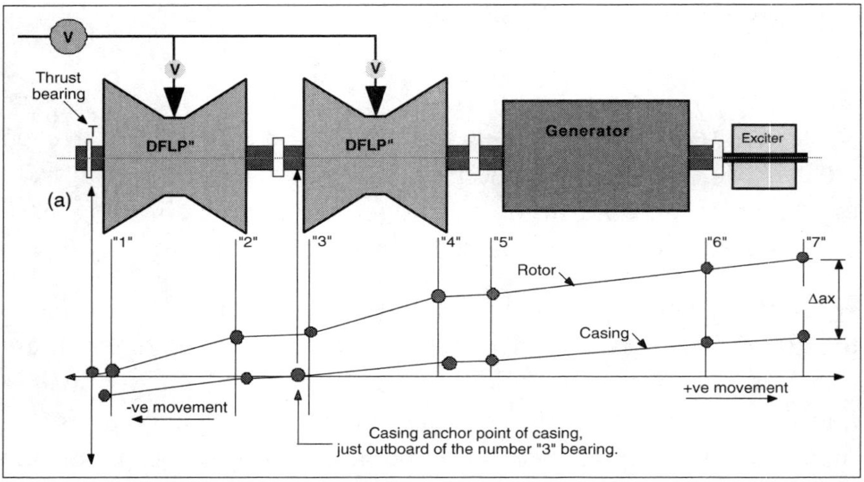

Fig. 2.4.6—The differential expansion of a geothermal unit comprising only low pressure sections. This unit has a rotor with a thrust bearing at the front pedestal, and the interconnected low pressure casings are attached to the foundation at a point near the number 3 bearing.

Figure 2.4.6 shows an arrangement suitable for a low-steam condition multi-flow unit—e.g., a geothermal application, where multi-flows are required for the larger outputs. In this design, the thrust bearing is located at the front end of the unit, typically in a pedestal containing the control mechanisms and the number "1" bearing. The "LPA" casing is anchored to the foundation in the region of the number "3" bearing, but just outside the casing region. The casings (normally fabrications) are interconnected through a shared bearing pedestal, containing both bearings "2" and "3." It can be seen that there is considerable negative movement of the "LPA" casing, and the same degree of positive movement in the "LPB" casing. Because of the lower temperatures in these units, the actual amount of differential is small, and this design splits the total movement so both double-flow sections can employ small axial clearances. The generator and exciter are shown in this figure but, as in the case of the unit shown in Figure 2.4.3, these are not too sensitive to differential movement and can be set with a reasonable degree of latitude.

Transient operating conditions

In addition to the requirements of normal operating clearances, it is necessary to consider thermal transients that can occur in the working fluid during operation. Under transient conditions, stationary and rotating portions of the unit are subject to steam temperature changes and accept and reject heat energy. This change in heat energy levels occurs at different rates between the various components and influences the relative positions they attain.

Considerations of transient conditions—and resulting expansions—are of considerable importance during the design phase and the setting of "cold stationary clearances." They affect the minimum and maximum differential clearance required to ensure that rubs between the stationary and rotating components do not occur during both normal operation, start up, and shutdown, when both "short or cold" and "long or hot" rotor conditions occur. The "long rotor" condition occurs when the rotor temperature is above those of the stationary portions of the unit by an amount greater than the normal operating temperature differential. "Short rotor" conditions exist when the rotor temperature is below that of the stationary portions by more than those achieved in normal operation.

"Short" and "long" rotor conditions occur during transients that are most often created by start up, shutdown, or when steam is being admitted to or cut off from the steam path. However, these conditions can also be introduced by boiler temperature excursion, when there are load changes, or in an emergency, when load is rejected from the generator and the main stop valves close.

These various transient phenomena introduce an imbalance or difference in steam path temperatures from "normal." Then, because of the difference in thermal inertia between the rotating and stationary portions of the unit, they will accept and reject heat at different rates, expanding axially by different amounts to the extent the running clearances must be modified from normal operating values.

In the following analysis the following symbols and suffixes are used:

- C axial clearance
- d axial distance adjustment
- a normal design operating position
- o off design operating position
- t turbine end of unit
- g generator end of unit
- n normal temperature condition
- h hot temperature condition
- c cold temperature condition
- r the turbine rotating portion
- s the turbine stationary portion

Long (hot) rotor. Occasionally, the steam path rotor temperature exceeds the temperature of the stationary portions by an amount in excess of the normal operating temperature differential. In such cases, there will be a period of time during which the differential expansion of the rotor relative to the stationary portions of the unit is larger than normal. Under these circumstances steam path portions of the unit must be examined to ensure there is no reduction of hot running clearances to the extent interference will occur. Interference would cause rubs between stationary and rotating portions of the unit.

Figure 2.4.7(a) shows a three-section turbine with high, reheat, and double-flow, low-pressure sections (the generator has not been included). Under normal operations, the differential expansion positions are shown, with the total differential expansion at the end of the high- and low-pressure sections indicated as "Δaxhpn" and "Δaxlpn" [Fig. 2.4.7(b)]. During periods of transient conditions—when the rotor achieves a temperature in excess of the stationary portions by a larger differential temperature—there will be changes in the relative positions between the components.

In the high and reheat sections, this results in an increase in the axial clearance between the stationary and rotating components "Cao" and "Cai" (Fig. 2.4.4) and a decrease in the clearance "Ci" between the rotating blade discharge and the inlet to the following stationary row. Normally, the clearances represented by "Ci" are significantly larger than the "Ca" clearances, so this is rarely of concern. However, in high temperature, high-pressure stages, it is always prudent to ensure "rubs" will not occur, particularly at the inner diameter of the diaphragm (clearances "Cxi" on the inlet and "Cxo" on the outlet sides of Figure 2.4.4), where there could also be both elastic and plastic deformation of the web.

The differential expansion changes give total differentials of "Δaxhph" at the high-pressure end and "Δaxlpn" at the generator end of the double flow low-pressure section [Fig. 2.4.7(c)].

There are other considerations in the double-flow low-pressure section. While axial clearances between stationary and rotating rows will increase in one flow, in the other they will also change as a consequence of the thermal difference—but the direction of the resulting movement will be reversed. It should also be noted that the normal differential expansion in the low-pressure sections are large because of the accumulated effect of rotor expansion and the introduction of such transients can aggravate the overall situation.

Figure 2.4.7(d) shows a diagrammatic representation of two stages—the first stages in the two flows of the double flow low-pressure section. We assume the unit is running at steady conditions rather than being subjected to start-up transients—*i.e.*, it shows the normal running clearance of "Caot" at the turbine end and "Caog" at the generator end flows, as defined by the equations on page 117. The values shown and defined in Figure 2.4.5 are the "normal hot running clearance."

Under conditions of the hot rotor transients, the turbine end will experience expansion. Because of the hot rotor condition, there will be an additional rotor growth ("dhrt"). Similarly, the stationary portions will expand in the negative direction (from a centerline position) by an amount "-dhst." These movements will give a final hot transient running clearance from "Caot" to "Coth" of:

$$\text{Coth} = \text{Caot} - \text{dhrt} - \text{dhst}$$

This analysis shows that during the long (hot) rotor condition, the expansion at the low-pressure section turbine end causes a decrease in the axial clearance of (dhrt + dhst), *i.e.*, the rows move, closing the axial clearance.

Similarly, at the generator end of the low-pressure section, the turbine rotor will have an additional positive movement ("dhrg"), and the stationary portions will have expanded by an additional amount (in the positive direction) of "dahg." This means the normal hot running clearance ("Coag") will have changed. This expansion gives a hot transient running clearance "Cogh" of:

$$\text{Cogh} = \text{Coag} + \text{dhrg} - \text{dhsg}$$

This analysis shows that during the long (hot) rotor condition, the expansion at the generator end of both stationary and rotating por-

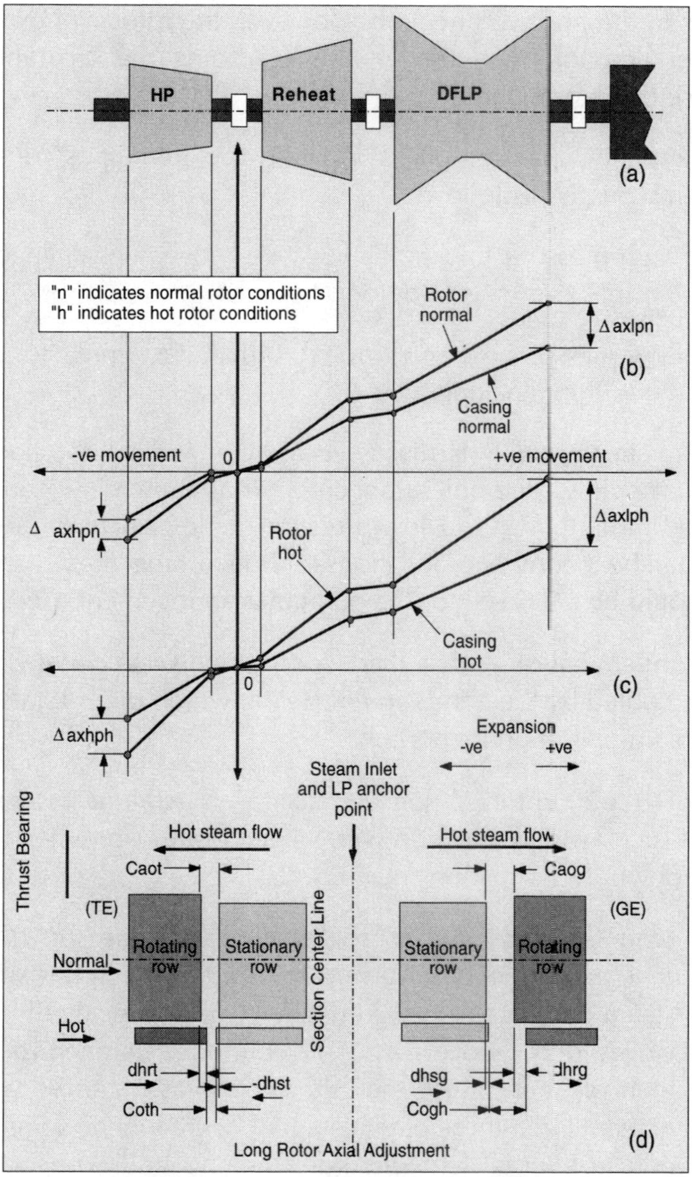

Fig. 2.4.7—The "long" or "hot" rotor. In (b) is shown the normal differential expansion, in (c) the effect of temperature reduction, and in (d) the modified operating position on the first stages in the low pressure section.

tions is in the positive direction. However, the rotor will expand by a larger amount, with the final effect being the clearance will increase by an amount (dhrg - dhsg).

The total "expansion effect" at the two extreme positions of the unit can be noted as follows:

- high-pressure end, total negative movement changes from normal ("Δaxhon") to hot ("Δaxhph")
- low-pressure extreme, from normal ("Δaxlpn") to the hot condition ("Δaxlph")

Note: In the analysis discussed above—and for the cold rotor condition below—the unit arrangement is as shown in Figure 2.4.7, with the thrust block and low-pressure section anchors as shown. There can be a number of locations for both arrangements, and each unit should be analyzed for the particular arrangement used.

A similar analysis can be made for transients that occur when the rotor is cooled by the admission of steam, which cools it faster than the containing stationary portion.

Short (cold) rotor. When the steam temperature reduces suddenly—such as load rejection with valve closure—or when the unit is shut down, two situations may result:

- *Unit speed does not reach emergency values (<110% normal) and the main stop valves do not close.* In this situation, steam continues to be admitted to the unit using the control valves to lower overspeed in a controlled manner. The unit is then reconnected to the grid as soon as it can be synchronized. Under these circumstances, there will be some reduction in internal temperatures gradients and a quenching of the materials of the steam path; however, with the readmission of steam, it is unlikely there will be significant metal temperature changes or that mean metal temperatures will drop to the extent there will be significant change in the axial expansion and location of the steam path parts

- *An emergency overspeed condition is reached,* i.e., *the overspeed transient reaches emergency values (>110% normal) or the rotor accelerates at a rate allowing emergency conditions to be reached.* Under these circumstances, the main control valves will close and the rotor speed will be reduced. Little steam is admitted to the unit before the valves close and unit speed will reduce until it can be placed on turning gear. During such an overspeed transient, a number of phenomena have the potential to affect the temperature of the fluid remaining in the unit. These include sections containing deposited moisture—this moisture will flash to steam as the pressure decays. There can also be frictional windage heating of the steam remaining in the sections, causing temperatures to increase to values much higher than normally experienced

Under circumstances of sudden cooling, the rotors—of smaller thermal mass and immersed in the cooler steam—will lose heat faster than the casings; rotor temperatures will tend to be lower than those of the casing. When this occurs, differential expansion curves will be modified—from those shown in Figure 2.4.8(a), with a hot-end differential expansion of "Δaxhpn" and a low-pressure end differential expansion of "Δaxlpn"—to those shown in Figure 2.4.8(c), with a hot-end total differential expansion of "Δaxhpc," and a low-pressure end differential expansion of "Δaxlpc."

Note: In low-pressure sections, design geometries can be selected so rotors will be of larger thermal mass, and cool slower than the casing, which can be a simple fabrication subjected to ambient temperatures. This must be considered in establishing the possible rotor conditions—particularly for units with multi-flow exhausts.

The "short rotor" condition will modify the clearance between the stationary and rotating blade rows shown in Figure 2.4.8(d).

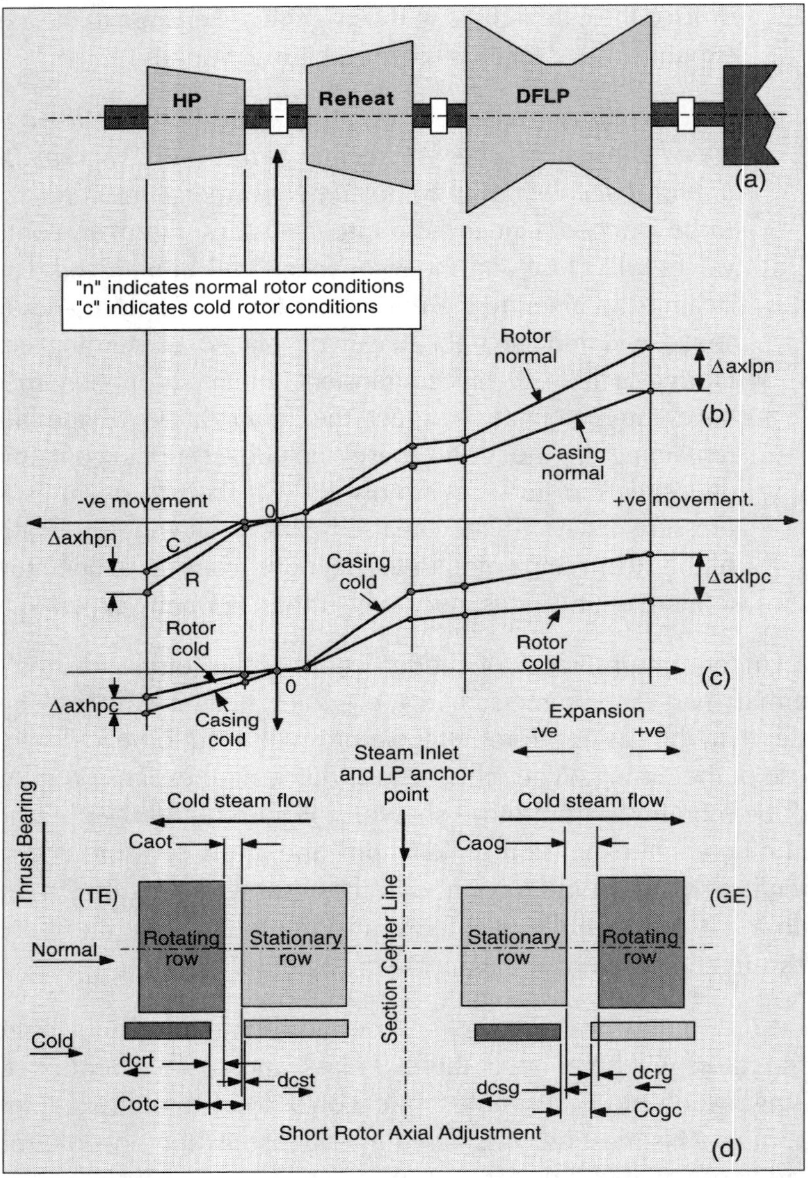

Fig. 2.4.8—The "short" or "cold" rotor. In (b) is shown the normal differential expansion, in (c) the effect of temperature reduction, and in (d) the modified operating position on the first stages in the low pressure section.

These normally run clearances "Caot" at the turbine end and "Caog" at the generator end and will be modified as shown to "Cotc" and "Cogc" at the turbine and generator ends respectively. Depending upon the rate at which the rotor cools relative to the casing, it is possible for the rotor to be at a lower temperature than the casing.

At the turbine end—and under conditions of a cooled rotor—there will be a rotor contraction towards the thrust bearing of "dcrt;" similarly, the stationary portions will contract, but move in a positive direction, towards the section center (anchor) by an amount "dhst." These movements will give a final cold transient running clearance ("Cotc") as follows:

$$\text{Cotc} = \text{Caot} + \text{dcrt} + \text{dcst}$$

This analysis shows during the short (cold) rotor condition, the contraction at the turbine end causes an increase in the axial clearance at the turbine end of (dcrt + dcst), *i.e.*, the rows move, increasing the axial clearance.

Similarly, at the generator end, the turbine rotor will experience a negative movement or contraction of "dcrg" towards the thrust bearing, and the stationary portions will have contracted by an amount "dcsg" in the negative direction towards the section centerline. Therefore, the normal hot running clearance ("Caog") will have changed. This contraction gives a cold transient running clearance ("Cogc") of:

$$\text{Cogc} + \text{Caog} - \text{dcrg} + \text{dcsg}$$

This analysis shows during the short (cold) rotor condition, the contraction at the generator end of both stationary and rotating portions is in the negative direction. However, the rotor will contract by a larger amount, with the final effect being the clearance will increase by an amount (dhrg - dhsg).

The design process considers and evaluates these "long" and "short" rotor conditions based on assumptions of the possible maximum and minimum temperature excursions that can be anticipated. This excursion review establishes minimum "cold stationary clearances" so there will be as few "rubs" as possible during operation. In addition to considering the clearance effect within the steam path stages, it is necessary to consider other locations within the unit where clearances are tight and rubs or excessive movement could affect total unit operation. Such locations can include oil seals adjacent to bearings, particularly at the low-pressure section where oil could be drawn into the condenser; also, at barring gear locations where misalignment of the gears can occur. Each of these axial positions needs evaluation and the cold clearance set, recognizing the damage potential these locations represent under the influence of transients.

The effects of thrust reversal

The steam path utilizes a high-energy fluid at various pressure levels that exert a thrust on each surface with which they are in contact. Around the surfaces of the rotating portions of the unit these pressures exert axial thrusts that must be balanced. On the inner surfaces of the casing these pressure thrusts are contained by the hoop stresses developed in the casings. Pressure "drops" also exist across the stationary blade rows, and the pressures developed between the blade attachments and the casings balance these thrusts. This fluid pressure is sufficient that in the case of diaphragms they are limited from moving in an axial direction by the force developed between them and the casing location. However, as will be discussed below, the diaphragms, as a consequence of this axial thrust, will deflect elastically; in certain circumstances, in a plastic manner as well. In the axial direction there are thrusts that need to be balanced.

In addition to the axial force developed by the steam pressure, there is also an axial thrust developed as a consequence of steam momentum change within the rotating blade rows. This thrust acts in the same direction as the pressure declines from inlet to exhaust.

As a consequence of steam momentum and fluid pressure within the casing, an axial thrust develops on each rotor, causing it to attempt to move in the direction of these axial forces to the greatest extent possible. To prevent (or limit) such unrestricted movement, the unit contains an axial thrust bearing. Such a bearing must contain faces intended to control and limit this axial migration while maintaining axial rotor position within close tolerances. Thrusts are in fact developed on the rotor in both directions, and the thrust bearing is sized to control these total differential effects and minimize damage that might result from excessive level or reversal of thrust.

If the unit is to operate as designed, the design engineer is responsible to evaluate thrust levels developed, and then to select design parameters to help ensure the net or resultant thrust is always in the same direction, and of manageable magnitudes. In addition, a margin is always designed into selecting this thrust bearing sufficient to allow for some level of operating damage and for deficiencies to be accommodated as they affect the thrust. Four phenomena, or components, affect the total axial thrust.

Piston thrust. At any position on the rotor where there is a change of section (increase or decrease in diameter), there will be an axial thrust produced. These forces are known as piston thrusts, and in magnitude are equal to the product of the pressure acting on the vertical face and the exposed axial annulus area. Such thrusts are present at both ends of each rotor and at other section changes. The end thrusts oppose each other. By judicious design they can be arranged to help balance thrust to a considerable degree.

Consider the four-wheel rotor shown in Figure 2.4.9(a). Here the radial dimensions are shown as diameters "D1" to "D4." [For convenience it is assumed that the rotor body diameter ("D4") is constant at all axial positions]. In the details of the rotor end positions

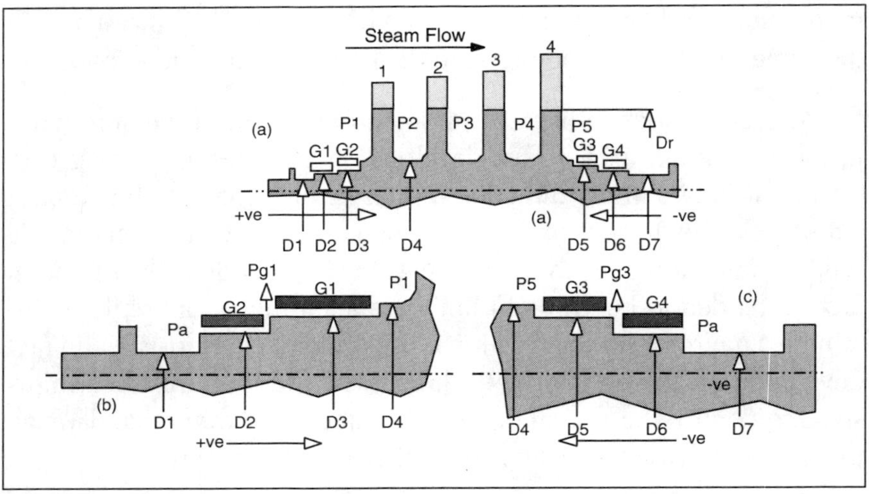

Fig. 2.4.9—The "Piston Thrusts" developed on the rotor ends.

[Fig. 2.4.9(b) and (c)], where flow-restricting glands ("G") are used, there are changes in shaft diameters in addition to axial position pressure changes. Consider the high-pressure end. Two gland rings have a pressure differential across the first ring ("G1") of "P1" to "Pg1" and on the second ring ("G2") from "Pg1" to "Pa." At the shaft position, where the steam enters the first gland ring, there is an exposed axial area of:

$$\frac{\pi}{4}\left(D4^2 - D3^2\right)$$

This annular area has a steam pressure ("P1") acting on it. Therefore, there is an axial thrust ("Ta") developed that is equal to:

$$Ta = P1 . \frac{\pi}{4}\left(D4^2 - D3^2\right)$$

Similarly, at the first "leak off" position between glands "G1" and "G2" there is a vertical exposed area of:

$$\frac{\pi}{4}\left(D3^2 - D2^2\right)$$

With a face pressure of "Pg1," this produces an axial thrust ("Tb") of:

$$Tb = Pg1 . \frac{\pi}{4} \left(D3^2 - D2^2 \right)$$

Similarly, at the second "leak off" position—steam escaping past gland ring "G2"—there is an axial thrust "Tc" of magnitude:

$$Tc = Pa . \frac{\pi}{4} \left(D2^2 - D1^2 \right)$$

In this case, "Pa" is shown as atmospheric pressure being equal to 14.7 psia. At the shaft end, another small thrust ("Td") is produced by atmospheric pressure. Its magnitude is:

$$Td = Pa . \frac{\pi}{4} . D1^2$$

Because the final "leak off" is to the atmosphere, thrusts "Tc" and "Td" can be combined. However, for purposes of demonstrating methodology, these are best considered separately.

The total axial thrust developed at the high-pressure end ("Thp+") can be found from:

$$Thp+ = Ta + Tb + Tc + Td$$

The implication of the "+" sign is to indicate the same convention of considerations or phenomena acting to the right are given a positive notation, and those to the left a negative notation.

At the low-pressure end the same reasoning is applied. This gives an axial thrust "Te," in the negative direction at entry to the leak off "G3" of:

$$Te = P5 . \frac{\pi}{4} \left(D4^2 - D5^2 \right)$$

At the leak off between "G3" and "G4," there is a thrust ("Tf") that again is negative, and has the value:

$$Tf = Pg3 . \frac{\pi}{4}\left(D5^2 - D6^2\right)$$

Using the same logic, the thrust ("Tg") at the atmospheric "leak off" is:

$$Tg = Pa . \frac{\pi}{4}\left(D6^2 - D7^2\right)$$

The axial thrust developed on the shaft end must be considered in terms of the thrust developed on the coupling flange; the fact that air exists on the coupling face (both sides), and the coupling flanges (when joined) are of the same diameter. It is also necessary to consider that the total thrust ("Thp-") is negative. Therefore "Thp-" is given by:

$$Thp- = Te + Tf + Tg$$

This axial thrust ("-Thp") represents the total thrust acting to the left in Figure 2.4.9(a) and (c). Taken together with any thrust developed on the coupling flange face—normally zero—the total axial piston thrust ("Taxp") developed on this rotor can therefore be found from:

$$Taxp = (Thp+) + (Thp-)$$

The gland systems shown in Figure 2.4.9 represent a relatively simple arrangement. In fact, the gland systems at both high- and low-pressure ends of the rotor are considerably more complex than those shown in Figure 2.4.9. Figure 2.4.10 is a sealing system in which one "leak off" (between gland rings "G2" and "Gx") goes to a sub-atmospheric gland-sealing condenser. In this case, the total thrust ("Thp+") would be determined from:

$$\begin{aligned} Thp &= Pa.\tfrac{\pi}{4}\left(D4^2 - D3^2\right) + Pg1.\tfrac{\pi}{4}\left(D3^2 - D2^2\right) \\ &+ Px.\tfrac{\pi}{4}\left(D2^2 - Dx^2\right) + Pa.\tfrac{\pi}{4}\left(Dx^2\right) \\ &= \tfrac{\pi}{4}\left[Pa\left(D4^2 - D3^2\right) + Pg1\left(D3^2 - D2^2\right) Px\left(D2^2 - Dx^2\right) + Pa.\left(Dx^2\right)\right] \end{aligned}$$

This arrangement is significant because the pressure "Px" is sub-atmospheric, and has therefore a lesser effect on the total axial thrust.

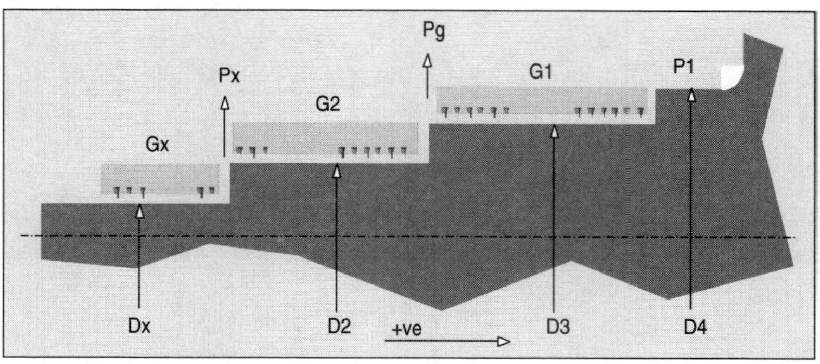

Fig. 2.4.10–A shaft end sealing system with a sub-atmospheric leak off at "Px".

Other considerations of total piston thrust are those related to reverse flow steam paths, where either the high-pressure expansion is broken into two separate and opposite flows, or where a combined high and reheat steam path are contained on a combined rotor. Consider the following arrangement: In the high/reheat section of a portion of a combined rotor shown in Figure 2.4.11, the high-pressure section has a shaft diameter of "D4a" and the reheat section, a shaft diameter of "D4b." At the crossover position—with a seal section "Gc"—the rotor has a diameter of "D8." Therefore, there is a negative thrust ("Th") of:

$$Th = Pa.\frac{\pi}{4}\left(D4a^2 - D8^2\right)$$

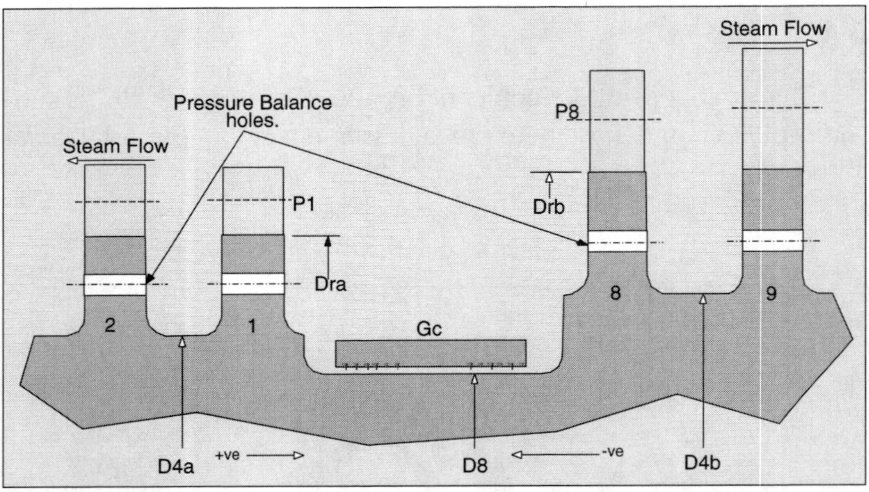

Fig. 2.4.11—Showing the piston effects at a reversal point in a high/reheat section.

A similar positive thrust ("Tr") develops on the reheat section. This is given by:

$$Tr = Pa.\frac{\pi}{4}\left(D4b^2 - D8^2\right)$$

The resultant thrust ("T") at this crossover position is equal to the difference in the two values given by the previous equation and that following.

$$T = Tr - Th$$

This piston thrust is a major contributor to the total thrust the axial bearing must accommodate. By selection of rotor diameters and arrangement of the steam path elements, this thrust can be reduced to a value the bearing can carry without exceeding acceptable values.

Wheel face thrust. If a pressure differential exists across a rotating blade row, there must also be a pressure difference (and therefore axial thrust developed) on any wheel face or vertical face that is part of that row.

Many stages are designed with the blade elements carried on the outer rim of wheels that are either formed integral with the central rotor or shrunk onto it. With this structure, thrust develops across these wheels, which act in the axial direction of steam flow and in effect represents another form of the piston thrust. Consider the four-stage rotor shown in Figure 2.4.9. Its four wheels each have a blade root diameter of "Dr." Total thrust developed on the wheels is the sum of the individual wheel thrusts—which is again the product of the pressure differential across the wheels—and their vertical face area. For the rotor shown in Figure 2.4.9, the axial wheel thrust ("Tw") is given by:

$$Tw = \frac{\pi}{4}[Dr^2 - D4^2].\left[(P1 - P2) + (P2 - P3) + (P3 - P4) + (P4 - P5)\right]$$

In the "impulse" design stage, the magnitude of the reaction at the vane root section is relatively small—and normally just positive—although negative reaction stages have been designed to meet other requirements. This means the pressure differential across each stage is relatively small. However, many impulse wheels have a large axial area, and can therefore develop significant levels of thrust in the steam flow direction. To minimize the possibility of this thrust causing excessive axial loads, "pressure balance holes" in these wheels allow equalization of pressure, helping to minimize the axial thrust developed across them. These holes also allow for the effects of transients or mechanical damage that could introduce some maldistribution of stage pressures. This allows them to equalize themselves without allowing high levels of thrust to develop.

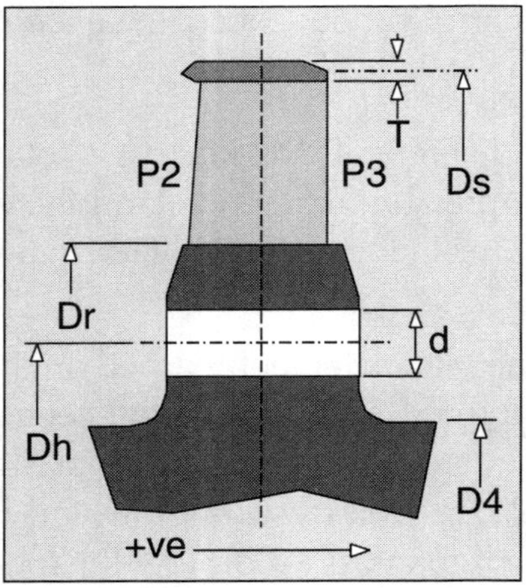

Fig. 2.4.12—The pressure balance hole at a diameter "Dh".

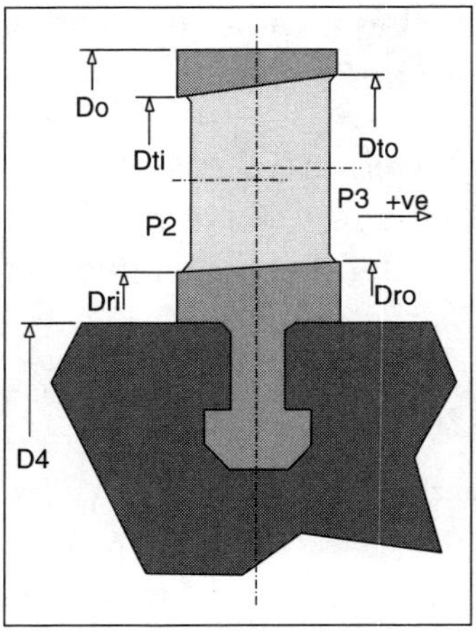

Fig. 2.4.13—A single reaction stage on a drum rotor.

Consider the wheel shown in Figure 2.4.12. A pressure balance hole of diameter "d" has a mean diameter "Dh." At this mean diameter there is no pressure differential across the wheel, as pressure is equalized by flow through it. However, at the diameter of the blade root "Dr," some small pressure differential exists due to the root reaction. If this root section pressure drop is defined as "dP," then a linear reduction from "dP" at "Dr" to zero at "Dh" can be assumed to exist. Then the wheel thrust "Twh" in this case can be taken to be:

$$Tr = 0.5 . dP.\frac{\pi}{4}\left(Dr^2 - Dh^2\right)$$

Another portion of this thrust that can be determined is normally of such small magnitude it is ignored for all practical purposes. This is the thrust "Tcv" developed on the coverband, and is equal to:

$$Tcv = (P2 - P3) . \frac{\pi}{4} . T . Ds$$

Where "P2" and "P3" are the pressures existing at the tip section, they are not of the same magnitude as those present at the root section; a vortex calculation must be performed to determine their values. The mean diameter of the coverband is "Ds" and its radial thickness "T."

In the case of a reaction stage carried on a drum rotor (e.g., the rotating blade row shown in Fig. 2.4.13), a pressure drop develops across the rotating blade and acts on the root section—from diameters "Dri" to "D4," and at the tip or outer diameter, from "Do" to "Dti." If it is assumed that these pressures vary with radial reaction, then the thrust "Twr" can be found from:

$$Twr = \frac{\pi}{4} \left\{ \left[\left(Dri^2 - D4^2 \right) . \left(P2r - P3r \right) \right] + \left[\left(Do^2 - Dti^2 \right) . \left(P2t - P3t \right) \right] \right\}$$

Note: This equation is developed from considerations of pressure differentials from inlet to discharge rather than total thrusts in the upstream and downstream direction. This equation also ignores the effect of sloped sidewalls and the axial component of any thrust developed on them. (See also the equations on page 138.) It also takes into account:

- *the effect of blade damage*—if the blade rows sustain any form of mechanical damage, it modifies the row discharge area and affects the pressure distribution throughout the steam path (and therefore, the axial thrust). In the majority of incidents, this will be of no consequence; however, there could be circumstances in which this can increase the load on the thrust bearing to unacceptable levels, and corrective action must be taken

- *the effect of chemical deposits*—deposits will often form as a consequence of impurities carried over with the steam. These impurities will often be deposited on the blade vanes and will tend to close the throat. Under normal circumstances, these deposits will build to a thickness of 0.005" to 0.010" *i.e.*, they will close the throat by 0.010" to 0.020." However, circumstances can arise in which much thicker deposits will form. This will influence the pressure distribution and therefore, the axial thrust

Blade annulus thrust. A pressure differential exists across the rotating blade annulus. This totally axial force is the product of the rotating blade row pressure drop and the annulus area across which it acts. From the inlet to the row, the pressure is not of equal magnitude along the entire radial height, because of the radial flow component of the steam velocity. However, at the discharge surface, the pressure is sensibly constant at all radial heights. For blades of relatively small radial height (height/mean diameter <0.20")—as shown in Figure 2.4.13—the pressure at the mean diameter (P2 and P3) can be used with sufficient accuracy that the axial thrust "Tba" on the row can be determined from:

$$Tba = (P2 - P3) \cdot \frac{\pi}{4}\left[(Ds\text{-}T/2)^2 - Dr^2 \right]$$

Equation 2.4.13(a) is suitable for use with an impulse blade row (Fig. 2.4.12) with a relatively small pressure drop developed across it. However, in the case of a reaction stage having a high degree of reaction, it is necessary to evaluate the entire stage, including those developed on the inner and outer sidewalls of the vane itself. Figure 2.4.13 shows a reaction rotation blade row that has "sloped" sidewalls. In this case the total "Tba" thrust consists of the following terms:

$$Tba1 = P2\frac{\pi}{4}\left\{\left(Do^2 - Dti^2\right) + \left(Dti^2 - Dri^2\right) + \left(Dri^2 - D4^2\right)\right\}$$

$$Tba2 = -P3\frac{\pi}{4}\left\{\left(Do^2 - Dto^2\right) + \left(Dto^2 - Dro^2\right) + \left(Dro^2 - D4^2\right)\right\}$$

The outer and inner sidewalls carry a mean pressure 0.5(P2 + P3).

$$Tba3 = -\frac{P2 + P3}{2}\cdot\frac{\pi}{4}\left(Dto^2 - Dti^2\right) + \frac{P2 + P3}{2}\cdot\frac{\pi}{4}\left(Dro^2 - Dri^2\right)$$

The total thrust ("Tba") is therefore the sum of these three components:

$$Tba = Tba1 + Tba2 + Tba3$$

Note: The equation contains those terms of the previous for the wheel thrust "Twr," but has assumed a constant pressure along the radial height of this rotating blade row at both inlet to and discharge from the row. This introduces no great error for small radial height elements. The first equation has also ignored the sloped sidewalls and any axial component of thrust developed on them. Examine all equations in this section for blade annulus thrust.

For larger radial height blades, it is necessary to determine the pressure distribution along the inlet height of the vane. From this can be determined the total thrust, either by considering the mean thrust, or determining the thrust over small radial sections, and adding their individual values to determine the row total.

Steam momentum thrust. The change in momentum as steam flows between the blade vanes is shown in Figure 2.4.14—the force diagram on a blade due to the change of steam momentum. The total thrust component "AD" produces a component in the axial direction equal to "DE." This value must be determined for each stage in the unit and added to the thrusts developed by the other mechanisms present in the unit. This axial force is directed downstream in the direction of steam flow.

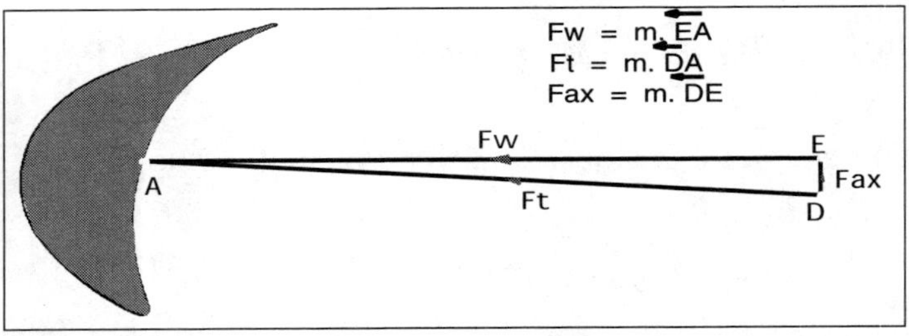

Fig. 2.4.14—The force diagram and its components on a turbine blade. Vector ED is the axial thrust component.

For any blade row, the thrust due to the change in steam momentum can be determined from consideration of the vector diagram and the change in the axial component of velocity that is achieved. For the vector diagram shown in Figure 2.4.15, the change in axial component of velocity "ΔCax" is given by—

$$\Delta Cax = \left\{ C1.\mathrm{Sin}\,\alpha 1 - W2.\mathrm{Sin}\beta 2 \right\}$$

—so for the axial thrust "Tdv" for a stage flow of "m" #/sec, the total thrust can be determined as:

$$Tdv = \frac{m}{g} \left\{ C1.\mathrm{Sin}\,\alpha 1 - W2.\mathrm{Sin}\beta 2 \right\}$$

Pressure balance pistons

Reaction stages have an axial thrust component that is considerably larger than equally-sized impulse blades. This means single-flow sections normally require "pressure balance pistons" in order to reduce thrusts to a level that normal-sized thrust bearings can accommodate the loads developed in normal service.

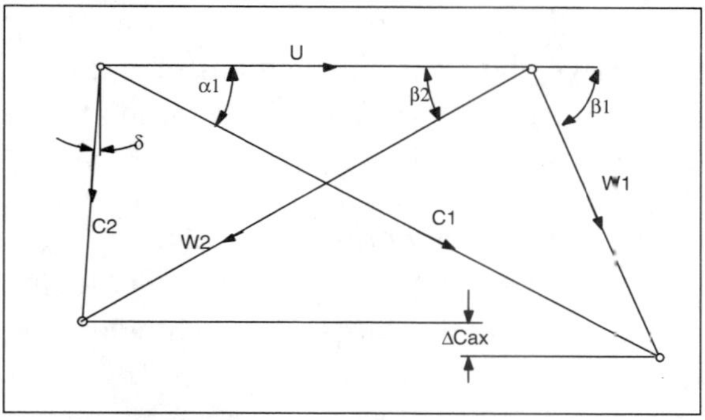

Fig. 2.4.15—The vector diagram of a single stage, showing the variation of velocity in the flow of steam through a stage.

The balance piston consists of a wheel machined onto the rotor. It is designed to operate with a pressure drop across it sufficient to balance the portion of the axial thrust (which is in excess of what might normally be developed) and capable of being carried by an acceptably-sized thrust bearing. Figure 2.4.16 shows a typical high-pressure, single-flow reaction section. The expansion consists of an impulse control stage with a pressure drop across the rotating blades from "P1" to "P2," and then a series of rotating blades with a pressure reduction from "P2" to "P3." Piston thrusts also develop on each end of the rotor. As a consequence of the reaction blade pressure drops, a large thrust develops towards the exhaust end. To counter this, a balance piston is incorporated into the design, with a gland seal system above it ("G4") to limit leakage. The thrust developed on this piston ("Tp") acts in the opposite direction to the blade reaction thrust, and is equal to:

$$Tp = \frac{\pi}{4} \cdot \left\{ P1.[D8^2 - D7^2] - P5.[D8^2 - D9^2] \right\}$$

Using the pressure balance piston makes certain variables available to the designer to help select a suitable geometry. These include:

- *the point in the total thermal cycle to which the pressure "P5" is connected.* The steam leaking past this dummy piston is at a high energy level and should be returned to the unit, preferably on the steam path at a pressure as high as possible
- *the dummy piston outer diameter "D8."* This diameter can be varied within the range of what can be contained within the casing. However, the larger this diameter, the greater the leakage area, and the greater the energy wasted
- *the selection of diameters "D7" and "D9."* These are the diameters controlling the piston thrust area around the dummy, and can be varied within limitations set by stress and critical rotor speed, to control the available area

The total axial thrust developed on the rotor is the sum of these four individual components each of which, for normal operating conditions, is determined as part of the design process. Where possible, parameters and overall steam path geometry are selected to minimize the required size of the thrust bearing. Unfortunately, it is possible that during operation, transient situations can occur to cause an increase in or reversal of the normal thrust. In the majority of situations, this will only be a short duration phenomenon, which will correct itself—but during an incident, it will be necessary for the thrust bearing to accommodate the change. While the majority of these transient situations are of short duration, conditions can arise that result in permanent damage to the blade system introducing a permanent change in the thrust level. This can result in unit shutdown so corrections can be made.

As stated, the design process determines the magnitude of the total axial thrust of the coupled sections and arranges for a thrust bearing of sufficient capacity to carry this load. However, in an effort to minimize the size of the required thrust bearing, the designer will utilize certain options to help ensure this total thrust occurs at acceptable levels. The design options available include:

- arranging individual sections of the unit so their axial thrusts are in opposite directions—where they tend to cancel
- using intercept valves to shut after emergency stop valves (ESVs). If these ESVs close too early, the high-pressure section thrust will be unbalanced momentarily and can cause thrust bearing failure
- reversing flow in individual sections after partial expansion, so the net thrust is reduced. This flow reversal also has other advantages
- modifying rotor diameters to increase or decrease total piston thrust effects. When making such modifications, it is also necessary to consider the effects on the critical speeds of the rotor
- employing "pressure balance pistons" on reaction design stages to counter excessive blade reaction and wheel thrust
- using "pressure balance holes" in impulse wheels to minimize wheel face thrust
- arranging extraction locations in double-flow sections—where steam is extracted for regenerative feed heating—so net thrust induced on the rotor is in the most advantageous direction to reduce the total thrust

Thrust bearing failure

It is uncommon for a thrust bearing to fail—but it can occur. When it does:

- the oil wedge between the thrust bearing and thrust collar on the rotor will break down and cause the white metal to overheat and melt. This can cause extensive damage to the blading, which, depending upon axial clearance, will make "rubbing" contact

- this failure of the oil wedge allows axial movement of the rotor, causing heavy contact between the metallic surfaces of the thrust collar and bearing pads

A potential consequence of this contact failure is axial movement of the rotor—in either direction—so clearances from thrust collar to the active and inactive faces must be controlled; this has the potential to modify the axial position of the rotor. It may only occur for short periods but cause extensive damage to the rotor and bearing, possibly introducing "rubs" that embrittle the rotor. This could cause local hardening that may be irreparable unless it can be machined off. This depends upon unit geometry. The blades may require replacement.

Differential expansion with flexible couplings

The rotors discussed in previous sections have been considered to be the solidly-coupled type, and so it is possible to consider the total rotor "string" as a single, solid forging. In such a design, the axial thrust developed in one section is transmitted through the couplings to a single thrust block. It is therefore necessary to consider all expansive movement of the rotor, in both directions, occurs from the active face of that thrust bearing.

When flexible couplings connect rotors, the differential expansion produced in one section must be carried by a thrust bearing in that section. The thrust developed in each section must be transmitted to an axial bearing located in that section, *i.e.*, each section requires an individual thrust bearing capable of locating the rotor within the casing surrounding it. Therefore, conditions established by a solidly-coupled rotor do not exist with the flexible design, and tighter hot-running clearances can normally be accepted.

Modern turbine units tend toward the use of solidly-coupled sections. While this may introduce certain problems of accommodating large axial movements, on balance these are found to provide advantages that certainly outweigh the disadvantages.

Different forms of flexible couplings are available, and there are advantages to each of them. If a tooth coupling is used, for example, there is often a tendency for it to lock if transmitting higher torques than it was originally designed for.

RADIAL EXPANSION OF THE STEAM PATH PARTS

Steam path components exhibit growth in the radial direction during operation. This growth is due to two separate phenomena affecting all rotating components to a degree. One of these will also affect the stationary components to a lesser effect. These phenomena are detailed below.

Thermal effects

As the unit heats with the admission of steam, the individual components accept this heat and expand in all directions. There are natural laws expressing the extent of expansion in terms of material properties (linear coefficient of expansion) and the degree of temperature increase. Consequences of this thermal growth in the axial direction have been considered in the previous section, and the modifying effects of thermal transients and other phenomena will be considered in subsequent sections.

In the radial direction, the effect of thermal radial expansion is evident in both stationary and rotating portions of the unit. However, because the casings (stationary components) are slightly cooler than the corresponding rotating portion (due to heat radiation), the casing will expand to a lesser extent than the rotating portions at the same axial location.

The casings carry the stationary blades or diaphragms and radial seal portions of the unit. These thermal effects cause the components to expand and contract radially in response to changing temperature environments. It should be noted that as the unit heats the casing will radially expand outwards. Any blades or diaphragms they carry and locate will also expand in the radial direction, maintaining their total alignment. Stationary blades located in casing grooves will move with the casing because they are locked in the radial direction by the root portion of the blade. Diaphragms, however, are located within grooves machined into the casing and will move radially, but because they are not constrained by it, they may move at different rates relative to the casing and their temperature ramp rates. This is discussed in more detail below.

Radial stress effects

Mechanical stresses are induced in the rotating portions of the unit by their own centrifugal force. These stresses have a component that induces radial growth (or extension) in both the blades and rotor. These stresses are proportional to the square of the speed. Because the stationary portions of the unit are not subject to centrifugal loading, these will not grow in the radial direction due to any rotational effects. There is, however, a degree of radial growth due to the internal pressure, causing a "hoop stress" in the casing walls. This will cause some small amount of radial stress growth. Also, it is not uncommon for this internal pressure to cause the casing to become slightly elliptical due to the greater stiffness at the horizontal joint.

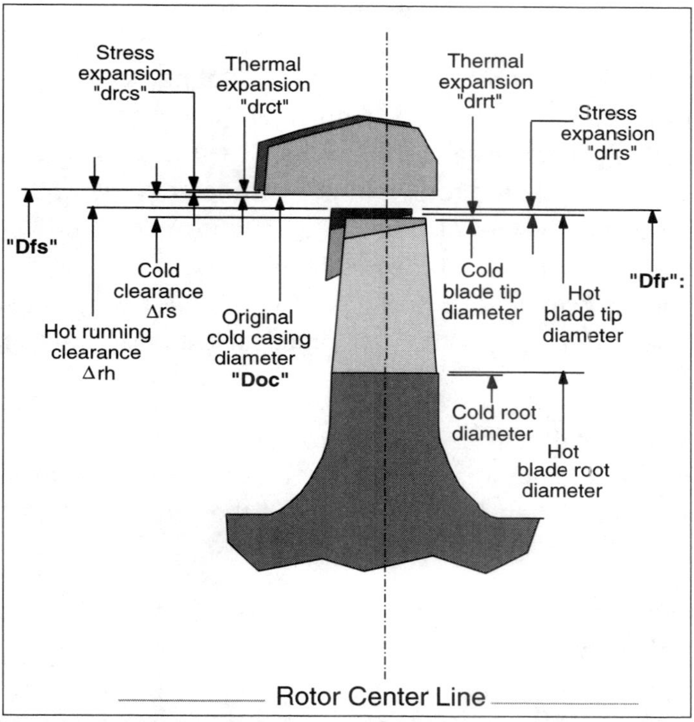

Fig. 2.5.1—The thermal and stress effects influencing radial growth in a turbine stage.

Combined factors of radial growth

The radial growth elements for a single stage are shown in Figure 2.5.1. To the right of this figure are those effects that influence the wheel diametral changes; to the left, those that influence the casing.

The rotating blades and rotor will grow as a consequence of both stress and temperature. Consider first the effects of temperature on the rotating portion of the stage. For this wheel (including the rotating blades) the effects of radial growth are shown as "drrt," due to temperature. During operation, all components of this stage will reach a temperature that is at some variation from inlet to discharge edges, depending upon the degree of reaction. In fact, a mean temperature can be determined for all elements of the rotating stage (Fig. 2.4.2). This can be taken as the mean temperature between inlet and discharge, and all elements can be considered to grow radially as a consequence of it. While the rotor can be considered to be at the same mean temperature as the rotating blade, it is possible that it will also be influenced to some small degree by heat soaking along the rotor from the hotter to the cooler positions, and from its cooler side to the lower temperature stages. For all practical purposes this can be ignored.

The centrifugal extension of the rotating portions of the stage is shown to give a total radial growth of "drrs." This growth represents the sum of the growth in the wheel and the growth in the blade, including vane and root attachment.

Therefore, the final operating diameter of the rotating blades ("Δfr") is given by an extension beyond the cold stationary position ("Δor") by the expression:

$$Dfr = Dor + 2(drrt + drrs)$$

On the left of the stage are shown the radial changes that occur in the casing. The original inner casing diameter is shown as "Doc." At the normal operating condition, the casing expands radially ("drct" represents the total growth due to thermal influences; "drcs" represents the effects of the casing's radial expansion due to the internal steam pressure). Therefore, the hot operating diameter "Dfc" of the casing is given by:

$$Dfc = Doc + 2(drct + drcs)$$

The cold internal clearance "Δrc" is found from:

$$\Delta rc = \frac{\text{Blade tip diameter cold} - \text{Cold internal diameter of the casing}}{2}$$

$$= \frac{Dor - Doc}{2}$$

Taking into account the radial growth of the stationary and rotating components, the hot running clearance "Δrh" can be determined:

$$\Delta rh = \frac{Dfc - Dfr}{2}$$

$$= \frac{[Doc + 2(drct + drcs)] - [Dor + 2(drrt + drrs)]}{2}$$

These clearances are calculated on the assumption that the seals and/or wall above the blades are formed by the casing inner surface. If this is not so and seals are carried in an extension of the diaphragm, these equations may not be valid. This is considered in the following section.

In the case of a casing-carried seal, the designer selects the internal casing diameter to accommodate the blade tip radial growth with cold clearance sufficient to ensure a lack of running interference between the parts during operation. As discussed above, the stages can be subjected to thermal transients. They will heat various stationary portions of the unit, expand at different rates, and be influenced by the mass of the horizontal flanges that can heat and cool, thus attaining constant temperature at a slower rate due to the large volume of material at these locations.

In the case of low-pressure sections, the casings produced by fabrication can heat and cool faster than the rotating portions. This also has to be considered.

Influence of stationary blade locating method

Stationary blade rows are located radially to provide guided flow from the stationary to rotating blade rows. This is achieved by ensuring adequate inner and outer lap in the hot rotating position. To ensure this condition occurs, consideration must be given to the manner in which these stationary blade vanes are located and how temperature will affect them. There are two distinct methods of locating these blades:

- The blades are located in grooves machined into the inner surface of the casing, or a blade carrier that locates and carries several rows. Upon expansion or contraction in response to temperature changes, the blades will accept heat and expand, so there will be no increase in gaps between individual blade segments
- The blades are located in a diaphragm produced in segments (normally two), each of which has a 180° arc. Diaphragms are not exclusively located in two arcs, and designs of 6-10 segments are in use

Associated with these different constructions are considerations relating to retaining radial positions as the unit heats, expands, and is subjected to steam loads:

- For stationary blades located in a casing, the blades will move with the casing (and therefore change their diameters) as the casing moves radially outwards upon heating or radially inwards upon cooling
- When the blades are carried in a 180° segment, they are often bolted together at their horizontal joint, and will be less affected by casing movement. The casing simply supplies axial location and mechanisms to hold the diaphragm at the correct radial location at its center. The diaphragm will assume the lowest possible position due to its weight. If it is

supported at the casing centerline, weight has less influence on radial location

- Diaphragms constructed of a number of individual segments carry the possibility of vane radial movement with temperature change. It depends upon the method used to locate the diaphragm outer ring in the casing. Each design is considered separately

Effects of operating transients

It has been shown that thermal transients affect axial clearances between the stationary and rotating portions of the unit, and these must be taken into account when setting cold axial clearances. In a similar manner, it is necessary to consider the effect of these transients on radial settings—considerations influenced particularly by the manner of construction. As in the case of axial expansion and contraction, considerations must be made for the weight of the various components and the rate at which they receive and reject heat from the surrounding steam. Those considerations include the following:

Load rejection. The effect is triggered when the unit is disconnected from the grid. Under these circumstances, the rotor will be accelerated to higher rotational speeds, increasing the radial stress and therefore the radial extension. The designer must consider these factors when establishing the cold setting radial clearance.

In the event of load rejection and main valve closure, both the pressures and temperatures within the casings are reduced. In this situation, the lower temperatures will reduce the existing thermal radial growth of both the rotating and stationary components, and reduce the total radial expansion effects. These diametral corrections will not occur suddenly, but require time for the temperatures to reach an equilibrium condition. A corresponding reduction of internal pressure will result in a reduction in the casing hoop stress extension that will occur immediately. An internal pressure reduction will have no effect on rotor stress extension. Therefore, as the casing cools and contracts the rotor will increase in radial growth due to the increase in speed. The designer setting the cold stationary clearances must consider these effects.

Overspeed transients. When load rejection occurs, and the unit is disconnected from the grid (or other driven device), the unit enters an overspeed transient. The extent to which speed increases above normal operating values depends upon a number of factors—including the load at the time of valve closure; the speed with which the control and stop valves close, and (in the case of a unit with free moisture in the various sections) the extent to which this moisture can flash to steam with pressure decay and continue to expand through the steam path, contributing to an overspeed increase.

There are two considerations of overspeed transients and the extent to which these will affect the radial clearance:

- *The unit attains a "normal overspeed"—i.e.,* the unit does not increase its rotational speed to the extent the main stop valves close. In this situation, steam will continue to be admitted to the unit in such a manner that internal temperatures do not rise excessively due to windage (in the case of a main power generating unit, it will continue to carry station auxiliaries and can be reconnected to the system as speed reduces to normal operating values)
- *The unit reaches emergency overspeed values, or accelerates at a rate that would cause a trip anticipatory device to close the main stop valves.* In this situation, clearances must be evaluated in terms of the anticipated temperature differences that could arise

Note: The "trip anticipatory device" is designed to sense the rate of rotor acceleration in the event of an overspeed. If this device determines the overspeed will reach excessive values (in excess of normal overspeed, or about 108-110% normal), this mechanism will trip the unit to close the main and intermediate stop valves.

In the event of overspeed, the direct centrifugal loading of the rotating portions of the unit will increase in proportion to the speed of rotation squared. This increase in radial stress will cause the rotating blades and wheels or rotor to expand radially. The casing unaffected by this increase in speed will not increase its inner diameters. In fact, as pressure decays (with valve closure limiting steam admission to the unit), the internal pressure on the casings will decrease, reducing the radial expansion caused by the hoop stress induced in the casing. In general, these diametral changes of both the stationary and rotating components are small, but should be considered in the design phase, particularly for the larger blade elements where stress levels are high.

Temperature up-ramps. This pertains to some occasions when boiler temperatures rise. During operation it is normal for the steam to be attemperated and the temperature of the fluid entering the turbine to remain sensibly constant at the design value. However, if the temperature does manage to exceed the design values at some location within the steam path, then local expansion will occur in the radial direction.

Temperature down-ramps. Temperature reduction can occur in the steam delivered from the boiler superheater or reheater. This is not too common, and may only continue for short periods. However, during the cooling period, both sets of components can move and running clearance changes should be considered. This is unlikely to be a major cause for concern relative to other factors, but in a situation where other unpredictable factors are present, it could compromise running clearances.

Water ingestion. In the event water is ingested into the unit, it will have a quenching effect on various components, depending upon which portions of the unit suffer the cooling. This can effect rotor concentricity and casing alignment, both of which have the potential to effect running clearances.

The consequences of radial growth

The design process anticipates radial growth of the stationary and rotating portions of the unit and makes allowances for them. In establishing the normal radial clearance, several considerations need to be factored into the selection of the cold setting clearance, as they will impact upon the successful running of the unit and the maintenance of clearances.

Lifting of a rivet attached coverband. For those stages having a coverband attached to the blade's outer diameter by the riveting of a tenon, centrifugal loads—induced by their own weight—act on the coverband. These tend to make it deflect radially outward [Fig. 2.5.2(a)].

With this form of coverband attachment, the tenon will nominally hold the coverband in place, and it will be rigidly attached to the blade's outer diameter. However, under all rotation conditions, there will be sufficient force induced by the centrifugal loading of the coverband itself to make the inlet and discharge edges (and the overhung coverband at the batch end edges) "curl." This is shown in Figure 2.5.2(a)—a radial outwards movement of "ri" at the inlet and "ro" at the discharge. Curling extension at batch end positions could be even more severe.

It is not reasonable to expect this radial outwards movement would introduce rubbing under normal operating conditions. There is the possibility [shown in Figure 2.5.2(b)] of the coverband rising on the tenon (particularly as the unit ages) and for total outward movement to increase. This adds an additional extension ("da") as the riveting becomes less effective and increases total outward movements to "ri+da" and "ro+da" at the discharge. Also, if the unit goes into an overspeed condition—and particularly if the overspeed is accompanied by higher than normal levels of rotor vibration—it becomes possible for rubbing to occur. This may only result in wear of the seal strips and some gouging of the coverband and does have the potential to cause some degradation in the performance of the unit.

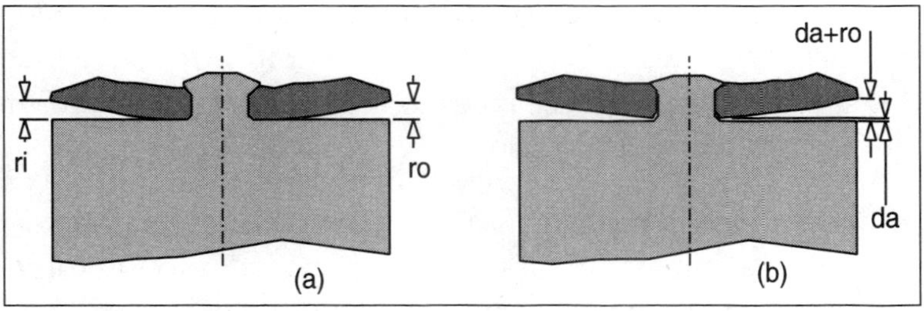

Fig. 2.5.2—The cover band attached by riveting. In (a) is shown the curling effect at the inlet and discharge edges. In (b) is shown the effect of time creep in the radial direction.

Lifting of a formed integral coverband segment. If the stage contains a coverband that has been formed integral with the blade vane, then these individual segments will also "curl" outwards under the influence of their own centrifugal force. However, in this case, the amount by which these segments will deflect radially are influenced by the following:

- *Form of the integral coverband.* The coverband portions will curl outwards under the effect of centrifugal loading depending upon their unsupported sections. As shown in Figure 2.5.3(a), there is an overhung portion ("a") of the lozenge shaped coverband that will tend to curl because of the distance from the profile. In this case, the extreme position ("u") is farthest from the profile and, therefore, the radially most deflected section moving outwards by an amount "rs." It is also that portion of the coverband most likely to rub, if there is some surface or seal above it

- *Position of the center of gravity of the coverband segment.* In some designs, the shape of the coverband is dependent upon the method of manufacture. When plunge milling forms the vane, the coverband will also have end shapes that are consistent with the profile face. Figure 2.5.4(a) shows a coverband segment with a joint face consistent with the pressure face of the vane. Large overhung sections at both inlet and discharge edges (an amount "a") will induce a large curling stress and cause these faces to move radially outwards under the influence of rotation

- *Machining thickness.* Many stages with integral covers are trim machined after assembly of the blades to the rotor. This trim machining—on surface "z-z" of Figure 2.5.4(b)—achieves a concentric surface suitable for sealing. It is normal for this trimmed diameter to be determined by design, allowing for a suitable leakage clearance at the tip. Should there be an accumulation of manufacturing errors, the trimmed thickness of the coverband ("Tk") can be produced too thin. While this will reduce the centrifugal load (and therefore the stress) of the coverband segment, it will also cause a reduction in the section modulus of the coverband, making it more susceptible to bending deflection. This will produce a larger radial movement than design had intended, and could cause excessive outward deflection. This effect of trimming is shown in Figure 2.5.4(b).

Slanted outer sidewalls. Some stages have rotating blades with "slanted" outer tip sections (Fig. 2.5.5). These stages are often used without coverbands. In these stages, clearance must be considered in both the radial and axial direction when the rotor is in the "long" or "short" positions. These extremes represent critical positions, as

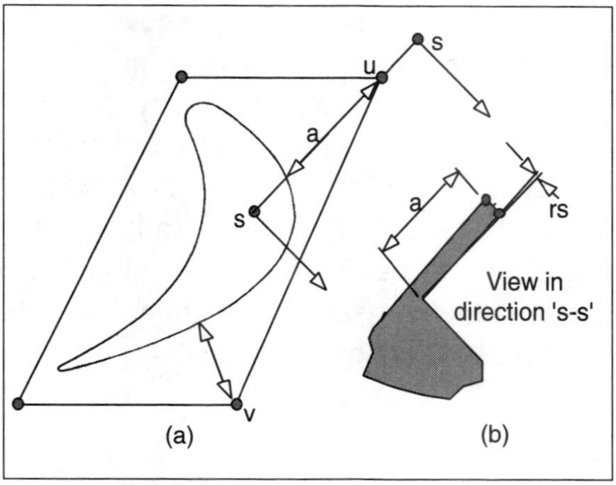

Fig. 2.5.3—The lozenge shaped integral cover showing the radial outward movement, under the effect of centrifugal loading.

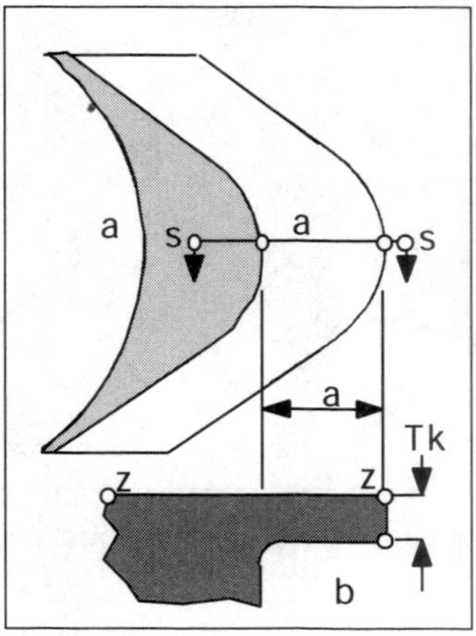

Fig. 2.5.4—The shaped cover, and the effect of trimming

there is positive ("+ve") and negative ("-ve") movement. Axial differential expansion could cause the radial clearance to be consumed and the blade tips to rub on the casing. Many designs utilize this slanted outer surface on low-pressure stages and are most likely to experience large axial expansions in a two-flow design. Here, the blade tip is thinned to a "squealer" section to allow close clearances between the blade tip and casing. If rubbing should occur in this design, excessive amounts of heat are not generated, and the blade will not bend because of the rubbing effects.

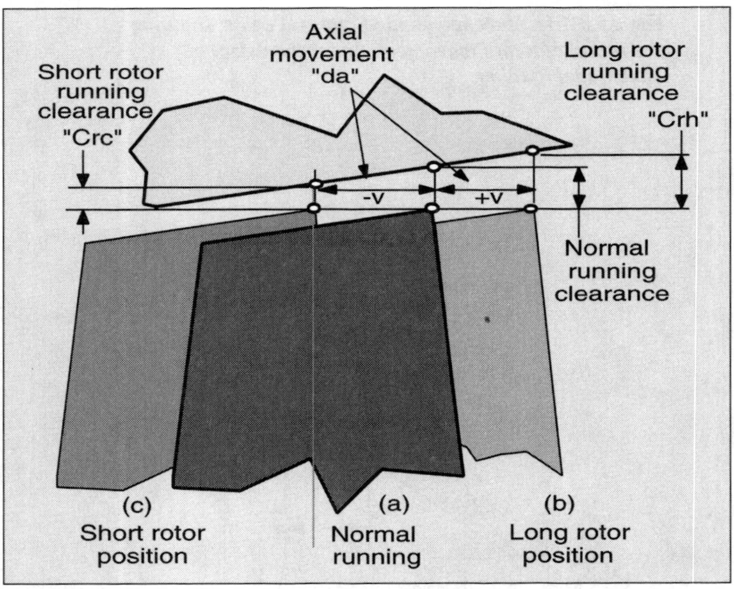

Fig. 2.5.5—The started outer side wall, and its effect on radial clearance.

DIAPHRAGM DEFLECTION AT PRESSURE AND TEMPERATURE

During operation, the stationary blade rows are subjected to a pressure differential from inlet to discharge sides. This pressure differential is applied to impulse and reaction designed stages. However, because the pressure drop across an impulse stage is much higher, and because the inner surface of the stationary blade or vane is normally attached to the inner web of a diaphragm, the consequences of this condition can be much more severe on these impulse stages. Both impulse and reaction stages can also be operating at high temperatures, which modifies the mechanical strength of the material.

Two stationary blade cases must be considered in terms of the amount of axial deflection: the diaphragm type stages (as used in impulse designs) and some low-pressure stages, which have high levels of reaction on them. There is also the need to consider the stationary blading of high reaction stages, where blade roots are assembled into a blade carrier or casing.

High pressure, high-temperature diaphragms

The high pressure, high-temperature diaphragm consists of an outer ring, located either in a groove produced on the inner surface of the casing or on a ligament carried from the casing. Because of this location, the outer ring cannot deflect. However, the blade row and the inner web develop high pressures across them, and will deflect axially.

This effect is shown in Figure 2.6.1. The pressure drop across the vanes will deform by some amount shown as "dxv" and the inner web—which acts as a pressure barrier between stages—will experience a pressure gradient in the downstream direction deflecting the diaphragm towards the following row of rotating blades (which at the inner diameter can be taken to be "dxw"). The total inner deflection is therefore:

"dxd = dxv + dxw"

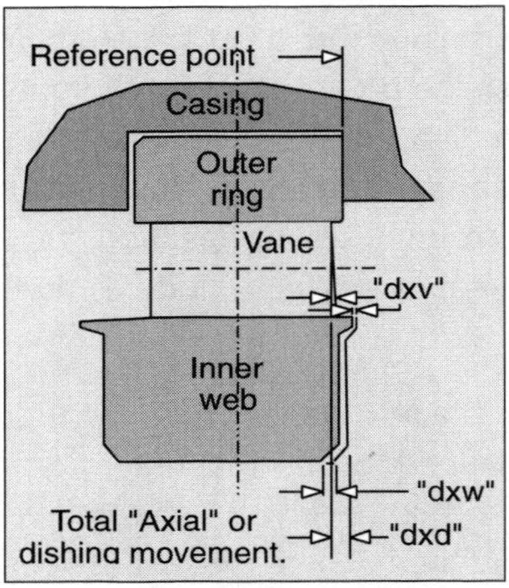

Fig. 2.6.1—Showing the axial deflection.

For the lower temperature, lower pressure stages, the pressure differential is not as large. Normally the surface area exposed to the pressure is greater, and there will be some small level of deflection.

In any stage, there will be a deflection downstream as pressure is applied to the unit diaphragm. This initial deflection takes an "elastic" form and will recover and return to its original axial location as the load is removed. Long-term exposure to this pressure at an elevated temperature will be sufficient to make the diaphragm creep in the axial direction. This creep deflection is termed "plastic" and represents a permanent deflection. The diaphragm will not recover this portion of the total deflection when the pressure is removed from the unit.

The design-stipulated cold clearances must consider the creep deflections and make adequate allowance in the initial setting. This ensures that, as the unit ages and there is an accumulation of deflection, there will not be running interference. *i.e.*, the axial clearance "Cxo" (Fig. 2.4.4) will not be totally consumed and rubbing will not occur.

Stationary blades

In the case of stationary blades, located in a casing or blade carrier with little or no thick inner web, there will be an axial deflection "dxv" as shown in Figure 2.6.2. This refers generally to a reaction stage with a lower pressure drop than a similar capacity impulse diaphragm vane. However, the reaction stage—because of space limitations—tends to utilize blades of smaller axial width, and these tend to have less resistance to axial deformation. One manufacturer brazes these stationary blades into groups, which supplies extra resistance to deflection.

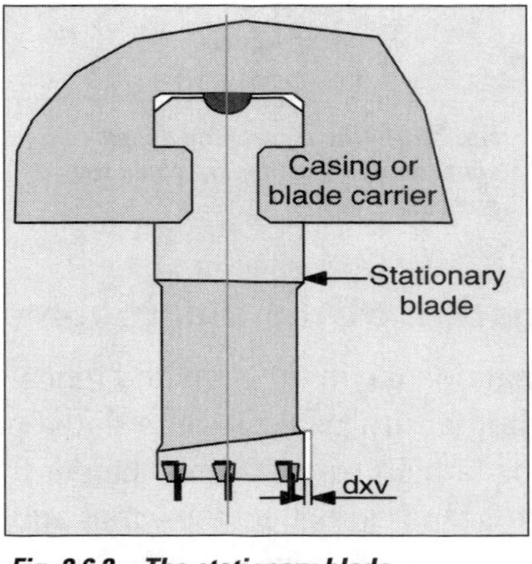

Fig. 2.6.2—The stationary blade.

A major consideration of the stationary blade is that it does not have an inner web, and is not subject to the considerable axial thrust that can develop across it. However, if these individual blades are not located adequately in the casing root slot—either as a consequence of poor assembly practices, poor control of tolerances on the vane access slot width, or root tolerances not maintained—there can be elastic and plastic deformation, which can increase the possibility of rubs under certain conditions of blade tilt. Critical axial widths that should be controlled are shown in Figure 2.6.3. In addition, the caulking that is often applied to the blades must be sufficiently tight to cause good contact between the blade root underside and the casing to prevent the blade tilting.

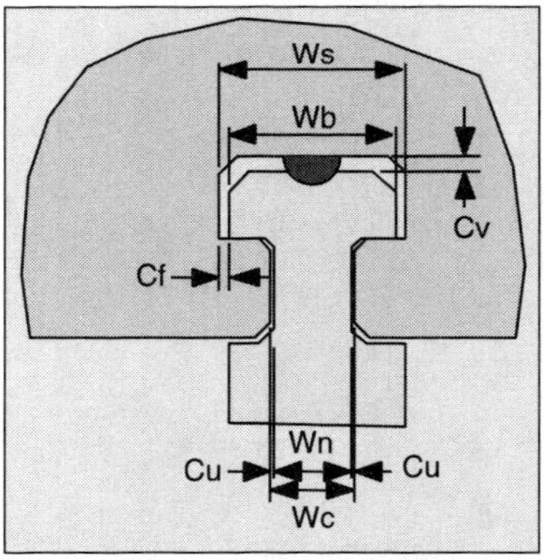

Fig. 2.6.3—The controlling dimensions around a stationary blade root and casing slot.

Creep deformation of the diaphragms

Diaphragms that operate in the high and reheat sections of a unit are subject to high intensity axial pressure differentials while simultaneously operating at high temperatures. Under these conditions, a continual axial stress tends to deflect the vanes and inner web downstream. (Creep deformation is covered in greater detail in chapter 5).

During operation, two forms of axial deflection occur—the deflection that is recovered when the axial pressure differential is removed, and the deflection that is not recovered (plastic and permanent). This second, plastic extension is termed *creep deformation.* The higher the stage temperature, the greater the degree of deflection per unit load, and the more susceptible to this form of deformation the element becomes.

Metals creep at all temperatures when loads of sufficient magnitudes are applied. However, the amount is often so low in the lower temperature stages that this effect can be ignored. However, if temperature and stresses are sufficiently high, metal will creep until rupture occurs. When such combinations of stress and temperature are severe, the "creep rupture effect" must be accounted for in the component design. "Creep rupture" is the term used to identify the mechanics of deformation and failure of metals under steady stress at elevated temperatures.

The creep deflection of metals can be considered on the curve in Figure 2.6.4. This curve is drawn to show the material deformation ("Et") that is the amount "dxv" and "dxw" of Figure 2.6.1 under a constant load at a given constant temperature. This curve represents a plot of material extension (or deformation) as a function of time. The diaphragm deflection under load "L" falls into four distinct stages:

- *Immediate extension "A."* Under the action of load "L," this extension is elastic and the total deflection will be recovered if the load is removed at any time
- *Primary creep "B."* The creep continually decreases as a result of the strain hardening the material that accompanies the elongation
- *Secondary creep "C."* Creep ordinarily remains relatively constant because it represents a balance between the strain-

hardening ability of the material and its susceptibility to thermal softening

- *Tertiary creep "D."* Creep is elongation to rupture, and is the final stage. The elongation rate increases because of local necking down, crack formation, and structural changes to rupture at condition "F"

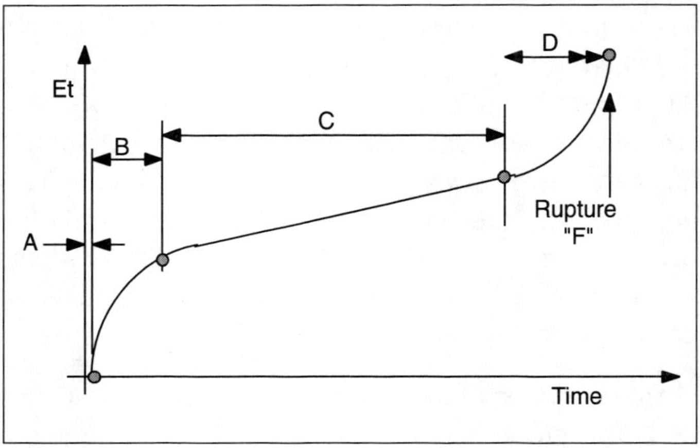

Fig. 2.6.4—The time phases for creep deformation to failure "f".

The minimum creep strength of a diaphragm is defined as the stress required to produce a specified amount of secondary creep ("C") in a given time—*e.g.*, 1% creep deformation in a time period of 10,000 or 100,000 hours, at a particular temperature under a constant stress. A material's rupture strength is usually expressed as the stress required to produce failure, by rupture, in a stipulated time (e.g., 100,000 hours at a particular temperature).

This creep rate is very sensitive to temperature at the higher values at which the diaphragm is operating. Families of material creep properties are available (Fig. 2.6.5), where "T1">"T2" >>"T5." These curves allow the total deflection ("dxd") to be predicted for a stage as a function of its temperature, stress levels, and hours of operation.

The creep strength of any particular diaphragm is dependent upon its resistance to the creep rupture effect, and the ability of the designer to calculate its probable strength. These include:

- the relative strength of the various parts of the diaphragm, (vane and inner web)
- the variation of the strength throughout the assembly due to the changes of metal cross sections
- difference in the material components, and their different creep resistances at different positions
- uncertainty of the distribution of pressure and temperature over the component being considered

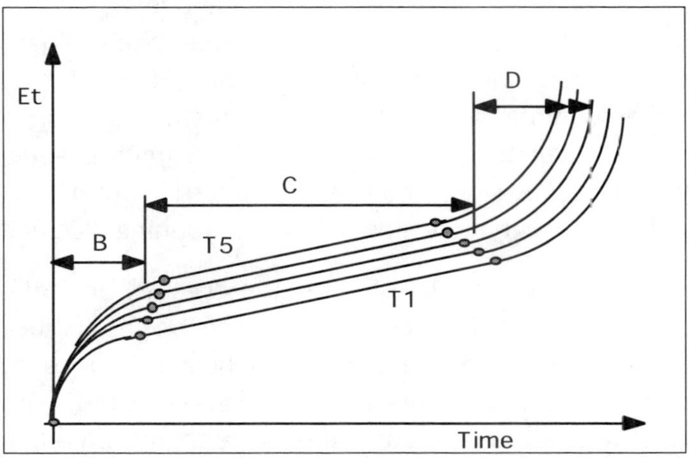

Fig. 2.6.5—A family of creep curves for temperatures "T1" to "T5".

UNIT "END-TO-END" LATERAL ALIGNMENT

The total alignment of the steam path components requires they be "in-line" at all operating conditions and in all directions. This includes "end-to-end" alignment. To achieve alignment it is necessary for the person responsible for this operation to achieve side-to-side measurements sufficient to ensure all stationary components are equidistant from the center on those surfaces needing interaction with the rotating portions of the unit.

The first action step to achieving total alignment is normally to place the bearings at the correct radial height (discussed above and drawn for a multi-section unit in Fig. 2.3.7). The bearings are the major consideration in the total alignment. Once they are set, it then becomes necessary to align the other components (diaphragms, packing heads, oil seals, etc.) so they follow the same catenary at all axial positions along the length of the sections. Figure 2.3.9 shows some small amount of lateral adjustment required to account for the oil shift ("dh"). Normally, this is a relatively small amount—below the level that special provision should be made for it—but it will cause some small adjustment in the side-to-side hot running clearances.

With bearing heights adjusted, it is necessary to adjust the remaining components. This requires that their height first be set so casing horizontal joints are at the correct height. Various components are then adjusted side-to-side in their correct radial position laterally and vertically. These components are then locked in place.

The methods of establishing the center position for gauging each of these three parameters employ a means of establishing a straight line from the front standard to the back-end generator or exciter. This alignment requirement is considered in greater detail below.

METHODS OF FIELD ALIGNMENT

Methods are available for achieving an alignment of the stationary components within the individual casings of the unit and the sections relative to each other. It is necessary to achieve this alignment so the efficiency of energy conversion and unit reliability is maximized within unit capability.

Stationary component alignment

Within the casings there is a need to assemble the individual stationary elements (or their carriers) in the radial and axial position. (In this section only radial adjustment is being considered.) Figures 2.8.1 and 2.8.2 show the basic settings of the steam path components in the cold stationary position for impulse and reaction stages. (These requirements are considered in greater detail in later sections of this chapter.) These figures indicate the diameters and clearances that should be achieved for the complete circumference of the blade rows to assure a level of performance consistent with design predictions.

Figures 2.8.1 and 2.8.2 show the basic steam path diameters ("Dot" and "Dor") at discharge from the stationary blade rows. These figures also indicate there is some means in the impulse stage to make adjustment by modifying the dimension "Crx" to modify the actual position of the diameters. However, this adjustment does not modify the diameters themselves, but simply their location within the steam path. In the case of the reaction stage in which the stationary blades are located within a casing or blade carrier, the diameters are established by the diameter of the locating root shoulder ("Ds") and are not adjustable. If these blades are carried in a blade carrier rather than a casing, they carry some level of adjustment by relocation of the carrier within the casing. However, any adjustment made to one row will affect all stages located within that carrier.

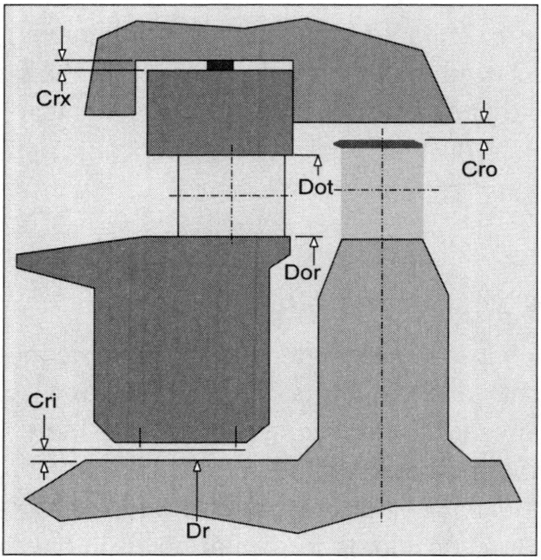

Fig. 2.8.1—The radial alignment requirements of a wheel and diaphragm constructed stage.

When making any adjustment to the discharge diameters "Dor" and "Dot" in both impulse and reaction rows, there will be an effect on the clearance from the inner stationary blade diameter relative to the rotor diameter "Dr," which influences the radial clearance "Cri." Normally, seals will be fitted at this location with no likelihood of rubs between the massive material pieces. However, adjustment can cause damage to the seal strips and represent a loss in stage efficiency, depending upon the sequence of assembly.

The dimensions shown in Figures 2.8.1 and 2.8.2 are for all circumferential locations and, as shown in Figure 2.8.3, must be met at all positions.

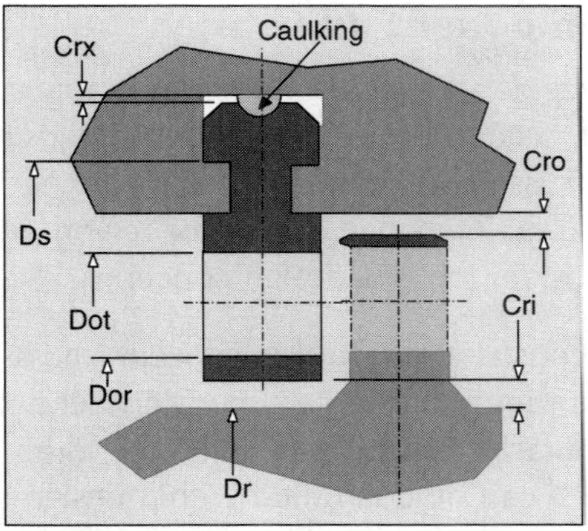

Fig. 2.8.2—The radial alignment requirements of a stationary blade inserted into a casing or blade carrier.

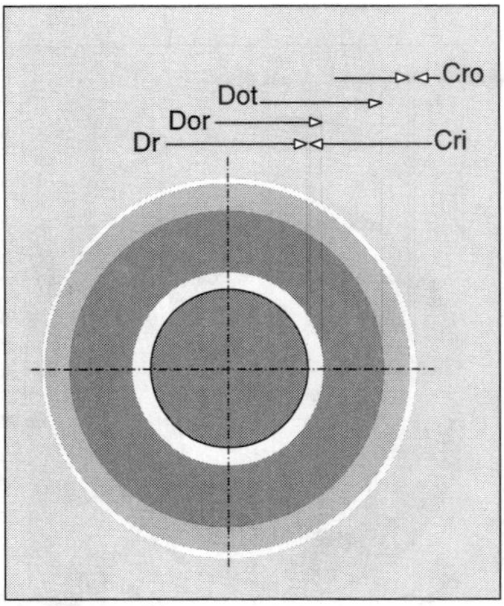

Fig. 2.8.3—The radial clearances and alignment.

Means of component adjustment

A certain level of radial adjustment is available to manufacturers and operators of steam turbines. It represents adjustments that must be made at initial assembly in the field, and after each component is removed during a maintenance outage for rework, repair, or simply for cleaning. Consider the two design concepts:

Diaphragm construction. In this concept, the stationary blades are located within inner and outer rings and dependent upon their method of manufacture. The values of the discharge diameters ("Dot" and "Dor") can be controlled within design specified limits. Figure 2.8.4 shows two diaphragm halves located at some arbitrary vertical distance apart ("G") before closure on assembly. In the final assembled condition they are in intimate contact. Keys are used at their interface to locate them correctly relative to each other. However, these keys are intended and dimensioned to produce axial rather than radial alignment.

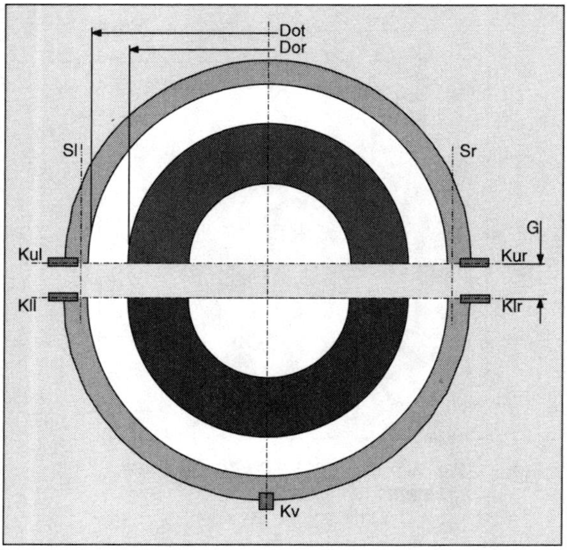

Fig. 2.8.4—The locator key positions and adjustments.

Movement of keys "Kll" and "Klr" achieves adjusting for the lower half in radial alignment. Support is also provided by the vertical key "Kv," which is intended to assist in alignment and prevent diaphragm "sag" at high temperatures under the effects of its own weight. The key "Kv" also takes the torque reaction developed in the stationary blades and so maintains transverse alignment. Triangulation of these three keys is undertaken at assembly to allow close clearance on final assembly.

The upper and lower halves can be assembled by two distinct methods. First, the upper half can be located in the upper half-casing with the horizontal joints of the inner web and outer ring being common or in line with the upper half- casing half-joint. When the upper half-casing is lowered into position over the lower, the diaphragm upper half locates on the lower. The joint keys are engaged to provide both location and, to a certain extent, provide a steam seal. An alternative design allows the upper half-diaphragm to be bolted to the lower half using studs (shown at positions "Sl" and "Sr"). The upper half-casing is then lowered over the assembled diaphragm. A steam seal is formed between the casing and diaphragm outer ring by the steam pressure acting across it. In each case, the steam path discharge diameters ("Dot" and "Dor") can be achieved within design limits.

Inserted blade design. This design locates stationary elements within a casing or blade carrier. Their diameter is adjusted relative to the rotating elements to provide an acceptable steam path. The radial location of these stationary blade elements, assuming the base dimensions of the individual blades are correct, depends entirely upon the shoulder diameter of the root groove in the casings or carriers, which is shown as "Ds" in Figure 2.8.2. The steam path discharge diameters remain at their correct location relative to the rotating blades (assuming the casing remains concentric).

Other stationary components. Within the steam path are a number of components (packing heads and seal components, in addition to the stationary blade elements) required to guide and limit steam leakage. Keeping foreign materials out is as important as keeping steam in. Oil used in the bearings, for instance, is not allowed to enter the steam path, particularly in the low-pressure regions where sub-atmospheric pressure exists. Under certain circumstances, such pressure could suck oil into the steam system along the central shaft.

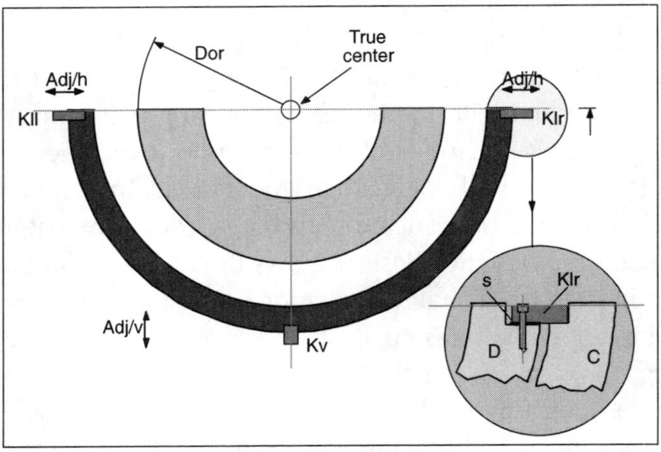

Fig. 2.8.5—Details of key support and adjustment.

The component adjustment process

When a lower half-casing is set in a horizontal position (relative to the bearing pedestals and those bearings supporting the rotor within that casing), it becomes necessary to adjust the various stationary portions within the casing that locate about the rotor to ensure concentricity. Consider the diaphragm lower half shown as Figure 2.8.5, which is held in position by keys "Kll," "Klr," and "Kv." Each of these keys has adjustment. Keys "Kll" and "Klr" can be adjusted in the horizontal position by amounts "Adj/h." They can also have their vertical height adjusted by change of the shims "S" (shown by detail in the inset to this figure). This vertical height is adjusted until the half joints of the diaphragm ("D") and casing ("C") are level. The vertical key ("Kv") is also adjustable in the vertical position by an amount "Adj/v," which may require modifying, or changing the key.

Methods of alignment, casing to casing

There are currently two methods available for the alignment of the components to ensure their bore and (in the case of blading) their discharge diameters "Dor" and "Dot" are at design values. These methods are described below:

Piano wire method. This method requires a tight piano wire be placed on the casing at the midpoint between the horizontal surfaces. This establishes the "true center" of the major component—the casing. The requirements of this method are that a wire of known diameter and material properties be suspended between centers of known length, with a weight suspended from one end.

Figure 2.8.6 shows a wire of known properties stretched by use of a weight "W" between centers "C-C" over a length "L." The wire is tensioned and has uniform stress along its length. At the mid-span position is a maximum defection ("D."). At any other position ("X"), the deflection ("d") can be determined from the expression for a catenary. More often a "sag chart" is available to determine the diameter, mechanical properties, and weight used over a variety of lengths ("L.")

With knowledge of the wire sag at any axial location, the individual stationary elements can be aligned within the casings. The casings can be set relative to their neighbors, with allowance made for the various bearing settings required to account for the various phenomena (as previously discussed).

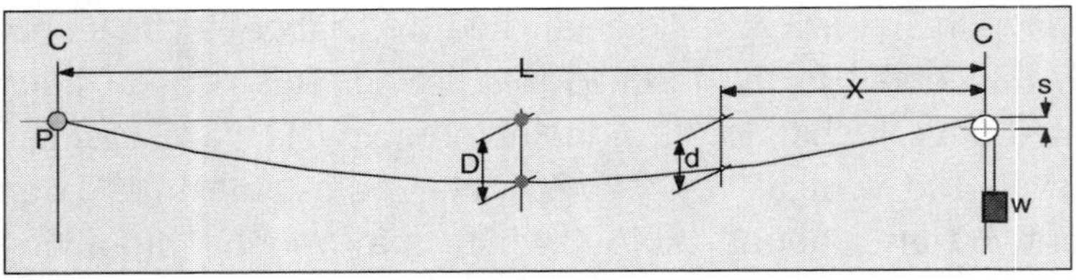

Fig. 2.8.6—The "Catenary Sag" of a piano wire.

Laser optical alignment. A more recent development is the use of laser optical alignment techniques. A beam is established between two reference points and the various components are aligned to this beam. Figure 2.8.7 shows the process of lining out a section, showing a bracket being rotated to various positions around the inner surface to be adjusted. Using readings from these various positions adjustments are made to achieve concentricity.

Fig. 2.8.7—Laser optical alignment.

Checking with leads

A means often used to check alignment and clearances is to place lead strips at various axial positions along the section. The rotor is then lowered into position, with the weight of the rotor crushing the leads locally so their thickness occupies the radial clearance at the top and bottom dead center. When the rotor is removed, the thickness of the lead is measured, providing an accurate reading of the cold setting unbolted clearance. This is also often performed with the bolts at the horizontal joint, both bolted and unbolted, to achieve "true" readings.

UNPREDICTABLE FACTORS AFFECTING DESIGN CLEARANCE

Thus far we have considered predictable factors that influence the hot running axial and radial clearances within the steam path. These are factors the design engineer examines during the various stages of design, to the accuracy with which components can be set at cold stationary conditions. The goal is to enable the unit to operate with a minimum of risk of running interference, and the performance of each stage maximized (within the limitations set by mechanical considerations).

There are, however, *two* groups of considerations influencing the actual running clearances. The first have been considered; the second group is those factors that are unpredictable. These phenomena occur normally as a consequence of operating conditions, operating excursions, or major transients of inlet steam parameters. There are, however, other considerations to these unpredictable phenomena that occur as a consequence of stress relief, extended exposure to the high temperatures and pressures of the steam, and possibly other factors of design or manufacture.

Unpredictable phenomena can modify the shape of the components by various forms of permanent deformation, which as a consequence affects the initially selected dimensional settings. What is particularly difficult with these phenomena is that while the probability of their occurrence once or twice during the life of a unit is almost certain, and some allowance can be made for them, their extent is unpredictable. The severity of the deformation or the modifications that result is normally dependent upon the type and severity of the condition causing the modification.

Various phenomena can occur, but the most common form of deformation or distortion follows:

Casing ovality (bowing)

Casing walls are not of uniform thickness at all tangential locations. There is a massive heat sink at the horizontal joint where the thick flanges can cause temperature differentials under conditions of loading and temperature transients. For these reasons various locations of the casing accept and reject heat to the steam at different rates while operating at the same temperature and under different operating conditions and situations. These temperature effects—together with the possibility of residual stresses—can cause the turbine casings in most instances to assume an oval form.

Figure 2.9.1 shows a schematic section through a turbine half-casing. The casing has an internal diameter of "Dx" and a vertical half height of "Vx = Dx/2." This is the design specified requirement. The halves are held in their operating position by a series of bolts at the horizontal joint. The boltholes are also shown. Under the action of high temperature steam, these dimensional conditions can

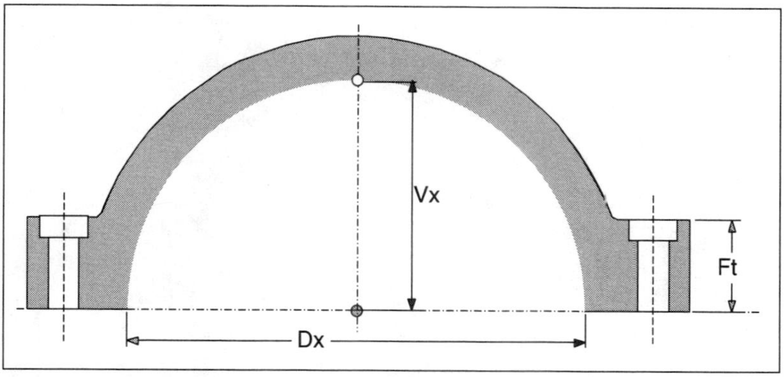

Fig. 2.9.1—The design conditions at a section through an upper half casing.

change, and the design roundness can be lost. There are two conditions that can occur:

A reduction of the horizontal joint diameter "Dx" to "Dx - 2.dRx." Under these conditions, the vertical half height ("Vx") will increase by some amount ("+dVx"), as shown in Figure 2.9.2(b). This condition can be damaging to the extent the casing will close down on the rotor at the horizontal joint, and in the case of severe distortion, can make it difficult to lift the top half. It is also possible the horizontal joint studs will bind in their holes and become necessary to be bored out rather than removed by normal means.

In some instances these thermal effects cause residual stresses to be induced in the casing halves at a level that will allow the rotor to be turned while the casing halves are coupled. However, as soon as the studs are released, the casing moves in response to the residual stresses and clamps on the rotor. These are situations that must be evaluated in terms of removing the top half, and the potential this poses for damage to the rotating blade rows.

An increase in the horizontal joint diameter "Dx" to "Dx + 2.dRx." In this situation, the vertical half height ("Vx") will decrease by some amount ("-dVx"), as shown in Figure 2.9.2(a).

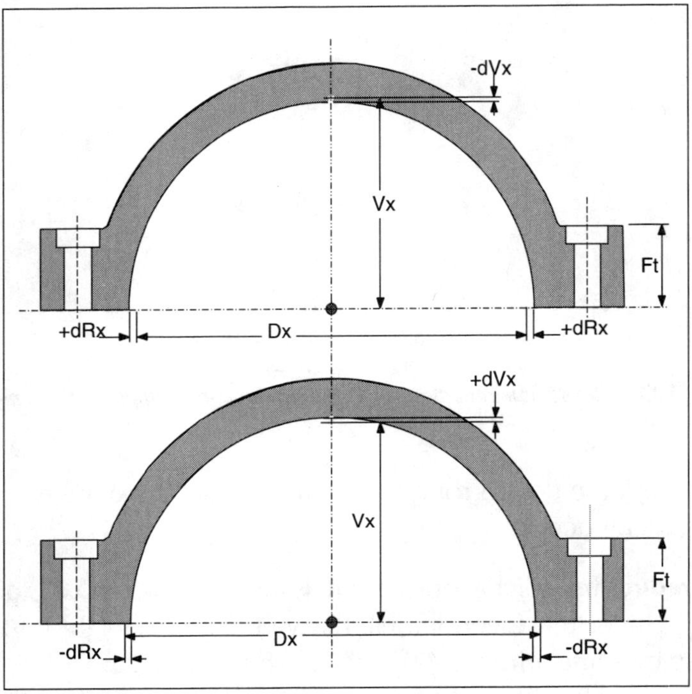

Fig. 2.9.2—Casing ovality. In (a) the casing has an increased horizontal joint width, and in (b) the horizontal joint is reduced.

This again is a situation where the casing can clamp the rotor, either at low rotating speeds, at standstill, or when the studs are released. However, this clamping is in the vertical direction. While this can cause rotating blade damage by bending them—particularly as the unit comes to rest—and the deformation becomes a maximum—this form of deformation will not cause a horizontal clamping of the rotor and prevent disassembly. However, the studs may need to be bored out if the horizontal movement of the casing is sufficient to bend them in the casing halves on release of the tightened studs.

If either of these deformation movements of the casing occur, it is possible for the radial sealing arrangements to be damaged if the movement is large and occurs while the rotor is turning at any speed.

Figure 2.9.3 shows grooves cut into the casing at the horizontal joint by the rotating blade seal as the casing has moved radially (in Fig. 2.9.2(b) condition). This situation could become sufficiently severe that blades could be bent and the rotor "grabbed." This will have the effect of destroying the sealing capability at the rubbed location, and depending upon the method of seal construction, can have significant long-term effects on the efficiency of the unit, if they cannot be replaced.

Fig. 2.9.3—Casing damage produced by heavy rubs of seal strips carried on the rotating blades. This introduces grooves on the casing at its horizontal joint.

With the condition shown in Figure 2.9.2(b), it is possible ovality can worsen when the horizontal joint bolts/studs are released. At that time, the rotor will "move in" at the horizontal joint and two situations can arise:

- Threaded components at the horizontal joint bind in the casing holes, making it difficult to lift the top half casing. This normally requires the bolts/studs be drilled out

- The rotor is grabbed by the inner surfaces of the casing. This would make it difficult to lift the blades that are in tight contact with the casing by bending, without damage

Ovality can be corrected (to the extent the unit can be reassembled and returned to service) by the controlled application of heat. The means to make such corrections are discussed in greater detail in chapter 7.

Casing humping (hogging)

A second distortion mode that can effect a casing is "humping"—a vertical displacement upwards from the original flat position. This can result as a consequence of certain operating phenomena (from rubs to residual stresses), possibly induced by excessive positive or negative temperature ramps. However, the most common cause is the result of the rotor being brought to rest and left stationary while hot gases (steam or air) remain in the casing. Under these circumstances, hot gases in the casing rise and overheat the upper half, which expands relative to the lower. Upon cooling, the casing horizontal joint will not reassume a totally flat position. The effect of this on steam path alignment is to make the rows of stationary blades rise and become misaligned relative to the rotor. Affected blades are located either in the casing, a blade carrier, or diaphragm—particularly near the center span. Methods of correcting this condition are discussed in chapter 7.

Horizontal joint leakage

As a consequence of operational distortion of the casing as discussed, there is always the possibility that leakage will occur between the surfaces of the horizontal joint. This situation must be corrected, to prevent steam bypassing the blade rows and lowering efficiency, and to prevent steam reentering the steam path, possibly causing a two/rev stimulus.

The most severe form of damage that can result to the horizontal joint surfaces is when water is present in the steam and a washing-erosion material loss occurs. In this situation, it is normally necessary to make a weld repair. If the opening has been present for some time, it is possible there could be deposits on the surfaces (Fig. 2.9.4). These deposits can in fact help seal the gap and help reduce the leakage quantity, but they do not correct the situation. Figure 2.9.5 shows the leakage pattern of superheated steam leaking past a stationary blade row. There are also deposits present on this surface.

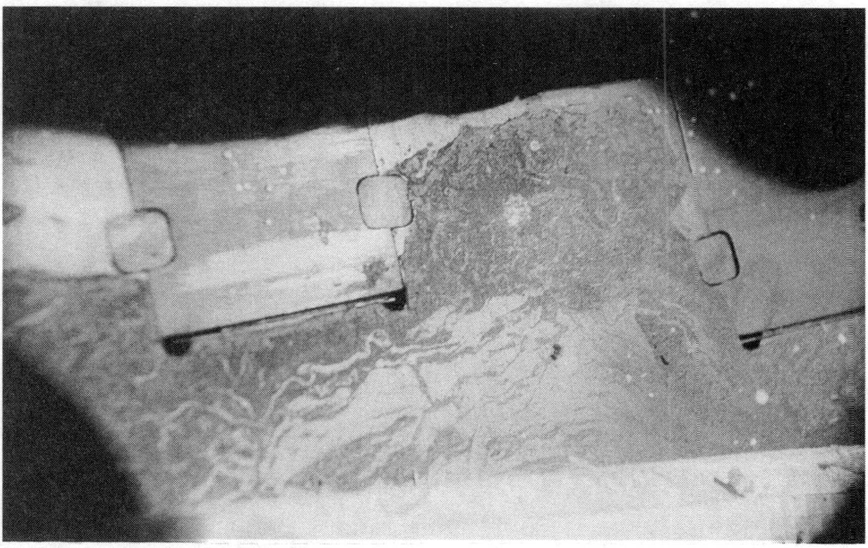

Fig. 2.9.4—The leakage path between the horizontal joint of a high pressure casing. Had the steam contained water there would have been washing erosion at these locations.

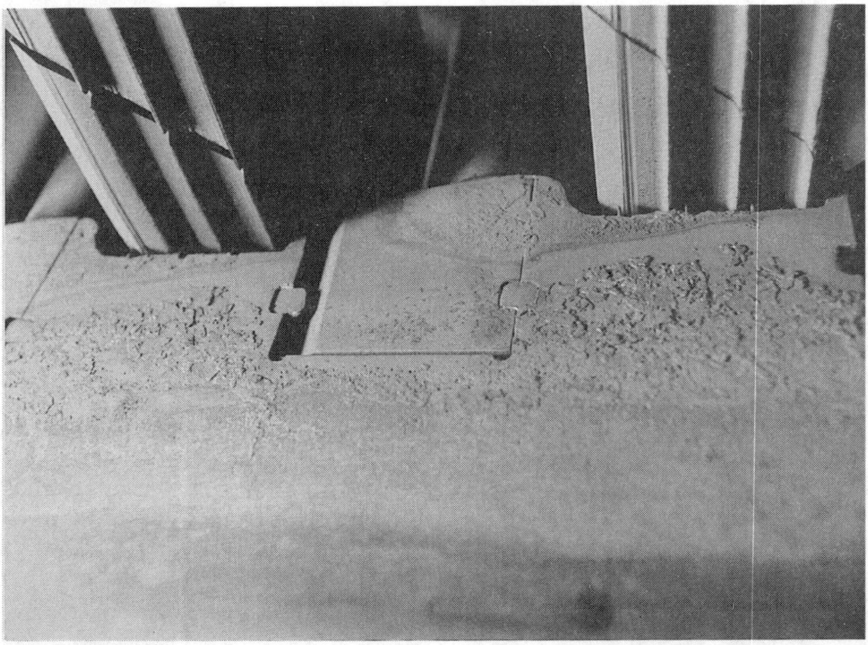

Fig. 2.9.5—Surface deposits on a horizontal joint surface where there has been an opening for an extensive period, and the unit has operated with high carryover from the boiler.

Casing boring after adjustment

Even after heating corrections, many of the damaging conditions that occur in the casings will retain a degree of "out-of-roundness" in the shells. While much of this may be acceptable, it may sometimes be necessary to undertake corrective machining and modify the form of the components located in the shell. Such decisions to machine must be made only after careful evaluation of the total requirements. Figure 2.9.6 shows a boring bar mounted to a high-pressure inner casing. In this case, the bar is machining the cylindrical faces to return the section to a fully concentric condition.

Fig. 2.9.6—A boring bar being used to machining a casing and achieve local concentricity.

Diaphragm dishing

This condition occurs when the diaphragms assume a degree of "dishing"—an axial deflection of the vanes and inner web in the direction of the pressure gradient. This effect is generally the result of either:

- excessive pressure at high temperatures, causing creep deformation (as discussed previously and shown in Fig. 2.6.1, with a total deformation of "dxd"). This deformation may not be to the same axial extent at all tangential positions

- the effects of a hard axial rub between stationary and rotating portions of the unit while it is in operation

The net effect of this deflection is for the diaphragm to deflect axially downstream (in the direction of the pressure gradient). This effect consumes the axial clearance between the diaphragm face and the following rotating row wheel.

Dishing can also modify the sealing clearance and tooth setting between the rotor and diaphragm inner seal diameter. If "hi-lo" teeth are used, this can cause the seal strips to rub (*i.e.*, it will modify the tooth settings). This is discussed in chapter 7, and is illustrated as dimensions "g" and "h" in Figure 2.12.25.

In fact, all of these effects are discussed in chapter 7, and Figure 7.7.2 shows deposits from a rotating wheel on a diaphragm inner web after a sustained heavy rub. This diaphragm had been hardened to about BHN 380 in the region of the deposited material. This diaphragm deflected downstream and also took an elliptical form [see the following section, *Axial deformation of stationary (body) blades*].

In some earlier designs (older units still in service), the selection of the weld prep, filler materials, and welding process (fabrication sequence) has had an effect on local strength. If uneven around the half ring, it allows a different degree of axial deformation, which contributes to an uneven dishing effect.

Figure 2.9.7 shows the measured values of axial deformation in a diaphragm half, which has operated for 150,000 hours at high temperatures. The deformation ("dxd") was measured on the hook face of the diaphragm seal-locating slot (which was the cleanest surface available) and had been made free of deposits. Note that the deformation is greater towards the horizontal joint. This form of deformation can be corrected by heat, with axial forces applied to return the element to its original position. It may be necessary to undertake some "skim" cutting after heat treatment, particularly on the seal surface on the packing ring.

Note: While dishing from creep deformation can be corrected by various means, creep is a metal failure at the grain boundaries, and correcting this condition will not reestablish the grain boundary strength. Such correction is therefore considered temporary.

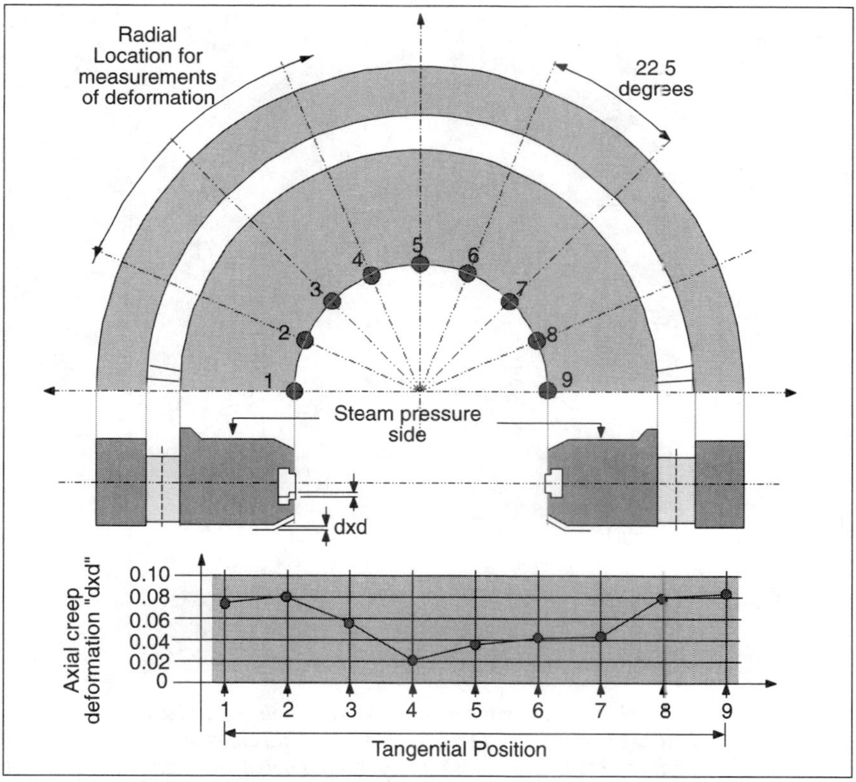

Fig. 2.9.7—The measured values of axial deformation "dxd" at the seal slot on a diaphragm which has considerable exposure to high temperature and pressure differentials.

Axial deformation of stationary (body) blades

In the event stationary blades inserted in a casing are deformed to the extent that "rubbing" contact could occur, it is normally far more cost effective to replace the blades rather than attempt to straighten them.

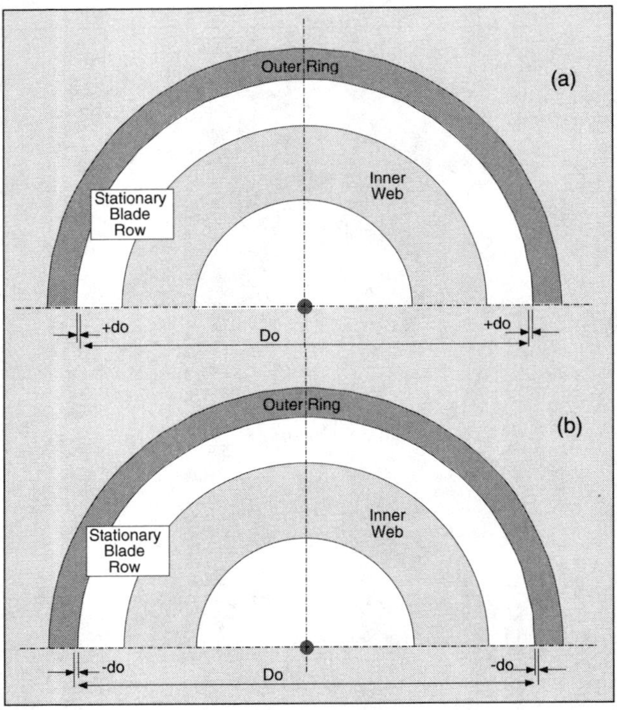

Fig. 2.9.8—In (a) is shown a diaphragm, in which the horizontal joint has opened by an amount "+do" on both sides. In b) the horizontal joint has closed by a similar amount "-do" on both sides.

Diaphragm concentricity

Just as the casing's horizontal joint diameter will change under certain load conditions, pressure, and temperature, so the diaphragm can creep "in" or "out" at the horizontal joint. If this deformation occurs, it will "carry" the stationary blade row to other than the original diameters and cause the steam path to loose concentricity. Under extreme conditions it can also produce rubs of the seal strips onto the rotor body, coverband, or other location.

This type of distortion—both closure and opening of the horizontal joint—can be corrected by jacking the joint to the correct diameter (Fig. 2.9.8). The halves are then subjected to a stress relief cycle, normally in a vacuum furnace. This should be done after the steam path, the steam seal surface, and the horizontal joint have been protected by a suitable compound (one that is inert and will prevent oxidation). It may be necessary after this treatment to skim-cut the steam seal face and horizontal joint to achieve an acceptable seal. Alternately, it might be necessary to make a surface weld deposit to build up the joint material. This will require machining and (possibly) stress relief. If the joint has opened, it can be closed mechanically using an external jack and then stress relieved as described above.

Fig. 2.9.9—The horizontal joint of a diaphragm which has deformed and moved radially "in". The horizontal joint key can be seen.

Figure 2.9.9 shows the diaphragm horizontal joint gap produced by two halves, after they have been removed from the unit and clamped together. It shows the effect of a diametral change that has occurred. In this stage, the horizontal joint has moved in, causing the horizontal joint to open at a position closest to the rotor. The key can be seen in the gap now existing between the top and bottom halves.

Rotor bends

Phenomena associated with the operation of turbines can induce bends into the rotor. Normally, if these are not too severe—in excess of about 0.006-0.009"—the rotors can be balanced and can continue to operate. However, with this form of operation, the rotating blade rows at and opposite the position of the bend maximum condition are no longer concentric with the stationary row, and the lap will be consumed at both the inner and outer diametral positions.

Operating with a bent rotor—even if adequately balanced—does not represent a satisfactory condition and should be corrected as soon as possible. However, it may be necessary to accept such operation for a period in order to return the unit to service.

STEAM PATH AREA REQUIREMENTS

In assessing dimensional and tolerance requirements of the steam path, it is necessary to consider the objectives of the blade system, relate these objectives to principal dimensions, and then consider their possible influence on various performance criteria. This process establishes blade geometry and certain radial alignment requirements, assists in the selection of various elements, and provides guidance in selecting arrangements suited to achieving these concurrent objectives.

In designing systems to achieve these two objectives, sufficient information must be generated concerning mechanical loads and thermal conditions within the steam path, so component selection is facilitated. Other concerns such as stress levels within the local environment and frequency effects, can also be investigated. After components have been selected and sized, manufacturing techniques are chosen and tolerances applied to ensure a satisfactory product.

For the purpose of this section, the information that the design process should generate and specify (and the owner should be familiar with) concerns the relationships existing at various locations and areas, and the rates at which it varies throughout the steam path. In defining the form the various components take, the designer selects or defines tolerances within which the individual components must be produced, how they are assembled, and the tolerances they provide to ensure area requirements are met.

The area relationships within the steam path that should be considered are:

- The area of the passage formed between stationary and rotating blade profiles, and its variation from inlet to discharge
- The discharge area of individual passages and its variation as a function of blade height
- The variation of discharge area from passage to passage within the row.
- The total stage discharge area
- The leakage area occurring at seal points between stationary and rotating portions of the unit

As stated, any unit's performance will be influenced by the design of these areas and the manufacturer's ability to consistently meet them within design specified tolerances. Normally, after steam path components are manufactured and assembled, the manufacturer will undertake a series of audits to determine whether the overall dimensional tolerances (both of the individual components and their assembled arrangement) are within acceptable limits.

When the unit is removed from service, damage is found, and repairs are undertaken. It is necessary to perform this work in such a manner that the mechanical integrity of the components is retained (or improved) and original area relationships are preserved. Although there is some latitude in these values, if it is known a repair procedure will cause a significant variation from the design-determined values, then a careful evaluation of the possible effects on the performance level should be made.

THE STAGE OPERATING DEFINITION

An important aspect of the design phase has the engineer establishing the dimensional requirements of the various components comprising the steam path, the tolerances within which they are to be produced, and the material to be used. In addition to these obvious requirements, surface finish must be defined, special process procedures possibly identified, and the total quality (conformance level) to be achieved.

During original component manufacture—when a component outside normal tolerances is produced, or when a nonconforming situation occurs during assembly—it is the design engineer's responsibility to evaluate the condition and provide a disposition that will help ensure the unit meets its expected performance levels.

A *nonconforming condition* is one in which the existing elements no longer meet design requirements, *i.e.*, individual components have deteriorated in their mechanical properties or surface and dimensional requirements. They have probably lost surface material, and no longer comply with the requirements established by design.

If deterioration is found at a maintenance outage, it is again the responsibility of a design or plant engineer to rule on the condition and develop an acceptable disposition. However, in this case, the disposition may not return the unit to the same condition it possessed as a new addition to the system. While such a disposition may return the unit to an acceptable condition—perhaps even possibly long-term reliability—it may also compromise efficiency. Often this has to be the case until a final disposition—possibly involving the provision of replacement parts—can be implemented.

Subsequent evaluation of these operational nonconformances may reveal two (or more) possible scenarios for correcting the situation and returning the unit to an acceptable level of performance. Some of these options may depend upon there being available suitable techniques for either repairing or refurbishing the condition, others may require the use of replacements parts. It is advisable for the maintenance engineer to remain aware of these techniques, the limits of their capability, and how (and if) they could be applied in the situations being considered.

The dimensional requirements discussed in the following sections represent conditions to which the components were originally manufactured, and should preferably be returned after they have been rebuilt and/or adjusted during the maintenance repair process. In the design-specified condition, elements are arranged to ensure steam expends its thermal potential energy and is converted to rotational kinetic energy within the rotor. The stages are also arranged so the steam flows through them with a minimum of loss, and the efficiency of energy conversion is maximized.

During the original manufacture of turbine steam path components (and during their repair and replacement), certain spatial relationships in the axial, tangential, and radial directions must be met, while others (by preference) should be met to assist in optimizing unit performance.

Purchasers of steam turbine equipment must ensure that suppliers adequately define manufacturing needs, and have an acceptable and controlled system for ensuring these are consistently met. The purchaser must also determine that a system exists for evaluating and disposing of nonconforming conditions at a level (within the manufacturer's engineering organization) competent to make such decisions.

Engineering tolerances are selected by design to help ensure performance targets are met without incurring costs that would make the equipment too expensive to purchasers. Where appropriate, components should be able to be disassembled for repair or replacement. Tolerances should also ensure components can be interchanged, both within similar units in a station and between stations.

STEAM PATH COMPONENT ARRANGEMENT (AXIAL/RADIAL DIRECTION)

Irrespective of the degree of refinement used to manufacture and assemble steam path components, these parts must be correctly aligned within the unit to permit the controlled expansion of the steam from one row of elements to the next. If this is not the case, then neither sophistication of design, nor quality of manufacture will produce a unit with an acceptable level of performance.

Consider the cross section of the two stages shown in Figure 2.12.1. An impulse stage is shown in (a) and a reaction stage in (b). These figures show the basic spatial requirements for "clearance," "lap," and "blade setting" in a plane through the stage.

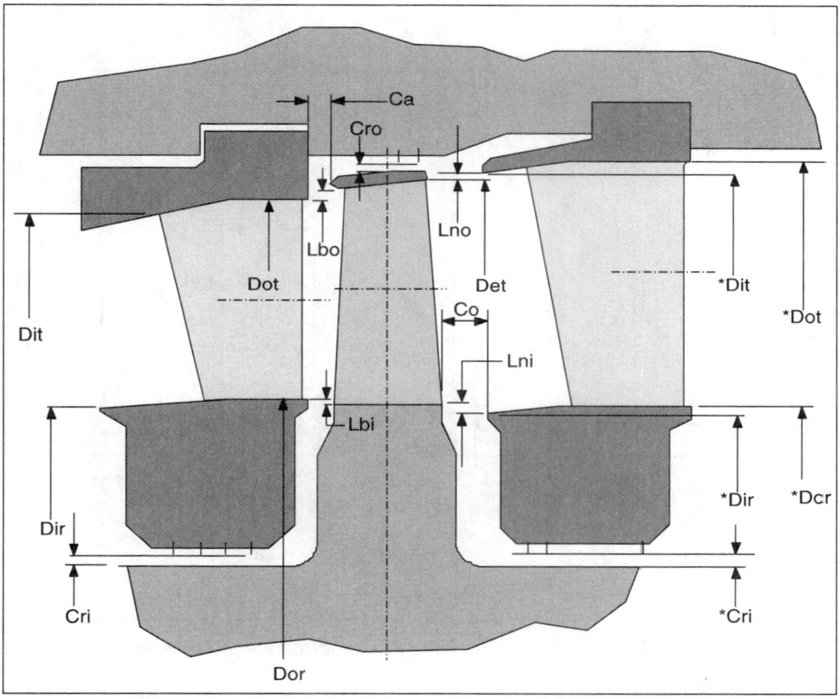

Figure 2.12.1(a)—The spatial requirements of an impulse stage in the axial/radial direction.

Obviously, these alignment and adjustment settings of the steam path components must occur in the cold stationary condition. Any relative movement of the parts that occurs during operation must be determined, and allowance for them made in establishing these cold settings. The principal factors influencing the hot running condition—and therefore the cold condition alignment—were discussed earlier.

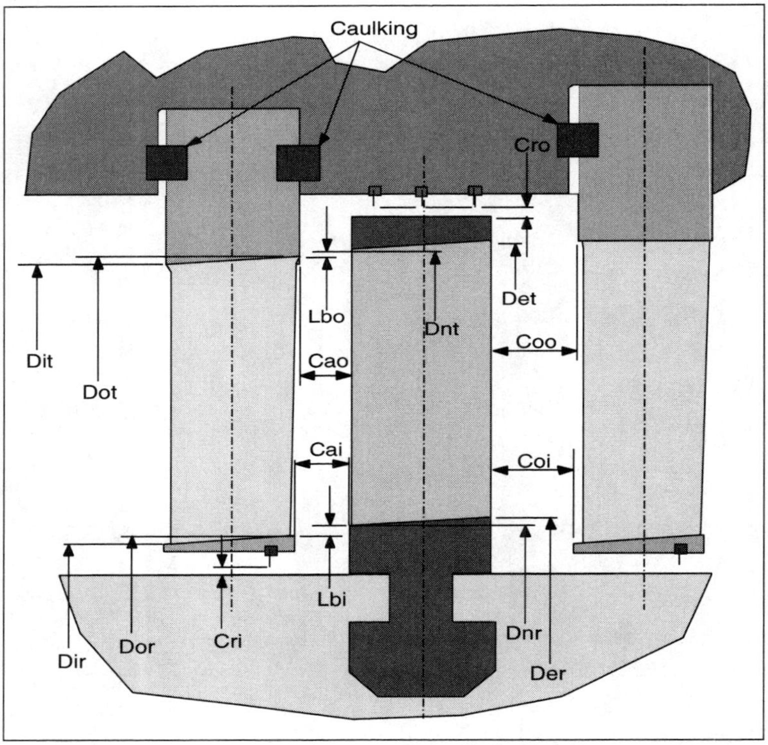

Fig. 2.12.1(b)—The spatial requirements of a reaction stage in the axial/radial direction.

The design process accounts for (and evaluates) the influence of the various operating phenomena and determines certain critical alignment or "setting values" in the components, as well as their required position relative to each other. Establishing cold settings ensures the clearances "C" and lap "L" dimensions are consistent with design requirements, and as predicted to occur during operation.

To assist (and promote) steam flow from one row to the next with a minimum of disruption, the inlet height of any row receiving steam is chosen to have a lap "L." This makes the inlet height of a stationary or rotating blade row marginally larger than the discharge height of the row from which it is receiving the steam. Because of differences in axial gap between stages—and to allow for the radial flow effect of the steam at exit from any stage—the value of lap will vary from row to row. The radial flow effect will require the outer diameter laps to be larger than the inner laps. However, if the outer lap becomes too large, there is a possibility of flow separation and excessive turbulence being induced at the entry to some rows.

Clearance requirements must be considered in both the axial and radial direction, depending upon stage location and arrangement. Critical clearances may, in addition, occur at the inner and outer extremities of the stage. The leakage areas occurring at any seal point provide an opportunity for steam to bypass the blade elements. This leakage constitutes a direct loss to unit output and efficiency. Efforts made to control lap and clearances at the design values can therefore represent considerable gains in maintaining unit efficiency.

The various components of the steam path must be machined and assembled so that during hot running conditions the various portions of each stage are in an acceptable alignment during normal and transient operation. If this is not done, then stage efficiency will be reduced because steam will not flow from one element to the next without incurring higher losses. Should rubs occur, they could lead to the opening of clearances. Under the most severe conditions, such rubs can cause "burning" of the surfaces in contact, which will affect material properties and could eventually cause failure of the affected parts.

During normal operation, the steam path parts must be so aligned that steam will flow from one row to the next and incur minimum aerodynamic losses due to steam path irregularities or misalignment.

Steam path efficiency is not an important consideration during transient conditions of load, temperature, or other cycle conditions that affect either clearances or lap. However, the clearances and lap for normal operation must be chosen so rubs and interference do not occur during these transients. This requirement will often mean the design process will specify values of clearance and lap to accommodate transient requirements—selected values that are larger than optimum for normal operating conditions. Although this does lead to relative inefficiencies of the steam path, such conditions cannot be avoided. To ensure acceptable alignment during operation, it is necessary to predict the operating conditions using them to calculate the "cold setting" requirements that will provide for the four major predictable influences of movement, and also make adequate allowance for the potential affect of transient operation.

To achieve acceptable "cold alignment," three major areas (or "regions of tolerances") must be controlled during the manufacturing and assembly phases. These areas must also be reviewed for compliance at completion of any repair/refurbishment actions, if these repair techniques are capable of affecting the basic dimensions.

These dimensional requirements represent conditions that must be correct in the individual components and assemblies before erection begins, or it will not be possible to achieve an acceptable unit alignment. These areas are:

- rotor machining and assembly
- diaphragm and stationary blade manufacture
- seal production and adjustment

The quality of the steam path is established before the unit is delivered to the site and field-erected. It is established by the design process, the method by which the design requirements are met, and finally, compliance with the three groups of considerations listed above. It is necessary to know the requirements and compliances needed to design these components so they can be confidently installed in the unit.

Rotor machining and assembly requirements (axial direction)

The rotor is an assembly on the main central spindle of rotating blades, coverbands, and tie (lacing) wires. It can also include shrunk-on wheels, coupling flanges, seal rings, and other elements. The rotor can also be that portion of the unit on which the thrust collar is located, and therefore the major component responsible for the total axial position of the unit during operation.

A typical high- or intermediate-pressure, double-flow rotor is shown in Figure 2.12.2. It indicates the major axial dimensions that must be achieved before the rotor is delivered for assembly.

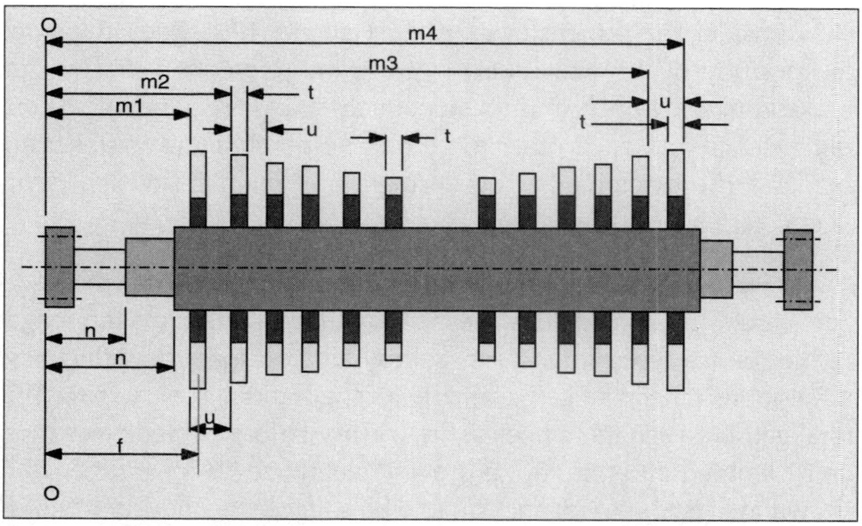

Fig. 2.12.2—The basic control dimensions on a rotor. In this example a double flow element, without a thrust block.

Certain dimensional requirements must be met to ensure the rotor can be assembled into the casing and fit among the stationary blade rows to achieve the clearances required by design. In defining these dimensions, a reference point must be established. If the rotor carries the thrust collar, then it is selected as the reference point. If (as shown in Fig. 2.12.2) there is no thrust collar, a coupling face is selected. Such a face is shown as position "O-O" in Figure 2.12.2 (an alternate location would be one close to the center of the rotor span). A coupling flange can be adjusted at erection by using spacer plates of various thicknesses between the flanges. The important requirement in Figure 2.12.2 is the relationship of the various dimensions to each other. When a thrust collar is used as reference, the relationship of these dimensions to each other is important, but since no site adjustment is possible, the absolute dimensions to the thrust face are also critical. Ultimately, all axial dimensions are checked and set relative to this thrust collar, whichever rotor it has been produced on

Consider the dimensions "m" in Figure 2.12.2. They reference the position of the individual stages to the reference face. Integral and shrunk-on wheels must be accurately spaced relative to the coupling flange and each other to permit the diaphragms or stationary blades to be assembled between them and maintain design clearances [Fig. 2.12.1(a) and (b)].

In Figure 2.12.2, the important wheel dimensions are shown as "m1-m4." This establishes the wheel pitching along the axial length of the rotor. These dimensions, combined with the wheel thickness "t," establish the axial gap available for the stationary blade row. This final gap between the wheels is "u–t." The stationary blade row must be assembled into this gap to provide clearances "Ca" and "Co" [as shown in Figs. 2.12.1(a) and (b)] and be suitable for "hot" and "cold" rotor conditions.

Note: When considering the "Ca" and "Co" clearances, it is important in those stages with an overhung coverband (Fig. 2.12.3

and 2.12.4) to normally define clearance as shown in these figures (*i.e.*, the clearance from the stationary blade rows to the knife edge of the seal). This clearance is set (the coverband is machined) after the rotor is assembled. It is also important to ensure the axial clearance to the blade inlet edge does not compromise the running clearance under the effect of transients.

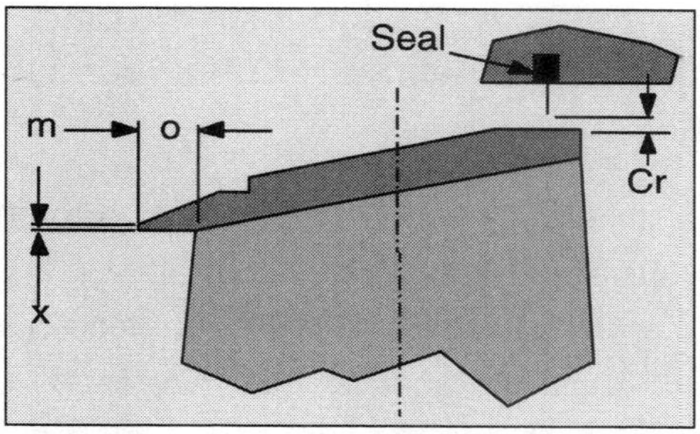

Fig. 2.12.3—An angled tip seal system employing one axial and one radial strip.

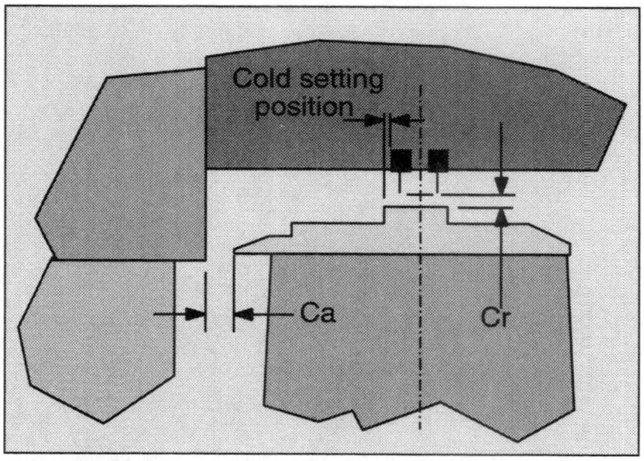

Fig. 2.12.4—A tip sealing arrangement for a conical tip section.

With the considerations mentioned above, the blades must be assembled to the wheel and two major dimensional requirements must be controlled. First, the coverband must be spaced in its axial extremities and set so interference does not occur. If the band contains an axial seal or provision for a portion of a radial seal platform, these must be spaced correctly (Fig. 2.12.3 is an axial seal coverband.) Here the band has an overhang portion "o" and a required knife-edge thickness "x." These dimensions are normally achieved by trimming after the band is assembled to the blade tips. If the band also contains a platform (Fig. 2.12.4), then the platform must be axially spaced so the seal strip above it during normal operation is in the correct position.

Figure 2.12.5 shows the root platform "a-a" which must be accurately placed. This is of considerable importance if a seal is produced in the root area, and requires a platform that is machined integral with the root. Such a radial seal is often produced on a stage with tangential entry roots, and is produced by machining excess stock left on the root form after the blades are assembled to the rotor.

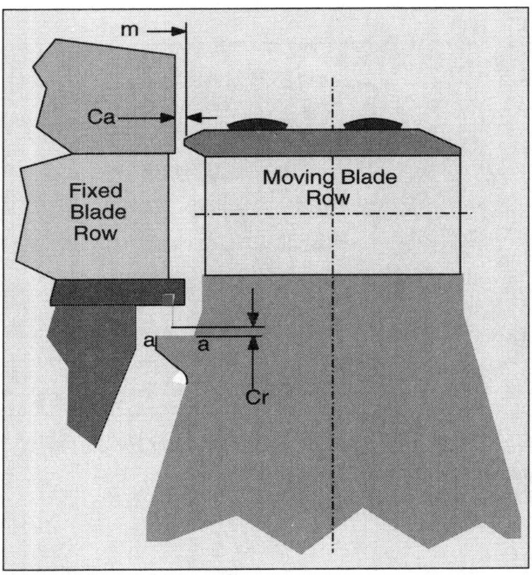

Fig. 2.12.5—A root sealing system used to minimize leakage to the inner seal diameter.

Some blade designs—particularly those required for close axial clearances in a reaction type stage—are produced and assembled to the wheel with excess stock or material remaining on them for final trimming. This is shown in Figure 2.12.6, where the "stock" is shown as an axial amount "s." This material is trimmed after assembly to provide the correct axial spacing ("m"). Care must be exercised in trimming this stock to ensure there is sufficient material below the vane to support it. The material should not be removed to the extent "k" exceeds the value shown as "k2." If the machining compromises the blade vane support, there could be an error in some axial dimension, possibly relating to the root groove position.

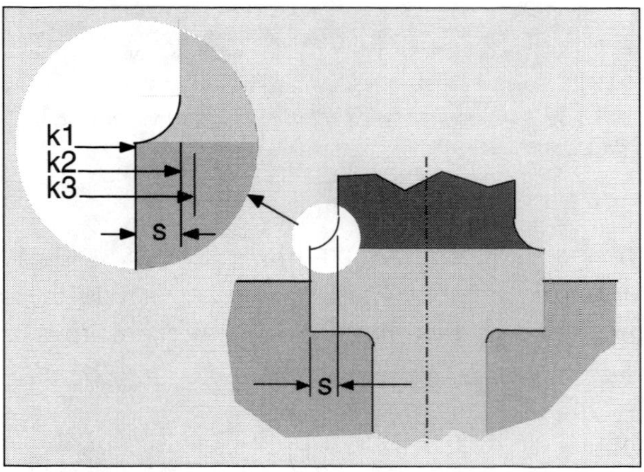

Fig. 2.12.6—Material trimmed from the rot platform of a rotating blade installed on a drum rotor.

Impulse rotor blades are located at the outer diameter of wheels, which in high and intermediate pressure sections are formed integral with the rotor central spindle. However, drum type rotors, because of the greater number of stages required to expand the steam efficiently, have their rotating blades located in grooves machined into the outer radius of the rotor center drum section (Fig. 2.12.7). In machining these grooves it is necessary to locate the groove at position "f" and also to control the neck width "g." This ensures the axial spacing can be achieved and there is no opportunity for the root block to twist in the groove and modify the vane-setting angle.

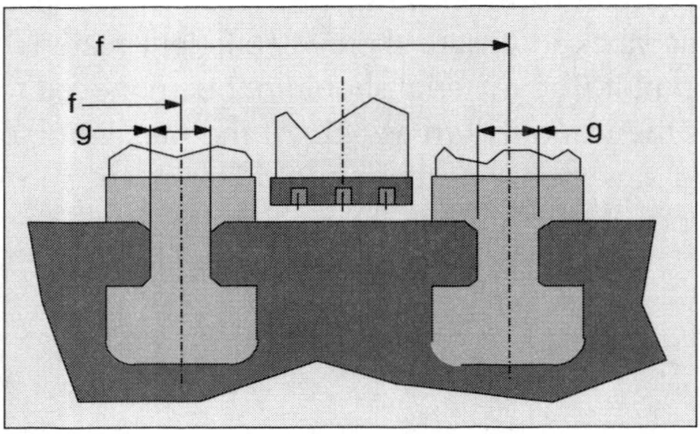

Fig. 2.12.7—Closely spaced rotating blade rows mounted on a drum rotor.

In establishing the correct axial position has been achieved for these drum-carried blades, it is necessary to gauge that dimension "f" has been achieved. This must be done by measuring from the reference face to the blade groove side (*i.e.*, to "f+g/2" or "f-g/2").

On many rotors there is provision for machining the seals region to ensure the steam sealing strips are correctly located relative to the stationary portions that normally carry the seal strips. Figure 2.12.8 shows the region of castellations required for "hi-lo" strips. For the seals to be effective, and to operate without interference during transient conditions, the shoulder location "n" must be correct, and then the individual pitch ("p") and castellation location ("w") must be correct relative to the shoulder position ("n"). This is of considerable importance on systems that employ the "hi-lo" tooth configuration, where teeth are alternately located on the upper and lower castellations of the rotating parts.

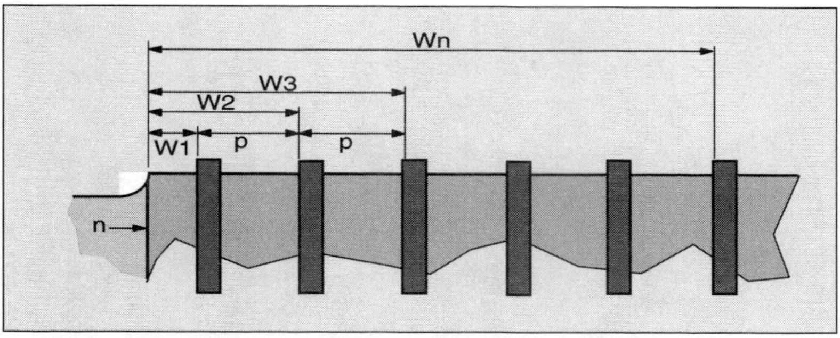

Fig. 2.12.8—The control dimensions at a castellated rotor location showing critical dimensions.

Some designs utilize seal strips that are caulked directly into the rotating portions of the unit. This is shown as Figure 2.12.9(a), where the pitching ("P") must be maintained at the design values located at a known distance ("j") from the reference face. The seals must be machined to a maximum diameter "Dk" to form the seal with radial clearance "Cr." Designs are also used where seal strips are caulked into both the rotating and stationary portion [Fig. 2.12.9(b)]. Here, the pitching is the same, and the clearance between the rotating and stationary seals is selected so rubbing interference will not occur during transient conditions. The teeth are arranged to have seal diameters "Dko" and "Dki." The radial clearance is set at "Cr" at the inner and outer seal positions.

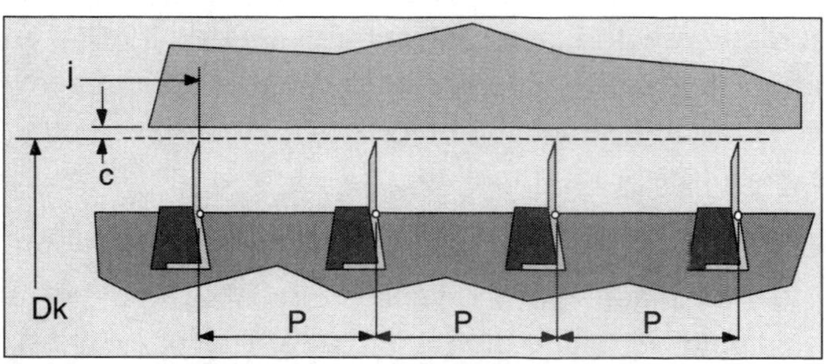

Fig. 2.12.9(a)—Seal strips mounted in a stationary portion of a unit, showing control dimensions.

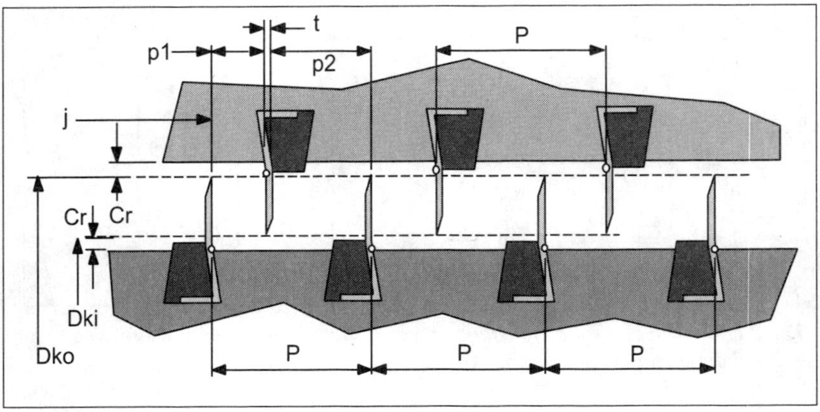

Fig. 2.12.9(b)—Inserted teeth mounted alternately in the stationary and rotating portions.

Rotor machining and assembly requirements (radial direction)

In addition to the axial requirements, the rotor also must be radially in the correct position. The more important diameters are shown in Figure 2.12.1(a) and 2.12.1(b). This requires that diameters must be set to achieve lap "L" and clearances "Cr" consistent with design specifications.

The considerations concerning radial clearances at many positions are shown in Figures 2.12.3, 2.12.4, and 2.12.5 as steam transfers from blade row to blade row, and in Figures 2.12.9(a) and (b) at shaft-end positions. The sealing requirements at diaphragm-to-shaft positions, and under stationary blade rows, are also essential to minimize leakage losses and maintain efficiency at the highest possible levels.

Consider the diaphragm shown in Figure 2.12.1(a). Here, the "Di" and "Do" diameters define the steam paths at inlet to and discharge from the stationary blade elements. These set the steam path height at inlet and discharge to the blade row for the stationary portion of this stage. Therefore, to achieve a suitable lap ("Lbi" and "Lbo" at inlet to the rotating blade row), the rotating blades must be located with inlet diameters "Dit" and "Dir" (Fig. 2.12.10). Similarly, at discharge from the rotating blade diameters, settings must be at "Dot" and "Dor" from the row to achieve the correct discharge area, and then the following stationary blade row geometry arranged to achieve laps "Lni" and "Lno."

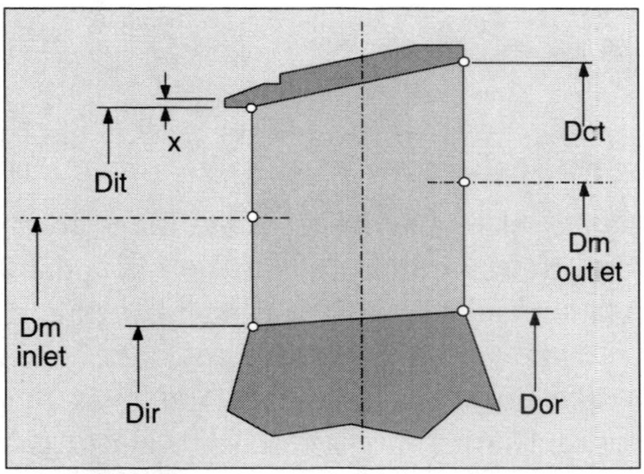

Fig. 2.12.10—The control dimensions for a blade at its inlet and discharge diameters.

For the rotating blade row to achieve the design lap in the cold stationary condition, it is necessary to achieve the inlet and discharge diameters as shown in Figure 2.12.1(a) and (b). These diameters are achieved by mounting the blade to the wheel or rotor and then, where necessary, using caulking to achieve a final diameter fix (Fig. 2.12.11). Other diameters—particularly those associated with producing platforms to form radial seals—are normally machined after assembly by the removal of stock material left on for that purpose. Such radial adjustment is also achieved in certain designs by rolling side grips to move the blade radially outwards under the action of the rolling force.

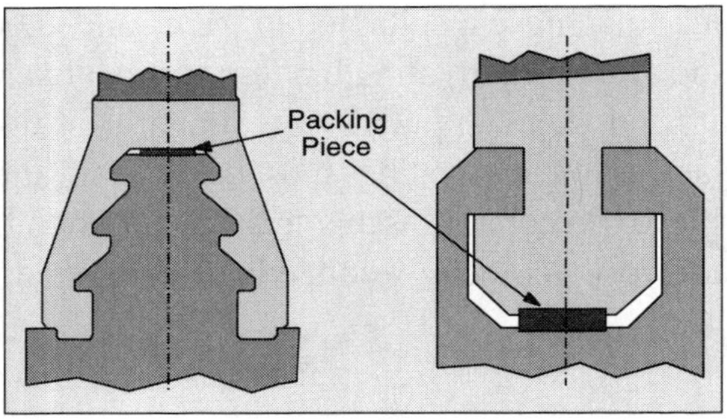

Fig. 2.12.11—A packing piece wedged between the blade root and wheel rim.

Discharge annulus form

Further diametrical requirements of the blade rows (stationary and rotating) are to ensure that the variation of radial height and placement of blade to adjacent blade is maintained at design tolerances throughout the complete row. Figure 2.12.12 shows the discharge portion of adjacent passages in a blade row, either stationary or rotating. These passages have a discharge height "h" and inner and outer design diameters of "Di" and "Do," respectively. This figure shows the tolerance bands "dDi" and "dDo" within which the inner and outer sidewalls must be located to ensure a smooth transition of steam flowing from one row to the following.

These tolerances may also be stipulated by the manufacturer as the permissible radial variation for the passage position ("dRo" and "dRi") from the design position. In addition, tolerances can be expressed for a radial variation ("dSi" and "dSo") from one passage to its neighbor. These "dS" passage variation values are not necessarily the same as the diametral tolerance bands, which apply to the total diametral variation within a row. The total requirement may also be expressed as a value "dh," within the diametral tolerances "dDi" and "dDo." Height tolerances refer to the total variation of the passage height, while diametral and radial tolerances refer to the passage position. In establishing permissible height variations, the design engineer must consider the effect such variations will have on the total discharge area, and set these height tolerances accordingly.

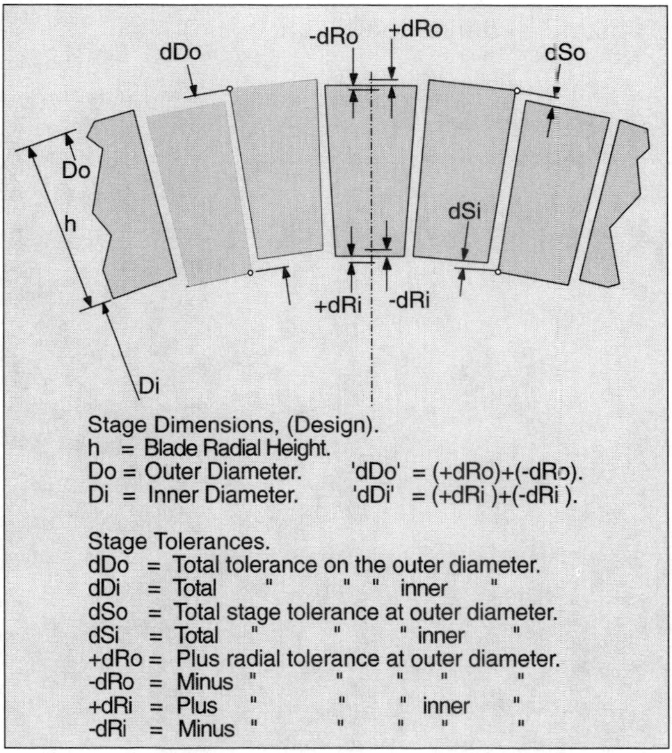

Fig. 2.12.12—The control dimensions and definition of the tolerances around a blade row discharge.

When any form of refurbishment is undertaken on either stationary or rotating blade rows, it is necessary to take no actions that will compromise area tolerances, or at least minimize those that may do so. In many older designs using cast iron or steel diaphragms, there can be a considerable radial variation of the inner and outer sidewall positions. It is also common to see these diameters affected by solid particle and water erosion. During maintenance outages, when these elements are being refurbished, it is often possible to take some remedial action that will reduce the total variation "dsi" and "dSo" within a row. Figure 2.12.13 shows a finishing process on a weld-fabricated diaphragm, where the outer sidewall is dressed to achieve the correct diametral dimensions.

Fig. 2.12.13—Finishing the outer side wall of a weld constructed diaphragm.

Many units, when removed from service, will be found to have lost material from their sidewalls. This loss most commonly occurs on the outer wall, and is aggravated by the radial flow effects on the steam and the eroding material that it is transporting. This condition is often seen in a last-stage cast iron diaphragm outer wall [Fig. 1.9.20 of chapter 1, Fig. 2.12.14(a)] where moisture impact and washing erosion has removed material. In Figure 2.12.14(b), material has been

Fig. 2.12.14(a)—The outer side wall of a cast iron diaphragm, showing material loss due to water washing and impact erosion.

Fig. 2.12.14(b)—The outer sidewall of a high pressure fixed blade row, where material has been lost due to solid particle erosion.

lost due to solid-particle erosion of an outer sidewall. These are conditions that must be assessed in terms of their potential effect on stage and unit performance, and decisions taken concerning the need for corrective action.

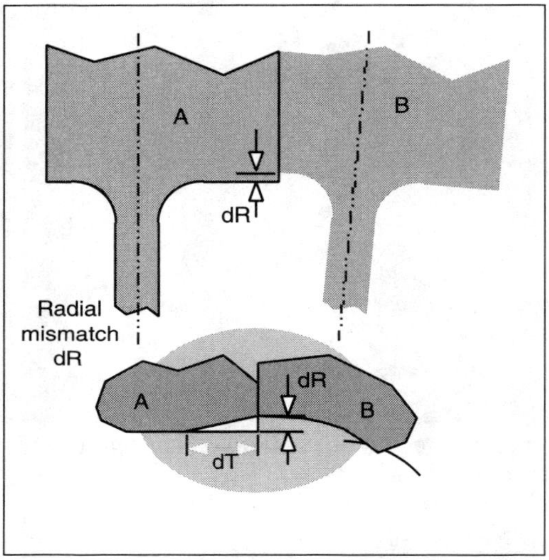

Fig. 2.12.15—Outer sidewall discontinuity and shaping to correct the passage shape.

When the abutment of individual stationary or rotating blade elements forms the passage, it can cause a discontinuity in the maintenance of discharge diameters and the form of the expansion passage. Such a condition occurs at one diametral location—either inner, or (as shown in Fig. 2.12.15) the outer. The effect of this discontinuity is to introduce turbulence in the expansion passage, causing vortices and eddies to be carried into the following row of stationary or rotating blades. This carryover has the potential to induce aerodynamic losses and introduce stimuli that can create vibration in the blade row. In terms of its effect on efficiency, this is more significant in shorter blade rows, where the radial disturbance represents a greater percentage of the total radial height. As shown in Figure 2.12.15, a mismatch such as "dR" should be blended to minimize the carry-over. However, the effect of the diameter change must still be within the tolerances for diameter, as defined by Figure 2.12.12.

There may also be diametral discontinuities at the inner diameters of rotating blades when it is necessary for root blending radii to meet at the root platform (Fig. 2.12.16). This is a condition that cannot be avoided. However, there should be some stated tolerance that controls the mismatch dimension "dr" to minimize the turbulence that is established.

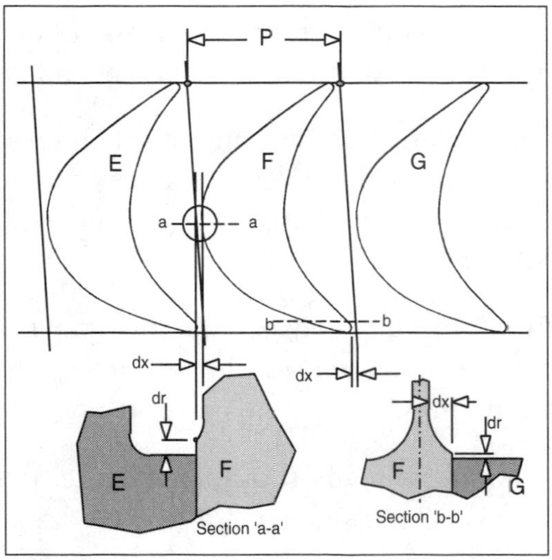

Fig. 2.12.16—The effect of a root blending radius on expansion passage discontinuity.

Rotating portion circumferential arrangement

In the circumferential direction, efficient and reliable performance requires correct pitching of the vanes. Vanes cannot be proud or recessed (see "da" values in Figure 2.12.17) from the mean inlet and discharge edge beyond those tolerances established by the designer. These requirements will be considered in more detail later. At this juncture it is sufficient to consider the two blade rows—one stationary, and one rotating, as shown in Figure 2.12.17—and consider the following characteristics:

Vane setting angle "ξ." The vanes must be set in the cascade so an expansion passage is formed between them. If the vanes are set with an error in the stationary row "ξs" or rotating row "ξr," the shape of the expansion passage is compromised.

Vane pitch "P." The tangential distance between adjacent profiles (stationary and rotating) must be set at the design value "P" within the tolerances "+dP" and "-dP." Variation of these pitches will modify the shape of the expansion passage beyond design limits.

Vane axial placement "a." The vanes must be set so the axial displacement does not exceed the design values "+da" and "-da." The actual error at inlet and discharge may be different if the setting angle "ξ" is also in error.

Axial gap "Ca." This gap between the rows can be compromised if the values of "+da" and "-da" for the stationary and rotating rows are beyond design limits.

Depending upon the method of manufacture, the vane settings may not compromise the clearance "Ca" if there are lips at the inner or outer diameters of the vane. This should be explored when these characteristics are examined on the unit.

Diaphragm and stationary blade manufacture and assembly

The axial positioning of the diaphragm is of critical importance, as this establishes the relative positions of the stationary and rotating portions of the unit relative to each other. It also establishes the axial clearances between these components under all conditions. Figure

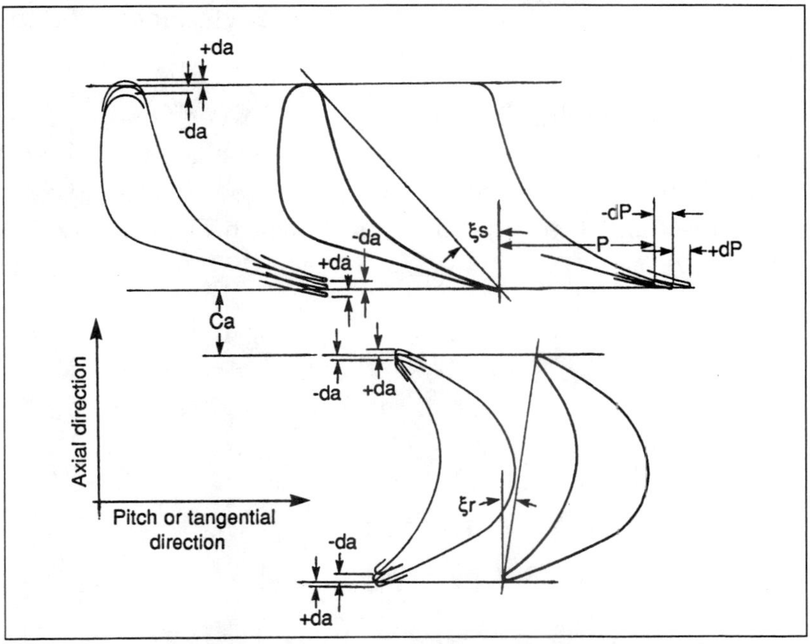

Fig. 2.12.17—Section through a stage stationary and rotating blading, showing the extent of spatial deformation and the possible effect on axial clearances.

2.12.18 shows the cross section through a diaphragm, in which the axial position of all vertical (axial) faces is set by their position relative to face "O-O." This face ("O-O") becomes the steam seal surface, and is located in the casing or blade carrier at a distance ("A") from some reference position, normally located within the cylinder. All axial dimensions such as a, b, c, and d of Figure 2.12.19 are then set by reference to face "O-O," and are in a known position within the total steam path, and relative to the anchor point of the casing.

Another important dimension shown in Figure 2.12.19 is the setback of the vane discharge edge from the "b" surface. This was discussed previously and is shown in Figure 2.12.17. At that time, the possible effect in terms of the vane axial displacement from the design condition by movements "+da" and "-da" on the axial clearance "Ca" was considered. In the case of this diaphragm, it is clear that dimension "b" will not impact on axial clearance, but could have an effect with some stage designs.

Also shown in Figure 2.12.19 are two radial seals that, on assembly, will form a constriction above the rotating blade row. These are machined to diameter "Ds" but must also be located axially to achieve the correct axial position as shown in Figures 2.12.3 and 2.12.4.

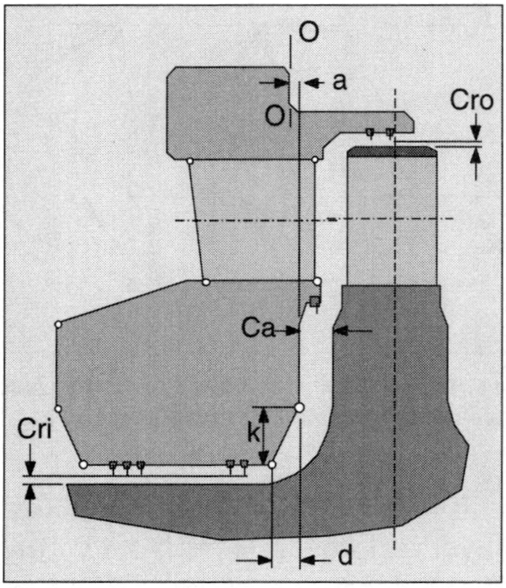

Fig. 2.12.18—Control dimensions for a diaphragm, axial dimensions shown relative to the steam seal face "O-O".

The inner portion of the diaphragm will normally carry seal strips to minimize leakage past the stationary blade row. These are shown in Figure 2.12.1(a) and (b) as having a radial clearance "Cri." At these inner positions the rotor is in close proximity to the diaphragm, and it becomes necessary to provide a chamfer at this location so that clearance from the diaphragm to the rotor fillet radius and from the wheel to central spindle portion is not reduced by operation. Such chambers are shown as ("d"- "k") on the discharge side.

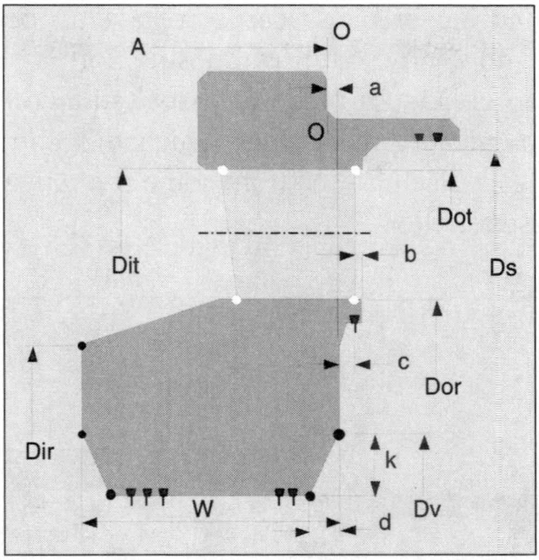

Fig. 2.12.19—Portion of a blade row showing the seal strips at the diaphragm inner diameter adjacent to the rotor.

Shown in Figure 2.12.18 is a detail of this region. The axial clearance between the diaphragm inner web face and the wheel face is shown as "Ca." This clearance is shown without defining its radial location. In the cold stationary—as a new condition—this clearance will be present along the entire radial face. It is also necessary to ensure this minimum clearance is maintained around the fillet radius of the wheel to the spindle section. It is known that certain operating characteristics will modify this value. It will change along the radial direction being consumed to a greater extent at the inner diameters and can also be affected by operating age.

Many rows of stationary blades are located directly into an inner casing, as are many diaphragm rows. For reaction-type stationary blades, this form of construction is used where the pressure drop on the stages is such that the stages must be placed closer together to conserve axial pitch, and where the pressure drop allows such a construction.

The inner casing or blade carrier of a unit locates within the outer casing, and such a form of construction is shown in Figure 2.12.20. Again, contact is made at a seal surface "O-O," and the outer casing provides grooves into which this carrier locates. The face "O-O" is at a distance "A" from some reference or datum position, as discussed earlier.

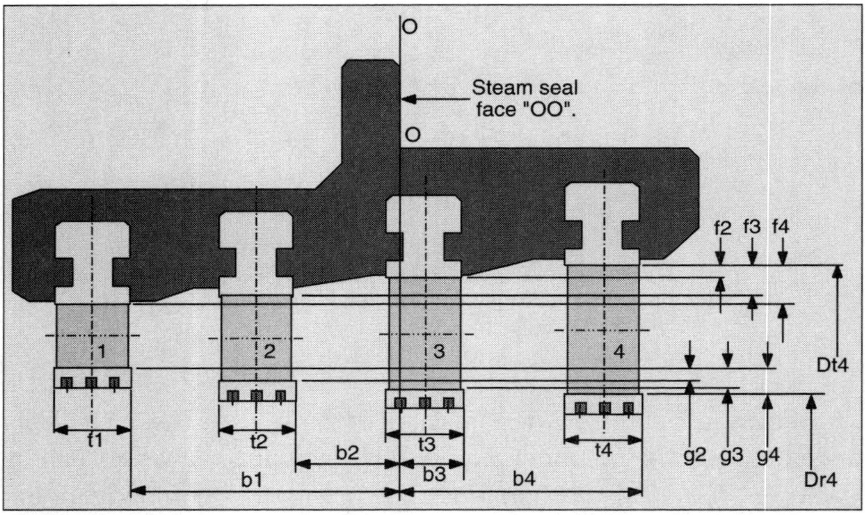

Fig. 2.12.20—A blade carrier locating four stationary blade rows, and showing the control dimensions from the steam seal face "O-O".

The rows of stationary blades carried by this casing have their axial and radial positions defined by dimensions "b1....b4" in the axial direction, and "g1....g4" and "f1....f4" in the radial directions. The blade rows also have an axial thickness defined as "t1....t4." The axial dimensions "b" and "t" define the axial space available to admit the rotating blades and maintain an adequate axial clearance. In the radial direction, dimensions "g" and "f" are selected to provide inner and outer diameters "Dt" and "Dr" for each stage sufficient to achieve the design lap when the unit is assembled.

In many longer vortex design stages, the blade axial width changes with radial position. In these designs, the axial clearance may remain relatively constant, but be axially displaced (Fig. 2.12.21). This diagram represents a reaction stage with the stationary blades located directly in a casing. Together with the width taper that exists on the rotating blade and the coverbands, there are axial clearances "Cao" at the outer diameter, and "Coi" at the inner. As shown in the inset to this diagram, the width and overhang of the covers bands can have an effect on the clearance and the "m" setting of the rotors. The cover overhang (shown as "da") is normally established by final rotor machining.

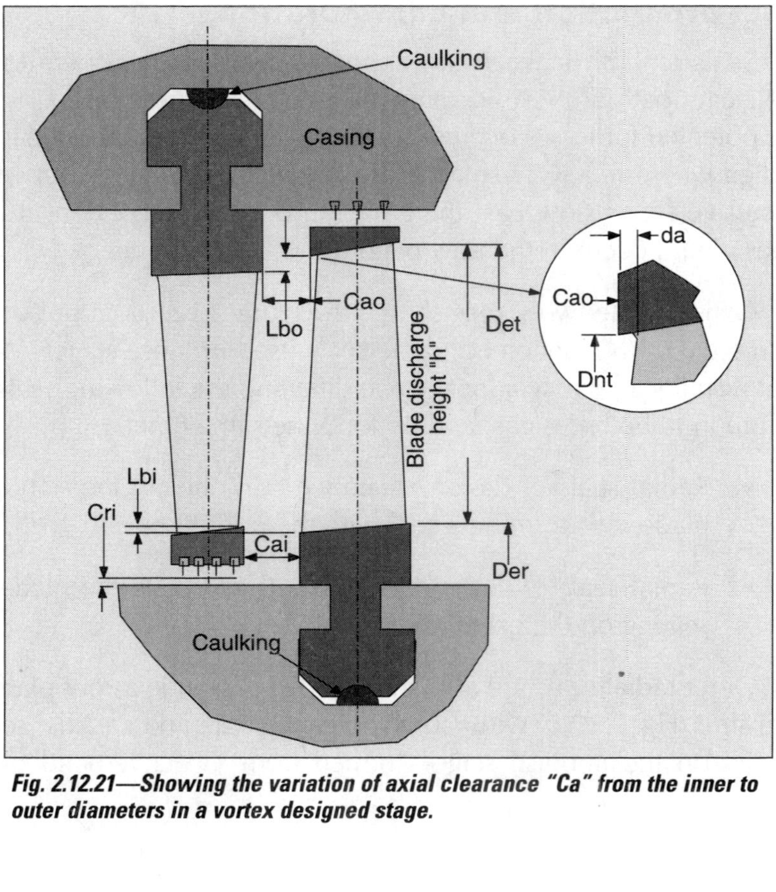

Fig. 2.12.21—Showing the variation of axial clearance "Ca" from the inner to outer diameters in a vortex designed stage.

Stage details

Modern units have many details incorporated into the individual stage design that help improve or maintain stage performance, by contributing to maintaining unit reliability or efficiency. Many of these details are associated with sealing systems designed to reduce wasteful leakage past the blade rows. There are too many such details to discuss in great detail. However, their presence and function must be recognized. Not meeting the design requirements may negate their value and degrade the total performance of the stage and unit.

Seal production and configuration

Seals to limit leakage are arranged—where space permits—in the steam path at those locations where a pressure drop exists and the potential for losses occurs. Seals may also be incorporated in the design where relatively small pressure differentials exist, but where excessive steam flow past the area can occur to the extent disruptions would occur in the flow pattern.

Consider the two stages shown in Figure 2.12.1(a-impulse) and Figure 2.12.1(b-reaction), and the seals used in those stages. While the overall stages may appear to be different, the following seals are common to both designs, and perform the same function:

- Radial seals of design clearance "Cri" below the stationary blade row, forming a seal with the rotor body
- Radial seals of design clearance "Cro" above the blade tip, sealing on the coverband

Other radial seals are also shown. These include a root platform seal strip (Fig. 2.12.5) with radial clearance "Cr" and an axial seal—shown on the impulse stage—formed from the coverband of the rotating blade row, and forming a constriction to the diaphragm outer ring, clearance "Ca."

In addition to these seals—which are an integral part of each stage design—seals are used extensively at shaft-end positions in all sections of a unit. This shaft-end packing is required to limit leakage either of high-pressure steam from the steam path (contained within the casing to atmosphere) or from the inlet between a high-pressure section and a lower pressure section of a unit employing the reverse flow design. In these reverse flow designs, the steam path of two distinct pressure range sections are mounted on a single rotor. In these designs it is necessary to limit the quantity of steam leaking from the high to the lower pressure section, which bypasses several stages and does no work.

Seals are also required at the shaft-end positions of low-pressure sections. They limit the quantity of air leaking into the unit, and minimize the quantity of sealing steam that must be supplied to each end of the shaft preventing the inward leaking air from gaining access to the low-pressure hood that is below atmospheric pressure.

Seal configuration can take various forms. The actual form employed at any location is a function of the space and material available for producing the seal, the most suitable method of attaching it to either the stationary or rotating portions of the unit, and the relative motion between them at that location.

The shaft-end seals are located at some known axial distance "n" from the thrust or coupling face. For straight-through seals where running interference cannot occur, this is the only dimension of concern. However, if the seals are of the "hi-lo" type (shown for the rotor portion in Fig. 2.12.8), the castellations are pitched at "p," having a set back distance "w1" to the first castellation. These dimensions are critical, as misplacement will often cause interference between the castellations and seal strips.

Seal geometry and arrangement

The geometry of typical seal systems should be considered for installation at those locations where there is sufficient room for the use of multiple series constrictions—for a small number, two or three—and where only single strips can be used. The possible arrangements should be considered in both the axial and radial direction, or in combination. At some axial locations (where the differential expansion is large), it is possible the total number of strips provided may not be effective. The particular strips that will be effective are dependent upon the differential expansion under a given set of operating conditions. The larger number of sealing strips is provided to ensure that sealing at all axial positions and under all loads is as effective as possible. The following possible arrangements for multiple seal strips should be considered:

Straight through. The straight-through design (Fig. 2.12.22) has minimum axial alignment requirements, though the only concern for clearance is in the radial direction. The axial placement requirements are that the first tooth ("k") is placed effectively at all extremes of axial expansion conditions of the rotor, relative to the stationary portions of the unit (and recognizing the requirement of the distance "n" as defined in Fig. 2.12.2). It is also necessary that at rotor diameter changes the space requirement for the last tooth ("q") before the diameter change is at a distance "r" from the shoulder, so interference with the shoulder will not occur.

This form of labyrinth seal is used where axial movement can be large from cold setting to final operating condition, and where large axial movements are experienced due to differential expansion during thermal transients. In this straight-through arrangement it is known that a portion of the kinetic energy generated at each constriction is carried through from one seal strip to the next (Fig. 2.12.23). This carry-through effect is unfortunate, but unavoidable.

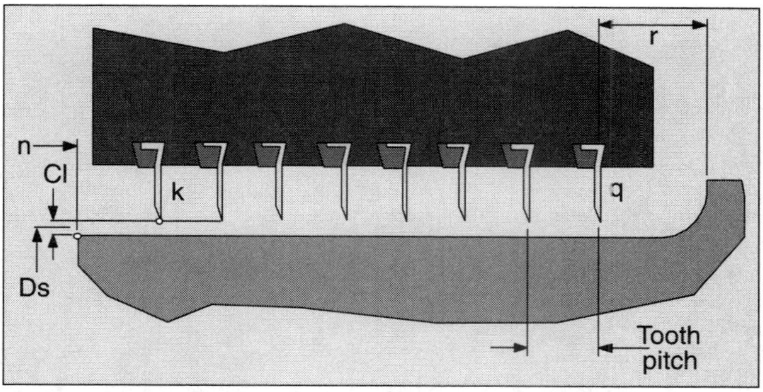

Fig. 2.12.22—The "straight through" design with inserted teeth.

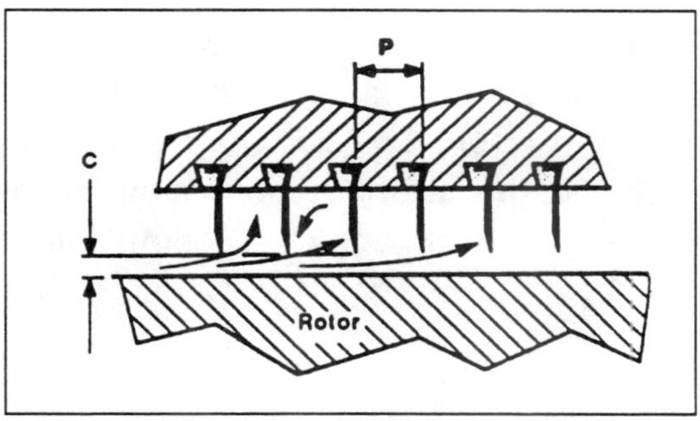

Fig. 2.12.23—Showing the flow patterns set up in the labyrinth seals.

Simple "hi-lo" configuration. It is normal to use the "hi-lo" type configuration at those locations where differential expansion is not large, and there are relatively small amounts of axial movement under all extremes of transients. Such an arrangement is shown in Figure 2.12.24, where alternate teeth form on the rotor and castellations are produced on its surface. The principal advantage of this seal is that thermal energy expended in expanding past a constriction—and converted to kinetic energy—is almost completely destroyed upon impact with either the castellation wall or the seal strip face.

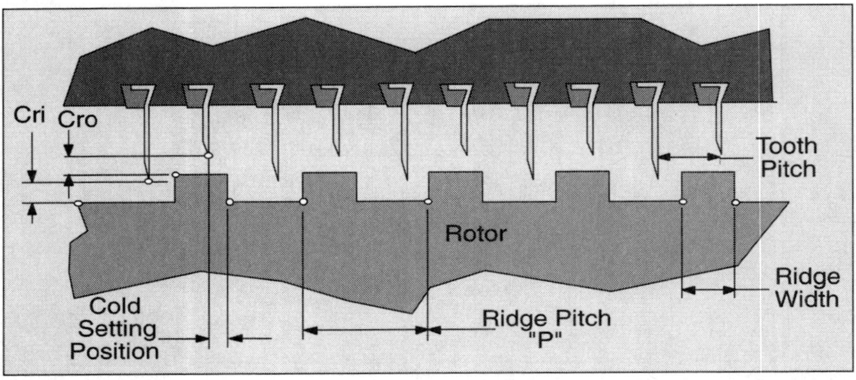

Fig. 2.12.24—The "hi-lo" teeth with a single "hi" tooth at each castellation.

The cold setting requirements for this design are shown in Figure 2.12.24, and the required axial clearances in the gap between castellation shown in Figure 2.12.25. The axial clearances "g" and "h" are set to design specification eliminating the risk of rubbing interference between the teeth and castellation vertical surfaces.

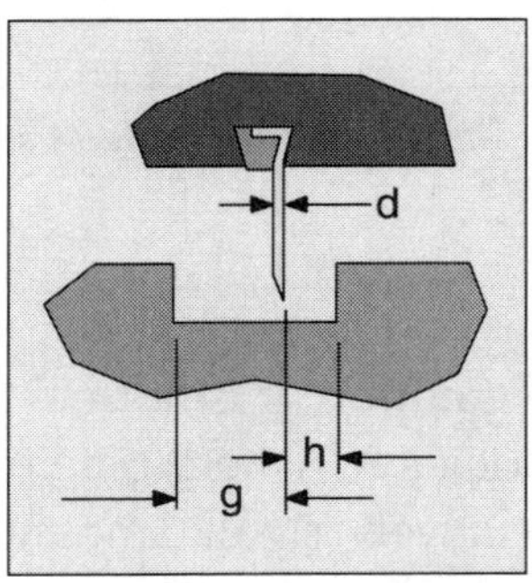

Fig. 2.12.25—Cold tooth settings.

The radial clearances above the rotor and castellation are shown as "Cri" and "Cro." It in normal to make these clearances the same, and this single clearance will be referred to as "Cr." These radial clearances are set by design at a minimum value sufficient to avoid "rubbing" during operation, to allow the phenomena described earlier, and possibly some rotor vibration, particularly during transient operations.

The "hi-lo" configuration for large differential expansion. At higher levels of differential expansion—but not so large that the straight-through design need be used—the design shown in Figure 2.12.26 can be applied. This design employs two high teeth for every low one. With this design, only one "hi" tooth is effective on the castellation at any time; the "lo" tooth is always effective. However, at the extremes of operating position, there is always one "hi" tooth effective. The cold settings are shown in Figure 2.12.26, and the pitching requirements between the castellation are shown in Figure 2.12.25. However, the actual pitch values will have increased because of the larger total pitch required between the castellations. Compare Figures 2.12.24 and 2.12.26.

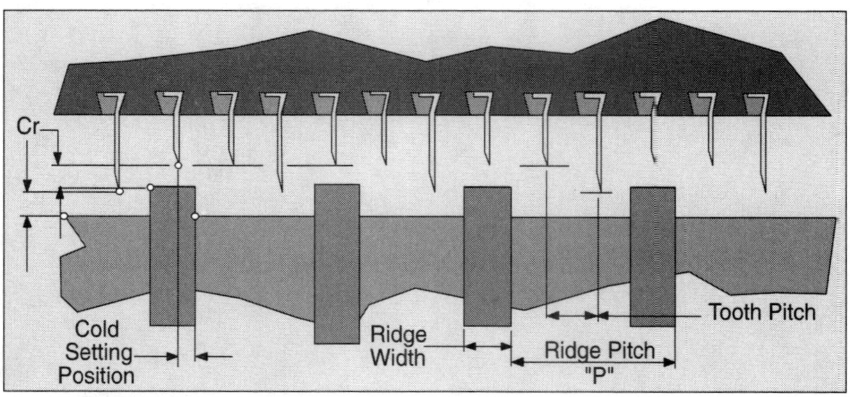

Fig. 2.12.26—The "hi-lo" configuration with two "hi" teeth to allow for large differential expansions.

The optimum point to convert from the use of the double "hi-lo" to the straight-through design is a matter for design evaluation. In Figure 2.12.22, there are shown eight effective teeth, and in Figure 2.12.26 there are nine effective teeth (only one "hi" tooth is effective at any time). It is a matter for the design evaluation to establish which will be the most effective design to use in the axial space available, recognizing the superior effectiveness of the "hi-lo" configuration in destroying the kinetic energy of the leaking steam.

The seal teeth shown in Figures 2.12.22 through 2.12.26 are of the inserted strip type. However, a commonly used form of tooth is that which is carried on an inserted gland ring (Fig. 2.12.27). These forms of sealing arrangements are considered in detail in chapter 11.

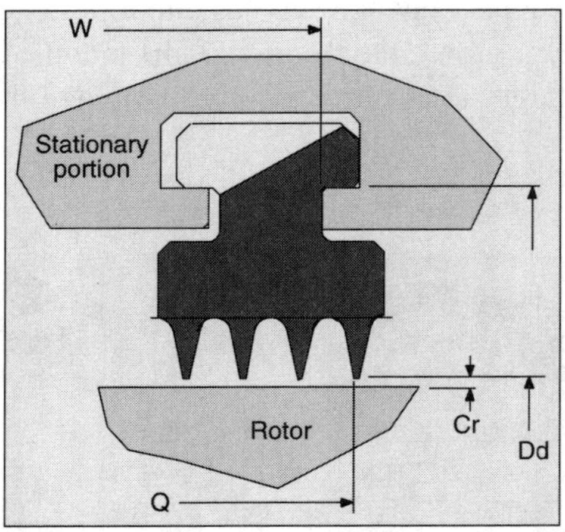

Fig. 2.12.27—The segmented seal ring, with control dimensions.

The selection of seal configuration

The seal configuration used at any location is dependent upon a number of factors acting to make one arrangement more suitable than others. Among these considerations:

- *The differential expansion at that location together with any other factors that might influence the axial position of the rotor relative to the stationary portion of the unit.* This can include certain requirements including the need to achieve axial shift for coupling adjustment of component disassembly

- *The axial space that is available to mount the seals.* There is no advantage to using more seals than can reasonably be effective. To place the seal strips too close will make them ineffective

- *The cost of producing the seals.* If the financial cost of producing and mounting seals is more than the cost that can be achieved, then the less effective sealing system will be installed

Consider the possible geometries and arrangements used at any location. There are three major areas to consider. These include:

At shaft-end seal positions. At shaft-end positions, the most commonly used seal at the high-pressure, high-temperature locations close to the thrust block is that shown in Figure 2.12.24. That is the simple "Hi-Lo" arrangement. At low-pressure locations where the differential expansion is considerably greater, the straight-through arrangement is most common. Each axial location is analyzed during the design process, and the most effective design consistent with manufacturing and fuel costs is chosen.

Under the diaphragms and stationary blade rows. In the high-pressure, high-temperature stages of impulse units (the seals at the diaphragm-to-rotor locations), the seal system is selected based on the same considerations as those governing the selection of the system used at the shaft-end positions. Under these higher condition diaphragms, there is normally sufficient axial space that multiple constriction seals can be used, and because of the high-pressure drop across them, it is necessary such systems be used to limit leakage.

Similarly, in reaction stage stationary blade rows it becomes necessary to provide seals at the inner diameter, between the inner blade tip and the rotor surface. The inner diameter will normally have an inner coverband, which is either attached by the use of a tenon formed on the vane, or with a coverband formed integrally with the vane. Figure 2.12.28 shows four possible systems for attaching such seals at this inner location. In (a) and (b), the seals are shown as part of an inner coverband that is itself attached to the blade tips using tenons. In (a) the seals are formed integrally with the cover and in (b) they are caulked in place. In (c) and (d), the inner coverband is formed integrally with the blade vane and the seals are mechanically attached. In (c), these are attached by caulking and in (d) by an attachment method that normally is welding or brazing. In each of these four cases, normally seals require trimming after assembly of the blades to the casing or blade carrier to achieve the seal diameter and clearance "Cri."

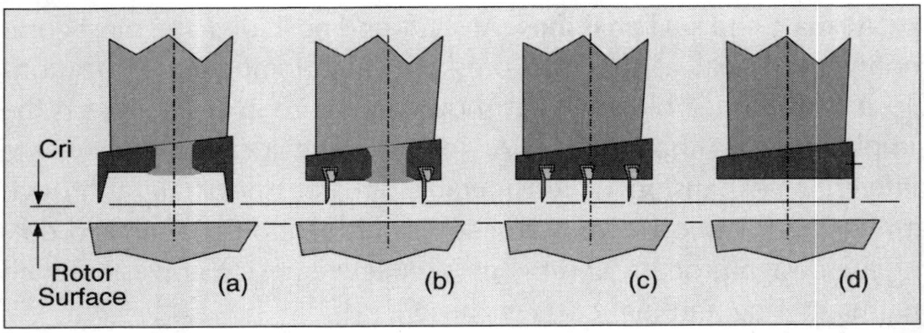

Fig. 2.12.28—Showing alternate seal arrangements at the inner diameter of stationary blades.

At rotating blade tips. The possible arrangements and design options at the rotating blade tips are far more varied than at the other locations. The options available allow these seals to form axial or radial constrictions, or a combination of both. Typical seal systems at the rotating blade tips are shown in Figures 2.12.3 and 2.12.4.

The seal strips used can either be produced integrally with the blade vane cover or be inserted into a portion of the unit—rotating or stationary. It is not common with rotating blades to caulk attaching seals to the rotating portions because there is the possibility of centrifugal effects detaching them during operation. Such designs do exist and have been used successfully, but considerable care is required in the caulking operation. It is most common for the seals to be caulked to the stationary portions of the stage.

From a maintenance and economic perspective, it is preferable to form the seals from separately attached components that are caulked or inserted into the stage. This allows their replacement when they have become worn (causing a larger leakage area and no longer sealing). Separate attachment is considered optimum, because this allows the seals to be changed without the need to replace the entire major component of which they are a part. Although procedures are now available to allow worn seals to be weld-rebuilt, the cost is higher than simply replacing inserted seal strips.

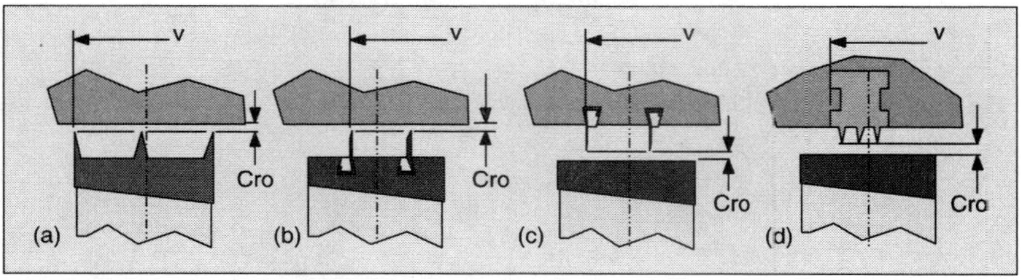

Fig. 2.12.29—Possible seal configuration at a blade tip.

Figure 2.12.29(a) shows radial seals at a rotating blade tip. These seal strips have been produced integrally with the blade cover, which can be attached or formed integrally with the blade vane. This type of seal is produced and the final form is machined after assembly to the rotating blades. Figure 2.12.29(b) shows a design in which seal strips are caulked into the outer surface of the coverband. As previously stated, this form of seal is not common, and can be a source of damage if the steal strips detach. Figure 2.12.29(c) shows a radial seal in which the seal strip is caulked into the stationary portion above the coverband. Figure 2.12.29(d) shows a seal in which a segment is inserted into a special groove produced in the casing. In each design shown, the axial control distance from the reference face is shown as "v" and the clearance as "Cro."

Figure 2.12.30 shows an arrangement that provides a seal in the axial direction. This is possibly the most common form of seal for impulse stages, in which a knife-edge is produced integrally with the coverband inlet edge. This type of seal is normally trimmed after assembly of the coverband to the blade tips, and establishes the axial clearance "Ca." This form of seal may not necessarily be referenced to a datum face, when the coverband is changed on maintenance.

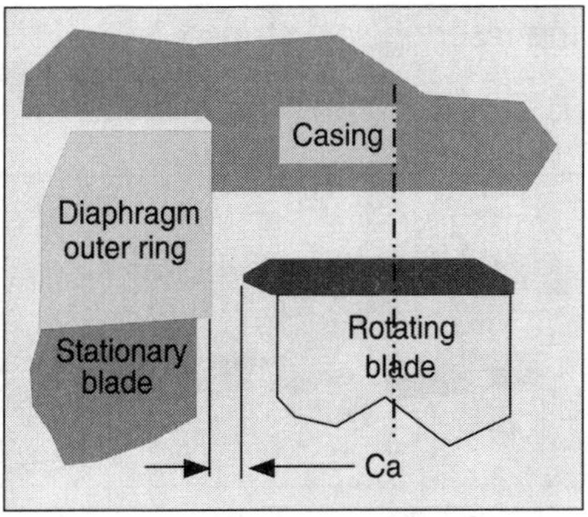

Fig. 2.12.30—An axial seal at the blade tip. This seal is formed integral with the cover band.

Figure 2.12.3 and 2.12.4 shows that there are designs combining axial and radial seals. There are many acceptable arrangements, and it is necessary to consider the spatial arrangements in terms of the particular design.

BLADE VANE AND CASCADE GEOMETRY

The method of calculating or determining the form blade vanes take is complex. In a future chapter we will review vane form from a theoretical perspective, making the assumption that the design specified form can be met without exceeding design specified tolerances. There are, however, certain characteristics of the vane—its form and placement on the root platform—which have a significant effect on the performance of the total unit. These geometrical requirements are considered in this section, together with efforts made to define which of these characteristics are important and cannot be relaxed without causing a deterioration in the performance potential of the unit. This section will also explain which parameters defining the total blade vane need to be considered when blades are manufactured.

Vane form requirements

The vane form is not critical in and of itself. The actual form of a blade is important only in terms of the shape of the expansion passages it produces between adjacent elements. In assembling blades (either stationary or rotating) errors in the profile form (or in the rate of change in vortex designs and its placement) can influence two expansion passages, and has therefore the potential to degrade performance.

Profile shape. The expansion passage must be considered in terms of the demand for and degree of convergence required. In supersonic rows, it must be recognized there is a need to produce a converging-diverging passage form. However, no thought has been given to the possible effect that manufacturing tolerances and errors could have on the passage, or even where tolerances need be applied. The profile is designed to achieve certain technical requirements. Among these are the needs to form an acceptable passage, and expand the steam with a minimum of losses due to form. The shape of the profiles will also control the total discharge area from the row. This will be considered.

In order to achieve its functions as effectively as possible, there are certain characteristics of form that must be observed. There are parameters that are defined by design, which specifies shape, and assign a tolerance band within which they are acceptable.

Fig. 2.13.1—A blade mounted in a shutter gauge for profile confirmation.

It is normal for the design specification to define an envelope of tolerances for the profile, providing a "plus or minus" (+/-) tolerance band for various locations on the surface. What such a tolerance band cannot establish is the extent to which such a tolerance can vary within the envelope. It is possible for the band to be (+) at one location and (-) at another adjacent location. Assessment of this condition calls for mature judgment on the part of the design engineer. Gauging the acceptability of a profile is normally undertaken using a "shutter" or "guillotine" gauge (Fig. 2.13.1) and a typical envelope of tolerances (Fig. 2.13.2) where variations from the defined shape are shown as "dT" and "-dT." Special tolerances are applied to the discharge tail.

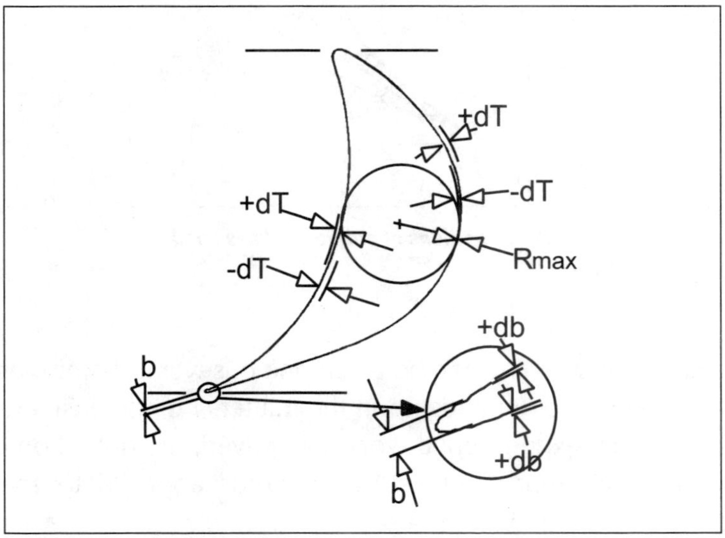

Fig. 2.13.2—The envelope of tolerances defining the profile form.

Chord "C" and thickness "T." From mechanical considerations, the more critical considerations of the profile are the chord and the vane thickness. These are best defined as the maximum thickness along a mean line traversing the median thickness (Fig. 2.13.2 and 2.13.3), where the maximum thickness is "2.Rmax," measured near the center of the profile.

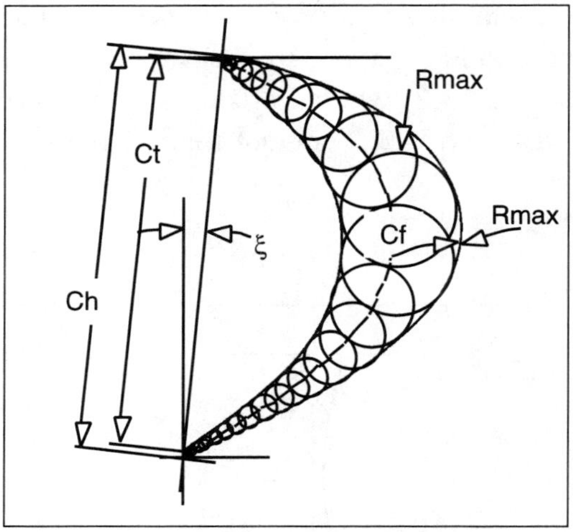

Fig. 2.13.3—Definitions of profile thickness and chord.

The chord of the vane can be given one of several definitions but the most appropriate depends upon the matters being considered. If discussion of any experimental work is involved, the definition used to derive any particular set of results is the most appropriate. The following definitions can be applied:

- *as the distance "Ch" across the horns of the profile.* This easily measured characteristic can be determined with relatively unsophisticated measuring instruments
- *as the distance along the center of the profile "Cf."* To determine this value, it is necessary to have it defined from com-

puter calculation. If only the profile is available, constructing the position of the center from drawn radii and then measuring must establish the chord

- *as a line drawn from the center of circles defining the inlet nose and discharge tail "Ct."* This is not a common definition but is most appropriately used when methods of calculating profiles determine these are positions to aid in the drawn construction of the profile

These parameters "C and T" influence the mechanical strength of the profile, since they influence both the profile area and the section modulus—a critical consideration when evaluating vibration characteristics.

Inlet nose. A profile's inlet nose has the potential to modify the streamline form of the steam as it passes between the vanes of a blade. Consider the three profiles shown in Figure 2.13.4. The center element ("S") has an inlet nose that divides the incoming steam and diverts it to the left or right to flow through one of two passages ("R-S" or "S-T") that it helps form. Steam entering the row with a high relative velocity ("W1") introduces a stagnation region ("s").

If the vane is manufactured to design there will be no discontinuities in the nose form, and the flow will be even in the passages (other than the possible separation bubble that can form). However, vane manufacture is a process that can involve a certain amount of hand polishing, and any uneven surfaces produced (as shown by flat "y-y" in Fig. 2.13.5) can magnify the disturbance that occurs and introduce significant turbulence into the flow.

The vanes shown in Figure 2.13.4 are typical of the rotating blades in an impulse type design. For stationary blades the velocity of the entering steam "C2" (see Fig. 2.4.15) is much lower, and will therefore cause less degradation in the stage efficiency.

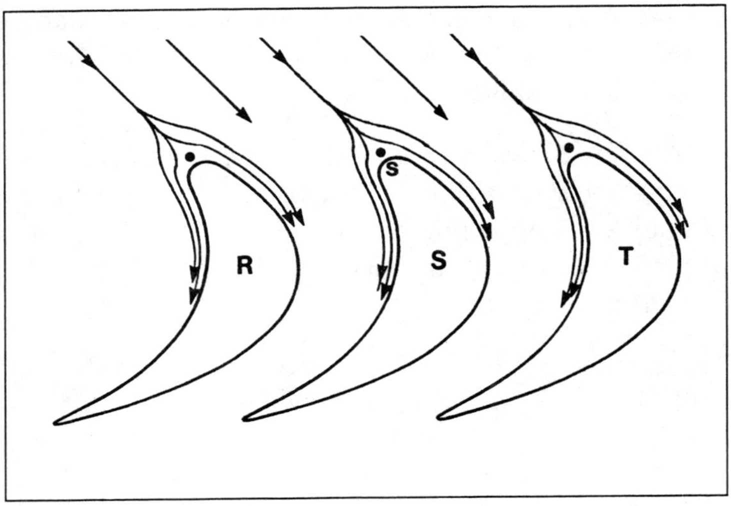

Fig. 2.13.4—Flow division at inlet to the stage.

Discharge tail. The discharge tail helps form the discharge point from the expansion passage. As such, it must provide the correct discharge area and angle (as explained previously). Stresses—both bending and direct tensile—are high in the tail region, and it is necessary to ensure the definition of tail thickness is observed. It is normal that the envelope of tolerances at the tail is limited to design with a (+) tolerance only. No undersize thickness is acceptable.

The tail can be finished to several forms as shown in Figure 2.13.6. In (a), the flat is maintained to a knife-edge. In fact, there is always a degree of rounding [as shown in (b)] and at times a complete rounding, as shown in (c). To assist in measuring discharge tail thickness, the most convenient method is to measure at some specified distance ("d") back from the discharge point to establish the thickness ("b"), as shown in Figure 2.13.7. Both "d" and "b" can be defined by design sufficient to permit accurate determination, allowing for the fact that the distance "d" is dependent to a considerable extent on the polished form of the discharge tail. Measuring too close to the discharge point can result in measurements made on the radii formed by polishing, *i.e.*, "q-q."

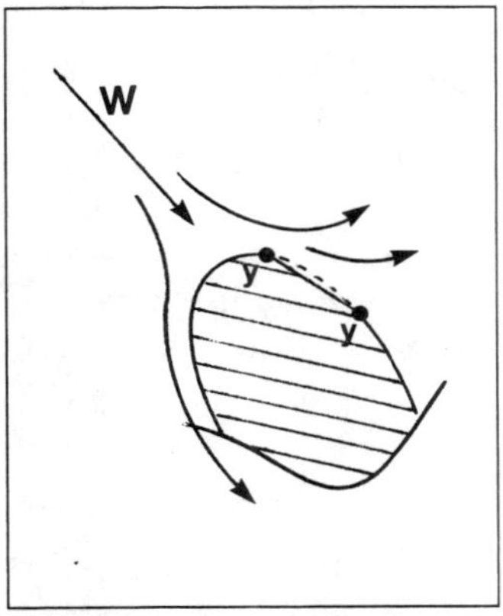

Fig. 2.13.5—A flat produced on the inlet nose of a vane, modifying the form of the profile.

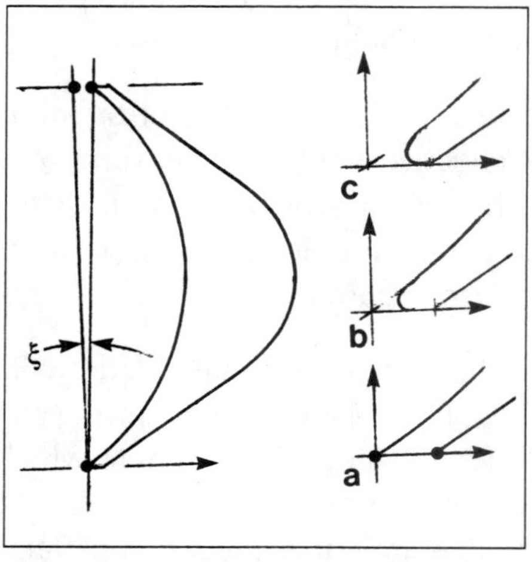

Fig. 2.13.6—Details of the discharge tail, and various forms.

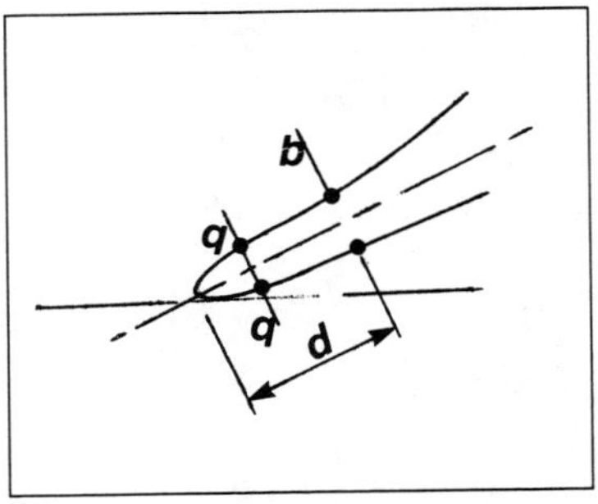

Fig. 2.13.7—Quantifying discharge tail thickness.

Vane setting angle "ξ." If a vane is manufactured correctly—meeting all requirements of form and finish—it is still unable to form an acceptable expansion passage if it is set within the cascade in such a manner that it is incorrectly inclined in the axial/tangential direction. As shown in Figure 2.13.8, to be correct, this setting angle must have its inclination "ξ." Modifying this angle by unacceptable amounts will modify the shape of the passage, and—depending upon the point in the profile about which the vane rotates—such work can modify the rate of convergence and modify the discharge throat by unacceptable amounts.

As can also be seen in Figure 2.13.8, errors in the setting angle can influence the effective width of the vane, causing errors of "+dW" or "-dW." The increase or decrease in width can occur at either of the edges, depending upon the point in the profile about which the "untwist" has occurred.

In service, this setting angle on vortex blades will become modified by the "untwist" of the blade vane occurring as a consequence of the centrifugal stiffening effect. The extent to which the blade will untwist is influenced by the stage hardware, such as the coverbands and tie wires that help restrain this effect to a certain extent.

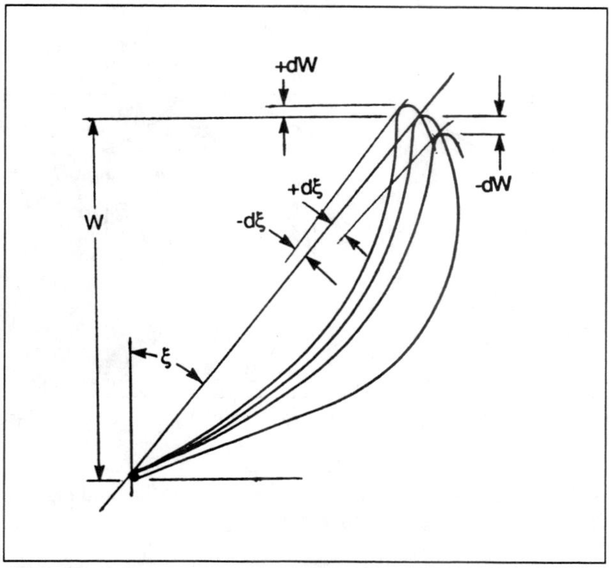

Fig. 2.13.8—The effect of setting angle errors "dξ" on row width "dW".

Pressure and suction faces. In the portion of row cascade shown in Figure 2.13.9, three profiles—"R," "S" and "T"—enable steam to enter the row at a velocity "W1" and discharge with a velocity "W2." Each vane forms one surface of two adjacent expansion passages (the suction face of one and the pressure face of another).

Figure 2.13.9 shows three profiles, or "faces." The center profile ("S") indicates these faces. The concave surface of the profile ("S") has a positive pressure relative to the mean pressure ("Pmean") in the passage developed on it due to the change in steam momentum. Similarly, the opposite face in the passage has a pressure negative to the mean developed on it. Therefore, the total force developed on the profile is the sum of the forces on both sides of the profile. This is shown in the pressure diagram, where pressure through the row reduces from "Pi" at inlet to "Po" at discharge.

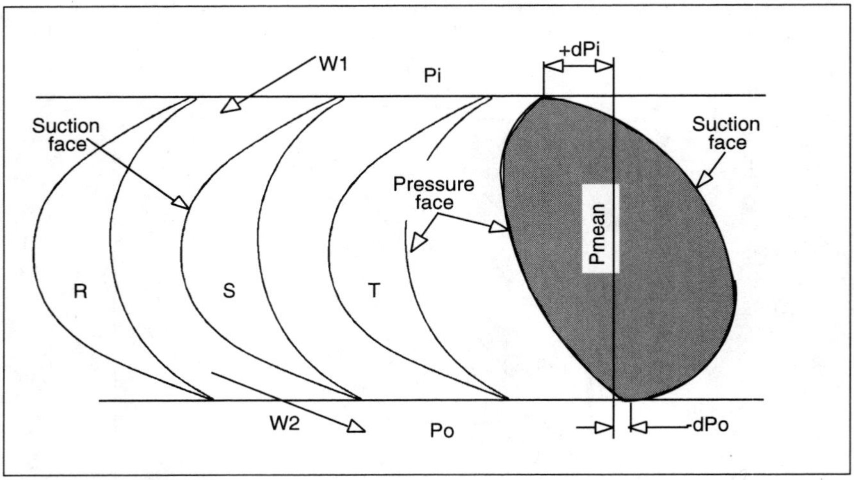

Fig. 2.13.9—The "pressure" and "suction" faces of a profile, and representation of the pressure distribution in the expansion passage.

Vortex form and profile change. The vane profile is a two-dimensional form. However, the vane of a blade is a three-dimensional body, which—depending upon the design philosophy used to establish vane requirements for that row—can have a profile that varies as a function of its radial position. This is done to accommodate the changing steam velocities and relative inlet angles to the row. Figure 2.13.10 shows the "stacking diagram" of a vortex blade, including the profile change at equidistant radial positions above the root section. The relative position shown for the profiles above the "stacking point" is selected to minimize the centrifugal bending stresses during rotation.

Various vane angles are essentially defined or selected by the thermal design process, and so in selection of profiles, the designer must meet and ensure the expansion efficiency guaranteed in the quotation to the client. Stage profiles are shown in Figure 2.13.11. The important angle considerations follow.

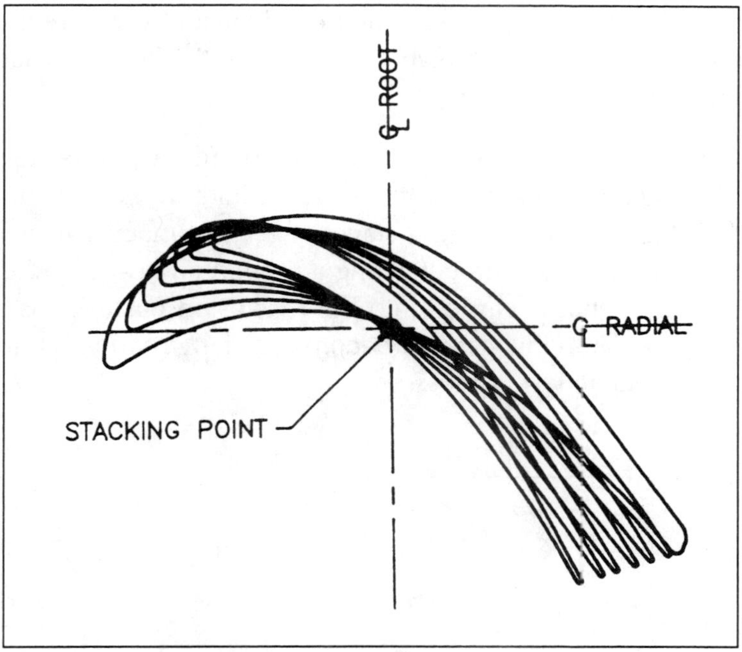

Fig. 2.13.10—The "Stacking Diagram" for a vortex vane.

Vane inlet angle "α0" or "β1." The vane inlet angle is selected to allow the steam from the previous row to enter the expansion passages with a minimum of incidence. The inlet nose of the majority of modern profiles is rounded. Experimental work has shown that larger inlet radii ("Rs" and "Rr" of Fig. 2.13.11) are better able to accommodate incidence with an off- design inlet angle. However, there are losses associated with all such incidences.

For the stationary row of an impulse design, the velocity entering the row is considerably less than that entering the rotating row. Therefore—because losses are velocity sensitive—these stationary rows are able to accept greater incidence than the rotating rows. In high-reaction stages, velocities entering the stationary and rotating rows are of the same general magnitude. This means that while both suffer the same order of losses due to incidence, the entering velocity in a reaction row is somewhat smaller than the impulse stage, and the level of losses are smaller.

Certain older designs of vane—particularly those used in impulse stage rotating elements—have an inlet that is not rounded, but flat (Fig. 2.13.6). In these stages, the flat surface will interfere with the inflowing steam, deflecting a portion and causing a degree of turbulence that is carried into the expansion passage. As a unit ages, it is possible this flat will become partially rounded, but there will still be some level of loss.

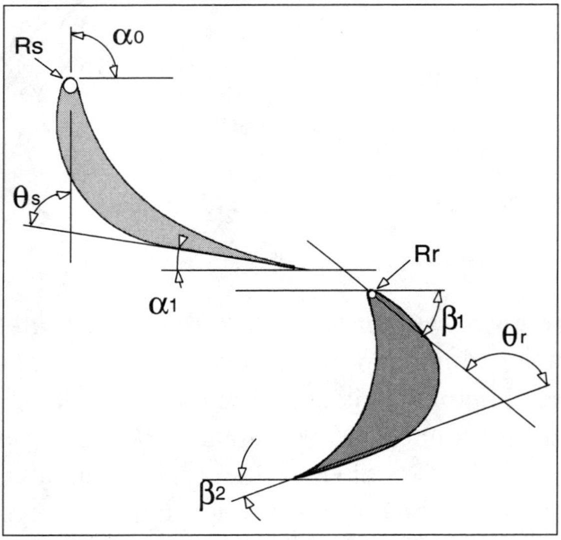

Fig. 2.13.11—Stationary and rotating blades showing basic characteristics.

Vane discharge angle, "α1" or "β2." The vane discharge angle is shown as the mean angle of inclination of the vane discharge tail. The value used in determining the angle requirement at discharge is the inverse sine of the ratio "O/P." However, the discharge portion of the vane must be set to support these geometrical requirements, and while there are minor differences, (as will be discussed in a further section concerning the pitch) the vane tail inclination assists in controlling the throat for any vane pitch at discharge.

Vane turning angle "θ." The vanes must be designed to admit the steam at an angle "α0" and discharge it at an angle "α1" on a stationary element, admit steam at an angle "β1" and then discharge it at an angle "β2" on a rotating element. This means the steam must be deflected through an angle (known as the turning angle). This angle is given the designation "θ," being "θs" for the stationary elements and "θr" for the rotating.

The magnitude of these turning angles can be established from the relationships:

$$\text{For the stationary row: } \theta s = 180° - (\alpha o + \alpha 1)$$
$$\text{For the rotating row: } \theta r = 180° - (\beta 1 + \beta 2)$$

In many stages, the profile actually used may not meet precisely the requirements of inlet, discharge, and turning angle "θs" or "θr." This is because profiles are expensive to design, and situations often require that existing profiles be employed rather than newly developed for any specific application.

Axial placement

Figure 2.12.17 shows the theoretical layout of the vanes for both stationary and rotating rows of a single stage. This figure also indicates that vanes can be proud by an amount "+da" or recessed by an amount "-da." Before discussing these errors, it is necessary to establish a definition for plus (+) and minus (-) deviation from the theoretical line of the vanes. A suitable nomenclature that will be used is:

- in a *positive deviation,* the vane projects beyond the constraints of the theoretical boundary line. This can represent a movement "up" or "down" stream in the axial direction, depending upon whether the row being considered is stationary or rotating, and whether this is at the "inlet to" or "discharge from" the expansion passage
- in a *negative deviation,* the profile is recessed from the theoretical boundary line. This negative movement, like the positive, is a function of the vane and its position relative to steam flow direction

The inlet and discharge edges are defined as the theoretical position of the vanes at inlet to and discharge from the row. These are shown for a stationary row in Figure 2.13.12, where both proud and recessed edges are shown at the design limits of "+da" and "-da." This figure indicates that if the blades are set with the correct setting angle, a blade proud to one edge is recessed at the other. However, errors in the setting angle may cause both edges to be either proud or recessed.

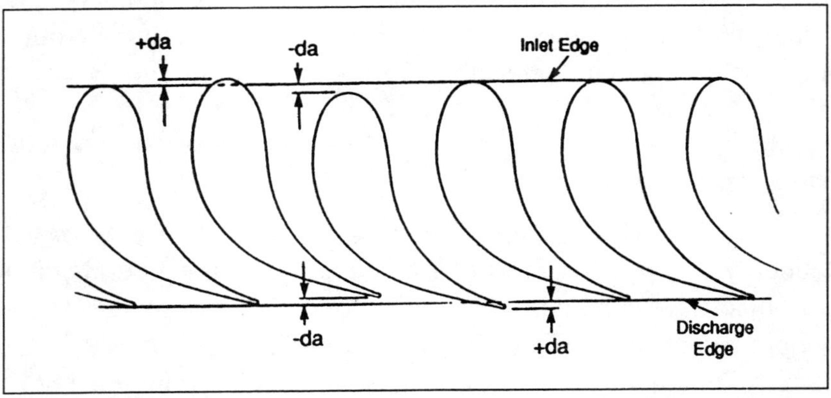

Fig. 2.13.12—Blade cascade showing "proud" and "recessed" blades at the inlet and discharge edges.

Inlet and discharge edge. The inlet and stationary edges shown in Figure 2.13.12 define the design-specified axial limits for the row. Also shown are the positive (+) and (-) limits at inlet to and discharge from the row. The limit for a proud or a recessed edge is set so the stage will perform at the design-specified level. The tolerances are set dependent upon the methods of manufacture.

Stationary blade elements. The limits established for stationary blades are normally set at values a little more liberal because of the manufacturing methods. This is particularly necessary when vanes are located in cast or fabricated rings. These blade elements can also be adjusted to achieve an improved geometric arrangement.

Rotating blade elements. The rotating blades are normally set with limits that are more restrictive than those applied to the stationary row. This is necessary because misplacement of the profiles will induce stresses in the vane due to bending as the blades rotate.

In general, the stationary row—particularly for fabricated diaphragm stages—has a greater tendency for error because of their method of manufacture.

Tangential placement (pitch)

The design process sets the tangential placement of one profile relative to others in the row, so the throat is produced on the suction face of the adjacent element. For curved-back profiles, tolerances may need to be more restrictive because of the changes errors introduce in the discharge angle.

The cascade

The cascade is a series arrangement of blades that are placed forming a series of expansion passages. This arrangement is selected to achieve two concurrent objectives:

- To provide a discharge throat at the tail position. This throat, when integrated along the radial height of the vane, provides a discharge area for the single throat that when integrated around the row provides the total discharge area required for that row
- To provide the discharge angle required to direct the expanding steam into the following blade row

To produce a satisfactory cascade fulfilling the requirements of area and angle the following geometric and spatial characteristics must be met:

Vane placement. The vane must be placed and secured so its position within the cascade is correct and meets design requirements. Shown in Figure 2.13.13 is a rotating blade pair, of a typical high reaction stage, and the development of the passage width as a function of the axial position. This represents the design optimum form of the expansion passage, and should be met within close tolerances.

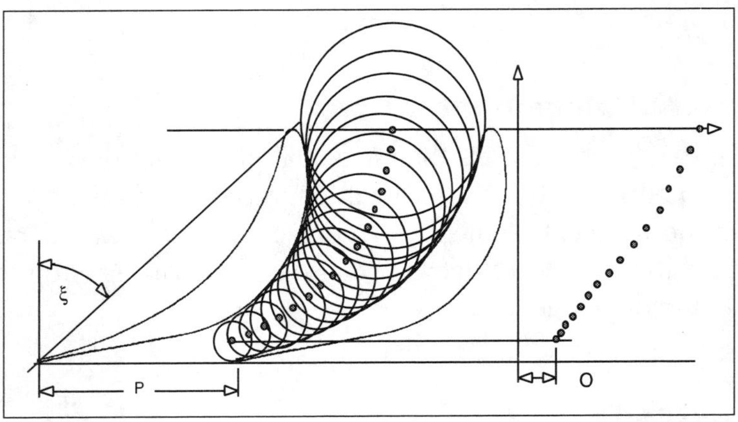

Fig. 2.13.13—Rotating blade pair for a high reaction stage showing the developed expansion passage.

Vane twist. The thermodynamic optimums of a blade row are a function of blade height, therefore the vane should change its geometry along the radial height to accommodate these variations and minimize losses.

In many small radial height stages, there is often insufficient predicted variation of the inlet, discharge, and turning angles for the cost of twisted or vortex blade to be justified. The vortex blade is, in general, more expensive to manufacture, but fuel cost saving can often easily justify the additional expense of producing the vortex profile. (However, with multi-axis milling, this production cost differential is relatively small, and only the complexity of milling small radial-height blades with integral covers will induce suppliers to use the cylindrical form.)

A general, but not absolute rule is if the ratio of vane height to mean diameter ("h/Dm") is greater than 0.356, then a vortex vane should be used. However, many manufacturers can justify the use of a vortex vane from a cost/performance perspective when the "h/Dm" ratio exceeds 0.10 to 0.15, depending upon fuel costs and predicted unit load factor.

Expansion passage form. Figure 2.13.13 shows the passage form between adjacent vanes. In this cascade, the throat width reduces from inlet to a throat of "O" at discharge. The pitch at discharge is "P."

The ideal passage shape is designed to turn the steam through the correct angle, accepting the inflowing steam at an angle "β1" and discharging it at an angle "β2." The passage should diverge at a rate sufficient to form the correct discharge area. It is also necessary to ensure there are no sudden changes in the radius of curvature of either the pressure or more importantly the suction surface, as this effect can cause premature separation of the boundary layer.

Throat. The throat between the discharge tail and the suction surface of the adjacent blade is one of the more critical characteristics defined by the design process. The product of the mean throat along the radial height of the vane and its height will give the discharge area from the passage. The sum of all the individual throat areas will give the discharge area of the stage that in turn defines velocities and pressures at discharge from the stage.

Pitch. The pitch is the circular distance between adjacent profiles, and if there are "Zb" profiles in a row, the theoretical pitch shown as "P" in Figure 2.13.13 can be found from:

$$P = \frac{\pi \, . \, D}{Zb}$$

where:

"D" is the diameter at which the pitch is being considered

Note: When calculating pitch, it is normal to use the circular pitch as determined from this equation. However, even when a relatively small number of blade elements are employed, the "chordal" pitch is often used, as it is more accurate.

Throat/pitch. The ratio of "throat/pitch ("O/P") establishes the steam discharge angle from the passages. In many rows, an audit is made of the values of "O" and "P" after assembly (plus any needed adjustment), and tolerances are set on these values.

Typical of the tolerances applied to rows are:

Parameter	Tolerance Range
Throat "O"	4.0 - 5.0 %
Pitch "P"	4.0 - 5.0 %
Throat/ Pitch "O/P"	2.0 - 2.5 %

Actual tolerances for any stage are dependent upon whether they are stationary or rotating, vane radial height, and method of manufacture.

Discharge area "Ad." The total discharge area from any row establishes the pressure at the discharge point and, therefore, the energy on the stage. In establishing this total area, consider the area of the individual passages ("ad"). Figure 2.13.14 shows the variation of throat ("O") along the radial height of a vortex blade vane (either stationary or rotating). It can be seen from this figure that the throat does not vary linearly, but is some function of the vane twist and setting angle variation. Therefore, the discharge area of the individual throat is:

$$ad = \sum_{o}^{h} o.dh$$

and the total discharge area "Ad" for the row is:

$$Ad = \sum_{o}^{z} ad$$

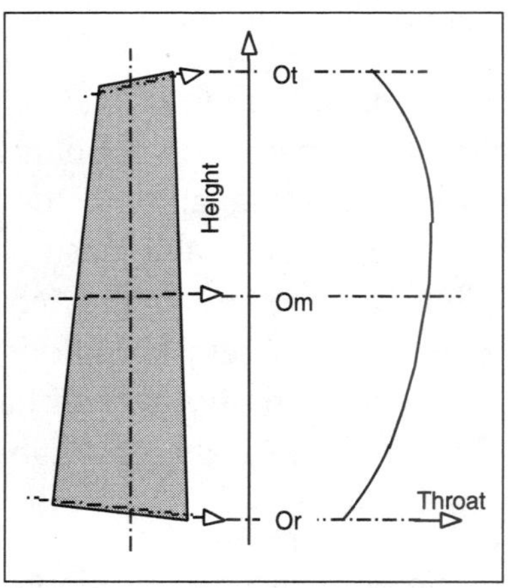

Fig. 2.13.14—Showing the radial variation of throat on a vortex blade.

Swallowing capacity. A blade row consists of a number of elements, each placed relative to adjacent profiles sufficient to complete the 360° or design-specified portion of the total arc for some stationary rows. If each vane is correctly placed, then each of the expansion passages will accept an equal amount of the fluid in the row—expressed as the total flow divided by the number of expansion passages comprising the row. The quantity of steam admitted to individual passages is defined as the *swallowing capacity* of that passage. The sum of the swallowing capacity of the individual passages comprising a row is the total flow in the row less any quantity of steam that bypasses due to leakage. This is taken up in the following section.

THE EFFECT OF VANE PLACEMENT ERRORS

We have considered the blade cascade and the characteristics required to define it, as well as how the parameters of the cascade are able to influence the total performance of the stage and directly impact the efficiency of energy conversion.

However, because various manufacturing techniques are employed to produce steam path components—and because of varying engineering tolerances available by design to allow the components to be produced at acceptable costs—variations in axial setting ("δa"), pitch ("δP"), and setting angle ("$\delta \xi$") are expected as part of a normal manufacturing process. Variations within design tolerances are anticipated, and the design should have sufficient margin in the predicted performance so the unit can reach guaranteed efficiency levels.

It is inevitable that components will be manufactured and assembled exceeding these tolerances. It is the responsibility of the design function to evaluate and recommend corrective actions, if these are considered necessary. It is also possible operating problems will modify the geometry of the stages and stage components. When it is available for inspection it is normally the responsibility of the operating engineer to consider these and their potential effect on the unit.

Various forms of errors can be introduced into the steam path, and it is necessary to consider the effect and extent to which these are acceptable. Design tolerances of the profile and the cascade must be based upon what can be "tolerated" before the quality of the cascade energy conversion ratio deteriorates by an unacceptable amount. Tolerances must not be set to accommodate manufacturing techniques and what can be achieved. If a particular technique cannot produce adequate accuracy, the technique must be changed—not the tolerances.

Unfortunately, there are no simple rules that can be formulated to determine acceptability. The extent to which errors are acceptable depends upon a number of factors, including the form of the profile, the turning angle, and a number of intangible factors specific for the row. The three most important are discussed below:

Axial placement error

The effect of axial placement has been considered, and Figure 2.13.12 shows proud and recessed blade vanes within both the stationary and rotating blade cascades. For both rows, it shows acceptable tolerances of "+da" and "-da." What have not been considered until now are the consequences of these errors on the parameters of the cascade, and what would cause such errors to occur. The most common conditions to induce errors follow:

Vane position on the root platform. A rotating and some stationary blades can be considered to consist of two principle portions—the vane and the root. They are connected through an integrally connected portion known as the root platform. It is necessary during the manufacturing phase to ensure the vane is located on the root platform in the correct spatial position relative to the adjacent elements, forming an acceptable expansion passage.

Figure 2.14.1 shows a vane drawn relative to the root platform. If it is assumed the platform is located correctly above the root load bearing or load transfer surfaces, then the vane must be positioned so the distances "dy1" and "dy2" from the vane inlet and discharge edges are correct. Therefore the vane is axially correct. In these positions the vane will have a total width of "Wv," and the root platform a total width of "Wr." It is possible that on many roots the distance "Wv" is equal to "Wr." Errors can still be encountered, though normally to a lesser degree.

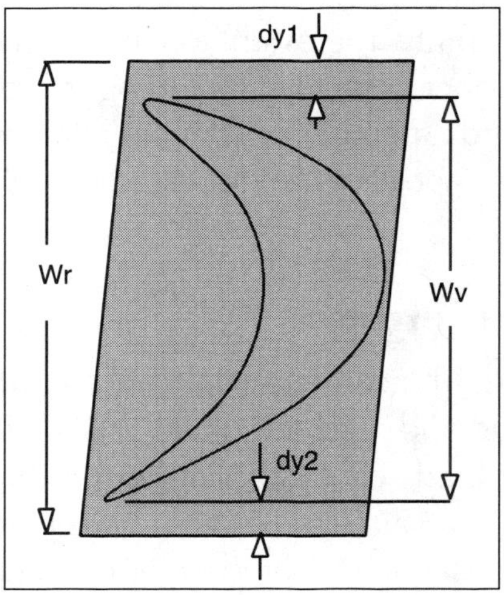

Fig. 2.14.1—Vane profile placed on a lozenge type root platform.

If however, the vane "S" is axially misplaced by an amount "e" (shown in the partial cascade of Fig. 2.14.2, with three vanes: "R,"

"S" and "T"), and the neighboring vanes "R" and "T" are correct, then the expansion passages formed will be incorrect. Throats formed with both adjacent vanes "Ors" and "Ost" are incorrect. Their total discharge throat (and therefore, their area) may be the same and within design tolerances, but their discharge angles will be modified (because the throat has changed) while the pitch remains substantially correct.

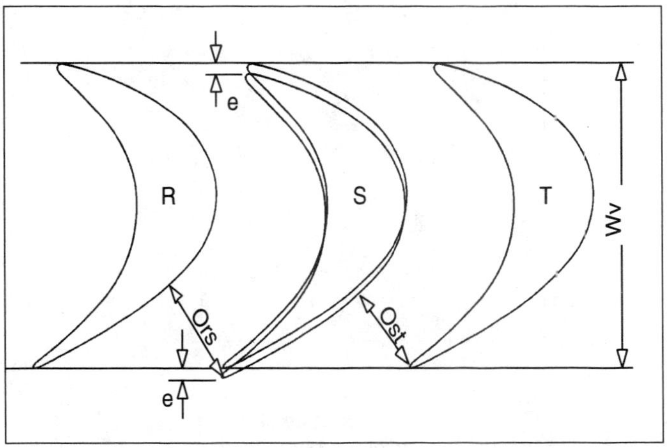

Fig. 2.14.2—Three vanes "R", "S" and "T" placed in cascade, the center vane "S" being misplaced axially.

For stationary elements, it is possible a larger tolerance can be accepted in terms of the vane movement from the theoretical axial position (inlet and discharge edges), provided all elements in the row are affected or in error to the same degree. However, with rotating elements, excessive axial tolerances cannot be applied and accepted, even if all elements have the same degree of error and would produce an acceptable expansion passage with adjacent elements on assembly. These more stringent tolerances on rotating elements are necessary because vane displacement on the root platform will introduce a centrifugal bending moment into the blade, leading to increased root and vane stresses. This effect is shown in Figure 2.14.3, where the center of gravity of the vane "Gv" (a distance "Y" from the discharge edge) has been axially displaced on the root platform by an amount "dY." This introduces a bending moment "M.dY," where "M" is the total centrifugal load of the vane.

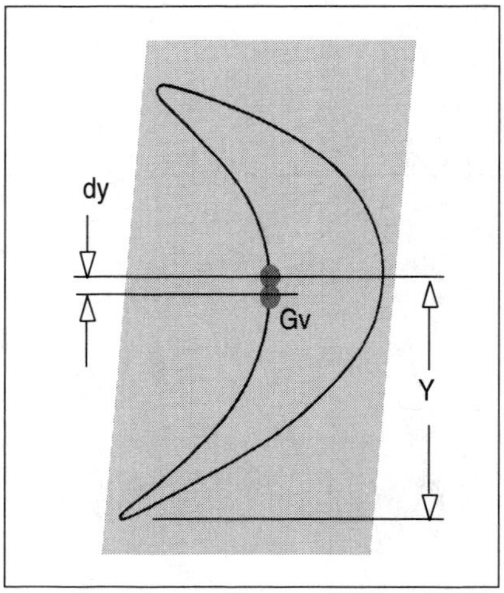

Fig. 2.14.3—Showing vane center of gravity "Gv" displaced from the root platform.

Root form position. The axial relationship of the root load-bearing surfaces to the center of gravity of the vane "G" is shown for a tangential root form in Figure 2.14.4. A theoretical line ("G-Gv") is shown passing through the center of the root profile and then through the vane "Gv -G." In Figure 2.13.10, it is shown that the vane profiles are "stacked" above their common center of gravity ("G"). This position ("G") lies on the radial line "G-G," shown in Figure 2.14.4 as "Gv."

In the manufacturing of the blade, the root form is produced with distances from the line "G-G" to the three pairs of load-bearing surface center points. "Lul" to "Llr" is the position at the upper "u," middle "m," and lower "l" radial positions on the left "l" and "r" of the centerline. These distances must be controlled, which is relatively uncomplicated as the root form is cut with a form cutter producing all six surfaces simultaneously. What must be assured, however, is that the vane is positioned correctly, at a maximum distance specified as "dax."

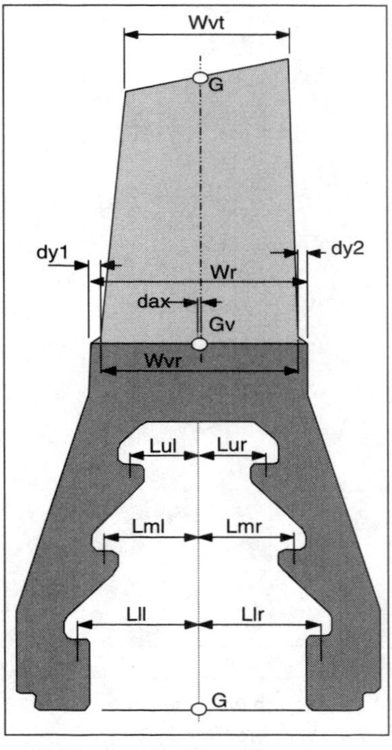

Fig. 2.14.4—Showing the relationship between the vane and root load bearing surfaces to a line through the center of gravity.

This correct root profile placement is assured by ensuring the root form cutter is positioned correctly as it passes through the root block. This root form is normally produced before the vane is machined and in fact is used to locate and hold the material block during vane machining. However, if machining the vane first produces the vane portion—which can happen with envelope or precision forgings—then care must be exercised in positioning the vane before cutting the root.

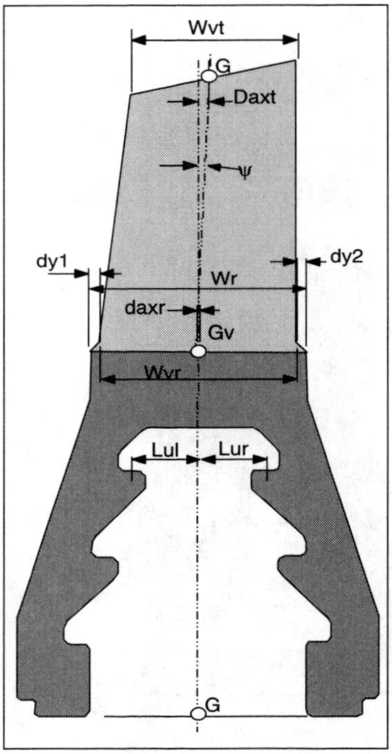

Fig. 2.14.5—A blade with the vane tilted in the axial direction by an amount "Dax" at the tip section.

Vane axial lean. In addition to vane misplacement on the root platform, it is possible for the vane to be inclined in the axial direction. This effect is shown in Figure 2.14.5, where the vane—having the same axial location at the root section and being offset in the axial direction by an amount "dax"—is also tilted at an angle "ψ," giving a total offset at the tip of "Daxt." The root offset is defined as "daxr" and the tip movement in the axial direction is "daxt-daxr." The net effect of this tilt can be to make the root axial position entirely satisfactory at the root section but beyond limits—or certainly with a greater total axial displacement at the tip.

This axial lean will also produce a bending moment in the blade increasing stress levels. Depending upon the extent, axial lean will induce higher stress levels in the blades and, in the majority of vanes, will tend to cause a greater level of distortion to the expansion passage forms. Each leaning blade affects two passages.

Vane untwist. As the unit starts up, the speed and centrifugal loading of the vane increases. Figure 2.14.6 shows two sections of a vane mounted in their final manufactured position above the center of gravity (or stacking position "G" at axial "A" and tangential "T" locations). The effect of rotor rotation is to make the vane "untwist" in a counterclockwise direction ("M"), so the tip section "Ti" rotates about "G" in an effort to achieve a position above the root section "Ro." If the vane is unconstrained, the extreme position of the tip section discharge point would move from coordinates "X," "Y" by amounts "dX," "dY." In untwisting, the vane will experience a turning moment in the vane equal to "M." This untwisting will modify the tip form, causing a modification of the throat and therefore the throat/pitch ratio that defines the discharge angle.

With freestanding blades, the only constraint to this modification of section position is the resistance of the blade material to the bending moment set up by vane rotation. The untwist is calculable. The vane is therefore calculated with pre-twist built in, so that during operation the profiles along the vane height will rotate to the position required by design to give the throat distribution required as a function of height.

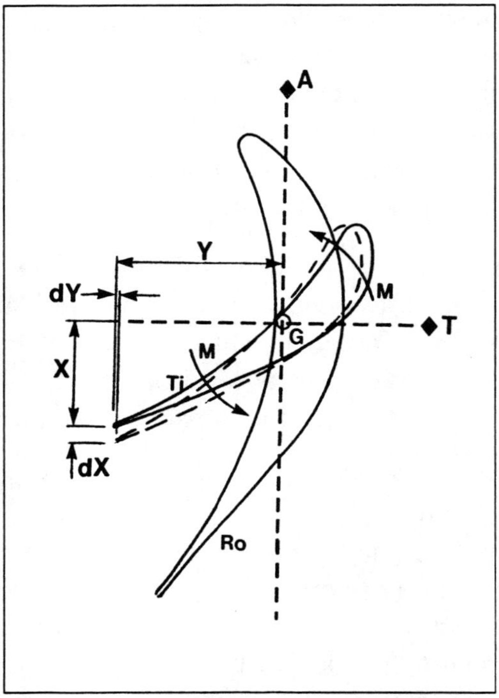

Fig. 2.14.6—A vane with tip intwist, about its center of gravity.

For those blades with coverbands (or tie wires) at their tip—attached to the vanes in such a manner they resist untwist—the situation is somewhat different. The untwist will experience considerable resistance and the vane will move only by relatively small amounts. Therefore, the blade can be manufactured with the throat distribution required during operation.

Coverband distortion. As we just mentioned, the coverband can resist untwist of the blades, reducing the bending moment and the stresses in the vane due to this constraining effect. However, it is also necessary to recognize that the coverband can also induce movement or displacement in the vane radial position. Various effects can cause this displacement:

- *Tenon hole in coverband misplaced*—The production of the tenon hole in a coverband (attached by riveting) is an operation that can be done on-site if the bands are not pre-punched. This tenon hole production is undertaken using a "punch and die." Any error in locating the hole will require the vane to be distorted by some small amount to facilitate assembly. In small radial height blades it is unlikely the vane can be adjusted, but in the longer elements this is entirely possible
- *Vane pitching*—The vane tips on certain elements (in stages using tangential entry blade roots) may have their pitches removed from the design value to facilitate blade window closure. If pre-punched coverbands are used, it is possible the vanes be distorted to allow this assembly. Again, considerations of the effect of the vane and its ability to be deflected allowing assembly must be considered
- *Incorrect tenon position*—The production of a tenon at the blade tip is normally a separate manufacturing operation. If some condition exists that will allow the production of a rogue tenon on one blade element (due possibly to the inclusion of other debris in a jig), this can cause error in the positioning of the tenon

Coverband distortion is not a common occurrence, and often occurs when clearance between the tenon and coverband hole assembly is achieved. However, it must be remembered that as the tenon is formed by the riveting operation, there will be some degree of vane distortion as the gap between the tenon and hole is closed.

Method of vane location. The method of locating the vane within the steam path can influence the axial location. When vanes are attached to a root through a platform, it is relatively easy to ensure the correct axial position within design tolerances. If, however, the vanes are located as part of a fabrication or casting process (as often occurs on many stationary row elements), then axial misplacement is common. When establishing tolerances with these forms of construction for stationary rows (the diaphragms), the designer recognizes these vanes are of a form permitting adjustment after completion of the assembly or manufacturing process.

Quantifying "acceptable" axial placement errors is extremely difficult. In addition to the requirements of achieving a throat that will contribute to and control the discharge area, there are considerations of the expansion passage divergence. In general, the greater the turning angle "θ," the more precise the axial pitching must become. Consider the three profiles "R," "S," and "T" in Figure 2.14.2. With a large turning angle, as profile "S" moves downstream, the throat at the entry to the expansion passage (formed between profiles "R" and "S") will close down at the inlet, causing a passage that is divergent throughout its length. This is unacceptable.

Pitch error

The theoretical pitch at any diameter can be determined from the diameter being considered and the number of blades in the row (See equation on page 246). The actual pitch at any position on a blade row will often differ from the theoretical value. It can be expected and accepted if within tolerances.

The effects of pitch error on an individual cascade needs to be considered in terms of their effect on the passage shape and the discharge area from the individual throats. It is obvious the total pitch around a row must be equal to the sum of the tangential pitches of the individual blades. A "plus error" in one pitch must be countered by a "negative error" in another—or in a series of "others."

Consider the three vanes "R," "S," and "T" in Figure 2.14.7. The normal pitch "Pn" is shown for the three profiles. The design value for profile "S" is shown in the design position. As a pitch error occurs, the pitches modify to "Pr-s = Pu," and "Ps-t = Po." The sum of these pitches remains the same as the original pitches. However, the distribution has changed. In fact, in a complete blade row there may be no pitches at the design value, but their mean must always be equal to the design.

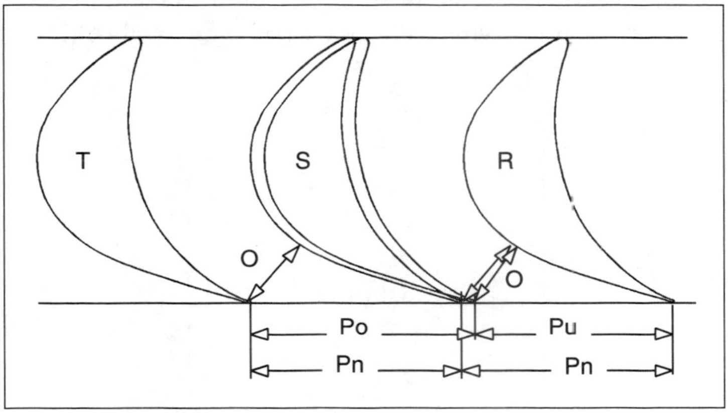

Fig. 2.14.7—Three vanes with the center vane off pitched.

The throats formed from discharge tail to adjacent suction face will also vary with pitch. The form of the throat and the ratio "O/P" will depend upon the shape of the suction face. It can be seen that the point on the suction surface at which the throat originates provides a clear indication of the steam discharge angle. Consider the two forms of discharge tail suction face profile that can be used—straight suction face and curved suction face—and their effect on the discharge throat and angle:

Straight suction face. The suction face at the point of minimum discharge width is formed on a straight surface, as shown in Figure 2.14.8. In this case the value of the discharge angle ("αo" or "$\beta 2$") is unaffected by pitch changes. There is, however, a small change in the ratio O/P, when the effect of the tail thickness "b" is taken into account. However, the value of "b" is sufficiently small that change can often be ignored.

Large pitches are the one exception to this constant "O/P" ratio. In these cases, when the throat is formed of the radius "R1" for the value of pitch, the throat no longer follows the linear relationship. This condition will often occur on constant profile vanes (cylindrical profile) at tip sections where the pitch has increased sufficiently.

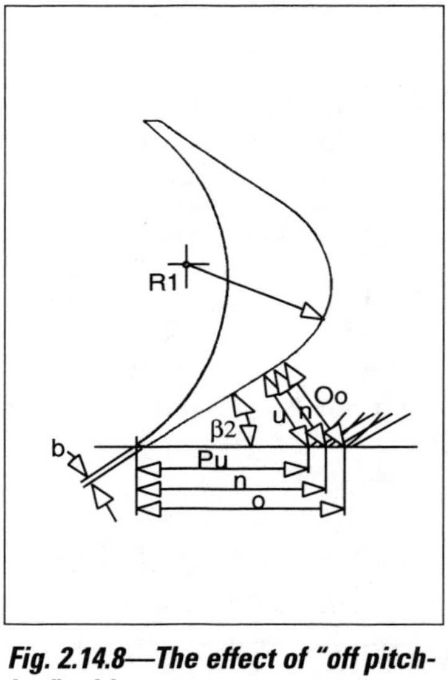

Fig. 2.14.8—The effect of "off pitching" with a

Curved suction face. If the discharge portion of the suction face is curved (Fig. 2.14.9), the throat is formed at a different discharge angle, providing an angle varying from "β2u" to "β2o" as shown by the tangent lines. Therefore, the effective throat and the ratio "O/P" are functions of the pitch, for all values.

This form of vane is therefore more susceptible to the discharge angle variation and is dependent upon the accuracy of manufacturing pitch. It can also be seen by inspection that the discharge curvature of the tail suction surface should be kept as large as possible minimizing the rate of angle change.

There are a number of contributing factors causing pitch error. These may occur individually or in combination. The most common of these are listed here.

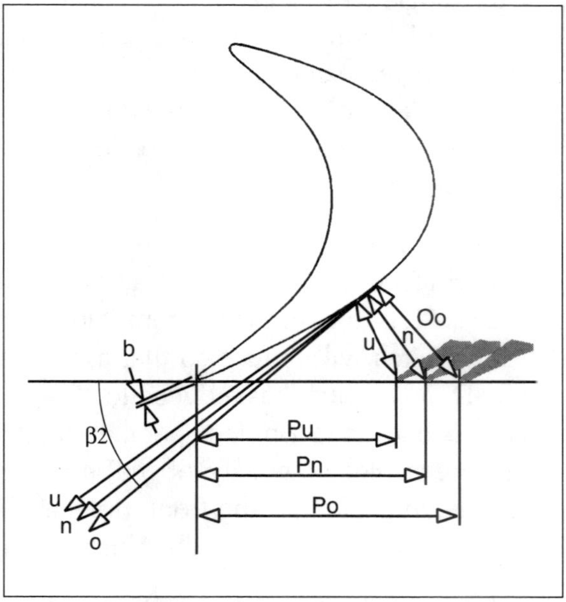

Fig. 2.14.9—The effect of "off pitching with a curved tail blade.

Root block thickness-tangential entry roots. For blades that enter the wheel in a substantially tangential direction, and are contained on root blocks that are in face contact on the wheel, the tangential pitch of the block will establish the pitch at the root section. Therefore, if the vane and root are correctly aligned, the pitch will be correct along the radial height.

Root block thickness-radial entry roots. In addition to the tangential entry blades, some radial entry pinned roots can have small tangential clearances between the blocks, or alternately mate to adjacent surfaces. There is the possibility that pitch errors can be introduced if the pitches of the individual blocks are not correct, or if having a small tangential pitch, the gap between the blocks is not maintained at design values.

Root block wedge angle. For tangential entry blades, the vanes are produced on a root platform, which has a plan shape suited to joining the vane to the greatest extent possible over its entire perimeter. In general, the root platform of these tangential entry blades is manufactured so there is adequate contact between adjacent surfaces when they are assembled to the wheel.

Consider a section through a root block (Fig. 2.14.10). Here the pitch at the root platform face is shown as "Pr," and at the root block has a total depth "Hx." As the distance from the outer surface decreases, the effective pitch will reduce so that at the base of the blade root diameter "Da," the pitch has reduced to "Pa." It is normal for the block to have its back (suction) face parallel to a line "G-G" that passes through the center of gravity of the vane profiles. To achieve the correct wedge angle "χ," the front (pressure) face must converge to a reduced pitch at the base of the block.

The value of "χ" can be determined from the number of blades in the row.

$$\chi = \frac{360}{Zb}$$

where:

Zb is the number of blades in the 360 of the row

It is important to consider the required accuracy of the wedge angle to ensure the blade pitch is not adversely affected by any errors. In assessing what this tolerance should be, it is necessary to

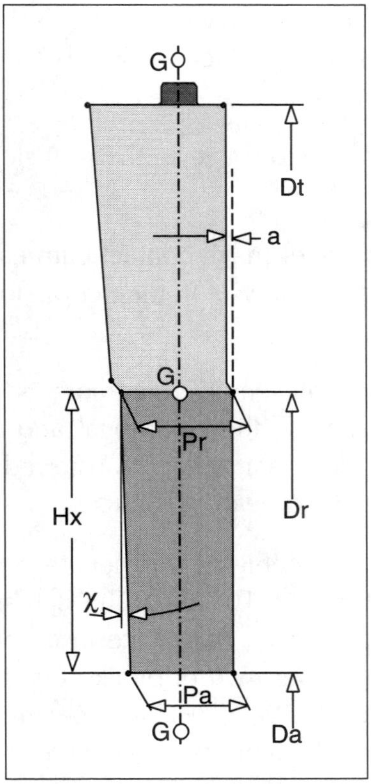

Fig. 2.14.10—A section through a blade vane and root block showing the alignment between the vane and block.

consider the functions of the root block relating to blade spatial requirements. There are four basic requirements to be considered:

- The blade vane be held in the correct radial alignment relative to both adjacent elements, and the radial line describing its center of gravity passes through the center of the rotor
- The root block be held at the wheel or rotor position so that it is unable to make any movement in the tangential direction.

Such movement occurs when there is inadequate contact between the root block faces—there is contact at one diametral position only

- The root is unable to rotate in the root slot by unacceptable amounts
- The root block faces make contact eliminating vibration and reducing the possibility of fretting corrosion in the interfaces positions

In addition to these placement requirements, it is also essential the root block be placed so the centrifugal and other loads developed in the vane during operation are transferred to the rotor evenly over any number of load bearing surfaces.

To achieve these requirements it is necessary for the root block angle "χ" be held at the +/- 0.10° level. This can appear to be excessively tight, but it represents a difference in the values of "Pa" and "Pr" of only +/- 0.0017" per inch of block height between the two. Therefore, if the pitches "Pr" and "Pa" are specified to a tolerance of design value of +/- 0.001"—a normal tolerance for blade pitch—then the root angle should be acceptable.

As an example, consider the blade root block shown in Figure 2.14.11, which is one of 120 blades in a row. The design diameters are shown as Da=32.14" and Dr=36.84." The blade root block is therefore 2.35" high.

At the extremes of tolerance, the wedge angle "χ" will vary between:

$$\chi_{max.} = \tan^{-1} \frac{0.9655 - 0.8404}{2.35} = 3.047^\circ$$

$$\chi_{max.} = \tan^{-1} \frac{0.935 - 0.8424}{2.35} = 2.950^\circ$$

The actual permissible angular variation in "χ" is from design 3.047° to 2.950°. These values are well within the engineering tolerance band of +/- 0.1° from the 3.0° design value.

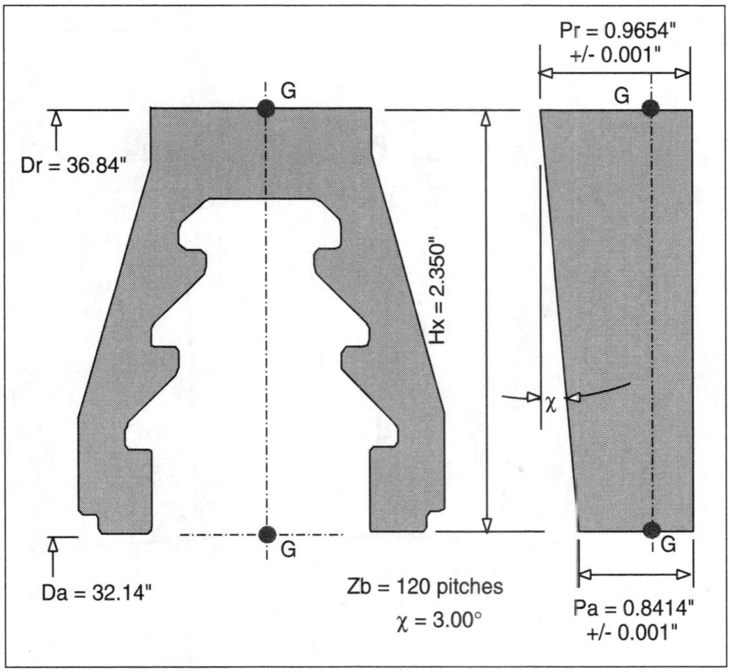

Fig. 2.14.11—The root geometry and block form.

Setting angle. The setting angle of the vane can have minor effects on the pitch between blades. This is not a major factor, but if there is a variation of setting angles between adjacent elements because of the method of manufacture, their pitch can be affected.

Vane position on the root platform. The vane can be misplaced on the root platform. This is not common, but for certain methods of manufacture there can be errors effecting tangential placement.

Vane lean. In addition to having the possibility of being misplaced on the root platform, the vane can also be produced with a lean. This lean can occur in either the tangential (pitch) or axial direction.

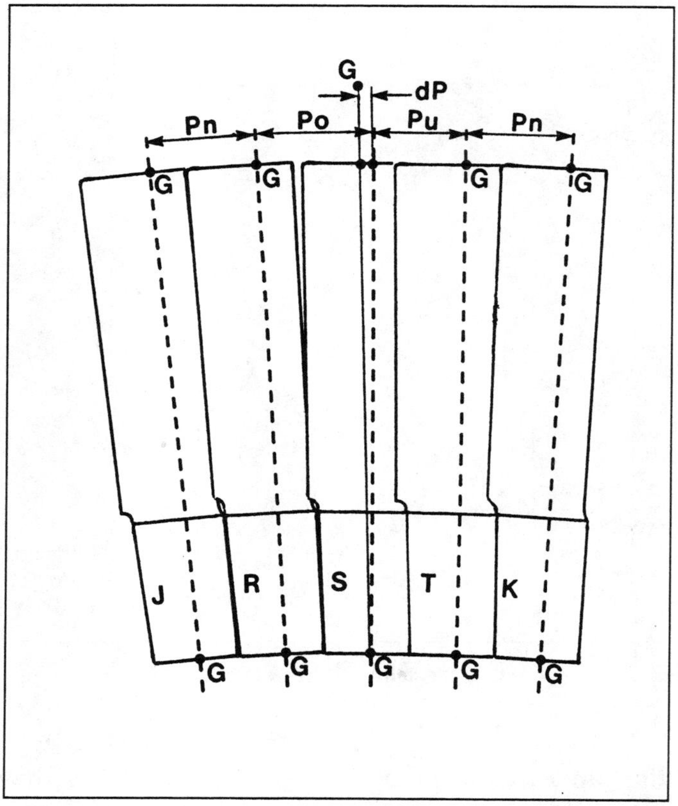

Fig. 2.14.12—The effect of vane lean on blade "S". The variation of tip pitch can be seen as error "dP".

The effect of this tangential lean is illustrated in Figure 2.14.12. Here, three blade vanes ("R," "S," and "T") are shown in their final positions as mounted on a rotor. Also shown are two additional blades ("J" and "K") adjacent to this group and pitched correctly. Of this three-blade group, the two outer vanes ("R" and "T") have their vanes in a correct radial alignment ("G-G"). However, the center vane ("S") has been produced with a tangential lean towards "T" by an amount "dP" at the tip, as shown. This lean produces tip pitches of "Pr-Ps=Po" and "Ps-Pt=Pu." The effect of this off pitching is shown in Figures 2.14.7 and 2.14.8. Figure 2.14.7 shows the effect on discharge throat and angle for vanes having a curved discharge suction face. In addition to the effect on passage shape, additional bending moments are introduced and the consequential stresses into the vane.

When it is determined that vanes are out of radial alignment after assembly, considerations of this error need to be reviewed. These considerations may be of no consequence, however, they can have the potential to impact on the performance of the unit and require detailed evaluation. These consideration include the following:

Pitch error vs. height. The vane can be off-pitched at one radial location but may also be bent. The implications of this are that the pitch can be within tolerance over a major portion of its radial length and then be outside over the remaining portion. Therefore, when an off-pitch condition is found, it could be advisable to consider this possibility, particularly with longer blades formed by manufacturing methods that induce bending and twisting, or blades that have employed a field repair involving the use of heat, which causes some forms of distortion.

The possible effect of the coverband. A coverband can cause pitch distortion, which can be localized over a portion of the total length. This condition occurs when a blade has tip distortion and the vane is bent to allow assembly. This is a complex issue, as the vane will have bending stress induced in it to facilitate assembly. However, the vane would also be stressed by the off-pitch condition and possibly in operation the bending stress could be reduced as the vane attempts to attain a true radial position. There is normally sufficient margin in the factor of safety that this should not be a concern.

Vane untwisting if straightened. If a vortex vane is found to be bent in the pitch direction, the effect of straightening it may also cause the vane to untwist and affect the setting angle and expansion passage form, as the vane will bend in the tangential direction by untwisting about its axis. Therefore, bending to achieve coverband assembly should be limited by the amount of untwist this introduces.

In an actual design, the designer is aware of the inevitability of vanes being off-pitched, and allows for a margin in both stresses and efficiency guarantee. However, there are limits placed on this effect and should elements be outside these after assembly, then it is necessary for the designer to examine and rule on their acceptability or define the corrective action required.

Method of manufacture (casting and welding). The methods of manufacture can have a considerable effect on the final pitch between the various elements, as well as the methods of correction available to conditions that exceed design tolerances. The rotating blades are normally manufactured with considerable care, and the possibility of errors after assembly is minimized to the greatest extent possible. Stationary elements manufactured for direct assembly to a blade carrier or inner casing—and manufactured to the same dimensional criteria as rotating blade elements—tend to meet design criteria within close tolerances. However, those stationary elements—particularly for diaphragm stages constructed and attached to inner and outer bands—often employ a casting or weld fabrication process. These methods tend to introduce greater variation in the pitch, but they also allow a greater degree of corrective measures after completion of the primary manufacturing process.

The steam path dimensional audit

At completion of manufacture and assembly to the rotor or casing, it is often useful to undertake an audit of the major blade dimensional values that have the potential to affect the performance (efficiency and structural reliability) of the unit.

Figures 2.14.13 (a), (b), and (d) show the measured values of pitch and throat at the tip and mean diameters and radial discharge height of 14 elements in a blade row, having a design radial height of 12.08". In (a), the measured values from the tip section of these blades are shown.

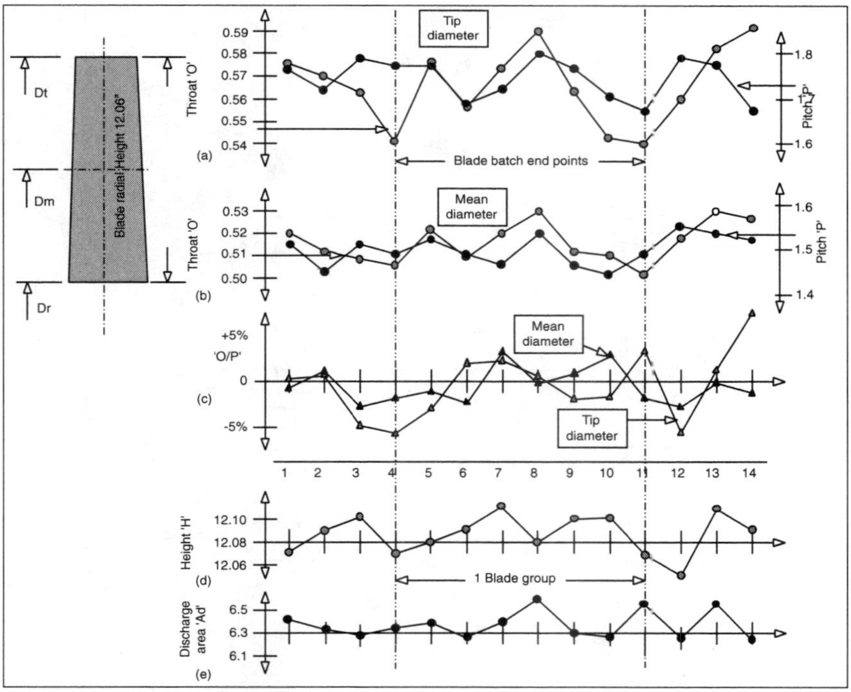

Fig. 2.14.13—The results of a partial audit on 14 throats formed between 15 blades in a rotating blade row.

They are grouped by a coverband connecting blades and having coverband discontinuities, batch end points, at throat 4 and 11, the throat and pitch are shown. In (c), the ratio "O/P" for this tip section "t" is shown. It can be seen that there is considerable variation of this ratio well outside what might be considered an acceptable range. Similar values of throat and pitch are shown in (b) for the mean diameter "Dm," and again in (c) the ratio "O/P". These values are more realistic, indicating these blades may have become bent or twisted in their outer sections.

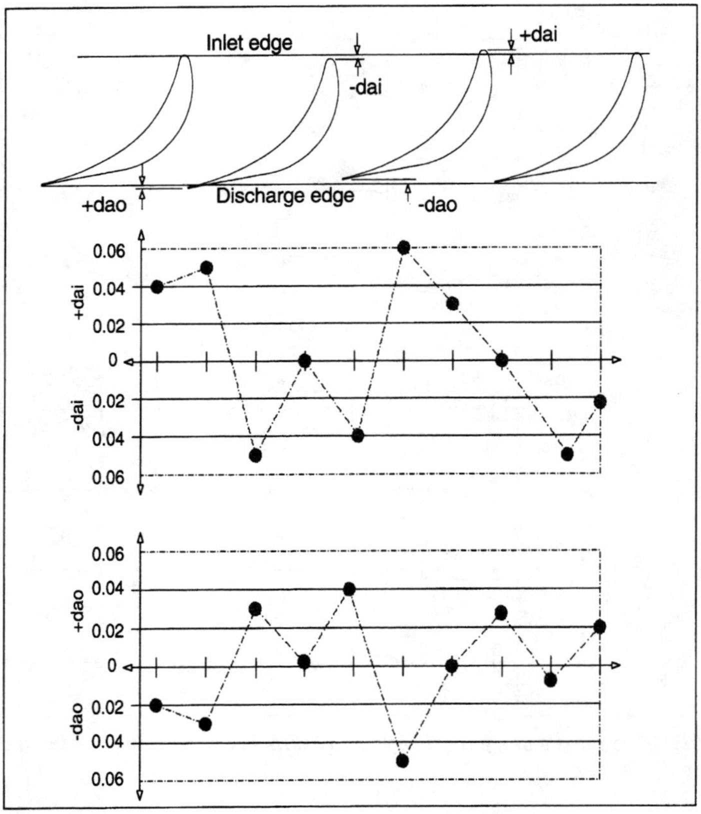

Fig. 2.14.14—Measured values of blade axial position relative to a mean inlet and discharge edge.

A similar audit can be made of the inlet and discharge edges, showing the possibility of "proud" and "recessed" blades. This effect will exist to a certain extent on all blade rows, but is normally significant only on those longer elements where the ability to, and probability of distortion exists. Figure 2.14.14 shows the measured values of ten 17.27" blade elements in a freestanding row. These measurements were taken at the inlet and discharge edges at a distance about 1.0" below the tip. In a normal audit, it is not known where the actual edges should be located; therefore, the measurements of "+/-da" are relative to an undefined edge. Such an audit can give unquantified information on the quality of blade production but the manufacturer normally has standards for "da," which should be met.

Such audits are not expensive to conduct, and are required only if a visual inspection suggests problems and excessive variations from design tolerances are present. In the case of a rotating blade with an attached erosion shield, it is possible some deformation will occur as a consequence of the heating process as well as the actual position of the shield on the vane.

Setting angle error

The setting angle "ξ" shown in Figure 2.13.3 is a critical characteristic of the row, as it defines the form of the expansion passage and therefore the potential efficiency of the row. Unfortunately, the setting angle is a characteristic which is relatively difficult to gauge because of possible small errors that may occur in form at the inlet nose and discharge tail. It is at these locations where relatively small changes of form can indicate an error in the setting angle of a vane which is, in fact, properly set. However, there are errors within the individual stage components that can occur and do affect the setting angle producing an expansion passage that may not meet design requirements within tolerance.

Many of the types of errors discussed previously also affect the setting angle "ξ" and introduce errors exceeding the design requirements. There are also other considerations, which include the following:

Root block twist. There are root forms that are entered into, and which during operation are surrounded by the wheel. These internal root forms are of a generic class often called the inverted "T" root. These roots have a central stem that has some small clearance from the stem to the wheel groove sides (Fig. 2.14.15).

The design specification would indicate there is equal clearance on both sides of this stem, and that this is maintained during operation. In fact, when the root block form is a lozenge, there is a tendency for the root platform to rotate and tilt the blades in the groove (Fig. 2.14.15). This twisting has an effect on the total spatial relationship of the cascade formed between the vanes, affecting the setting angle "ξ," the ratio "O/P," the discharge area "Ad," the row width, and many other characteristics of the row.

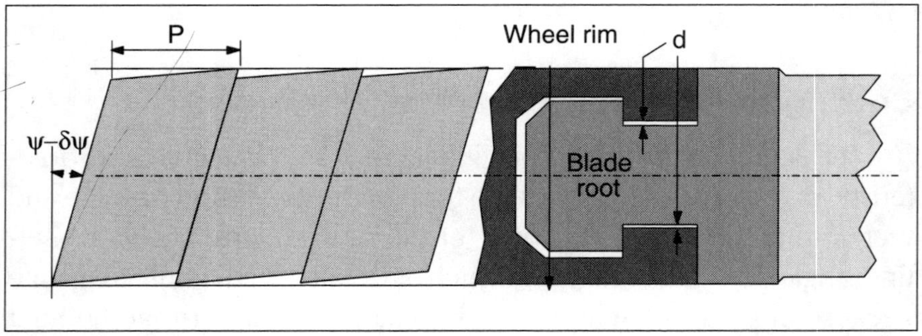

Fig. 2.14.15—Showing the effect of a lozenge root untwisting in the wheel due to gap "d". This untwist will rotate a blade through an angle "δΨ".

Blade untwist. The effects of blade untwist were considered previously, and the possible effect on pitch is shown in Figure 2.14.6 (assuming the vane were to rotate about its center of gravity "G").

However, no consideration was given to the effect of this untwist on the setting angle and expansion passage form.

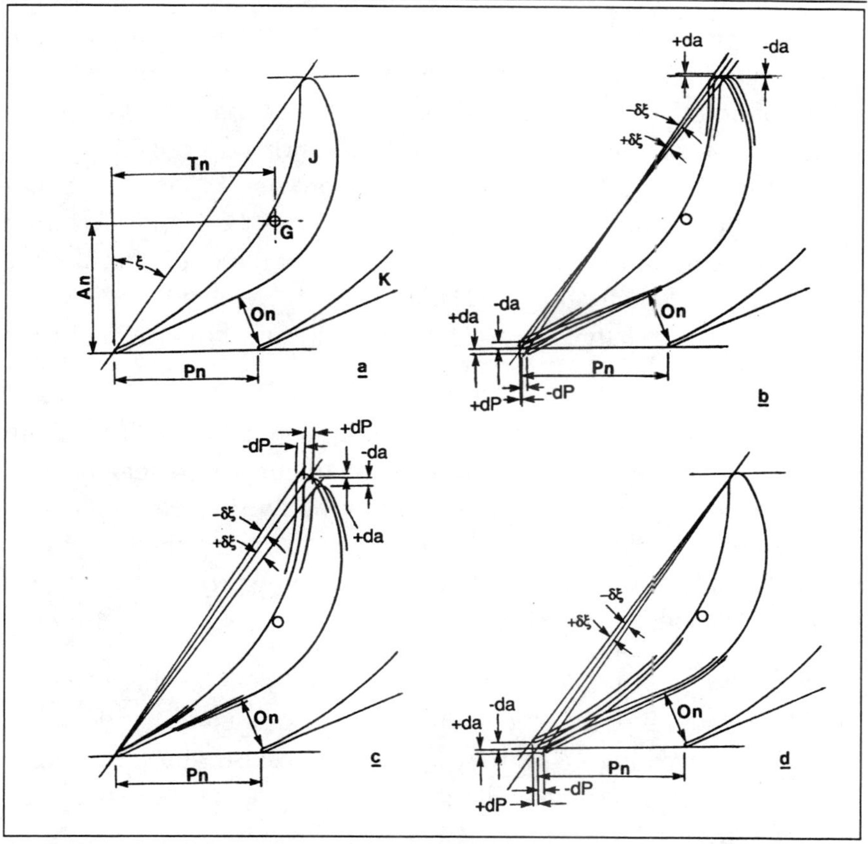

Fig. 2.14.16—The various forms of error which can occur in a cascade. Showing in figure 10.6.16(a) is the design condition.

Figure 2.14.6 considers one such profile. However, when examining the effects on setting angle, it is necessary to consider the effect of multiple blades with untwists, and if the untwist is caused by some phenomenon such as a heating process. In each case, there is no reason to believe each blade will untwist to the same degree. The effect of untwist on the setting angle is somewhat more complex.

Conditions can only be considered in general. Each condition will tend to be different and dependent upon the untwist of two vanes.

In terms of expansion passage form, convergence rate, design-setting angle, and discharge area and angle, it is necessary to consider the point on the profile about which the vane has rotated. This point of rotation can affect all of the above parameters. Figure 2.14.16(a) shows the design-specified condition of profile "J" and its position relative to the discharge tail of profile "K." These profiles have center of gravity position "G" at a distance "An" and "Tn" from the discharge point on the profile. For this condition, the form of the expansion passage is shown in Figure 2.13.13, which is convergent from inlet to discharge with a final throat "O" shown as "On" in Figure 2.14.16.

If the vane "J" rotated about some point within the cascade, there are an infinite number of variations of the major parameters that can be achieved, due to the position of both the twist in profile "J" and also movement in profile "K." Consider the following three cases, in which considerations are simplified by assuming profile "K" remains in its design specified position:

- *Profile "J" rotates about the centroid "G."* In this situation [Fig. 2.14.16 (b)], there will either be an increase or decrease in the setting angle by "dξ." In either case, the form of the expansion passage will modify, causing a change in the pitch and throat
- *Profile "J" rotates about the discharge tail.* In this case [Fig. 2.14.16 (c)], there will be a change in the throat, but these changes will have a smaller effect on the throat value. The width will either increase or decrease by an amount "da," depending upon the direction and extent of twist
- *Profile "J" rotates about the inlet nose.* In this change, rotation occurs about the inlet nose [Fig. 2.14.16 (d)], the conse-

quence of which can be an exaggerated change in the throats for any degree of twist. Again, the width will change by amounts "+/-da," depending upon the direction of twist

In establishing the extent to which twist is acceptable, the designer should avoid any condition which does not ensure that the expansion passage remains convergent from inlet to discharge (unless a converging/diverging form is required by design). The extent to which twist is acceptable is established by the design specification.

The profile shown in Figure 2.14.17 is for a low-reaction vane, normally defined as an impulse design having a small degree of reaction at its root section. In this design, the ratio in the inlet throat "Oe" to discharge "On" is equal to 1.40. In such a cascade, any relatively small change in the setting angle has the potential to modify the passage shape and convergence to the extent there is a considerable loss in stage efficiency.

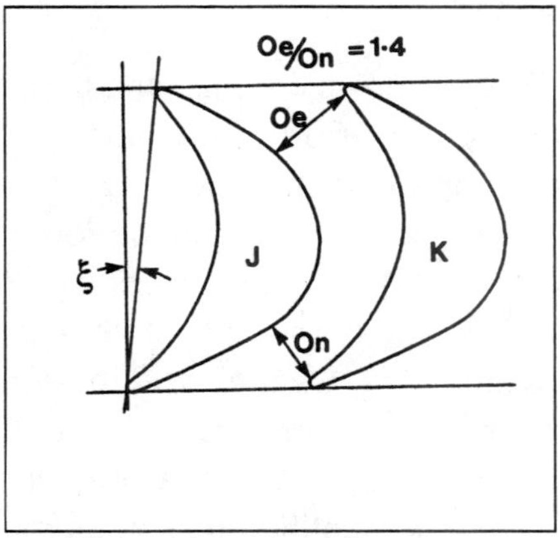

Fig. 2.14.17—The divergence ratio in a blade row.

Vane lean. The effect of blade vane lean was considered earlier in terms of its possible effect on axial and tangential (pitch) distortion. It is also possible for lean in the axial and tangential direction to have an effect on passage form. In general the greater the vane turning angle "θ," the less it is able to accept lean in either direction.

- *Axial direction*—In the axial direction, the lean will modify the discharge throat and, therefore, the area. If a rogue blade exists in a total row, the increase in throat of one passage is countered by the closing of another. However, the most common situation is a majority of blades in a row proud or recessed from the theoretical discharge edge (Fig. 2.13.12)
- *Tangential direction*—Lean in the tangential direction will produce a condition in which the passage throat will increase or decrease, depending upon the direction of lean. This form of error can distort the form of the expansion passage to the extent a diverging/converging condition could exist. The greater the vane turning angle—and the throat inlet to discharge ratio approaches 1.0 (Fig. 2.14.12)—the greater this extent can be

These three forms of error—axial placement, pitch, (refer to previous text concerning pitch error), and setting angle—have been shown to affect many important characteristics of the cascade blade row. It is not possible to manufacture a blade row eliminating these errors at an acceptable cost. The design function recognizes this, and establishes manufacturing and assembly tolerances within which variations are acceptable. After assembly, if these requirements are not met, it is the responsibility of the design function to evaluate and rule on acceptability of corrective action. Corrections that are made must be fully evaluated, and qualified personnel must undertake the evaluation process.

The discharge area and angle

The discharge area and angle are determined in terms of the physical dimensions of the vane and their setting at discharge from the expansion passages. These characteristics of the rows must meet design specification or the performance of the unit will deteriorate.

The audit is a meaningful method of checking compliance, but unfortunately, in rotating elements it is not possible to take corrective action. Therefore, the level of compliance is dependent upon the expertise and effort taken to manufacture the individual blade elements. This also applies to stationary blades that are inserted in the blade carrier or casing. For stationary elements, adjustment is often possible (see chapter 7).

The swallowing capacity

The swallowing capacity of a blade passage was previously defined as the quantity of steam that would enter a specific passage, and in a blade row that was perfectly spaced, each passage would carry an equal quantity of the working fluid.

Should there be any misplacement of vanes within the cascade that affects the passage inlets, the quantity of entering steam will be modified. Also, in any row extending for 360°, any error in one profile will modify the swallowing capacity of two expansion passages.

Consider the profiles "R," "S," and "T" in Figure 2.14.18, in which profile "S" is misplaced. The steam enters the row at a relative angle "β1." This profile "S" can be misplaced by any of the conditions shown in Figure 2.14.16. Because the steam enters the row in what might be termed a "division by opportunity," the steam will divide at the inlet nose, with the largest quantity flowing into the largest inlet throat. Therefore, if profile "S" were proud by an amount "+da" (Fig. 2.14.18), the passage formed between profiles "R-S" would receive more steam than that formed by profiles "S-T." The quantities would be reversed if "S" were recessed by a similar amount.

Similarly, a pitch error "+/-dp" will influence the steam quantity entering the individual passages. Setting angle error will also have an effect—the magnitude being a function of the degree of error and the point in the profile about which the vane has twisted. If the profile shown in Figure 2.14.19 is considered (with the inlet angle "β1" shown as 90°), then the influence of proud and recessed blades is not a consideration. However, the pitch and setting angle will still influence the flowing quantity, as discussed, for the blade with a larger turning angle (Fig. 2.14.18).

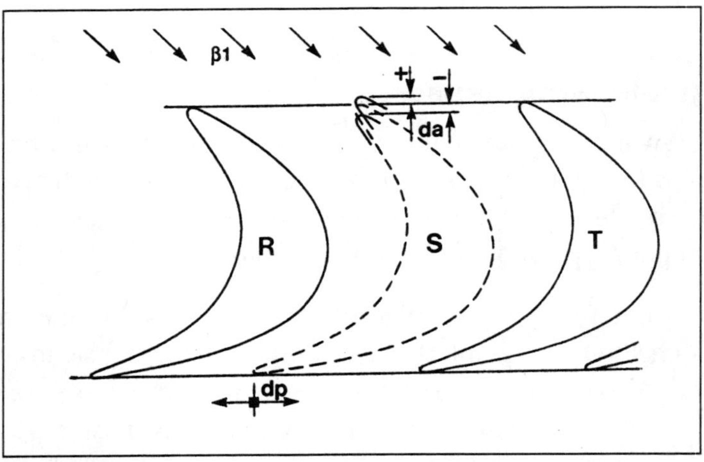

Fig. 2.14.18—Three blades showing the effect of a misplaced profile on swallowing capacity, for a vane with a large turning angle.

There are no general rules that can be developed concerning potential detrimental effects of swallowing capacity variation from design, and each situation should be considered separately. Suffice it to say, variation in the steam quantity entering a passage will influence the performance of a row, and there is a limit set by design establishing the tolerances within which variation is acceptable.

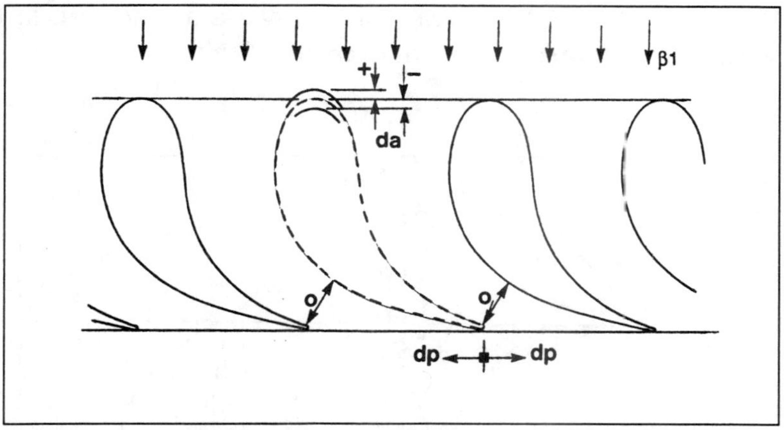

Fig. 2.14.19—Three blades showing the effect of a misplaced profile on swallowing capacity for a typical stationary vane.

Consider the profiles in Figure 2.14.20, in which profile "K" is misplaced as shown (chain line) relative to profile "J." This misplacement can be a combination of errors—"pitch dp," "axial setting da," and "setting angle dξ." These are sufficient to make inlet areas different. With the misplacement shown, area "Ajk" between profiles "J" and "K" is larger than area "Akl" between prof le "K" and "L." Therefore, passage "J-K" has the potential to admit a larger steam quantity. However, the discharge area "Ajk" is smaller. This total effect is to cause considerable variation of pressure and velocity distribution in the various passages.

There are several possible consequences on row performance associated with a variation in the individual passage-swallowing capacity that should be considered.

Steam discharge velocity. The quantity of steam discharging from passage "J-K" is different to that discharging from "K-L." If the discharge areas from both throats were the same, then the velocity from "J-K" would be larger. However, in addition to the larger flow quantity (Fig. 6.14.20), the throat of "J-K" has also decreased relative to "K-L." This reduction in area will also cause an increase in the velocity from "J-K" relative to that from "K-L."

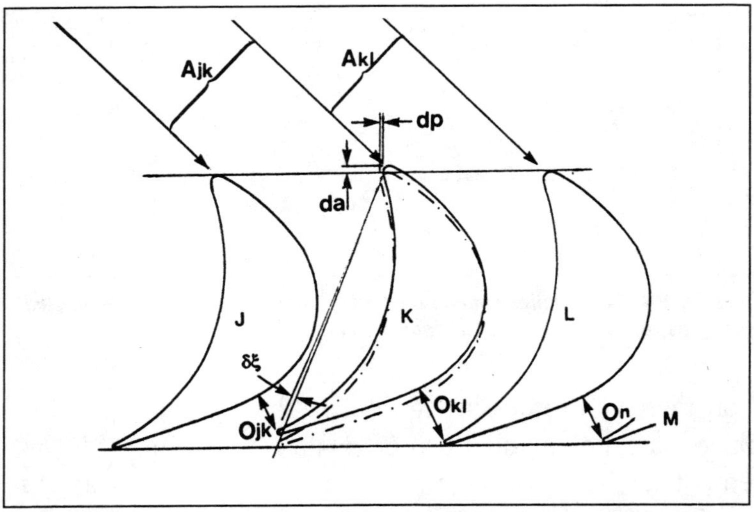

Fig. 2.14.20—A misplaced vane with the resulting stage parameters and inlet flow area.

Steam discharge relative angle. The value of the effective discharge angle is a function of the ratio "O/P." It can be seen that both pitch and throat can vary from the design values, as a consequence of misplacement error, and therefore the discharge angle will also change.

In the situation shown in Figure 2.14.20, the passage "J-K" will have a larger discharge velocity ("Cjk") at possibly a smaller discharge angle than the design. Also, passage "K-L" will have a lower discharge velocity ("Ckl") at a larger angle. The effect of this on steam entering the following row is shown in the velocity triangle for row "Y" in Figure 2.14.21, where the effect of both velocity and incidence in entering row "Z" can be clearly seen.

Steam bending stress. The steam-bending stress in a blade is directly proportional to the quantity of steam flowing across it, and the angle through which it is turned. Therefore, the stress induced in

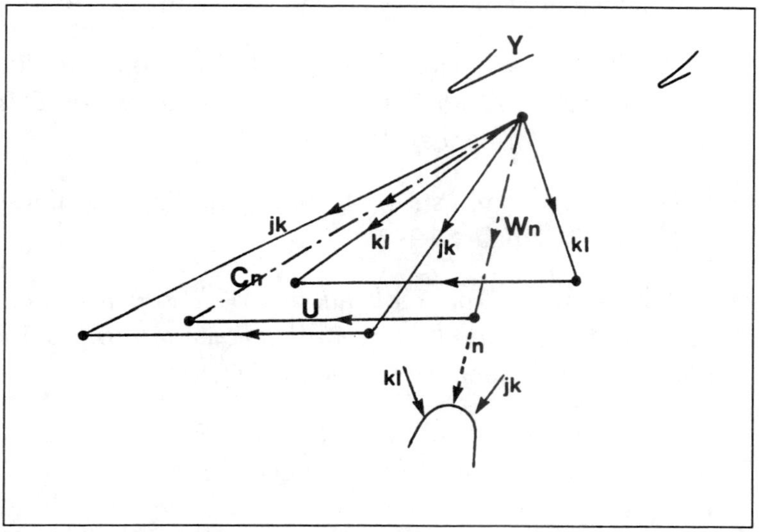

Fig. 2.14.21—The velocity triangles for expansion passages "J-K" and "K-L" of figure 2.14.20.

these blades will not be the same, because the flow quantity varies from one expansion passage to another, and discharge angles are also different from passage to passage. The effect of the discharge angle is relatively small compared to the effect of the changing flow quantity.

REFERENCES

1. Glatt, P., *Laser Measures Centerlines Quickly,* Economically, Turbomachinery International, January/February, 1994

2. Jordan, S., and M. J. Fraser, *Design Modifications and Repairs to Existing Steam Path Components to Improve their Existing In-Service Performance,* EPRI

3. Steam and Combustion Turbine Blading Conference, Orlando, FL, January 1992

4. Hollingworth, K., *The Application of Weld Repair Techniques to Reduce Costs and Outage Time on Steam Turbines,* Turbomachinery Maintenance Congress, Singapore, 1988

5. Rasmussen, D.M., and W.T. Durbin, *Steam Turbine Case Repairs to Extend the Operating Life,* ASME Paper 84-JPGC-PWR-44

6. Sanders, W.P., and I. Schulz, *Steam Turbine Rotor Restoration,* Eighth Turbomachinery Maintenance Congress, Kuala Lumpur, Malaysia, October 1992

7. Jones, G.T., and J. Gunning, *Causes for and Methods of Straightening Bent Turbine Rotors,* Eighth Turbomachinery Maintenance Congress, Kuala Lumpur, Malaysia, October 1992

8. Phonic, T.R., *Turbine Maintenance and the Straightening of Shafts,* Journal Of South African Institution of Mechanical Engineers, February 1953

Chapter

3

Steam Path Damage Induced by Water

INTRODUCTION

The steam that enters the turbine steam path has a high-energy content. In fossil-fueled cycles, this steam is normally superheated, *i.e.*, the steam contains a degree of heat in excess of that required to completely evaporate it. On expansion, this superheat is released in the steam path, converting its thermal potential energy to kinetic energy in the rotor. When all the superheat has been converted to kinetic energy, further expansion causes the remaining dry, saturated steam to give up a portion of its latent heat, which is converted to water. The flow then becomes a two-phase mixture of steam and water.

However, in the water-cooled nuclear cycle and in the majority of geothermal cycles, steam that enters the high-pressure section contains some small amount of moisture. This moisture portion of the mixture is increased throughout the initial stages of the expansion and the majority of the low-pressure stages.

Once water has formed, it must be removed or transported through the steam path by the parent steam. Because of the relatively larger mass of the moisture particles, they are unable to pass through the blade passages with the same degree of acceleration as the dry steam. These water particles must have work done upon them in an attempt to match steam velocities.

These steam velocities will act upon the water particles because the steam accelerates and changes direction in the blade passages. However, the water particles will be unable to accelerate and achieve velocity equality because of their larger mass. Therefore, the water particles will have trajectories within the steam path that are different from those of the steam. Many will collide with the surfaces of the steam-path components, upon which they will be deposited. After deposition, this moisture will be unable to re-enter the main steam flow, but will still experience large field forces from the flowing steam. This tends to make the resulting water film flow in the general direction of the steam particles. With moisture particles constantly being deposited on the surface from continued contacts, there will ultimately be a significant film of water flowing across the steam path surfaces.

It is this deposited moisture that has the potential to cause various forms of damage to the steam path components as it collects into pools and re-enters steam flow as larger droplets. In addition, that portion of the transported moisture remaining in the steam will reduce steam path efficiency, though its presence may not contribute to any mechanical damage to the steam path components themselves.

There are sources of water other than those formed by condensation—three principle sources of water within the steam path alone. The first two listed here are water-contained in the steam path, but restricted to existence in the saturated region. The entire list includes:

- water formed by condensation of the working fluid during its expansion through the stationary and rotating blade rows
- water sprayed into the low-pressure section exhaust to remove excess heat generated by frictional heating during part or light load operation (this will be considered in greater detail)
- any portion of the unit connected to other equipment within the steam cycle, *i.e.*, water re-entering the steam path from external sources

WATER CONDENSATION IN EXPANDING STEAM

As the steam expands, it first releases its superheat energy until it reaches the saturated condition. Then, with further expansion, a portion of the latent heat contained in the steam is released. This conversion of latent heat introduces a state where water is formed in the expanding steam. This water exists as fine particles (fog) that are transported by the parent steam through the steam path until it exhausts to the condenser, or impacts with a surface to which it adheres.

Consider such an expansion on the Mollier Diagram (Fig. 3.2.1). The steam expands from condition "A" to condition "D," at which point the steam is dry and saturated. It would be expected from the previous discussion that with further expansion, water would form in and be transported by the steam. However, for heat to transfer from the gaseous to liquid phase requires a finite period of time, and the expansion of steam in the steam path is extremely rapid.

Note: The elapsed time for steam entering a high-pressure section to expand through it, and through the reheat and low-pressure sections is about 0.2 seconds, if the time in the crossover pipes and the boiler reheater section is ignored.

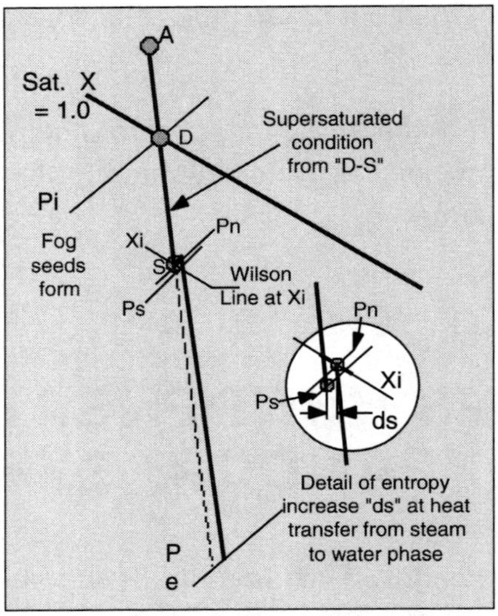

Fig. 3.2.1—Showing supersaturation of steam at the saturation line.

Because heat transfer cannot occur instantaneously, the expansion will continue as shown in Figure 3.2.1. Here, expanding steam crosses the saturation line at "D," and expands into the moisture region, achieving a super-saturated condition such as "S." At that point however, the transfer of heat will have been completed and the mixture will reach, or approach thermal equilibrium conditions, and moisture will form. This is the point at which a "fog" is formed, consisting of particles from about 0.5 to 1.0 microns in diameter. It is upon these "seed" particles that further deposition will occur in their passage through the steam path.

Once moisture particles have formed, various forces developed by the steam will influence them. This includes the drag force that causes the moisture particle to attempt to follow the steam flow velocity, being both accelerated and retarded through the blade passages. Because of their relatively greater mass, these moisture particles will not be able to follow the steam velocity, and will continue to extract energy from the steam, in an effort to achieve and maintain steam velocity. In addition, because of their greater relative mass, the water particles are unable to change direction as effectively as the steam, and will therefore assume a different trajectory or path in their flow between the blade passages.

The manner of water formation

There are three possible means by which water may form in the steam path:

- *Condensation on microscopic particles often impurities or contaminants contained in the working fluid.* However, modern methods of feed water treatment would preclude the contamination to any excessive degree. While there may be some impurities present, attracting a little condensation, the quantity is not likely to be sufficient to form a significant portion of the total moisture present at any time. Some of the impurities will combine with water, and this has other more insidious implications in terms of corrosion, but is inconsequential in terms of moisture formation and flow.

- *Condensation on the surfaces of the steam path components.* For steam particles present in the parent steam to condense on the surfaces of the steam path, they must contact it and be at a temperature that will cause condensation at the local ambient pressure. However, these metal surfaces are always covered by their boundary layer, which tends to be at a marginally higher temperature than the surrounding steam. This is because the velocity energy of the steam in the boundary layer has been reconverted back to thermal energy, raising the temperature of the steam. Therefore, the surfaces that the moisture particles contact are above the temperature of the surrounding steam, and condensation is unlikely.

- *Spontaneous nucleation after a limiting degree of super-saturation has been reached, and there has been sufficient time for heat transfer.* It has been shown by analysis that the only meaningful manner in which sufficient moisture can form in the steam is by this process.

It can be seen from these considerations that the moisture present in the steam path is essentially a consequence of spontaneous nucleation, and the formation of minute water droplets (which for convenience will be referred to as the water seeds) can be shown to have a size defined by the Kelvin-Helmholtz:

$$\mathrm{Ln}\left\{\frac{p}{po}\right\} = \frac{2.\sigma}{\rho R Ts r}$$

where:

p = Steam pressure
po = Vapor pressure corresponding to seed temperature Ts
σ = Surface tension
ρ = Seed density
r = Critical radius
R = Characteristic gas constant

Once these seeds have formed, they will continue to grow as expansion continues and further moisture is added. It is debatable whether droplets coagulate during their passage through the steam path, and it is not clear if smaller droplets (seeds) merge into larger drops (reference following sections on *Measuring moisture*

Distribution and Content and Water Removal from the Steam Path). However, other mechanisms for collection, present within the steam path, allow these droplets to agglomerate to much larger and more destructive forms.

Upon the formation of the water seeds, there is a small rise in the local pressure. This is shown in Figure 3.2.1 as an increase from "Ps" to "Pn."

RADIAL DISTRIBUTION OF MOISTURE

In the previous section the phenomenon of super-saturation and the delay in heat transfer from the gaseous to liquid phases was discussed. It is clear that if we consider the steam path of a low-pressure section in the region of the saturation line, then the point of moisture formation could be as shown in Figure 3.3.1. There would be a small axial distance after the saturation condition that would represent the under-cooled region of the steam path. This distribution assumes the pressure is constant at all radial positions along the axial length of the steam path.

Entry to the under-cooled region is pressure related, and originates at the point where the expansion crosses the saturation line. However, because the pressure gradient in the blade path is not radially constant at all axial positions, the position of the under-cooled region will not occur at a radially constant axial position. Figure 3.3.2 shows the measured pressure gradient in a low-pressure expansion. If the position of the onset of under-cooling is related to pressure, then this region will be of the form shown in Figure 3.3.3, with the moisture, as a fog forming as the heat transfer is completed at some lower pressure.

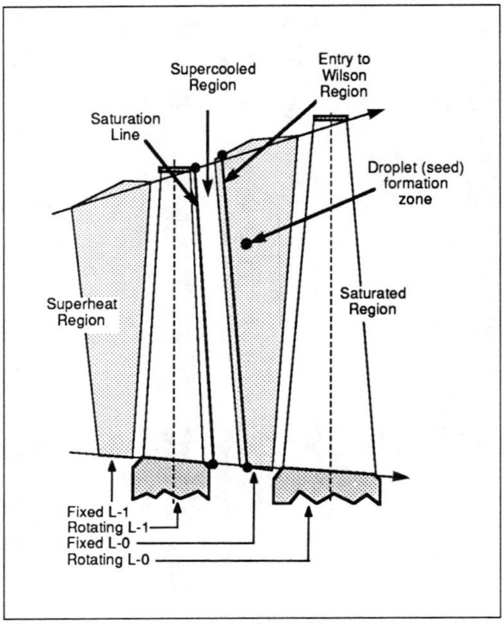

Figure 3.3.1—The position of the supercooled region with no radial pressure gradient.

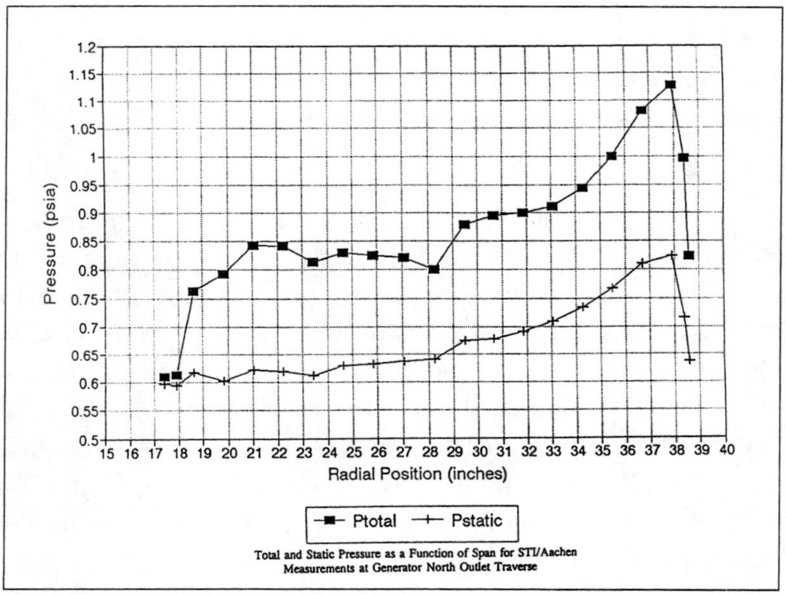

Figure 3.3.2—The pressure gradient in a low pressure stage.

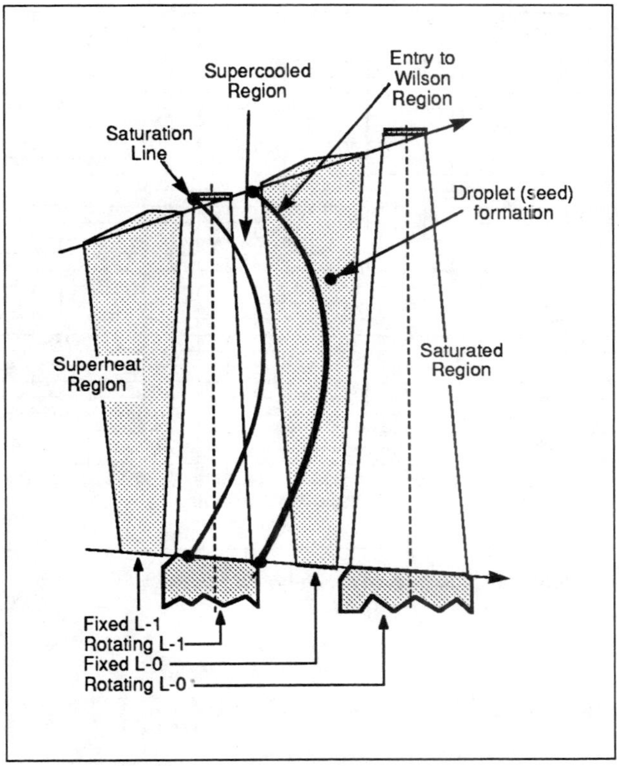

Figure 3.3.3—The pressure gradient in a stage with measured pressure distribution.

The axial position and entry to the under-cooled region can be established by measurement and calculation. However, it must be remembered that any measured condition applies to one set of steam conditions and flow quantities in any one unit. The position and form of the under-cooled region is influenced by the quality of steam delivered to the turbine from the boiler or reheater, the quantity of steam flow, and (for any design) the stage heat drops and vane configuration. Therefore, the regions shown in Figure 3.3.3 are steam condition and load sensitive. Actual measured values of moisture in the radial direction are shown in Figure 3.3.4.

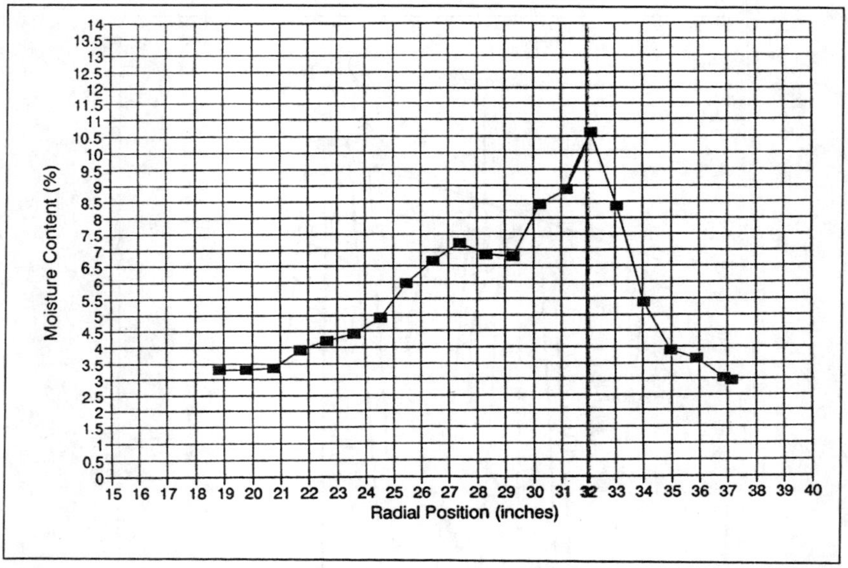

Figure 3.3.4—Measured moisture distribution.

MOISTURE DEPOSITION

Moisture droplets formed in the steam are unable to follow the direction of flow of the steam because of their relatively large mass. Therefore, the moisture droplets will be unable to change direction through the blade passages and will collide with, and be deposited upon the various elements of the blade rows. There are two main deposition areas to consider:

Deposition on the blade vanes

The path of dry steam flow between the stationary blade vanes is shown in its basic form in Figure 3.4.1. This illustrates the steam flow trajectories between a pair of stationary vanes. Shown are the streamlines and stream tubes of the steam flow between the pressure face of one profile and the suction face of an adjacent element. This stationary row is designed for a pressure drop to occur across it and for the steam to be accelerated at the expense of this pressure/enthalpy drop. Therefore, at discharge from the row, the velocity at "L" will be higher than at the inlet "K," and the pressure lower.

Note: The concept of streamlines is used to aid in the explanation of flow, and to represent a convenient means of anticipating the actual conditions that can exist at any position within a flow passage.

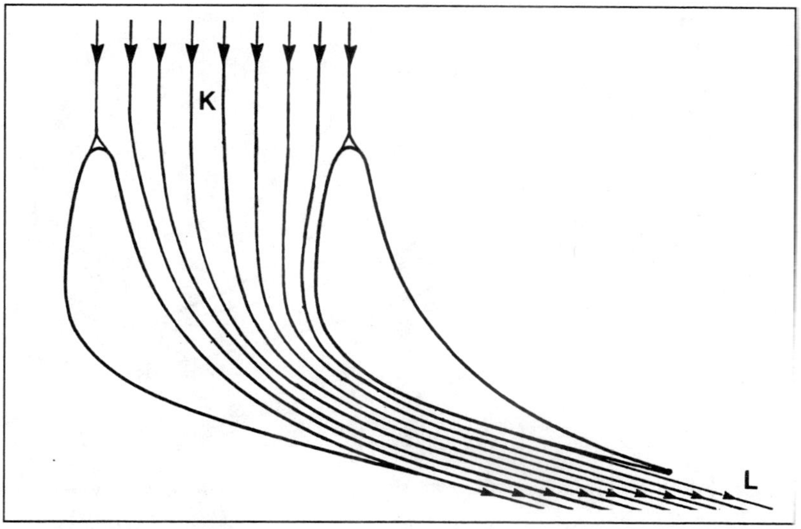

Figure 3.4.1—Steam flow lines through a fixed blade passage.

Now consider a moisture droplet at "G," shown in Figure 3.4.2, located on the streamline "G-G." As the steam enters the influence of the passage formed between the blade pair, the steam and the moisture particle are accelerated through it. However, because of its greater mass, the moisture particle is unable to accelerate at the same rate as the steam, and it takes a path as shown, "G-M" (position "M" being at the discharge point on the profile). The droplet will ultimately collide with the pressure face and, after collision, will remain on this face, unable to re-enter the main steam flow, and existing in the form of the original small particle. It can be seen, by examination, that all moisture existing from the profile inlet nose and position "G," is being transported by the steam, and will be deposited on the profile surfaces.

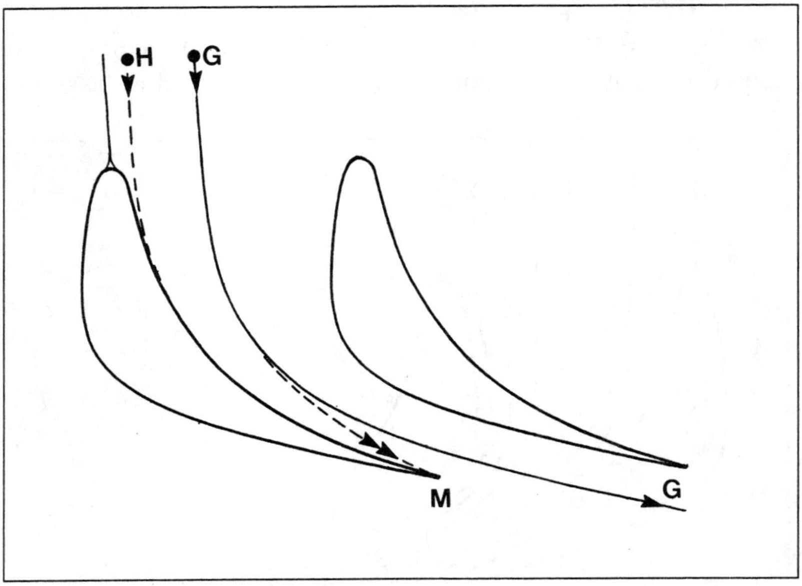

Figure 3.4.2—The trajectory of the moisture particle "G" which will collide with the discharge tail.

Due to the forces exerted by the flowing steam on the deposited moisture, this moisture will form a film that will flow across the profile in a substantially axial (downstream) direction, to the point "M" at the discharge tail. From this description, it is clear the quantity of water that exists as a film on the profile pressure face will increase as the distance from the inlet nose increases, and this is at maximum at the discharge point (or discharge tail) of the profile (*i.e.*, at position "M").

Before considering the water path at the discharge tail, consider the steam and moisture particle mixture that enters directly ahead of the profile inlet nose "Ni" (Fig. 3.4.3). Here, as the steam and any moisture particles it carries approaches the inlet nose, they must divide to flow over either the suction face "Fs" or the pressure face "Fn." The division is a matter of which side of the profile exerts the most influence on each of the individual steam and moisture particles. On the suction side of the profile, a portion of those moisture particles deflected around that side will be unable to accelerate sufficiently and will collide with the inlet nose as shown at position "T." This deposited moisture will then flow around the suction face, influenced by the same forces that act on the pressure face. However, the quantity on the suction face will not increase during the flow across this area, as there is little or no possibility of those moisture particles accelerating towards the pressure face impacting there. On the pressure face however, the moisture particles that are unable to avoid the inlet nose are added to those deposited (Fig. 3.4.2). These two sources represent the total moisture on the pressure face "Fn."

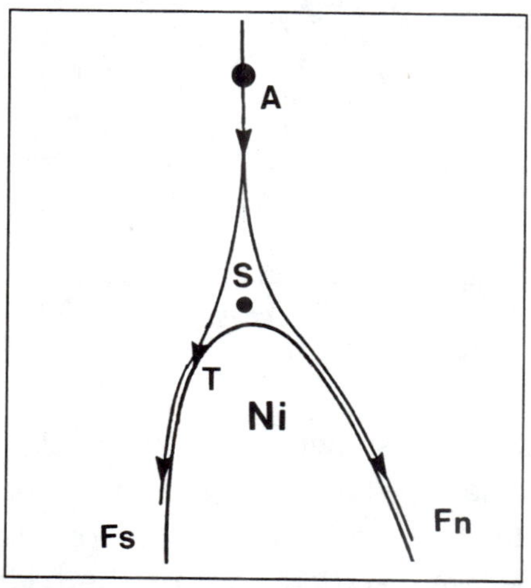

Figure 3.4.3—Flow division at the profile nose.

At the position on the inlet nose where the flow begins to divide, there is a slowing of the steam velocity as impact with the nose is partially avoided. At this point on the vane profile there is a general lowering of steam velocity, and with a lowering of velocity there is a small local increase in steam pressure, (velocity energy is reconverted to pressure energy). This condition creates the stagnation point "S." The Mollier Diagram for this position is offered (Fig. 3.4.4) with a steam inlet condition of "A" on the expansion line "A1-A2." It will be seen that accompanying the pressure increase "dp" (enthalpy increase "dH") is a small increase in the local temperature "dT" (1 to 1.5°F). In fact, the inlet nose of the vane will be heated to some small degree by this local increase in steam temperature.

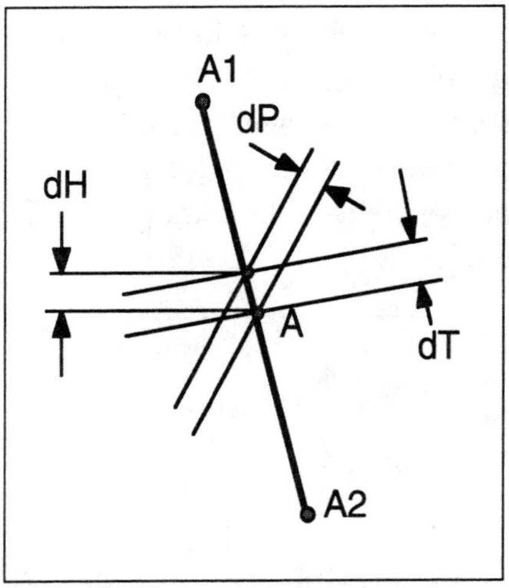

Figure 3.4.4—The steam conditions at row inlet.

At the discharge point (shown as "M" in Fig. 3.4.2), the moisture will accumulate as the quantity flowing down the pressure face continues to collect. However, the moisture cannot remain at this location because of the force exerted on it by the high velocity of the flowing steam. Consider the discharge tail (Fig. 3.4.5). The steam expands through the passage created between the profiles, and will eventually discharge from the row at the throat (minimum area), shown as "O-O," with a velocity "Cs." At this point the steam substantially leaves the influence of the profiles. At discharge from the throat, the boundary layer on the suction face will continue to adhere to the profile until some point is reached at which there will be separation; this is shown as occurring at point "Se." The void caused by this separation will shed vortices, which are carried over with the main steam flow into the following rotating blade row. Most significant in terms of the moisture film and accumulating moisture, is that the region of vortices is a low-pressure region, relative to that existing in the vane discharge region. This means that the moisture collected (or is collecting at the discharge point "M" of Fig. 3.4.2) will be drawn around the tail, where it will accumulate with the small quantity of moisture flowing down the suction face ("Fs" of Fig. 3.4.3) to form small puddles (or pools), "Q."

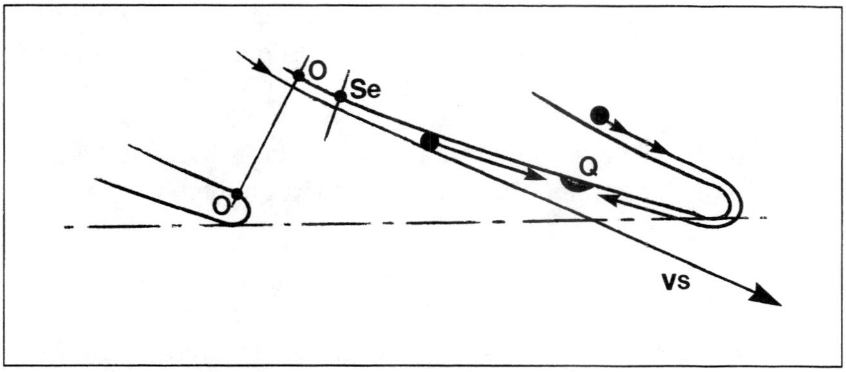

Figure 3.4.5—Moisture flow around a discharge tail, and collection on the suction face.

This moisture accumulation puddle "Q" will continue to grow in the vortex low-pressure region until it becomes too large to resist the drag forces of the flowing steam. These puddles will then be torn from the suction tail surface and re-enter the main steam flow as considerably larger drops. Because the drops torn from the discharge tail are no longer of a size that can be transported by the main steam flow, they break into many small droplets, but remain hundreds of times larger than the original seed droplets, and are of sufficient size to cause moisture impact erosion on the rotating blade inlet edge.

The manner of deposition in a rotating blade row is essentially the same as in the stationary blade row, with certain differences in the pattern of accumulation and flow. The physical considerations affecting these differences and causing the flow pattern to alter are listed below:

- Within the rotating blade passage there are large centrifugal forces acting on both the steam and moisture film. This centrifugal force causes the deposited moisture to have a large radial flow component on the surfaces of the vane
- The passage shape may be designed for very low degrees of reaction, and therefore the pressure drop may be lower. In nuclear high- pressure stages (the normal impulse design) the passage shape has a relatively small degree of convergence, and the degree of reaction is low
- In the rotating blade of an impulse stage, the velocity tends to decrease rather than increase through the passage

Deposition on the sidewalls

Expansion through the steam passages results in moisture deposition within them. We have discussed the deposition on the vanes, however, in addition to water accumulation that occurs on the vanes, there can be additional deposition on the sidewalls.

Consider the stationary blade row showing the lines of water flow with a radial component [Fig. 3.4.6 (a)]. The outer portions of the water droplets are deposited on the outer sidewall (shown at point "R"). After impact, this water will flow along the sidewall until it reaches the discharge point on the diaphragm, at which point it will either be detached and flow over into the rotating blade row, or flow out over the vertical face of the diaphragm and along the casing. A similar mechanism is shown for a diaphragm with slanted sidewalls [Fig. 3.4.6 (b)].

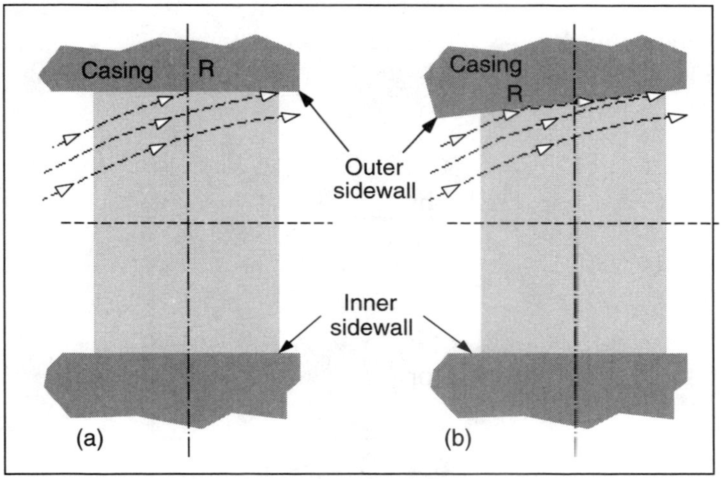

Figure 3.4.6—Moisture deposition and flow to the outer side walls of a fixed blade row.

Because of radial flow patterns, it is unlikely water will be deposited on the inner sidewall to any considerable extent. However, it is possible that any water flowing under the stationary blade row between the blade and rotor with leaking steam will migrate up into the blade row. This is a particularly unpredictable mechanism, and can only be determined to exist from flow patterns on the blades when they are removed for examination at an outage.

In the case of rotating blades, as soon as moisture is deposited from the steam flow onto the vane surface, it will experience high centrifugal loading, causing it to flow outwards across the vane faces. This effect is shown in Figure 3.4.7, where moisture is shown to be deposited at positions "R1...R4." From there, it flows outwards to the underside of the coverband band. At these locations, the moisture experiences a combined force from centrifugal and steam flow effects, causing it to have a substantially radial direction. At the coverband band, the water will flow in an axial direction, coming into contact with the coverband band under-face. This will continue until, at point "Z"—the discharge point—it will be centrifuged out to the casing or moisture-collecting device.

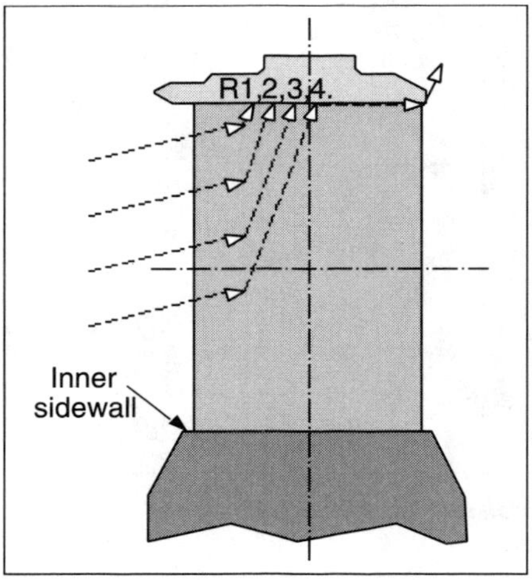

Figure 3.4.7—Moisture deposition and flow to the cover band in a rotating blade row.

Basically, no moisture is deposited on the inner sidewall from the steam flow, although it is possible some small amount could be present there (but this will originate from the leakage under the stationary blade row). As soon as this moisture is deposited on the blade lower surface or inner sidewall, it will experience large centrifugal inertia forces and flow outwards along the vane to the outer sidewall. In certain instances this water could detach from the blade root platform (particularly if there are any flow discontinuities on surfaces, such as the transition point from one blade platform to the next). In flowing outwards, the steam flow forces would influence the moisture, and it is not possible to predict any flow directions. However, this quantity would normally be small and of little interest in terms of predicting possible damage scenarios.

Water flow along the casing

Because of the radial component in both the steam and the water flowing in the steam path, centrifuged moisture continually adds a film of water on the casing inner surface. This film flows from the higher to lower pressure areas of the casing. Due to the gravity effect, it will also flow from the upper regions to the lower, from which points it can be drained. Another influence on its flow is the windage effect of the rotating blades. They set up forces that will attempt to force the steam onto the sidewalls, above the rotating blades, to flow in the same direction as the blades rotate. This is a small effect and considered to be of little consequence in considering damage mechanisms.

What is significant with this casing flow is when it migrates or drains to the lower regions of the unit, it must be removed, as it has the potential to cause significant erosive damage to the rotating blades.

MEASURING MOISTURE DISTRIBUTION AND CONTENT

Until recently, the actual measurement of water content or dryness in the steam path was particularly difficult to undertake, and reliable results hard to come by. However, recent technological improvements have introduced methods that provide data allowing this to be undertaken with acceptable levels of accuracy. It is anticipated that as technology advances, these methodologies will improve, and there will be even greater levels of confidence in the results.

Wetness probes can be placed in the turbine to traverse the radial direction and measure the moisture content at various locations. Figure 3.5.1 shows a radial probe that was mounted on the downstream side of a row of exhaust stage blades before replacing the casing. This probe [Fig. 3.5.2(a) and (b)] had alternative positions at "entry to" and "exhaust from" a large exhaust stage blade row.

Typically, a probe will use the "light scattering technique"—employing an argon-iron laser that projects a light into the flow path. This laser light is reflected off the flowing droplets, and reflected into a collecting device capable of counting/measuring their size. This device then integrates the total to determine the moisture present. By repeating this procedure at a number of radial locations, the total moisture distribution can be determined. The results from such a test are shown in Figure 3.3.4.

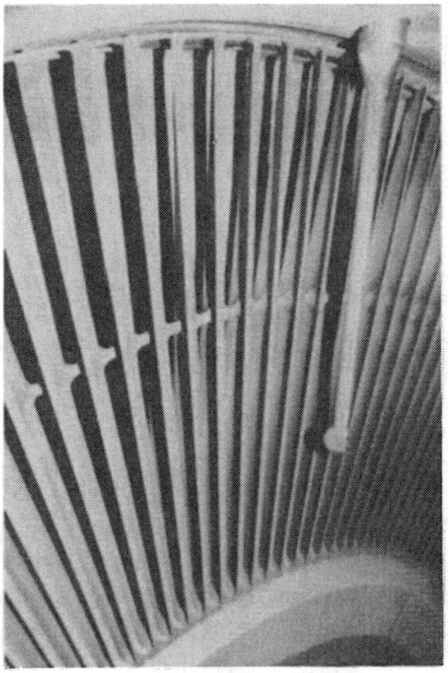

Figure 3.5.1—Laser probe at a last stage blade discharge.

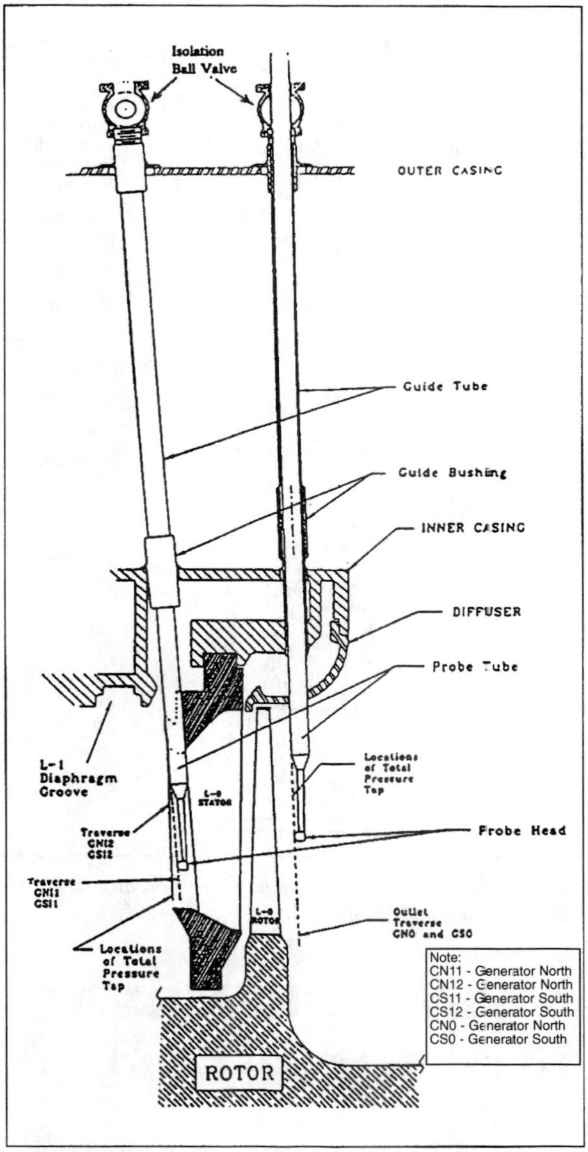

STI/EPRI

Figure 3.5.2(a)—The laser probe and its means entering the low pressure section.

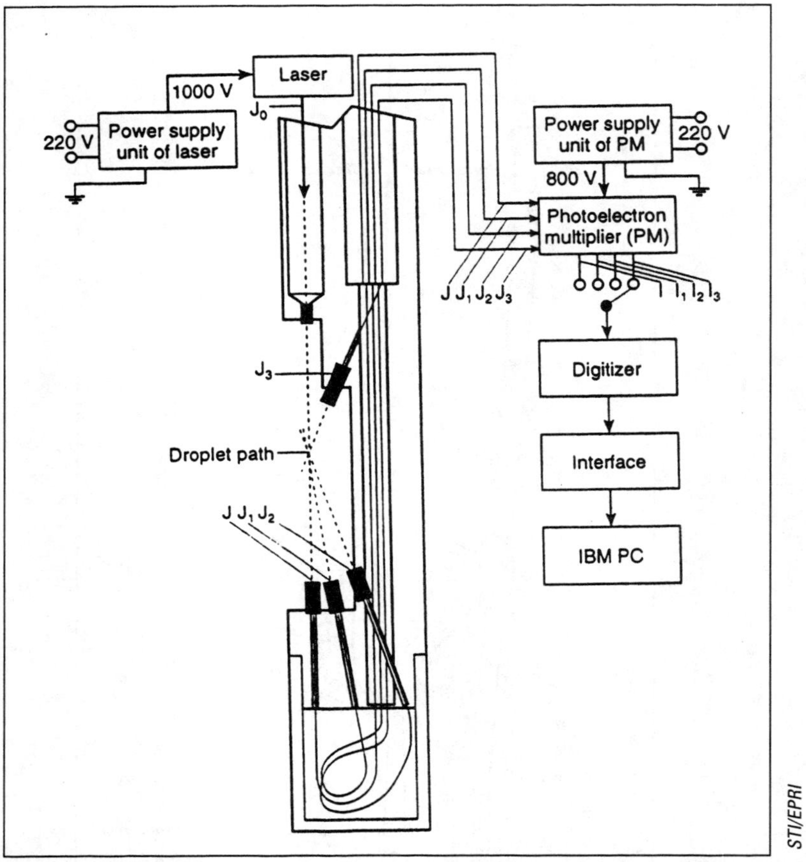

Figure 3.5.2(b)—The laser probe for moisture and droplet size measurement, using light scattering and attenuation.

WATER REMOVAL FROM THE STEAM PATH

All units in which steam operates with a moisture content have the need for the control and removal from the steam path as large a portion of this moisture as possible. It is also necessary for portions of the units not operating entirely or exclusively in the moisture region to have provision for moisture removal. This is necessary so that at a cold start, water present in the steam path can be drained away before high rotational speeds are reached. The majority of stages rely on the heating, evaporation, and carryover of the condensed moisture for its removal. This is a satisfactory method, but operators must be sure that sufficient heat has been added so all water has evaporated before high rotational speeds are reached.

There are three distinct methods of controlling the level of moisture in a unit:

- *By controlling the steam conditions.* This applies to both the superheated and saturated steam cycles. Saturated cycles include both nuclear and many geothermal
- *The removal of water from steam as it is transported from one turbine section to the next.* This is most effective in water-cooled nuclear cycles where the steam entering the high-pressure turbine already contains a small percentage of moisture or has only a very small degree of superheat so moisture is formed early in the expansion

  In cycles using saturated steam, such as nuclear cycles from water-cooled reactors and geothermal applications, the steam is dried before entry to the unit by similar moisture separating devices
- The collection and drainage of the deposited moisture from the individual stages

Internal collection and removal

For those stages that heating prior to start-up cannot effectively dry, and those stages having moisture formed in and deposited on them during operation, effective removal systems must be incorporated into the stage design.

To achieve drainage, it is necessary to collect this moisture as effectively as possible, pass it to a location where it can be drained, and then provide an adequate passage for its removal. This drainage must be completed without excessive steam blow down from the unit.

The majority of units employed on fossil-fuelled cycles have their low-pressure sections designed to collect and drain moisture. For units designed to operate on steam generated in the family of "water cooled reactors," moisture is present in even the highest-pressure stages. Therefore, it is necessary to collect and drain from the high-pressure section as well as from the lower pressure sections.

Moisture is collected at the casing inner surface by the incorporation of collection slots, or catchers, into the surface geometry. There are three sources of water in these casing wall locations:

- *that which is deposited on the casing inner surface from the radial flow effects of the expanding steam.* This is a relatively small quantity, but must be allowed for. In those stages where there has not been sufficient elapsed time, after the formation of water seeds, for these seeds to collect into large drops, this may be the only source of moisture

- *water deposited on the stationary blade rows—on both the vanes and the outer sidewalls—and flows with a large radial component out to the casing.* This is part of the total stationary blade row-deposited moisture that does not re-enter the main steam flow as destructively large droplets

- *water centrifuged to the casing from the rotating blades.* After impacting with the rotating blades, a portion of this moisture rebounds and will re-enter the main steam flow. However, the major portion will be centrifuged out to the casing inner surface

To collect as great a percentage of this moisture as possible, considerable care is taken in designing moisture collection slots. They must allow the moisture to enter, be retained, and then drained to the bottom half casing where it is removed. To achieve effective drainage, the collection slots must be dimensioned to collect moisture flowing along the outer sidewalls. The slot is optimally placed above the moisture that is centrifuged from the rotating blade rows. To achieve optimum axial placement, allowance must be made for the differential expansion occurring in the steam path. Differential expansion is load-sensitive, because the temperature of the various portions of the steam path varies with the valve settings, which are adjusted for the demand placed on the unit.

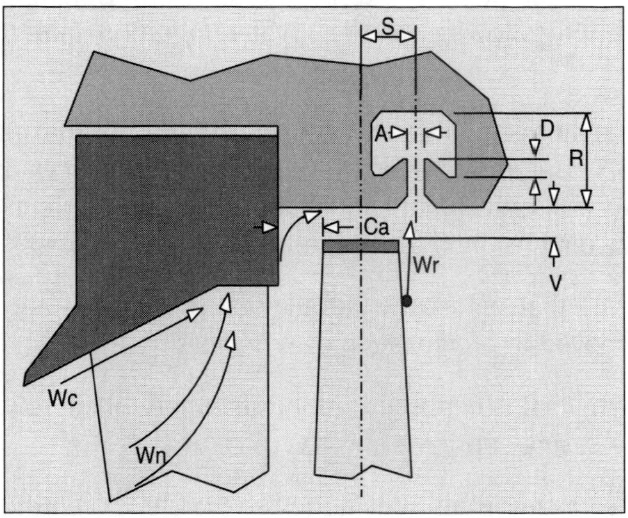

Figure 3.6.1—The internal water catcher belt.

As stated, the most common internal removal device, and that employed by all manufacturers, is to collect a large portion of the deposited moisture existing at the casing inner surface, and drain it to a suitable point in the cycle by means of a circumferential water catcher belt (often through a "U" tube). Figure 3.6.1 shows the basic arrangement of a water catcher belt placed radially outwards from a rotating blade. The method of constructing the collector belt can be seen. Here, at entrance to the catcher, an axial gap "A" is positioned at a distance "S" (in the operating axial position) from the rotating blade centerline to allow the moisture to enter. There is a relatively large radial distance "R" above the entry point to prevent water rebound, and then lips of depth "D" are produced at the belt entry providing a drainage path for the collected water to drain to the bottom dead center where it can be removed.

Note: The "U-tube" can, on load reduction, allow a portion of the moisture to "flash" to steam, which will re-enter the steam path. On load rejection it is possible a portion of the "U-tube" water will re-enter the steam path as "slugs" possibly causing significant damage (reference following section *Water Ingestion into the Steam Path*).

This catcher (circumferential belt) removal system, shown in Figure 3.6.1, is commonly employed in the latter stages of all low-pressure sections, where large quantities of free moisture exist. These designs are dimensioned to capture three separate sources of water:

- "Wc" that enters the stage along the casing, and was not removed in previous removal devices
- "Wn" that is deposited on the stationary blade row, flows to the casing and joins flow "Wc"
- The catcher is also positioned so that "Wr," centrifuged from the rotating blade, is captured

These designs are also arranged to have a steam blow-down of about 0.5% of the steam flow to help ensure effective removal of the water.

While this drainage design is used in the turbine low-pressure section, such a design can also be applied in the high-pressure expansion in saturated steam nuclear units. In addition to the geometry shown in Figure 3.6.1 an alternate arrangement, shown in Figure 3.6.2, can be used where the collector belt is located above the inlet edge of the rotating blades. In this design the inlet nose of the rotating blade has collection grooves, (shown in the detail) which capture the inflowing moisture particles "M," then provide channels for this moisture to be centrifuged to the casing, where it will enter the catcher throat "A."

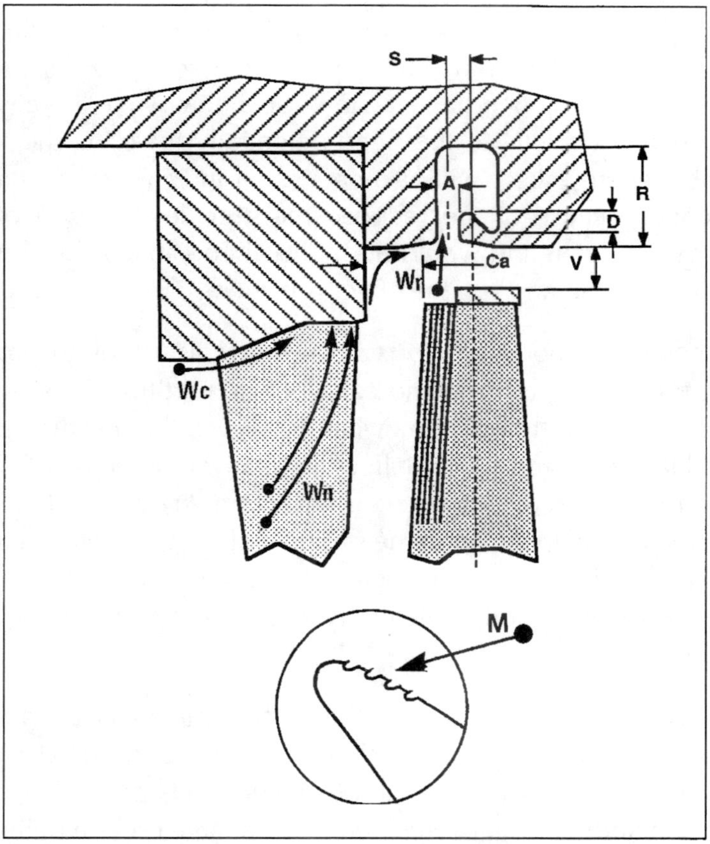

Figure 3.6.2—The water catcher belt, used with grooved inlet edge blades.

Moisture removal in the nuclear high-pressure sections is not undertaken to prevent erosion on those stages, rather it has the purpose of removing moisture from the expansion before it causes an excessive efficiency loss. This internal separation is also used as a preventative measure to reduce the duty imposed on the moisture separator in the nuclear unit intermediate system. While moisture quantities removed may be relatively small, they will in fact modify the shape of the expansion line as shown in Figure 3.6.3, where the enthalpy into a stage will increase by a small amount "dH" from "d1" to "d2," and there will be a small increase in the stage entropy of "dS." Examining one stage it can be seen that while there is a shift of the expansion line to the right, potentially lowering the level of available energy, there will also be a small increase in the state line efficiency because of the reduced moisture levels. Both of these changes are small and would not normally be able to be shown on the Mollier Diagram, but the effect is present.

Another method used to remove the moisture, is to provide a drain in the outer wall behind each nozzle partition. This has been shown to reduce erosion somewhat. However, the machining of the individual drain and bleeder lines is relatively expensive. Another method being used, but is too costly for many applications, is to bleed the boundary layer on the stationary blade partitions and draw away the moisture flowing there. This system has not been used extensively since results have rarely been found to justify the manufacturing cost.

At locations in the steam path where steam is extracted for regenerative feed heating, or other extraction purpose, if moisture exists in droplet form, this steam extraction acts as a very effective collection and removal device, removing a large portion of the moisture present in droplet form.

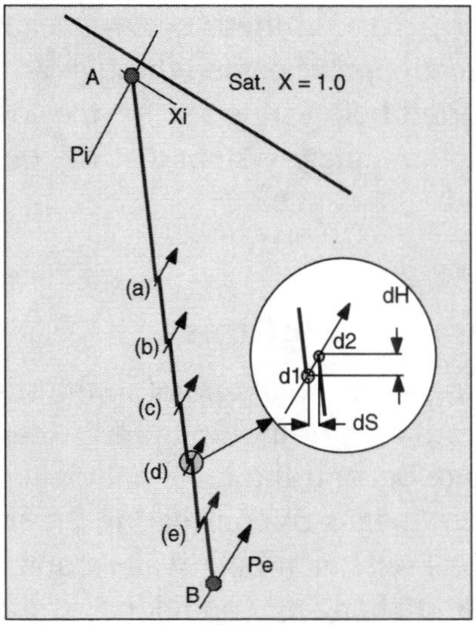

Figure 3.6.3—The effects of internal moisture separation in the high pressure section.

External removal (saturated steam cycles)

Cycles employing steam at or near the saturated condition upon entry to the turbine can, if the steam is to be removed during its total expansion, employ separating devices external to the turbine. There are various forms of separators all relying on the difference in density between the steam and water particles to effect separation. The methods of undertaking this separation are not a subject for this section. However, the effect of these separators on steam conditions is considered in greater detail in the following section, concerning saturated steam cycles.

To remove this moisture, the steam must be removed from the turbine after partial expansion, where it can be dried, and the water drained. External separation is found suitable only for these cycles where the steam when supplied to the unit contains transported moisture, as in some geothermal applications and those units used with water-cooled reactors. In these cycles, steam is removed from the high-pressure section, and passed through an "intermediate system," where it is dried before re-entry for the low-pressure expansions. It is in this intermediate system that the moisture is removed in a separator vessel.

Control of steam conditions

The moisture that exists at any point in the steam path is fundamental to the performance of the unit, and excess water quantities can result in damage beyond the level anticipated by the designer. The most convenient means of considering the influence of changing steam parameters on steam path water content is to review the steam conditions, and changing conditions, as they are represented on the Mollier Diagram.

The expansion lines for various cycles are shown in Figure 3.6.4. These lines indicate the moisture levels anticipated by the designer, and for which protective devices or water control systems are designed into the steam power cycle. Changes in initial, reheat and exhaust steam conditions will almost always affect operating moisture levels and modify the potential for damage. Therefore, if a unit has operated for a number of years with steam conditions within the cycle that are "off-design," through turbine, boiler, or condenser inability to maintain design specified operating parameters (*i.e.*, the inlet, reheat, and exhaust steam conditions could not or were not maintained) then the moisture levels will modify, and damage could result. Under these conditions the unit's mechanical condition could deteriorate, and should be examined for damage.

Superheated steam cycles. Initially superheated steam cycle conditions for a typical fossil unit are shown in Figure 3.6.5. In this figure the design condition and variations are shown as heavy lines. The initial design conditions are "Pd" and "Td," point "A." In the high-pressure section the steam would then expand to pressure "Ped" at point "B." Also shown in Figure 3.6.5 are variations from these initial design inlet conditions from pressures "Pu" to "Pl" and temperatures from "Tu" to "Tl." It can be seen that any variations within these ranges will not affect moisture, but will affect the conditions at "B," exhaust from the high-pressure section. At any flow quantity the inlet to discharge pressure ratios will be substantially constant:

"Pu/Peu" = "Pd/Ped" = "Pl/Pel"

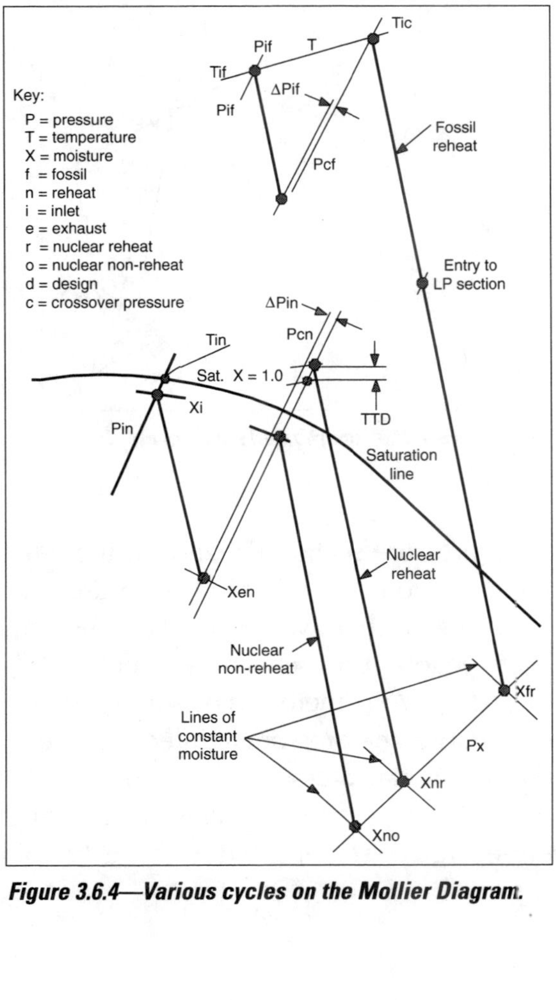

Figure 3.6.4—Various cycles on the Mollier Diagram.

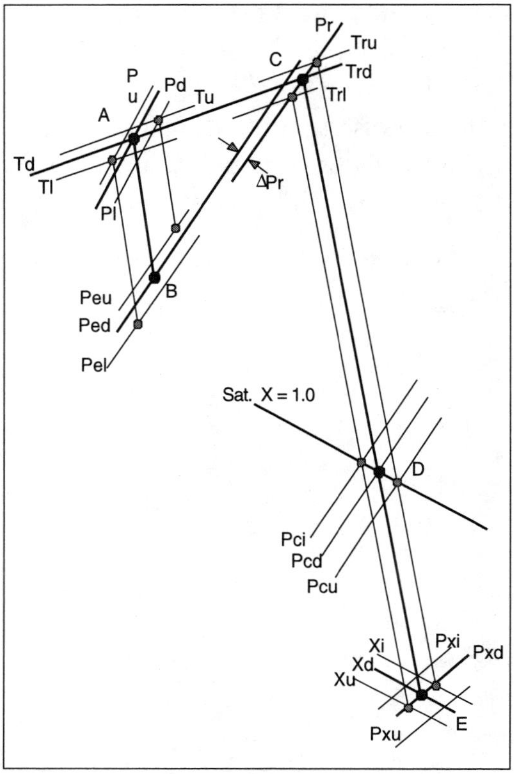

Figure 3.6.5—The superheated steam cycle.

From the high-pressure section exhaust, in the reheat system to inlet to the reheat section, there is a pressure drop in that system (piping, boiler reheater section, valves, etc.) of "ΔPr," and the steam enters the reheat section at a pressure "Pr," point "C." Depending upon the pressure drop "ΔPr" there can be some variation of reheater inlet pressure. Therefore, the pressure at inlet to the reheat section is dependent upon the initial pressure to the turbine at "A" and the pressure drop in the reheat components comprising the cycle. Therefore, the re-energized steam enters the reheat turbine section at "C" with a pressure "Pr."

However, despite these possible differences from HP turbine exhaust to reheat section inlet, the major parameter influencing low-pressure exhaust conditions, and therefore steam path moisture content is the reheat temperature "Tr" at "C." The effect of varying this temperature from "Tru" to "Trl" can be seen in its effect on the pressure at which the expansion line crosses the saturation line "D," the steam path and final moisture content at "E," when the condenser pressure is "Pxd." The variation of exhaust dryness content varies from the design-anticipated value of "Xd" by amounts "Xu" to "Xl."

Saturated steam cycles. In saturated steam cycles such as the nuclear and many of the geothermal applications, steam enters the turbine with initial moisture content (shown as condition "A" in Fig. 3.6.6, which represents the expansion lines for such saturated conditions). Steam produced by the water-cooled nuclear reactors has conditions of about 1,000 psia with small initial moisture content of about 0.25%, "Xdi." The moisture content before entry to the turbine having been reduced by the removal of any free moisture it might have been transporting from the nuclear boilers.

After admission to the high-pressure section of the turbine, the steam expands to the design selected pressure "Pi," at "B." This steam then exhausts from the turbine, and passes through an "intermediate system" to improve the steam conditions by drying it to condition "C," and then either reheating it to condition "D," or returning it to the low-pressure sections where it expands in the non-reheat cycle to condition "E." For the reheat application the steam is reheated using live steam extracted from the nuclear boiler (and possibly utilizing partially expanded steam removed from the high-pressure turbine). Reheating improves the steam condition to "D." Steam is then allowed to expand in the low-pressure sections to condition "F."

At entry to the high-pressure section of the turbine, the steam has a condition established by the performance of the nuclear boiler, and more importantly in terms of moisture by the performance of the moisture separator, which if not working correctly will lower the inlet conditions to pressure "Pd," moisture content "Xdl." The steam then expands, and will achieve the same pressure ratio across the high-pressure section. This will decrease the dryness content in the high-pressure section exhaust steam from "Xe" to "Xel."

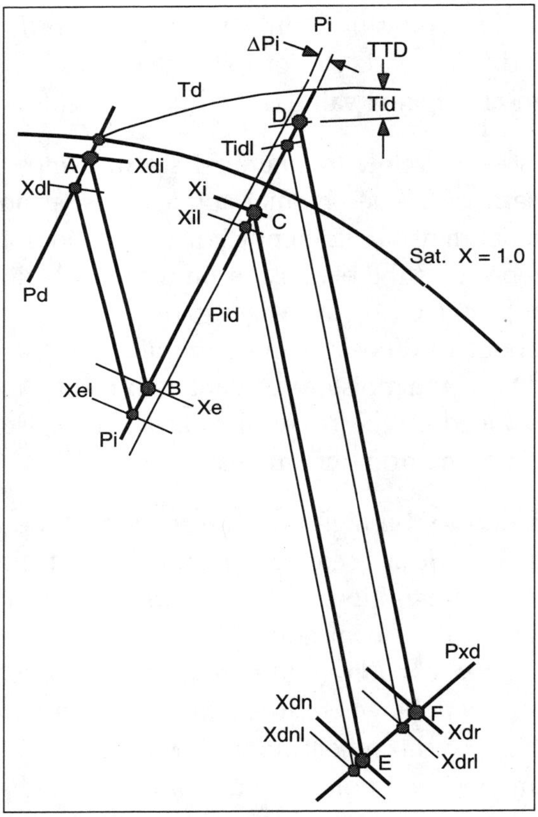

Figure 3.6.6—The nuclear cycle on the Mollier Diagram.

In passing through the intermediate system there will be a pressure drop "ΔPi." Also the performance of the external moisture separator will, if operating at design specification, improve the condition to "C," dryness "Xi." If however, the separator is not operating correctly then some greater moisture content at dryness "Xil" will be achieved. The effect of this on the non-reheat cycle can decrease the exhaust dryness from "Xdn" to "Xdnl."

If live steam reheating is to be used in passing into the reheater portion of the cycle, energy will be used undertaking the additional heating to evaporate moisture, improving the steam condition from "Xil" to saturated, rather than from "Xi" to the saturated condition. A degree of superheat will then be added by the reheating steam improving the steam temperature to "Tid," which is less than the saturated steam temperature corresponding to inlet pressure by an amount termed the "terminal temperature difference" (TTD). The maintenance of the TTD will help maintain the exhaust dryness at "Xdr." An increase in the TTD will decrease the dryness from "Xdr" to "Xdrl."

In the simple geothermal application, shown in Figure 3.6.7, steam will enter the unit with a condition "A," pressure "Pi," and dryness "Xi." (The inlet pressure on the majority of geothermal units is in the range 100psia+/-.) This steam will expand with normal efficiency to condition "G," pressure "Px," and at this condition, the dryness will be "Xgd." In the geothermal cycle, the major influencing factor is the effectiveness of the moisture separator ahead of the inlet valves—if not operating with design effectiveness, could give a dryness to the turbine of "Xiu" at entry, causing an increase in the final moisture content from "Xgd" to "Xgu."

Exhaust steam conditions. As the condenser reduces exhaust pressure, there will be an increase in the available energy, and the last stage discharge moisture content will increase with decreasing pressure. The extent to which damage occurs—as a consequence of this additional moisture that forms in the last stage due to the reduction of exhaust pressure—is dependent upon the location of the moisture and the particular form of damage.

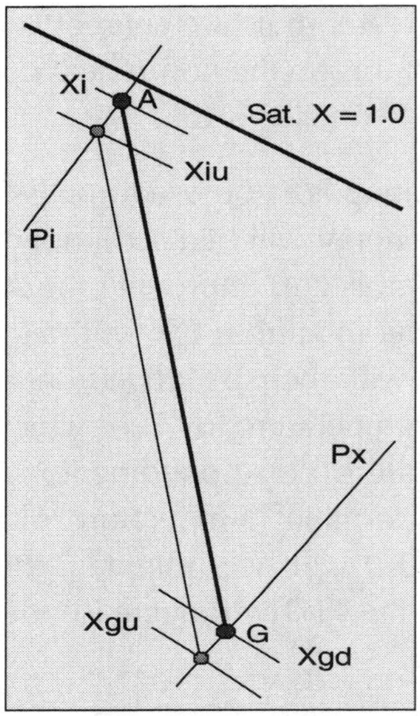

Figure 3.6.7—The geothermal cycle.

Damage, and its dependence upon moisture content, will be considered in greater detail in sections following.

It is clear from this discussion of steam conditions, these will have a considerable effect on the moisture content in the various portions of the unit. It is also clear there are pieces of equipment contained within the cycles that will also have a significant impact on the final moisture content, and the moisture throughout the steam path.

The effects of internal moisture removal on steam path conditions

The means of moisture removal by collector belts was discussed earlier, and this action of collecting and removing was shown on the Mollier Diagram Figure 3.6.3 for a high-pressure section, where the removal was undertaken from efficiency considerations. In the low-pressure section the quantity of moisture removed is more substantial, and the effect on steam path conditions more pronounced. The effect of such removal is shown in Figure 3.6.8 for a low-pressure section.

The effect of internal moisture separation is to reduce the moisture content of the working fluid at the stage position. This is indicated on the Mollier Diagram by expansion line discontinuities. On the expansion line shown in Figure 3.6.8 the steam enters the low-pressure section with a condition "Pm/Xm," for a fossil or nuclear unit. In this low-pressure section there are three stage positions where moisture is removed—"n1, n2, and n3." At each stage position the pressure remains unchanged, but the level of moisture is reduced. The net effect of this three-stage separation is to reduce the exhaust, or final dryness from "Xf" to "Xe."

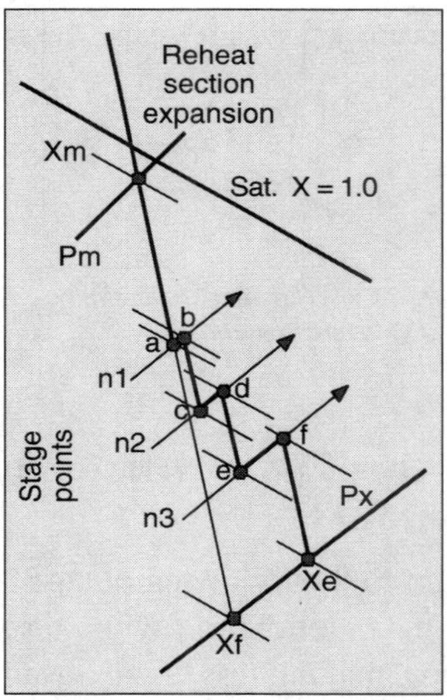

Figure 3.6.8—The effects of internal moisture separation on steam conditions in a low pressure section.

In the first removal stage the condition is increased from "a" to "b," drying the steam. A similar effect can be seen for the other two stages. These stages will have a total effect of modifying the final dryness from "Xf" to "Xe," and having the same effect on steam conditions as discussed for Figure 3.6.3.

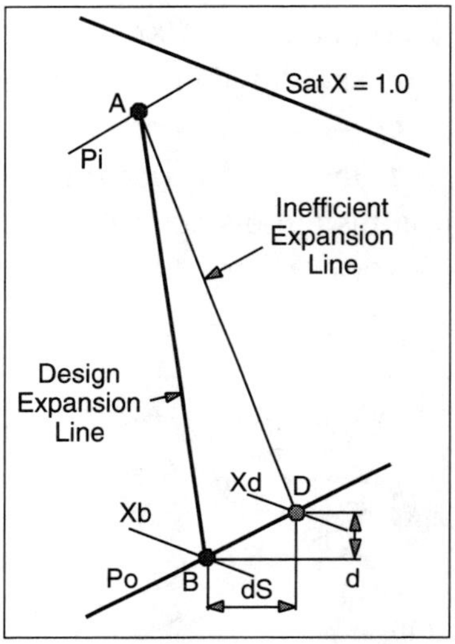

Figure 3.6.9—The effect of inefficiency on moisture content.

The effects of steam path efficiency on moisture content

From its initial start to the following outage, steam path efficiency will deteriorate. This deterioration will cause an increase in the expansion line entropy, making less energy available for conversion to work. However, as shown in Figure 3.6.9, for an expansion from pressure "Pi" to "Po" this deterioration will also increase the final dryness form "Xb" to "Xe." This drying is a reflection that the additional heat generated by frictional heating (inefficiency), will be returned to the working fluid, and will in effect evaporate some of the water contained in the steam. Similarly any increase in steam leakage quantities (which reduces efficiency), will mean the energy not removed from the leaking quantity will mix with and raise the total energy level of the working fluid. Both effect drying the expanding steam.

Similarly if any stage suffers mechanical damage, its efficiency will degrade. This effect is shown in Figure 3.6.10, where steam expanding from pressure "Pa" to "Pb" in a damaged stage will increase the stage discharge entropy by an amount "ds," and lose enthalpy for conversion to work by an amount "δh." In doing this the expansion line will shift from "a-B" to "a-b-D," and the end effect on moisture content will be to dry the steam from "Xb" to "Xd." The effect on final moisture content can be seen on the Mollier Diagram. The result of any frictional losses will be a small increase in the stage reheat.

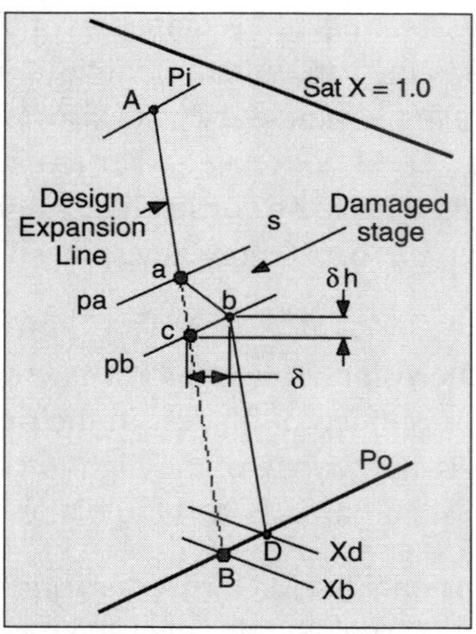

Figure 3.6.10—The effect of stage mechanical damage on moisture content.

The loss in efficiency of the entire expansion shown in Figure 3.6.9, or a single stage shown in Figure 3.6.10 can be found from:

$$\Delta kW = \frac{dH \,.\, m}{3412.14}$$

$$\Delta kW = \frac{\delta h \,.\, m}{3412.14}$$

where:

ΔkW = The power loss in kilowatts

m = The steam flow in #/hour

MOISTURE-INDUCED DAMAGE

Moisture in the steam path is capable of causing extensive mechanical damage to the various components that comprise it. This damage can both degrade efficiency and also force the unit from service for extensive periods to repair and/or replace parts. There are four major forms of damage to be considered, caused either by water formed in the steam path, or intentionally sprayed into it for cooling purposes:

Moisture-impact erosion. This is a form of erosion occurring on the inlet edge suction surface of blades, in their outer regions. This damage occurs as a consequence of the high impact force developing between the moisture particles and the blade material.

Trailing-edge erosion. This is a form of erosion that occurs on the discharge portion of last stage blades near the root section. Damage is a consequence of re-circulation of steam and water carried into the blades from the exhaust casing.

Wire-drawing erosion. Damage occurs when moisture, on one side of a metallic joint is forced by a high-pressure differential to flow between the joint faces, removing material by a washing/cavitation process.

Washing erosion. Washing erosion occurs as material loss from high velocity, possibly high-pressure steam flowing across a surface and removing material.

The following sections will consider these types of damage in greater detail, and where possible explain the material loss mechanism or process.

MOISTURE-IMPACT EROSION

This is the most common form of water damage in the turbine steam path. In order for this erosion to occur, the moisture must accumulate into considerably larger drops than exist from the formation of water seeds by nucleation. The damage results from the impact forces developed between moisture droplets carried by the steam and the blade.

During its passage between the blades, a portion of this transported moisture will be deposited upon the various elements of the steam path, and flow across their surfaces. This water flows as a film across the surfaces into regions where the film collects into larger puddles. They are eventually torn from the surfaces and re-enter the main steam flow, moving at considerably lower velocities.

This re-entrained moisture torn from the stationary portion (normally the stationary blade elements), causes "impact erosion" upon collision with the leading edge of the rotating blades. Typical of the material loss is that shown in Figure 3.8.1, where craters on the leading edge, and the extent of surface penetration back along the suction face is shown. Figure 3.8.2 shows a micro-section through a moisture impact eroded surface, defining the type of craters formed.

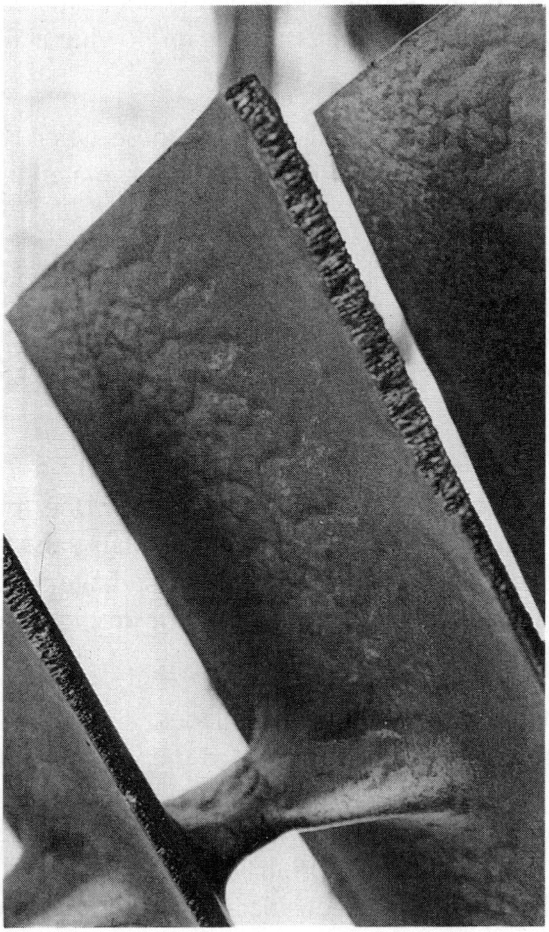

Figure 3.8.1—Moisture impact erosion at a blade outer sections.

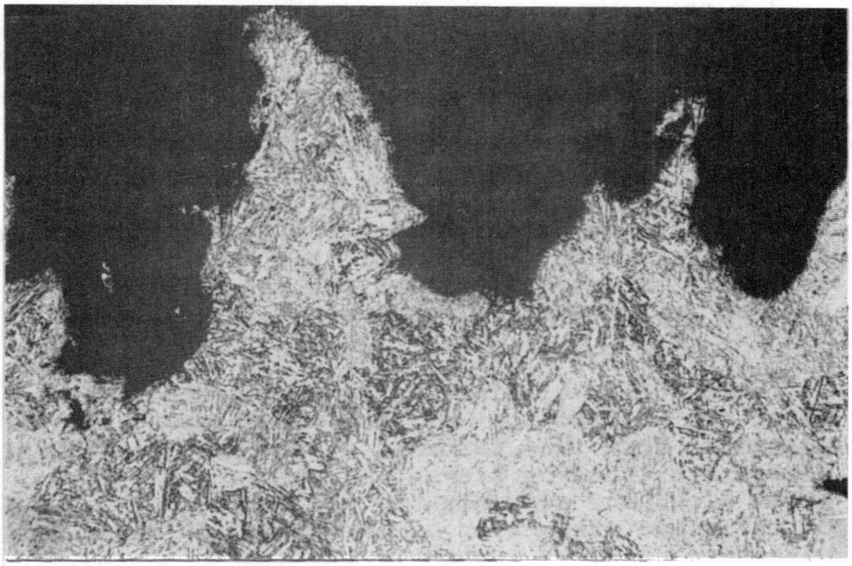

Figure 3.8.2—Micrograph of a moisture impact eroded surface.

The mechanism of impact erosion and material loss

The droplets formed in the steam path are not of themselves large enough to cause erosive damage upon impact with the metallic surfaces of the rotating blade elements. These small moisture particles must therefore aggregate to droplets of a much larger size before the impact forces developed between them and the rotating blades are sufficient to cause surface material rupture and loss. The previous section *Moisture Deposition* and Figure 3.4.5 shows the mechanism of water agglomeration on the discharge tail suction face of the stationary blades.

The water film flowing down the pressure face of the stationary blade is drawn around the discharge tail into the lower pressure region of the row. This is due to the separation of the boundary layer on the suction face of the profile, at or near the passage throat, and the lower pressure this induces in this region. At this point the film meets, and mixes with a small quantity of water flowing across the suction face. On the discharge tail suction face, the water collects into large droplets that grow until they are eventually torn from this location, and are carried into the following row of rotating blades (Fig. 3.4.5).

It is these large droplets, torn from the stationary blade trailing edge that cause the "impact erosion" damage on the inlet edges of the following row of rotating blade elements. Due to their greater mass, these droplets once torn by the steam forces from the suction face, accelerate from zero velocity, but cannot approach the velocity of the steam, which is high in the discharge region. Therefore, their impact with the rotating blade inlet edge is at an oblique angle, shown in the velocity triangle Figure 3.8.3. Upon impact large pressure forces are developed between the blade surface and water droplets.

Figure 3.8.3 shows the steam velocity triangle for the tip section of a large rotating blade, having a high tangential velocity. These steam velocity vectors have a suffix "s." This velocity triangle is typical of the tip region of a large rotating blade row. The blade has a high tangential velocity "U," and the steam discharges from the stationary blade row at a velocity "C1s," from a stationary blade with a discharge angle "$\alpha 1$." These velocities give a relative inlet velocity at inlet to the rotating blade row of "W1s," at a relative angle "$\beta 1s$." Superimposed on this steam vector diagram are the velocity components of the water droplets, suffix "w." The water droplets, starting from rest on the suction surface of the vane, have a relatively small velocity on discharge from the stationary blade row, being accelerated to a velocity "C1w" by the flowing steam. These droplets discharge from the stationary blade row at the same discharge angle "$\alpha 1$." The blade tangential velocity is unchanged at "U." The vector diagram for the water droplet is then shown having a relative inlet velocity to the rotating blade row of "W1w." The water droplets enter the rotating blade row with a relative "Shadow Angle ψ."

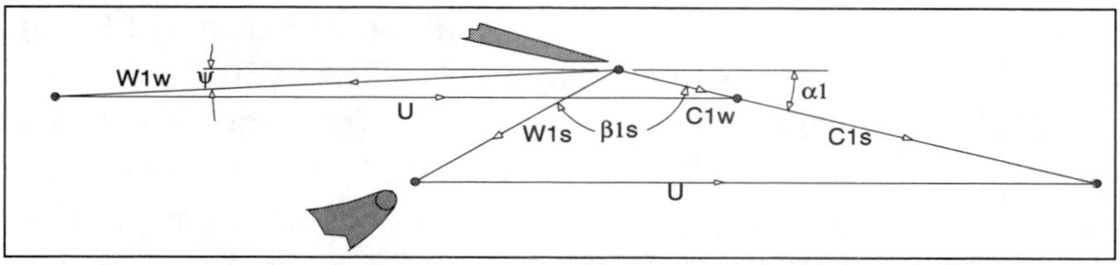

Figure 3.8.3—The velocity triangles for steam "s" and water "w" from a wet stage fixed blade row.

These large droplets torn from the stationary blade row discharge edge, are normally too large (about 4x10-3") to exist in the steam environment present at that location. Therefore, the drops break down into droplets to a size they are able to sustain. These droplets are still considerably larger than the seed elements (about 4.6x10-8 inches) formed by nucleation.

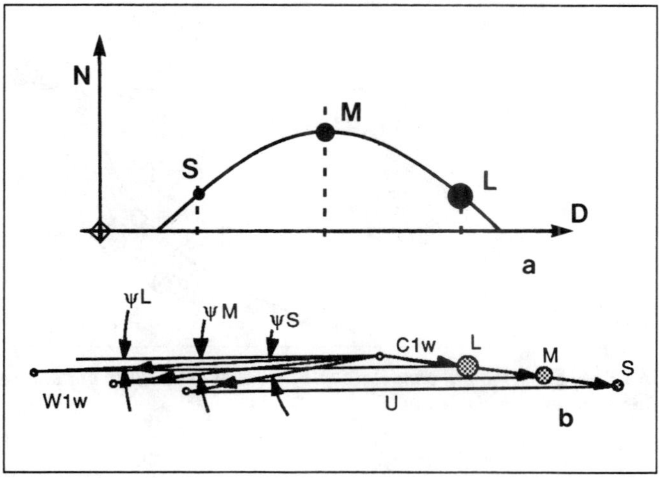

Figure 3.8.4—Water drop size distribution (a), and their velocity triangles (b).

Shown as Figure 3.8.4(a) is a diagrammatic representation of the variation of droplet sizes as they enter the rotating blade row. These droplets, having a spectrum of sizes, are accelerated to different velocities "C1w" by the transporting steam shown in Figure 3.8.4(b). Such droplets achieve different relative velocities "W1w" and shadow angles "ψ" depending upon their size. They impact with the rotating blades, but penetrate the blade cascade in an axial direction to different levels. In Figure 3.8.4(b) the velocity diagrams for three different drop sizes, "L," "M" and "S," can be seen. Their absolute velocities are "C1w," achieved after detaching from the stationary vane and being accelerated by the steam flow, which is in the same direction. Their shadow angles are "ψ," and their penetration in the axial direction is shown in Figure 3.8.5. As a consequence of their relative velocity "W1w," their penetration upon entry to the rotating blade row is dependent upon size, which influences the velocity "C1w," and the impact point on the rotating blade inlet edge.

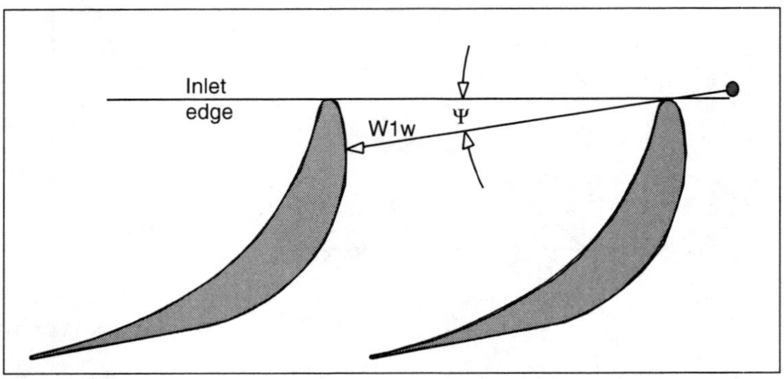

Figure 3.8.5—Showing the shadow angle "ψ" and the degree of penetration into the rotating blade row.

Upon impact between the water droplet and rotating blade inlet edge, there are two factors that can account for, or contribute to the erosive damage occurring. First, there is the initially high intensity pressure developed between the droplet and blade material at the impact point. Second, there are the high radial flow velocities of the water outwards from the point of impact.

Consider these two effects and how they would contribute to surface material loss. If the impact of the high velocity water particle and an elastic/plastic brittle material is considered, the material deforms to an elastic crater under the impact forces [Fig. 3.8.6 (a)]. The magnitude of the stresses induced in the material are sufficient to produce hairline cracks that form at and run around the shoulder of the indentation. After impact, the material recovers substantially retaining its original position. These hairline cracks will now have closed, but there is a circular fracture zone existing around the impact point. Shown in Figure 3.8.6(b) is a macro photograph of an aluminum target after impact by a high velocity mercury drop fired at the target. The circular crack zone can be seen.

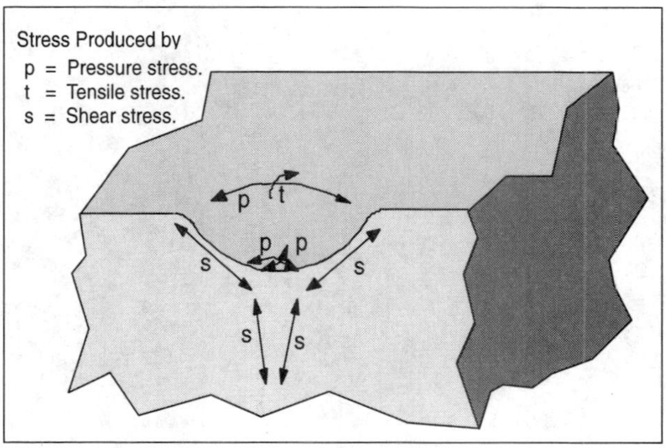

Figure 3.8.6(a)—The stress produced in a surface as a consequence of an impact.

As a consequence of this initial impact there is probably no, or minimal, material loss from the surface. However, subsequent impacts in the region of the initial impact will cause water to be present on the surface, again causing impact indentations. Immediately after these subsequent impacts, the water flows radially outwards. There are three aspects of this flow that can cause or contribute towards the erosive material loss. These are the high impact forces between the radially outflowing water and surface irregularities—many such irregularities having been formed at the hairline crack points. The high shear forces developed between the flowing water and surface material, and finally any cavitation effects of the high velocity water flowing across the profile face.

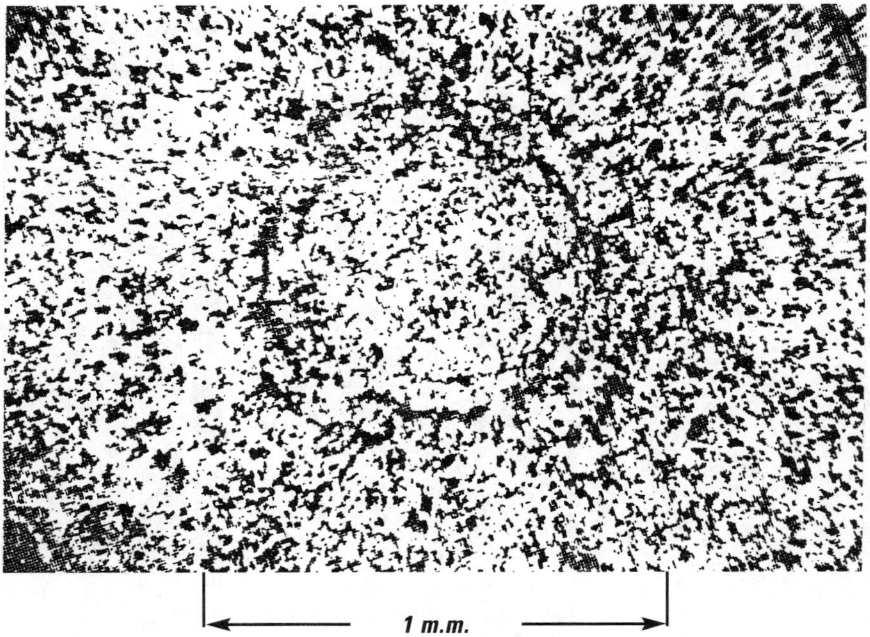

Figure 3.8.6(b)—The "fracture rings" after impact of a mercury pellet with a "hot pressed aluminum" target, at a velocity of 1400 ft/sec.

Methods of protection

The methods of protection currently used against moisture impact damage can be considered to employ two main principles that can be applied simultaneously:

- The collection and drainage of as much moisture as possible, which is present in, and deposited upon the turbine elements
- The deposition, production, or attachment upon the areas of the blade vane surface likely to be affected (a layer of hard resistant material that is better able to withstand the impact forces developed by the water droplets)

A third method is to increase the axial gap between the stationary blade discharge and the rotating blades inlet edge. This increased gap permits a greater distance, and therefore more time for the moisture droplets to be accelerated to the velocity of the steam, reducing their velocity difference relative to that of the blades. This method is not employed extensively, as the cost of increasing the axial length (compared to the losses in efficiency and the increased shaft length) more than offset any potential gain from reduced erosion.

Moisture removal. Moisture removal by internal (water catchers, etc.) and external means in the intermediate portion of the cycle were considered in the previous sections *Internal Collection and Removal and External Removal.*

Erosion resistant inlet edge. The most extensively used method of erosion protection, is producing on the inlet edge of the vane in the regions subject to erosion a hard resistant surface. There are several methods of providing and then attaching such a resistant surface. These include the following:

- *Braze attached resistant material.* The braze attached shield is the most common method of providing protection. The shield can be of many forms. Figure 3.8.7 shows four possible designs. The form of the shield is selected to provide protection, and more recent designs, to preserve the aerodynamic form of the vane at the point of attachment. Such contouring allows the shield to attach within a preformed recess and preserve the aerodynamic form of the profile. In Figures 3.8.7(a) and (b) the shields are of relatively simple design, and make no attempt to maintain the aerodynamic form of the vane. There will then be some degree of boundary layer separation, and efficiency loss with these forms. Figure 3.8.7(c) and (d) employ more complex shield forms, fitted into recesses, or specially designed access slots at vane the inlet. These shields maintain the nose profile.

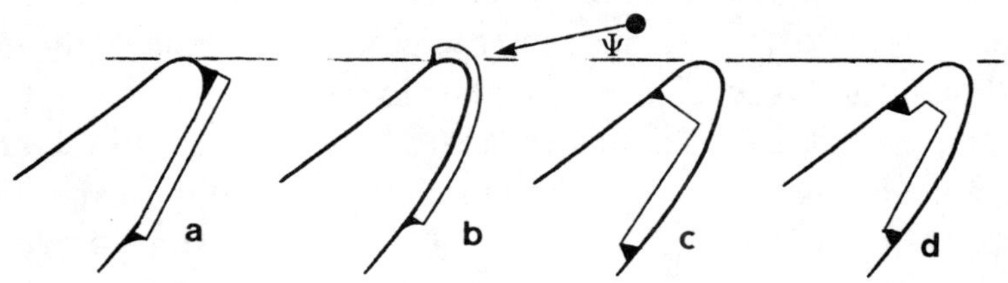

(a) and (b) Braze attached. (c) and (d) Weld attached.

Figure 3.8.7—Various nose forms of hard metal erosion shield.

Difficulty can be experienced with the braze attached shield, if the procedures for attachment are not followed in detail. The melting point of many suitable filler materials is close to the austenitising temperature of the blade material. Failure to control temperature can cause embrittlement of the vane at any point of overheating. This can ultimately lead to premature failure of the blade material.

For brazed attachment, the shield can detach in operation if the brazing joint is not adequate. Shield detachment leaves the blade tip material exposed, and therefore subject to a heavy rate of material loss.

Another difficulty encountered with the brazed shield has been that the softer braze material has been subject to heavy moisture impact erosion, and some undercutting from the moisture. It is possible however to design the shield so it is partially self-protecting at the braze point.

- *Weld attached resistant material.* Another method of attaching shields is by welding. Control of the weld attachment process is critical. Cracks have developed in the heat affected zone (HAZ) of the shield/blade interface, and while the majority of these cracks have "run forward" towards the inlet edge, some have regressed into the blade parent material. In the event a crack does develop some form of corrective action is required. This situation cannot be left uncorrected.
- *Thermal Hardening.* Thermal hardening is undertaken using either direct flame impingement or induction heating of the blade material. With such a hardening system, process control is critical, and stringent observance of procedures is mandatory. If this process is not controlled normally cracks can initiate in the HAZ, causing the outer portion of the blade to detach.

Figure 3.8.8 shows the section of a thermally hardened inlet nose displaying the effects and results of a hardness traverse. The base material had a design specified Brinell Hardness (BHN) of 260-275, and the material as seen from the base material readings fell into an acceptable range of values. At the transition HAZ, there was a dramatic change of hardness to the 400-415 BHN range (one reading taken too close to the profile edge gave a questionable reading). The hardened region ends abruptly at the HAZ.

- *Laser hardening.* A further development in the hardening of the blade vane material is by the use of lasers. This is not as common as other methods, but does promise to be placed into greater use as results justify its application.

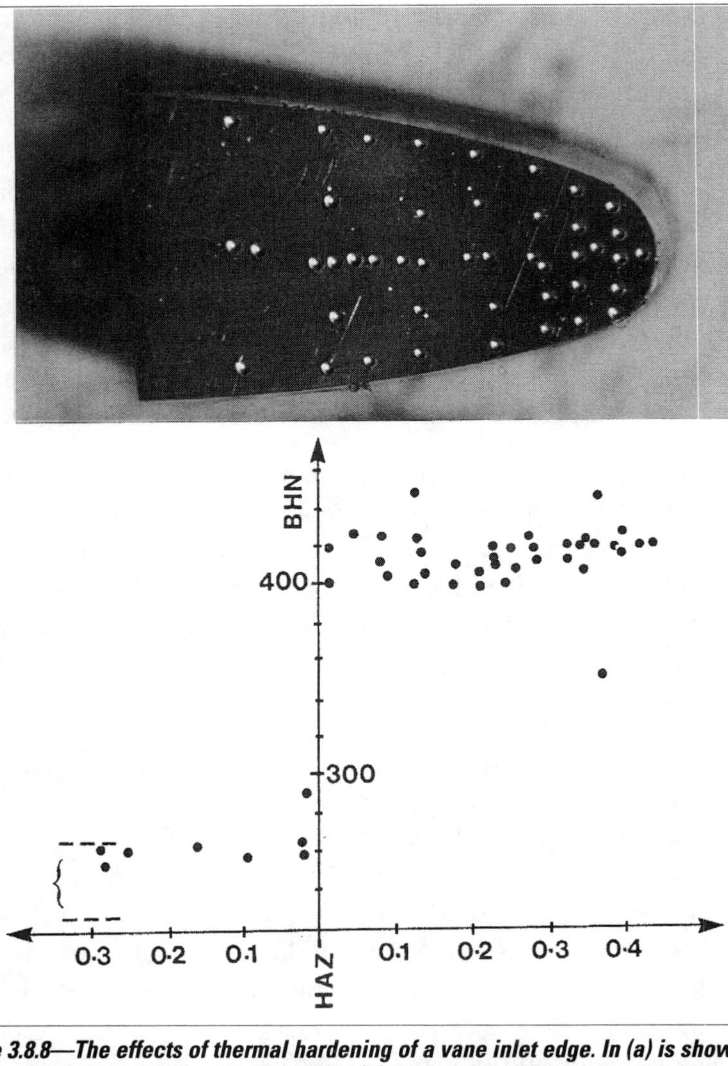

Figure 3.8.8—The effects of thermal hardening of a vane inlet edge. In (a) is shown a section through the hardened nose, and in (b) the results of a hardness traverses.

Secondary (concentrated) erosion

If a blade that has been in service in the moisture region for an extended period is examined, there are often regions where the local depth of penetration by erosive action is far greater than in the surrounding material. This localized heavier damage is caused by the occurrence of obstructions or collection points at different "up stream" locations within the steam path. At these points moisture can collect or concentrate until it is suddenly torn off by the main steam flow, or thrown off by centrifugal action, causing heavy localized erosion on the following rotating blades. Figure 3.8.9 shows this "concentrated" or "secondary erosion."

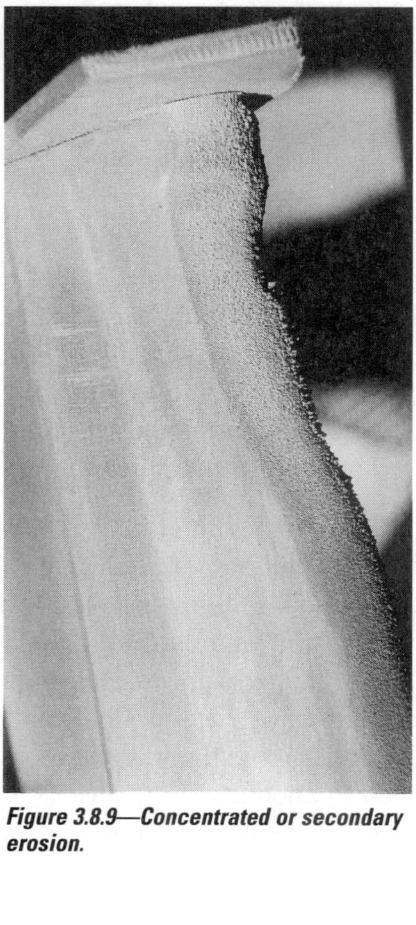

Figure 3.8.9—Concentrated or secondary erosion.

Careful attention to detail during the design phase can help recognize the potential for this form of damage, and help eliminate or minimize the potential for it to occur. Prototype units should always be examined early in their operating life, to determine if the early stages of such damage are present. If damage has occurred or there are indications of such concentrated penetration, the design should be reviewed, and remedial action taken before the damage has progressed to the point its effects are significant, and before components must be replaced prematurely.

In a well-designed stage, there should not be a need to replace blades during a normal operating life of 200,000 hours due to moisture impact erosion. However, concentrated erosion can force a unit from service for an extended outage, with the need to replace the affected row.

Off-shield erosion

During the design phase, the dimensional requirements of the erosion shield are predicted from an analysis of the stage operating conditions, and those anticipated under "worst conditions," including prediction of the shadow angles likely to be encountered. Using this data, a prediction is made as to the extent of erosion, the degree to which it is expected to occur on the blade, and therefore the level of coverage required by a protective shield.

Low-pressure blading is designed for a range of applications, a variety of stage inlet conditions, and a range of blade loadings (pounds of steam/water flow in the stage). The low-pressure exhaust blades can be subject to a range of moisture impingement probabilities. Last stage shield requirements are designed to accommodate this range of steam and water entering the rotating blade, and are normally designed to be adequate under the most severe conditions the blade is expected to encounter. However, it is a common operating experience that occasionally unique features of operation, or characteristics of a particular unit, can cause unusual patterns of material loss. These unique situations can cause the shield to be inadequate and erosion can be noted to occur to the extent the life of the erosion shield and/or blade can be placed in jeopardy.

Whenever a unit is opened for inspection and maintenance, blades should be examined for the depth of erosion, and also for any unusual "off-shield erosion." If it is considered the level of damage is progressing to the extent excessive damage can be anticipated and the reliability of the unit could be affected, the condition should be monitored either by measurements, making casts of the damage, or photographs. This monitoring should be done in such a manner the rate of penetration or extent of damage can be predicted and judgment made, even though subjective, of when corrective action could be required. Remedial action must be planned for and taken before the blade is destroyed. Destruction of the blade is not of itself catastrophic, however, consequential damage can be very expensive to correct.

The various forms of "off shield" erosion that a blade should be examined for consist of the following:

Beyond shield erosion. In this case the impact erosion extends back, beyond or behind the shield coverage into the base material. Such damage is shown in Figure 3.8.10. Often, this is not significant as the erosion may only be of a very light nature and therefore will not endanger the integrity of the blade. Should this occur however, it is advisable to note its extent and radial location and then monitor further deterioration of the vane. Should this type of damage be found early in the operating life of the unit—before 10,000 operating hours—and occur at one or two radial locations, it would be advisable to monitor the condition of the shield in this region, since this could be symptomatic of heavy localized, or secondary erosion, which will occur there.

Figure 3.8.10—Erosion beyond the erosion shield.

It is also possible with a shield attached by brazing, that early erosion behind the shield is simply removing the excess soft braze material. This type of braze erosion should not be confused with erosion of the blade material.

Below shield erosion. Normally, erosion does not occur below a blade threshold tangential velocity of about 800-850 feet/second (50" diameter at 3600 rpm). The shield will normally be designed to extend down the blade inlet edge until a point is reached on the profile at which this velocity does not occur. There are occasions when erosion is found below a shield and this progresses into the blade material to the extent the shield can be undercut and severe damage occur due to the stress concentration effect this material loss causes. Material loss in such a location can also modify the natural frequencies of the blades and move their frequency into a dangerous range. In such a situation shown in Figure 3.8.11, it is necessary to consider the extent to which the shield can be produced or modified to cover the lower radial portions. It would also be advisable to consider if by modifying moisture flow patterns some corrective actions on the preceding stationary blade row would help alleviate the problem.

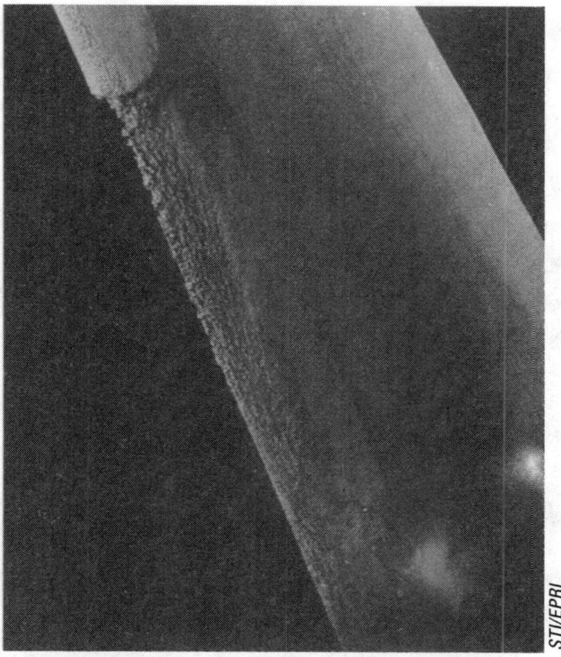

Figure 3.8.11—Erosion below the lower end of the erosion shield.

Erosion between shield segments. Many units, particularly those of an older design, employ an erosion shield produced from two or more individual segments. At the point where the shield segments butt, it is possible for erosion to occur and proceed into the softer braze or blade material. The shield is generally arranged in its radial direction so overlap of shield pieces occurs and the radial flow direction of the moisture particles tends to cause less severe damage. However, such damage does occur. Figure 3.8.12 shows the extreme of this type of erosion, where the outer segment of a shield has detached, and the resulting depth of penetration in the outer blade portion that was consequently exposed is being destroyed. In many instances, significant material loss can occur on the two ends of the shield, without its detachment.

If damage of the type shown in Figure 3.8.12 occurs, this can, if it is only on one or two elements, affect the dynamic balance of the unit, as the mass is being removed from the outer diameters of the vane where its centrifugal mass kinetic effect on balance, is more significant.

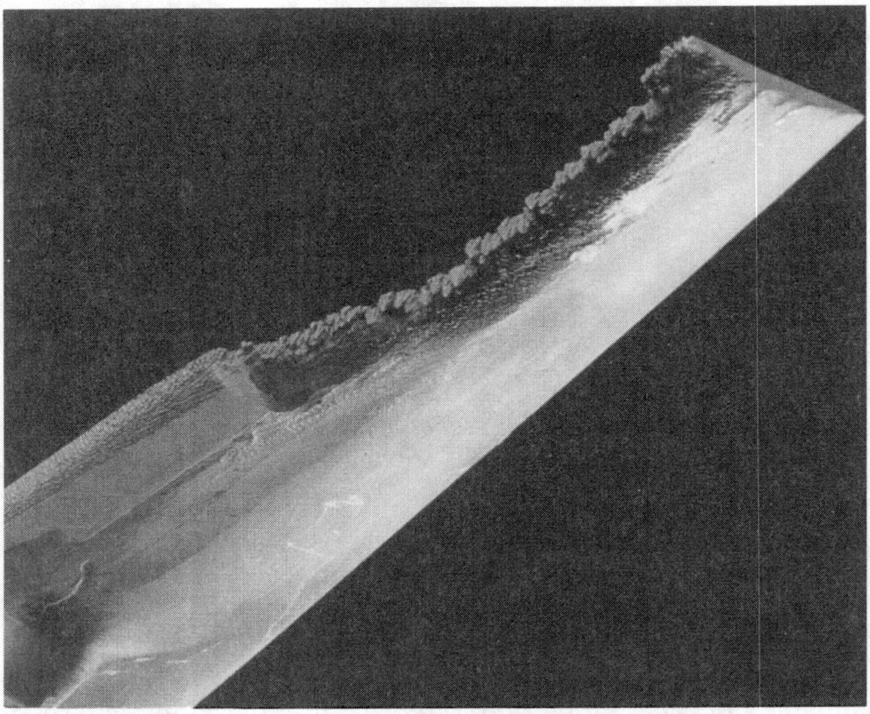

Turbine Blading (UK) Ltd

Figure 3.8.12—Heavy erosion into the blade base material after loss of the erosion shield. The shield having detached at the radial position where two segments butt.

Pressure surface erosion. When a stage has large quantities of moisture transported by the parent steam, it is possible for some particles of moisture to rebound after impact with the suction face inlet edge, to the extent they can impact on the pressure face of the vane of the adjacent blade causing material loss there. Such rebound effects can also be present from moisture particle impacting with a casing wall or other part of the steam path, rebounding into the blades. In general this condition is not dangerous. However, if it occurs accompanied by massive penetration into the blade material on the suction face, or losses in the shield material it can, depending upon the form of the shield, cause a weakening of the attachment to the extent the shield can detach leaving the surface of the base material exposed.

Erosion of the coverband. A form of secondary erosion, which can occur at the tip of the blade, is due to concentrated moisture. This moisture separates from the outer radius of the stationary blade or diaphragm. In this case the moisture droplets can flow into the rotating blade, and impact with the under surface of the coverband, or the leading edge of a coverband segment. In some stages, this can result in an undercutting of the tenons attaching the coverband. This effect can also cause a localized cutting of the coverband, as shown in Figure 3.8.13. Such damage has the potential to become severe to the extent it will require repair, or possibly replacement of the coverbands on the entire row.

If these phenomena are noted during early periods of operation, the effect can possibly be mitigated to some extent by rounding the discharge edge or corner of the stationary blade or diaphragm. This is effective only to the extent the steam is able to transport the moisture over the blade tip and past any leakage restriction points that are present. An action such as rounding the discharge wall edge, will have the effect of causing both the water and steam to flow towards the outer casing region, by-passing the blade, thereby reducing the level of moisture impact. (It can also introduce the possibility of tenon erosion, see following section.) Rounding the diaphragm outer ring discharge edge will also have an adverse effect on stage efficiency. However, the loss is relatively small. It is not possible or advisable to use the method of diaphragm edge rounding if the rotating blade has radial seals above it. This can cause additional efficiency losses, and possibly seal damage in the stage.

Figure 3.8.13—Erosion of the cover band.

Tenon erosion. In the event there is heavy moisture concentration above the tips of the blades, it is possible for tenons to be eroded (Fig. 3.8.14). This in itself may not be of consequence as the tenons could be held in place by the frictional fit between the tenon and sidewalls of the tenon holes in the coverband during the early years of operation. However, as this friction fit is normally lost or deteriorates as the

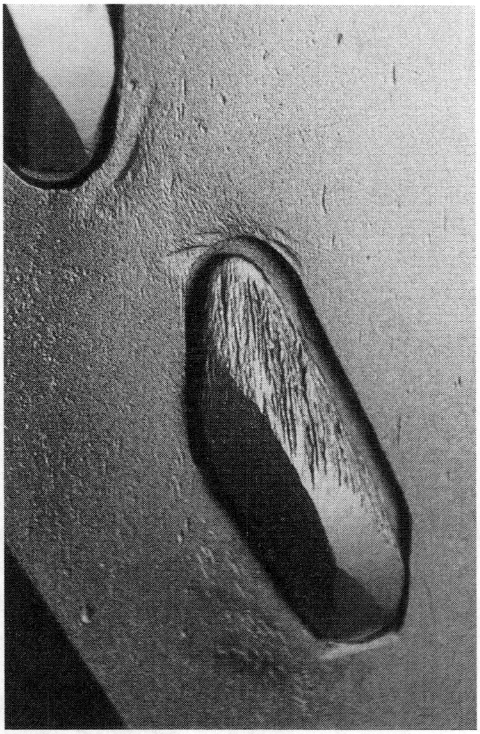

Figure 3.8.14—Tenon erosion.

unit ages, this is the time during which tenon material is lost. Tenon head material loss reduces the rivet head shear area, and will eventually have an effect on tenon integrity as the unit ages. In the event of a "fox-holed" tenon, the head material will not be removed as easily, as the head will be protected by the coverband.

Erosion of squealer tips. There are instances in the case of 12% chrome steel blades without coverbands, where squealer tips are used to limit tip leakage. These squealer tips will often suffer heavy material loss due to erosion of the thinned portion of the squealer. This is shown in Figure 3.8.15, where sufficient material has been lost—so excessive steam leakage will occur across the tip section.

This leakage will occur to the extent there will be a degradation of the stage efficiency.

Figure 3.8.15—Erosion of the "squealer tip."

Erosion penetration and time

A blade subject to moisture impact erosive attack, and therefore material loss, has damage that worsens as a function of time. However, this form of damage does not exhibit a linear relationship to time, but rather follows an exponential relationship.

The total "erosion-time" relationship is shown in Figure 3.8.16, plotting erosion penetration depth "e" as a function of exposure time "Y" years. The erosion penetration can be considered to occur in three distinct phases. The initial phase "A" of Figure 3.8.16 is commonly referred to as the "incubation period." This time is very dependent upon the material being eroded, being very short for 12% Cr steel, irrespective of its hardness, and relatively longer for a Stellite 6B material. This incubation period is very short. Normally units would not be opened for inspection when the material is being lost in this phase. Its only significance is the delay it can cause for the entry of the material into the secondary phase "S." During this secondary phase, there is a massive material loss rate over a period of about 7,000-10,000 operating hours. The erosion penetration reaches a depth "ε" during this period and the characteristic eroded surface appears.

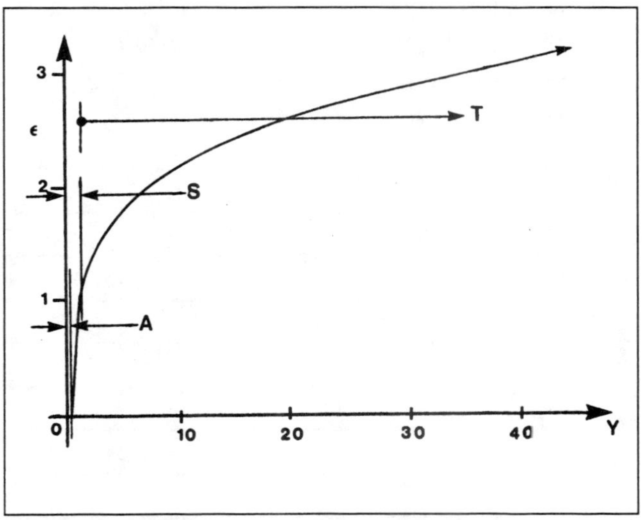

Figure 3.8.16—The relationship between erosion penetration and time.

The third or tertiary zone "T" is relatively slow, eventually approaching almost linear loss, during which the erosion depth of penetration "ε" doubles to "2ε" in about 7-10 years of further operation, if the operation continues at about the same load pattern and steam conditions as the first 7,000-10,000 hours. The erosion will treble to "3ε" in 35 to 40 years, again based on the assumption of the same load pattern and steam conditions.

Because erosion depth is time dependent, it is possible to produce a curve allowing a rough estimation to be made of future penetration depth predicated on the unit continuing to operate at the same steam conditions and with approximately the same load factors.

In measuring erosion depth of penetration "ε," it is important to ensure the depth is measured perpendicular to the direction at which the steam enters the row, *i.e.*, at the shadow angle "ψ" of Figure 3.8.5, where the shadow angle can be gauged from the position of the adjacent blade, and the depth "ε" is measured using a piano wire end to gauge this depth. This is an awkward and rather subjective method, but with experience the engineer can soon make a good and accurate assessment quite quickly.

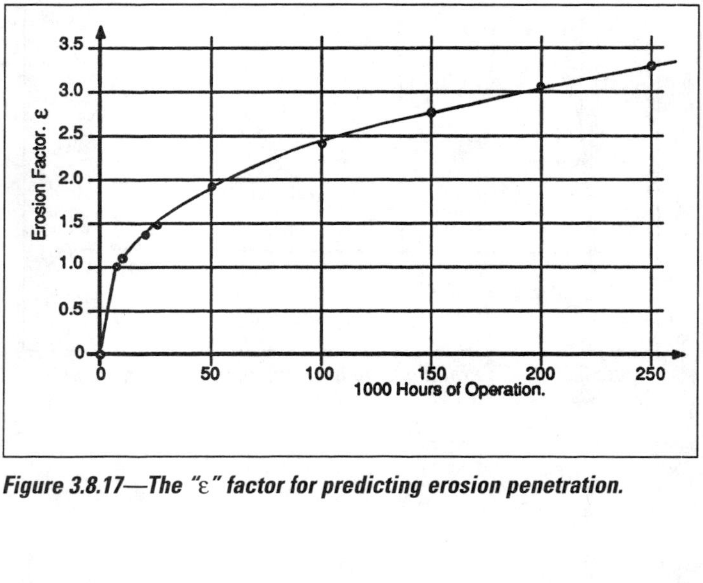

Figure 3.8.17—The "ε" factor for predicting erosion penetration.

The erosion/time characteristic for any 12% chrome steel blade material and forged Stellite 6B is shown in detail in Figure 3.8.17. Here the depth of penetration characteristic is set at "$\varepsilon = 1.0$ at 7,000 hours." This is characteristic of the first year of operation at a load factor of about 80%. Using the value of "ε" at any time, the depth for future operation can be predicted, assuming during this time the erosion does not penetrate the shield.

Classification of erosion severity

The classification of erosion severity is at best a subjective art. Terms such as "light" or "heavy" are meaningful only in terms of other parameters such as vane thickness, blade length, the period of hours the blade has operated, and even in comparison with other similar elements when examining rows.

In an attempt to quantify erosion, the seven levels defined in Table 3.8.1 are suggested together with an "0" level, indicating no or negligible damage. The description of this "0" level is at least 75% of the original inlet nose surface remains and is unaffected. There may be surface discoloration or chemical deposition on the surface but no significant material loss due to moisture impact. Figure 3.8.18 shows the profiles of blades illustrating the characteristics of the various levels of penetration.

The most significant measure of erosion damage is the depth of penetration into the blade or shield material. Often where large material loss has been experienced over an extensive radial distance of the inlet nose, measurements and estimates of the penetration are facilitated with the aid of a straight edge held at the original surface, preferably located from existing surface material. However, this may not always be possible in the case of extensive material loss.

Table 3.8.1 lists the categories, and in addition, attempts to equate them to a range of maximum penetrations in thousands

(0.001"). The physical description and surface characteristics are listed to provide some indication of surface appearance. It is not intended, or suggested, that all or any of these characteristics are present in every eroded surface grouped into the category. These are simply intended to act as a guide for recognition.

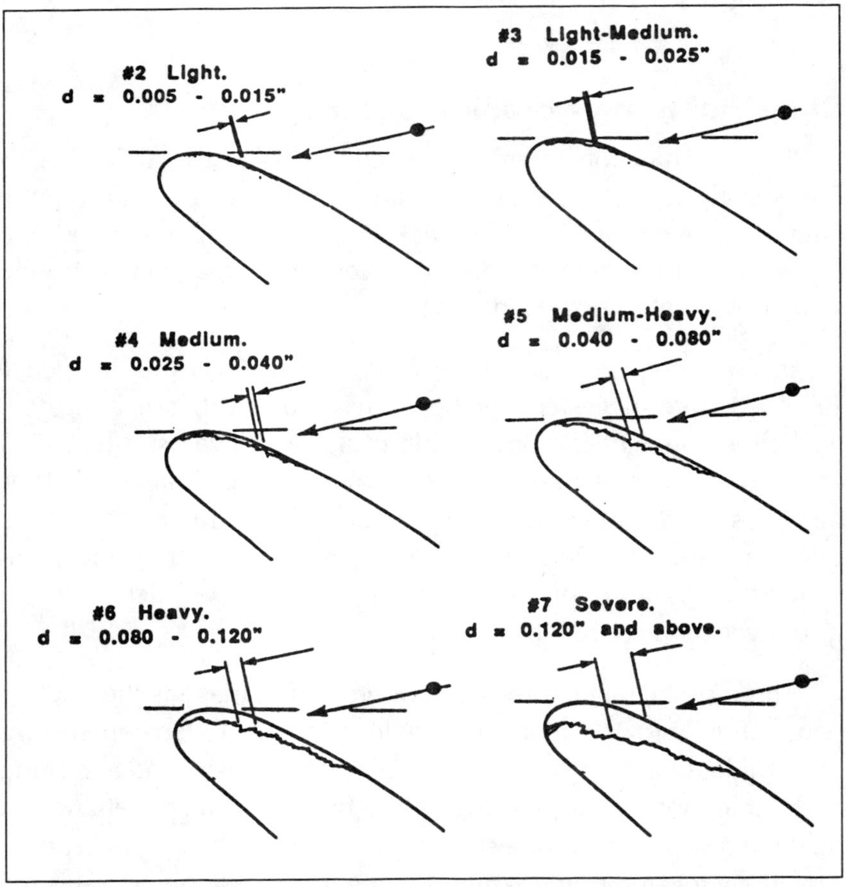

Figure 3.8.18—The inlet edge profiles of levels (2) to (7), illustrating some characteristics defining the extent of moisture impact erosion damage.

Indicator Range.	Surface Definition.	Description and Surface Characteristics.	Penetration in 0.001"
0		Surface unaffected by moisture - at least 75% of original surface remains.	0
1	Frosting.	A minimal amount of surface roughening, no perceptible surface material loss.	0 - 5
2	Light.	A small degree of penetration; original surface profile still evident; craters more than 0.015 inches wide, damage does not extend to the pressure face of the inlet edge, and damage is not visible from the pressure face.	5 - 15
3	Light-Medium.	Original surface still evident, but most removed. Craters more extensive; small particles of material eroded away. Erosion damage possibly visible from the pressure face.	15 - 25
4	Medium.	More extensive craters from 0.015-0.040 inches wide. No original surface material remaining, damage may have removed small amounted of material on the pressure surface also. 25 μ-inches particles missing.	25 - 40
5	Medium-Heavy.	Greater surface loss. Damage removing material from the pressure face to a greater extent. 0.040 inch particles missing. No evidence of profile shape on suction face.	40 - 80
6	Heavy.	Extensive material loss from the suction face. 80 μ-inch particles missing. No evidence of profile shape on the inlet edge or suction face. Craters possibly extending through any protective material, or shield, into the blade material.	80 - 120
7	Severe.	Massive material loss, no evidence of profile at inlet edge or suction face. Material loss on pressure face also. Moisture may be eroding back into inlet edge, exposing large ares of blade material on elements with protective shields. Larger than 0.120 inch particles missing.	120+

Note: The surface description given is intended to act as a guide in recognition of the extent of damage; few or even none of the characteristics listed may be present in an particular stage. The erosion loss pattern is more a function of the form and radial spread of the water droplets which flow from the stationary to the rotating blade row.

Table 3.8.1—Moisture Impact Erosion Scale, Used to Compare the Extent of Damage Penetration.

Monitoring damage

In general, moisture impact erosion in a well-designed blade will not of itself cause sufficient material loss that a blade will have to be replaced. Damage and material loss occur in the outer portions of the blade vane, but this should not affect the reliability of the blade. This is true to the extent the erosion occurs on the inlet edge suction face. There are however, certain aspects of secondary erosive material loss that can give cause for concern, and possibly require some form of corrective action:

- Excessive erosion below the erosion protective shield
- Erosive undercutting of tenons attaching a coverband
- Severe erosion beyond, or behind, the erosion shield
- Severe erosion endangering the attachment of the erosion shield

Moisture impact erosion is time dependent. However, the damage, or material loss does not vary linearly with time, and apparent severe erosion early in the life of a unit is not necessarily indicative of a poor protective system.

Localized or secondary erosion should be considered—examined to the greatest extent possible, and the cause explained. Blades subject to moisture impact erosion are elements that are tuned. These blades normally have high stress levels and their frequencies are sensitive to changes in blade section. This frequency sensitivity is such that it is normal for extensive material loss to occur before the blades require replacement.

An example of monitoring the last stage blades of a unit with predicted heavy erosion is shown in Figure 1.13.2. Here both the mean and maximum values of erosion penetration in the outer section are shown for one double flow low-pressure section.

Last stage blade erosion and condenser pressure

The extent to which moisture impact erosion occurs in a blade row, is obviously a function of the amount of water impacting the blade edge. The amount of water is indicated by the steam conditions from the Mollier Diagram. What is not so obvious is that the moisture content at exhaust from the stage—which is a function of the exhaust pressure produced by the condenser—has little or no effect on moisture impact erosion. Moisture impact erosion is a function of the moisture content at inlet to the stationary blade row of the stage, which will be collected and concentrated to discharge into the rotating blade row, causing the damage. Shown as Figure 3.8.19 is a Mollier Diagram indicating the steam conditions in an exhaust stage.

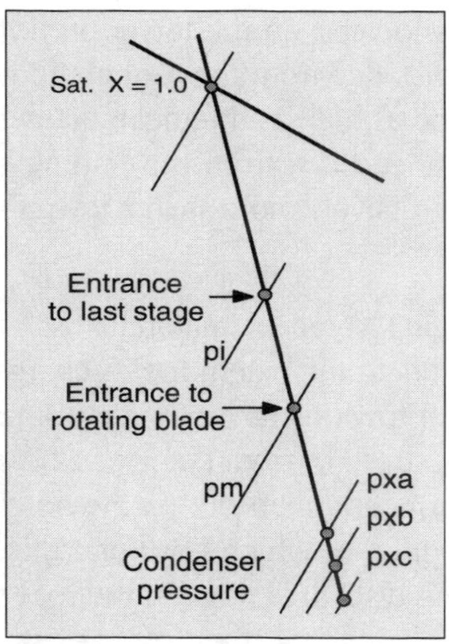

Figure 3.8.19—The effect of varying condenser pressure from "pxa" to "pxc", on the expansion line. The moisture impact damage is caused however by the water deposited in the fixed blade row, and will not be significantly influenced by the exhaust pressure.

As the condenser pressure "Px" reduces, there is a minimal change in the stage inlet conditions "Pi," and therefore moisture impact erosion potential will be unchanged. If blade inlet edge erosion represents a serious condition in any unit, there will be no advantage to increasing backpressure. This will not reduce the erosion damage rate.

BLADE TRAILING EDGE EROSION

A condition found on certain last stage blades, is erosive type damage of the blade suction face. This damage occurs on the discharge edge, and is located on the lower portion of the vane, often beginning a small distance from the root platform. This material loss can exist up the vane as high as the mean diameter. The damage initiating a small distance back from the trailing edge (about 0.3" to 0.5"), becomes progressively more severe towards the discharge (tail) position.

This is a "gouging" type of material loss, as shown in Figure 3.9.1. Damage occurs to the extent the discharge edge will become progressively thinner towards its discharge point. Such thinning will increase the direct stress by reducing load-bearing area, and also introduce a stress concentration due to the notch. This will become a stress sensitive region, at which position cracks and ultimate failures can initiate. The discharge edge in the root section of the blade is one where relatively high tensile stresses exist, and in addition will be subject to high order bending stress, as it represents the outer fibers on the blade material. The vibratory, or dynamic loading can be high in this region.

Figure 3.9.2 shows similar damage. In this case a crack has initiated is particularly difficult to see from the suction face, and even difficult to detect by non-destructive methods from this side. However, this crack penetrates through the thickness of the discharge tail, and can be seen by "red dye" examination from the pressure face (Fig. 3.9.3).

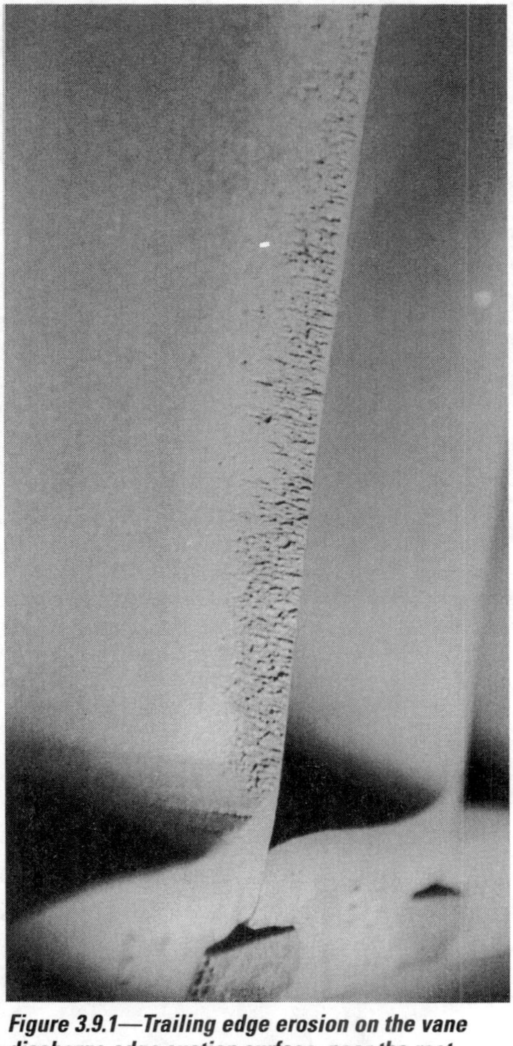

Figure 3.9.1—Trailing edge erosion on the vane discharge edge suction surface, near the root section of a L-0 blade row.

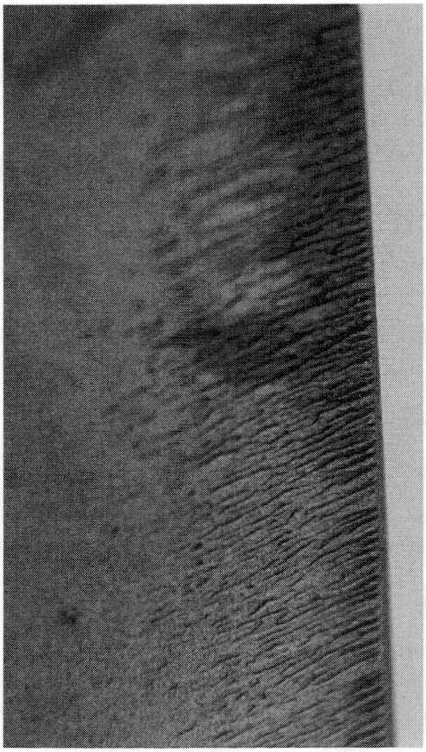

Figure 3.9.2—Trailing edge erosion on the suction face with deep "Gullies".

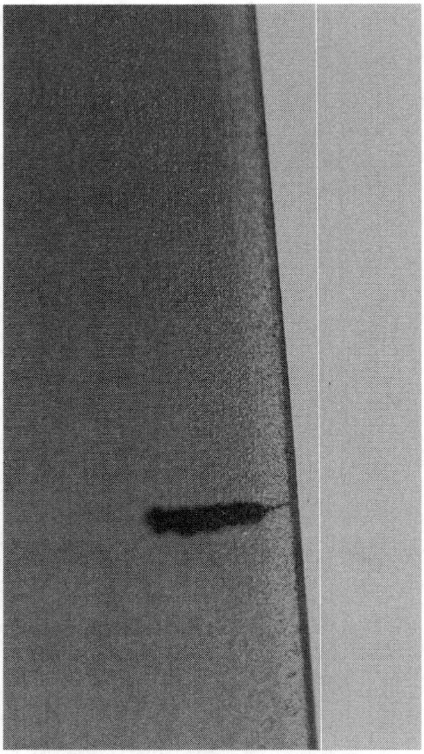

Figure 3.9.3—The pressure face of the vane shown as figure 3.9.2. In this view the crack can just been seen with "red dye".

Figure 3.9.4(a) shows the method of exposing this crack, as it was opened, and the failure surface is shown in Figure 3.9.4(b). Figure 3.9.5 shows detail of the cracked region.

The exact mechanism producing this gouging material loss at the trailing edge is not precisely understood. There is strong evidence to support the hypothesis that moisture particles re-enter the last stage blades adjacent to the root. This re-entry occurs only when the steam force is small and insufficient to inhibit such working fluid re-circulation.

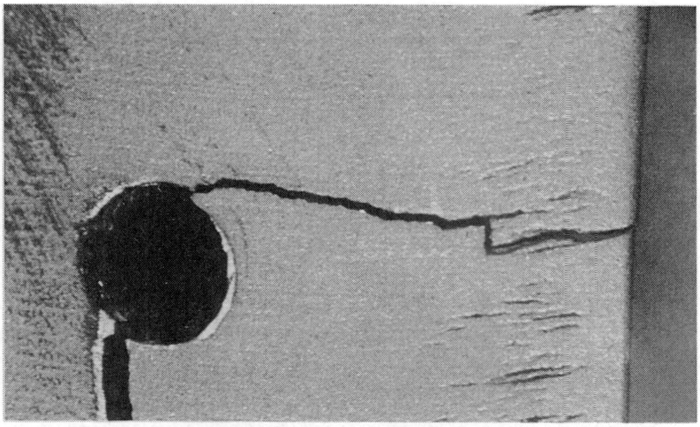

Figure 3.9.4(a)—An enlargement of the crack showing in figure 3.9.3 before exposing the surface. There a small hole has been drilled to allow the vane to be cut to expose the failure surface.

Figure 3.9.4(b)—The crack surfaces of figure 2.9.3 exposed.

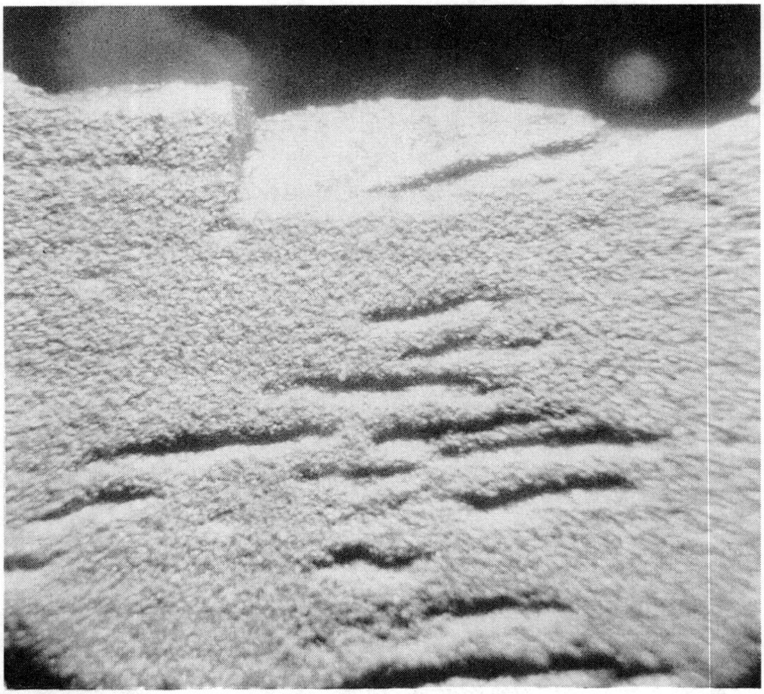

Figure 3.9.5—Enlargement of the crack initiation region.

Root section re-circulation

At full load the last stage blade discharge annulus will run full, and the streamlines can be depicted shown in Figure 3.9.6. This flow pattern will be fairly consistent, down to relatively lower part loads. However, the design flow density will influence the streamline position (pounds of steam per square foot of annulus area), and the exhaust pressure (specific volume). These two parameters establish the volumetric flow in the annulus. The flow density (back-end steam loading) is established by economic considerations in selecting the discharge area arrangement.

As load (and therefore steam volumetric flow) reduces, a point is reached at which the flow, under the influence of the radial flow component of the steam, will migrate to the outer flow regions. There will be no flow through the blade sections adjacent to the root or hub. At this and lower steam volumetric flows, a low-pressure region will be developed in the root sections, and a negative pressure will be present in the region of the rotating blade root section. The effect of this negative pressure will be to induce steam that is filling the exhaust hood, to flow back into the rotating blade annulus (Fig. 3.9.7). This steam, once it enters the blade passage, is then pumped radially out under the effects of rotation, and forced out of the blade annulus. These motions set up a re-circulation of steam in the lower section of the blade row—the lighter the flow loading on the blades, the greater the extent of this re-circulation will be.

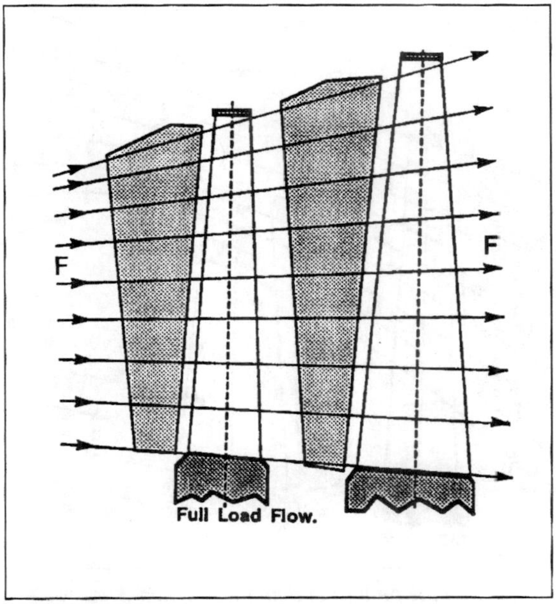

Figure 3.9.6—The steam stream lines when the low pressure section is flowing "full".

The total effect of steam re-circulation is the steam that re-enters the blade row has work done on it by the blades, causing the steam temperature to rise. The heated steam in turn heats the blades through which it is being recirculated. The final effect of root-rarefied flow is raising the temperature of the exhaust blades and the exhaust. This increase in temperature of the blade material modifies the mechanical properties that can—if the system heats sufficiently—change the natural frequency of the individual blades. Since all long last stage blade elements are tuned, it is necessary to provide a cooling system that will remove heat from the blades to prevent their failure. This cooling is accomplished by the use of water sprays providing a fine mist into the exhaust annulus to remove as much of the superheat present in the steam as possible.

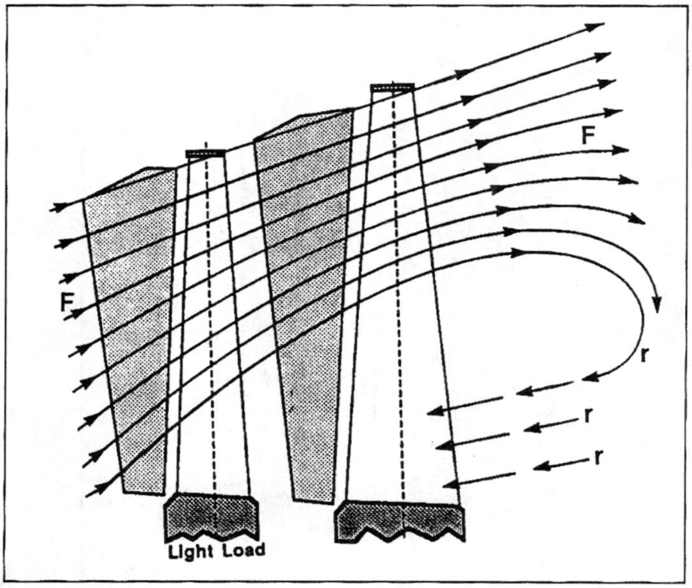

Figure 3.9.7—The flow pattern at light loads when there is steam recirculation into the root section of the last stage rotating blades.

Water sprays in the exhaust hood

To remove the windage heat that is generated by re-circulation, water is sprayed into the exhaust hood—preferably treated and not raw water. It is taken from some location within the steam cycle. Possibly the most common source is from the condenser hot well where there are unlimited quantities of relatively clean water (which could have some gasses dissolved in it). In spraying, the water is atomized, given as wide a distribution as possible, and a sufficient quantity is supplied to remove any latent heat from the steam.

Unfortunately, this cooling water—if supplied in excess quantities more than required to remove the superheat—is left free in the exhaust annulus. It is this free water that is captured by the re-circulating steam and is carried back into the rotating blades.

The damage mechanism

If conditions exist that steam is recirculated in the last stage exhaust, then any free water present can be drawn in with the re-circulating steam. As discussed above, water-cooling sprays are prime candidates for this water. It is also possible moisture can be entrained from other parts of the structural members of the exhaust hood, where it exists as a film on the surfaces. However, it is believed that the majority of any moisture re-entering the blades is carried in with re-circulating steam.

Figure 3.9.8 shows diagrammatically the effect of re-circulation in the exhaust hood, where water introduced through cooling sprays to remove the frictional heat re-enters the rotating blade root section. It is also clear that such re-entering moisture will not penetrate further up the steam path than the rotating blade elements.

The damage that occurs on the trailing edge cannot be explained by the relatively simple mechanism of the velocity of impacting moisture, since this damage occurs in the lower portions of the blade, where velocities are low and below what is normally considered to be the threshold values associated with material loss. It must also be considered that no material loss has been noted on the root portions of the blade. The root is often shielded, to a large extent, by diffuser plates and even bearing cones. In some instances, one would normally expect to find some form of damage however light in this root region if impact erosion were the only mechanism accounting for damage.

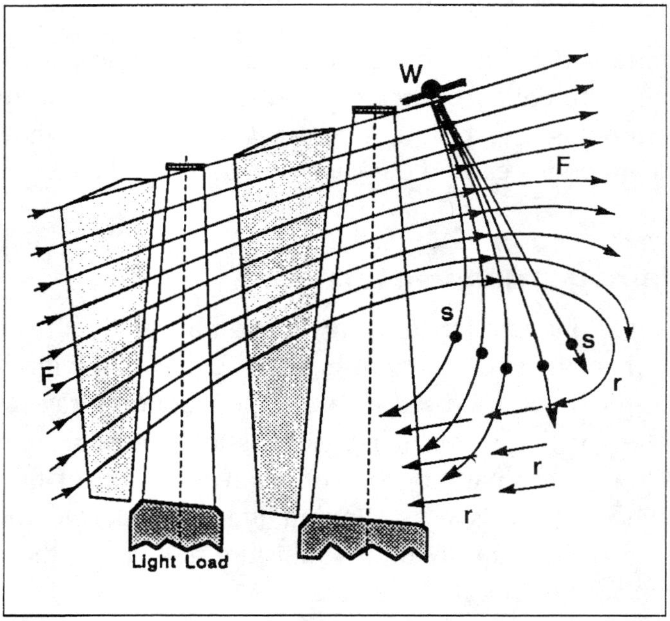

Figure 3.9.8—Water sprayed into the exhaust hood for cooling, has excess quantities carried back into the rotating blades.

This damage condition probably occurs as the result of three separate, but possibly interacting mechanisms, including:

- moisture impact type damage as discussed earlier. This form of damage could possibly occur in some locations, or at least contribute to the total material loss
- cavitation losses caused by the high velocity of the water droplets, which after impact with the blade discharge edges is pumped out from the impact point
- the interaction at the blade surface between the centrifugal and Coriolis forces at the point of impact and the water flow to the exhaust edge

Material loss model

A possible model that can be proposed for the material loss from the discharge edge, shown in Figures 3.9.1 and 3.9.2, is that moisture re-enters the blade vane from the exhaust annulus at the root section. The water carried in by the re-circulating steam is shown in Figure 3.9.8. Figure 3.9.9 shows the probable velocity diagram of the re-entering water at, or near the rotating blade root section. The water droplets have a velocity "Vw" approaching the steam velocity, and renters between profiles "J" and "K." The blade is rotating in the tangential direction at a velocity "U." Therefore, the relative velocity of the water into the blade annulus is "Vn," inclined at an angle "χ."

At this point of moisture re-entry, the spray moisture droplets are impacted by the blade surface at point "Q" (shown in more detail in Fig. 3.9.10). After impact, the moisture flows radially outward from the impact point "Q" at high velocities, up to 10 times the impact velocity. At impact the water flows outward from the impact point at velocities "u" towards the blade inlet edge, and "v" towards the discharge edge. The Coriolis forces deflect all the water towards the trailing edge, and accelerate it against the steam forces (which are in towards the vane inlet edge at this location and under these flow situations). As the moisture accelerates towards the trailing edge, it is

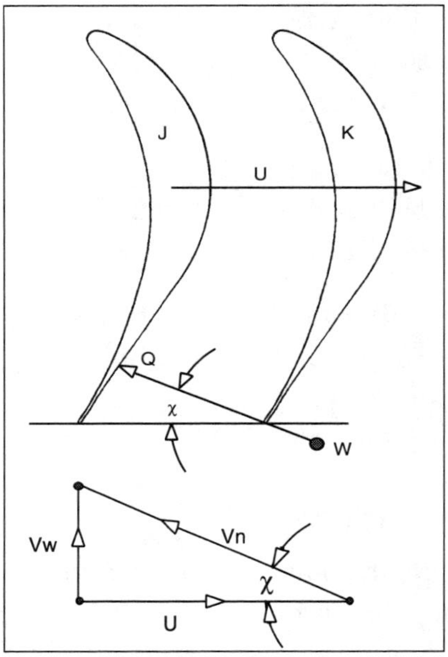

Figure 3.9.9—The probable velocity diagram of the re-entering water droplets.

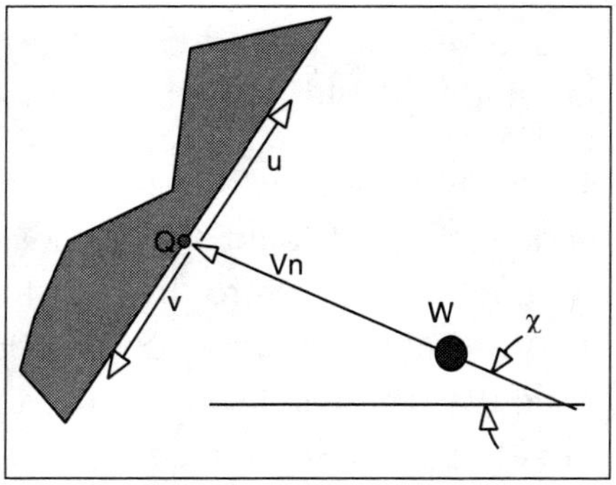

Figure 3.9.10—The water flow from the point of impact "Q".

possible some level of cavitation occurs, causing the blade material to fail at the surface. Subsequent impacts and high velocity flow will cause "gullying" and material loss. Because of the forces acting on the flowing water, the velocity "v" will be higher than "u," and these represent the forces discussed earlier, which cause the material loss.

It is entirely possible the impact at point "Q" will cause some plastic deformation of the vane material—which experiencing the same type, but not necessarily of the same magnitude of forces as at the inlet edge—will suffer the same form of elastic/plastic deformation and eventually the same form of material loss, but at a different rate.

As soon as there has been any material loss from the surface, a small crater such as "x" at a distance "z" from the discharge edge, will exist on the suction surface, shown in Figure 3.9.11. The form of the crater would be as shown in detail "A" after elastic/plastic deformation. As subsequent impacts occur at a distance "+dz," there will again be a preferential flow towards the discharge edge. This flow will cause high-pressure impacts on any vertical surfaces such as "a-a" and "b-b," causing a further material loss from this same crater, but predominantly a loss towards the discharge edge, which is the direction of the highest water velocity. By extending this reasoning, eventually there will be a small gully developed towards the inlet edge, which will act selectively to collect and channel the water outwards. It is possible to hypothesize the crater will eventually extend, and will be the beginning of a gully. Probably many such gullies will initiate at the same time and will run into each other—those receiving the largest flow quantities suffering the largest material losses. This effect of partial gullies joining can be seen particularly well in Figure 3.9.2.

Water for use in the exhaust hood cooling sprays is normally taken from the condenser hot well. While this water is essentially chemically clean, it could contain large quantities of dissolved gases, making the possibility of cavitation higher than if the water were extracted from some point above the de-aerating heater. It is

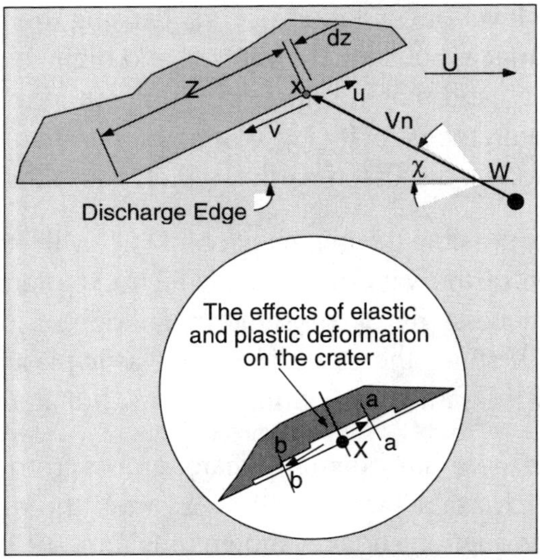

Figure 3.9.11—The possible mechanism of surface material fracture and loss upon subsequent impacts.

possible the high outward flow velocities after impact will lower the water film pressure, causing a portion of the water on the blade surface to "flash" to steam; this having the same cavitating effect.

The direction of the total forces acting on the deposited water after impact is the pumping force tending to make the water flow towards the discharge edge, and centrifugal forces tending to make the water flow towards the tip. Therefore, it would be expected the gullies would not be entirely axial, but have some radial component. This is confirmed from an analysis of the damage gullies shown in Figure 3.9.12(a) and a plot of the gully angle "υ" shown as a function of inclination to the axial position "A" in Figure 3.9.12(b), where it is noted the radial component is becoming larger as the radial distance above the root platform, and therefore centrifugal loading on the water film increases. The gullying angle "υ" is shown even more dramatically in Figure 3.9.2, where the damage shown is at a greater distance from the root section, therefore being influenced to a greater degree by the centrifugal force on the water after impact.

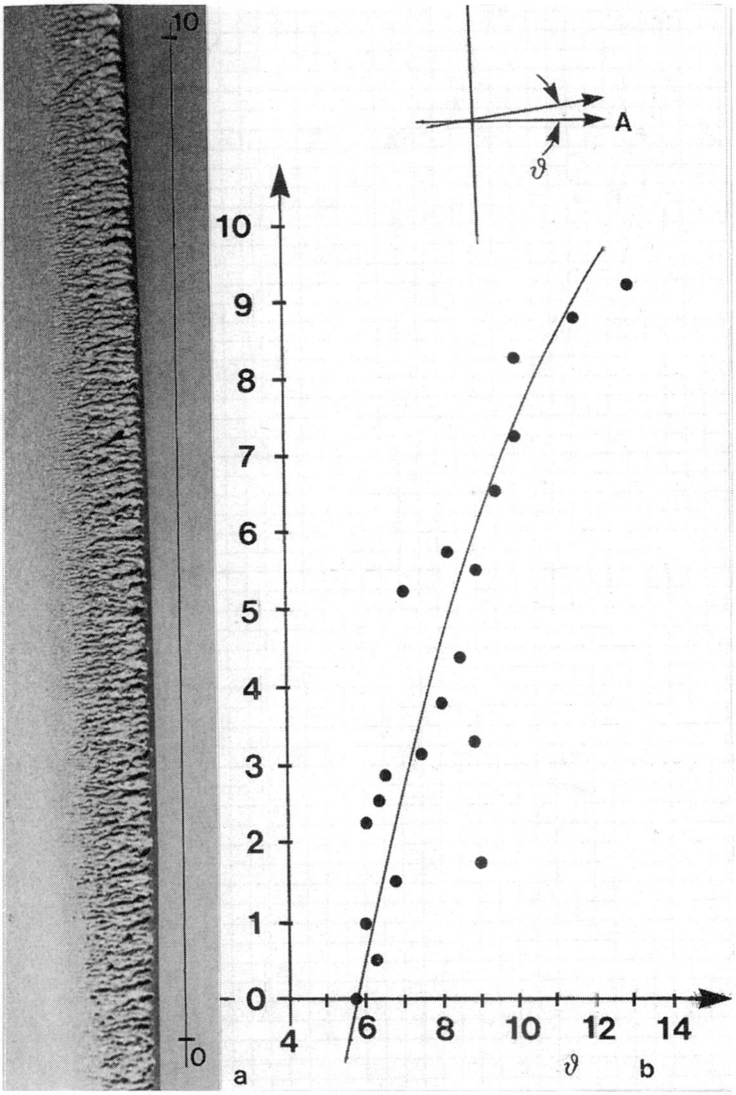

Figure 3.9.12—The "gullying angle" "υ", its variation along the radial height of the vane.

As this gullying continues, material loss will become more severe, and the gully depth "d" will increase, making the remaining discharge edge material thinner. As the depth of penetration increases, a depth is reached at which the remaining thickness of the discharge edge is insufficient to support the loads existing at this point and mechanical rupture will occur (Fig. 3.9.3). If this form of erosion is discovered in an operating unit it is considered advisable to monitor its depth "d" and the depth growth rate. Ultimately it could become necessary to replace the blades, as no reliable means now exist for weld repair in this region. Dressing the blades is inadvisable, as the removal of material in the lower locations can have a significant effect on blade frequency.

When such damage occurs it is difficult to formulate a rule that will indicate when the blades should be changed. A reasonable practice is to monitor penetration on the worst blade elements in the row, and be prepared to replace them when the depth "d," shown in Figure 3.9.13 reaches 0.050-0.060", or half the discharge edge thickness "t," whichever is reached first. This rule is no guarantee blades will not fail before such a depth is reached, as the gully represents a notch where stress concentration can be severe. Should a gully exists at a point where the vibratory stresses are high, rupture and consequently failure could initiate in relatively short periods of operation.

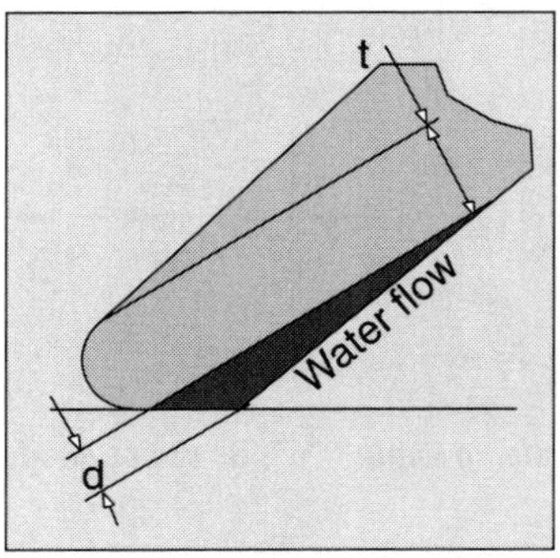

Figure 3.9.13—Measuring the depth of gullies.

Erosion penetration and time

With this form of damage, the depth of penetration is a function of the time of operation at light load, (below about 20%), with the sprays operative. It is therefore important the amount of time with this type of operation is limited, as this will deepen the gullies and reduce the load carrying section of the vane. Situations are relatively common where the unit is left in service with the sprays in continuous operation (this is normally inadvertent on the part of the operator). While undesirable from efficiency considerations, it should not cause any mechanical damage to the blades, as the annulus should be running full, and there will be no opportunity for re-circulation. Another situation that can occur is the exhaust pressure will rise, due to some fault in the operation of the condenser. It is then possible the sprays will become activated by the hood temperatures at loads higher than those at which the sprays would normally be used. This is because the volumetric flow will have been reduced and is principally through the outer section of the blade passage. Under these circumstances there can be re-circulation and damage.

If the instances of the use of water sprays are limited, the growth rate of damage depth "d" in Figure 3.9.13 should not be excessive, and be within the blades ability to resist crack initiation. However, this limited use of the sprays represents only a slowing in the rate of damage (gully depth) penetration. Once the gully exists, which is a notch-sensitive region, there will be some consumption of fatigue life. While the local stresses at the gully may not exceed the endurance limit of the material, as depth "d" increases to some extent each time the sprays are used, there could ultimately be blade failure. It may however take many years, and may in practice never become a chronic and continuing situation.

Proud blades

Just as the blades with a proud inlet nose at their tip sections suffer "sacrificial erosion," blades with a proud discharge edge in their lower regions will also suffer a sacrificial loss of material. In the situation of sacrificial loss at the inlet nose, this represents a loss of material and possibly a situation where a single blade may require refurbishment, the attachment of a new erosion shield, or even replacement. Unfortunately in the case of trailing edge erosion, the possible consequences of sacrificial erosion can be more severe. It can result in a single blade failing and causing the entire unit to be forced from service with subsequential damage to other blades in the row, if not detected, monitored, and corrected by replacement.

It is recommended that when discharge edge erosion is first noted, to determine if this is more severe on certain blades, and if these elements are proud. Any monitoring that is then undertaken should be on these elements, and to the extent necessary nondestructive testing be undertaken to ensure the unit is not returned to service with cracks present.

Monitoring damage

If a unit has operative water-cooling sprays in the exhaust hood, it is recommended the last stage blades always be examined at maintenance outages for this form of damage. If damage is found it is recommended the situation be monitored.

No general rules exist for estimating failure potential—the occurrence of cracks appears rather arbitrary—but is in fact the combination of a number of complex factors, including the depth of penetration, thickness of the remaining material, shape of the gully base, and amount of vibratory deflection at the gully location (which influences the vibratory bending stress at that location). Such an evaluation is beyond that of simple visual examination. To determine such detail, an "in-depth" finite element stress analysis is required. It is recommended the operator make such a detailed evaluation of the stress levels and failure potential, once the penetration "d" has reached a depth of 0.050" or half the vane thickness "t" at discharge, whichever is less. Preferably such an evaluation should be made prior to reaching a dangerous level of penetration. If an analysis proves the material loss to be dangerous, efforts should be made to limit operating at light loads, and if any control exists over cooling water spray quantities, this control should be exercised so excess water is not provided beyond that required to cool the blades. This is a very difficult judgment, and in trying to avoid one form of damage the operator could induce another. It is probable such blades will eventually require replacement.

When this damage is initially noted, it can be monitored by means of a plot of penetration "d" as a function of operation hours with the cooling sprays operative. Unfortunately this is difficult information to gather. While the depth of penetration is easier to define, it is far more difficult to maintain "sprays operative time log," which happens to be the major factor. Alternatively it might be possible to plot "d" as a function of the number of starts, or some other parameter reflecting the mode of operation of the unit. It is also recommended that photographs and even molds of the most severely damaged blades be collected to provide continuity between different observers.

Fortunately the last stage blades, and particularly the discharge edge can be inspected easily. When this type of damage reaches a critical level, such examination should be undertaken regularly, including the application of non-destructive methods. However, Figures 3.9.2 and the pressure face of the same blade in Figure 3.9.3 provide the most conclusive evidence that rupture has occurred. Therefore, it will be necessary to employ means that make the pressure face visible when the blade is still mounted on the rotor, making possible the inspection without the removal of the LP covers. The use of mirrors held between the vanes has proven successful, if difficult.

WIRE DRAWING EROSION

The phenomenon of "wire drawing" or "throttling" represents a free expansion of the water/steam mixture through a minute aperture (leakage path), and such an expansion takes place without the addition or rejection of heat. In a perfect gas, throttling produces no change in the temperature of a gas, but in steam or water/steam vapor it produces a small cooling effect, which is a measure of the fall in temperature per unit—fall of pressure under the condition of constant total heat. In steam, the throttling operation must have water phase fluid present at the beginning of the expansion to cause damage. If saturated steam as exists in the steam path were to expand, from any pressure, the final fluid would be superheated and also at a lower temperature.

For this operation to occur and cause damage in the steam path, there are three conditions that must exist:

- A two-phase water/steam mixture at an elevated pressure, at entry to the leakage path
- A pressure differential across the mechanical member or joint
- An initial or primary leakage path, through which the water/steam mixture can expand from the high to the lower pressure

Within the turbine steam path, all three of these conditions can be present as a relatively common occurrence. The first two are always present in a condensing unit, once the steam has crossed the saturation line. The probability of the third occurring, tends to increase as the unit ages, primarily in terms of the appearance or the development of a number of leakage paths. The structural material relaxes stress or twists and warps to a greater extent as a unit is put on cyclic operation with a greater number of stops, starts, and thermal transients. As this leakage expansion occurs it is characterized by two basic results:

- The leaking steam will achieve a high velocity in its passage through the aperture. This represents the conversion of the pressure of the mixture to kinetic energy, at constant total heat

- There will be a change in the physical properties of the steam phase of the mixture. If this expansion is examined on the Mollier Diagram (Fig. 3.10.1), it will show this is represented by a horizontal (constant total heat) line, which will move the mixture condition from "A" to "B"

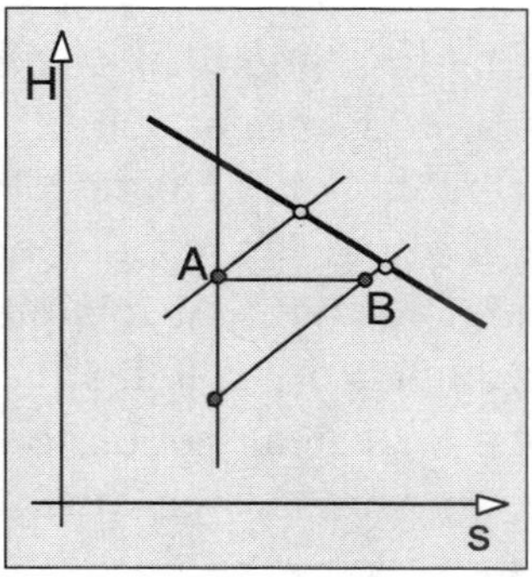

Figure 3.10.1—A throttling operation from condition "A" to "B" at constant enthalpy.

Within the steam path there are three possible consequences of this type of expansion that degrade unit performance:

- Steam that leaks past a joint will bypass the blade rows, and therefore not produce work in the stage. It is also possible, depending upon where the steam leaks, for it to re-enter the steam path causing a disruption to the flow of the steam that has expanded through the blade row, thereby introducing a further loss

- Expansion through the small aperture, the water phase fluid will cause the metallic surface to deteriorate and lose material by a process that breaks down the material surface structure. This surface loss will enlarge the aperture size increasing the leakage quantity, and therefore aggravating the situation

- The leaking steam will re-enter the steam path in such a manner setting up impulse loads capable of exciting dangerous levels of vibration in response to the frequency of the impulse. This is most common when leakage occurs at the horizontal joint of diaphragm halves, and the blades will experience a 2/rev stimulus, capable of becoming a high intensity impulse as the aperture enlarges

The extent of material loss, and final steam conditions (moisture content) after leakage, are dependent upon the steam conditions at entry to the leakage path. Shown in Figure 3.10.2 is a portion of the expansion line, where steam having initial conditions "Pa" and "Ma" throttles down to conditions "Pb." At this pressure the steam will have a dryness "Mb," which is dryer. If the final pressure, after throttling had been "Pc," the final steam would be dry with a degree of superheat "ΔT" above temperature "Tc" (the saturation temperature corresponding to pressure "Pc").

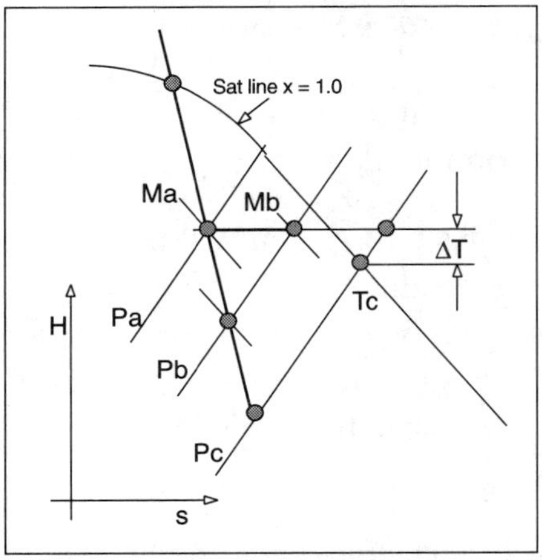

Figure 3.10.2—Throttling operations and the effect on the final steam temperature.

Leakage location potential

It is important to consider within the total internal portion of the turbine (including the steam path), where leakage apertures can develop, and where such damage can be looked for. The steam turbine contains a number of joints, both horizontal and vertical that are candidates for the appearance of these leakage paths. Most susceptible among these are the following:

The turbine casing horizontal joints. In the nuclear unit high-pressure casings the shells are normally produced from cast steel, but high-pressure differentials do exist, and if the joint is not "flat" or there is excessive warpage, then leaks can initiate.

Figure 3.10.3—A water/steam leakage path developed in a low pressure fabricated section.

In the low-pressure sections of both fossil and nuclear units the casings are most often produced from a steel plate fabrication. This plate is normally of a type that does not offer particularly good resistance to this form of erosion, and is therefore subject to damage. Shown in Figure 3.10.3 is the leakage path developed in a low-pressure fabricated section. In this case the leakage is from an extraction pocket to the space between the inner and outer casing. This type of damage will not interfere with the steam path efficiency by flow path interruption, but as the leakage path enlarges, there will be a waste of available energy—the escaping steam neither expanding in the unit producing work, nor being used for regenerative feed heating.

Some low-pressure sections use a cast iron or cast steel inner casing. Such designs are prime candidates for damage if not manufactured to specification. If the cast iron is of the non-weldable type such damage is difficult to repair.

Diaphragm horizontal joints. Diaphragms are by definition and design purpose intended to have relatively large pressure drops developed across them. They must also, to permit assembly, have a split joint in the horizontal plane. In the majority of designs, keys on this surface connect these joints. However, these elements are very susceptible to the wire drawing effect, and damage is quite common. Figure 3.10.4 shows the leakage path from the inlet side of the inner web to the keyway, and also the leakage occurring across the gland ring.

Split blades. There are designs of diaphragms that use vanes split at the horizontal joint, half of each vane being carried in the upper and lower half. If these blades are not joined with an adequate surface at the joint face, there will be leakage between the two halves. Shown in Figure 3.10.5 is the split vane geometry and the possible leakage path from the pressure surface "p," which has the majority of moisture deposited upon it to the suction surface "s." There is an initial small expansion gap "L" that will exist, and provide the initial leakage path.

This is not a common occurrence, as the pressure differential across the vane from the pressure to suction face is not large. However, in those cases where such leakage is present the leaking water/ steam mixture flowing into the suction face boundary layer can result in separation, causing localized turbulence adding a 1 or 2/rev low amplitude stimulus to the rotating blades.

Figure 3.10.4—Wire drawing erosion on the horizontal joint of a diaphragm. In this case the key has acted as a partial barrier to leakage, and the water has flowed to the sealing ring at the inner diameter.

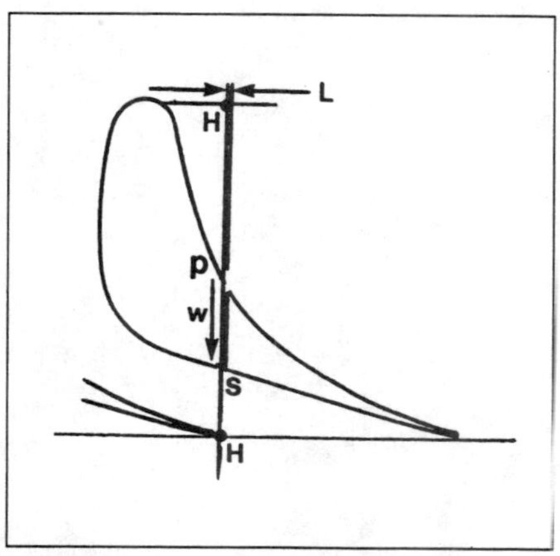

Figure 3.10.5—Flow across the "split" in a horizontal joint blade.

Between blade root blocks. As shown in Figure 3.10.6 leakage has occurred between the root blocks at the horizontal joint. This is a reaction stage in which the blade is located directly in the casing, or inner carrier. The blades are held in position and prevented from tangential migration by a screw attached key. This leakage is similar to that at the horizontal joint of a diaphragm, and can cause a 2/rev stimulus to develop. This form of erosive loss can also occur at other root block positions, and the production of the root block wedge angle is important in controlling the production of such a leakage path.

Figure 3.10.6—Wire drawing erosion on a root block at the horizontal joint.

Diaphragm and inner casing steam seal faces. These are the joint faces that form the major barrier where the diaphragms and blade carriers locate in the turbine casings. Such a steam seal face "O-O" is shown in Figure 3.10.7, as a blade carrier containing four stationary blade rows. The ability of this face to seal prevents steam bypassing four rows in this configuration. In Figure 3.10.8 is shown a similar steam seal face "O-O" for a single diaphragm. In each case these faces should be produced with a fine finish, 64-125√μ-inch. However, on the casing a similar finish is harder to achieve, and may deteriorate to 125-250√μ-inch upon final machining. This should however produce an acceptable leak tight joint under the influence of the pressure differential across the face.

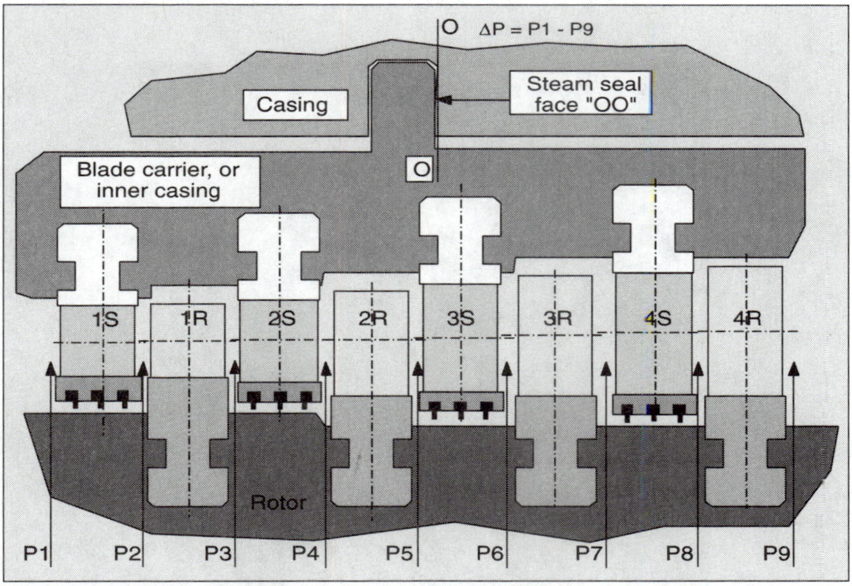

Figure 3.10.7—A blade carrier locating four stationary blade rows, and showing the control dimensions from the steam seal face "O-O".

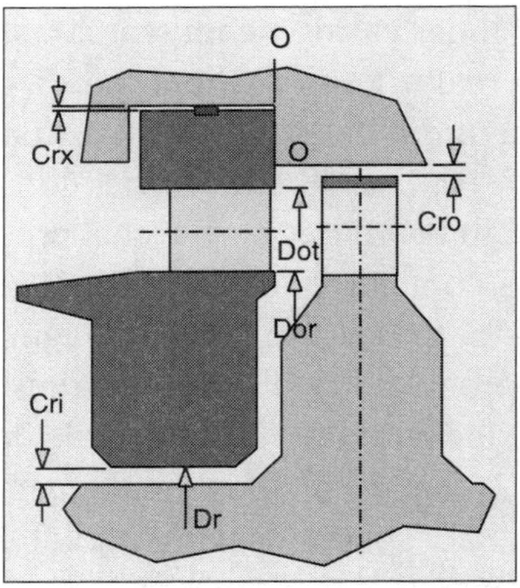

Figure 3.10.8—The requirements of a wheel and diaphragm constructed stage for sealing at face "O-O".

Figure 3.10.9—Wire drawing erosion on the steam seal face at a diaphragm locating slot of an older design cast iron casing.

The situation often arises, where a small leakage path develops across the faces "O-O" as the unit ages (there is the possibility of some small level of deformation of either component), then the possibility of leakage of a two-phase flow exists. Such a path will soon grow to unacceptable proportions. Normally the casing is a less resistant material and will suffer material loss to the greatest degree. Another source of the initial generation of leakage paths is some form of transient operation, where temperatures rise in the unit with the possibility of introducing temperature deformation.

Figure 3.10.9 shows a portion of a low-pressure casing at a diaphragm-locating slot, where there has been wiredrawing erosion on the seal surfaces of the cast steel casing.

Diaphragm gland ring locating slots. Those diaphragms that have gland rings located at their inner diameter forming a seal between them and the rotor, have the potential to erode on the seal surface, which is the surface forming the seal. This surface is shown in Figure 3.10.4. When the unit is opened for inspection, this surface should be examined and, if evidence of damage exists, corrective action should be considered. This leakage path can soon provide as great a leakage area as the labyrinth seal system.

WATER WASHING EROSION

Washing erosion is a form of surface material loss occurring as a result of the flow of water at high velocity over a metallic surface on the turbine internals. As the free water flows across the metallic surfaces of the turbine internals, there are mechanisms in effect that cause or contribute to micro-ruptures of the material surface. There are various considerations relating to this surface material loss that can include:

- initial irregularities on a surface, as the result of a number of operational events, and possibly including manufacturing processes. These can cause high impact forces to develop between the flowing water films. The surface will eventually suffer a material loss

- water film in the low-pressure sections of a turbine contains significant amounts of dissolved gasses, as evidenced by the need to continually use steam ejectors to remove these gasses from the condenser. As the environmental pressure reduces, these dissolved gasses "flash" out of their dissolved state, and in doing so cause micro-ruptures to be produced on the steam path surfaces

- the water may contain a small amount of corrosive ions that will attack the material structure, breaking down the bond between the metallic crystals. The subsequent washing of the surface causes these grains with a weakened boundary to break free

Washing erosion in the rotating blade row

The materials used to produce steam turbine blading, as used in stages having free moisture present in the working fluid, is a high quality 11-13% chrome steel and will resist the effect of washing erosion under normal circumstances. However, there are instances when this form of damage can occur:

- When the vanes are produced from an extruded section and a spacer piece of an inferior material is placed between the vane portions to maintain the correct tangential pitch, damage can occur. The effect of washing erosion on such a stage is shown in Figure 3.11.1. Here the spacer pieces, produced from 2.25% chromium steel, are eroding to the extent they have lost the form of the inner sidewall, and as a consequence will have destroyed the inner lap. This material loss also has the potential to introduce vortices and introduce turbulent flow into the steam discharging from the blades and flowing into the following stationary blade row

- In large nuclear units, where steam is produced in a nuclear boiler, it is necessary to "dry" the steam before it enters the turbine steam path. If the moisture separators (or driers) ahead of the high-pressure section are not sufficiently effective, large quantities of water can be carried over into the high-pressure steam path

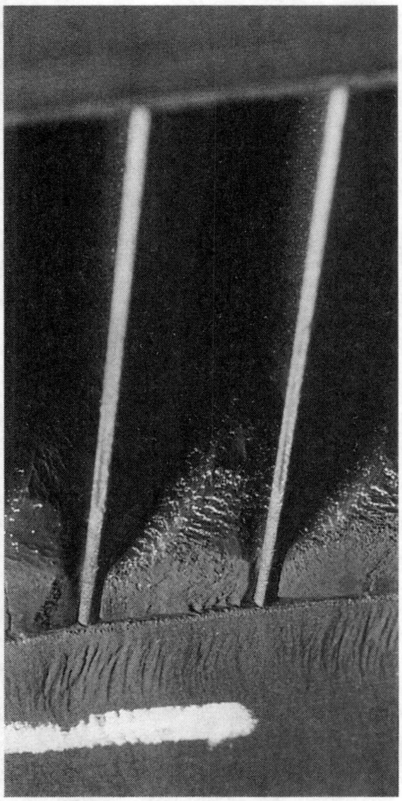

Figure 3.11.1—Washing erosion of the spacer pieces between extruded vane sections.

Figure 3.11.2 shows control stage rotating blades of a large nuclear unit, in which excessive free moisture has been carried into the high-pressure section. This moisture has accumulated at some point in the inlet system, flowed over into, and eroded the rotating blade root platforms. This material loss could eventually weaken the attachment, and adversely affect the reliability of the blade row. The discharge diameters would also be affected, which on a small radial height blade will have a significant effect on stage efficiency.

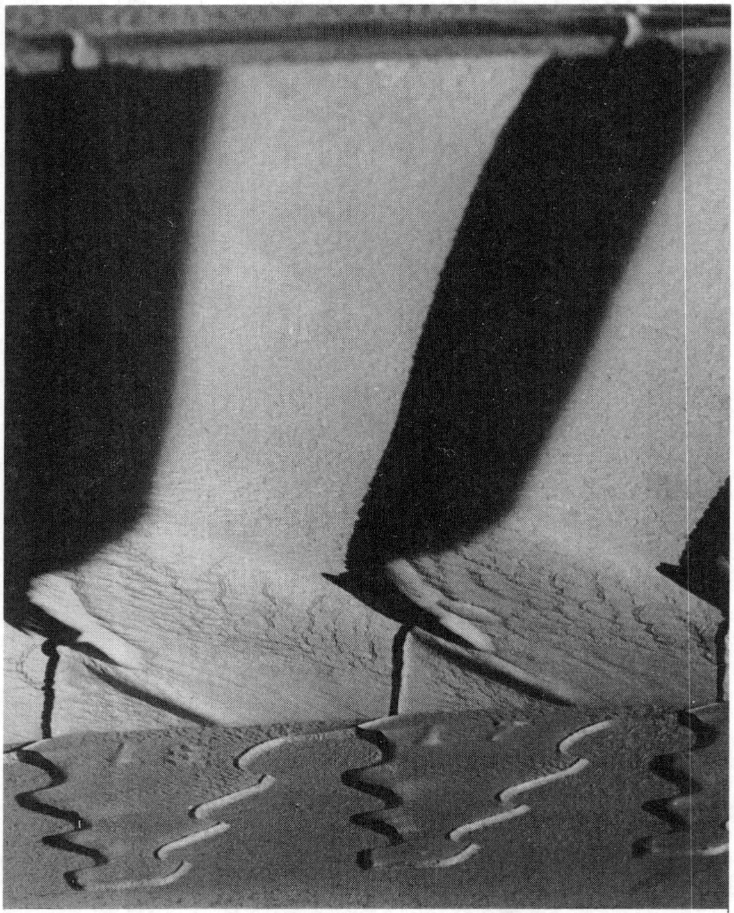

Figure 3.11.2—Washing erosion at the root section of a nuclear unit control stage rotating blade.

Washing erosion of the diaphragms and stationary blade rows

The stationary blades have moisture deposited on them from the main steam flow, and this moisture flows out across the blade surfaces in a chordwise direction, but also with a radial flow component introduced by the steam flow forces of the expanding steam. The actual flow patterns on the stationary and rotating blades can often be seen in the deposit pattern left to accumulate on the vane surface for a rotating blade (Fig. 3.11.3). Under normal circumstances the vane material is too hard to be influenced by the effects of washing.

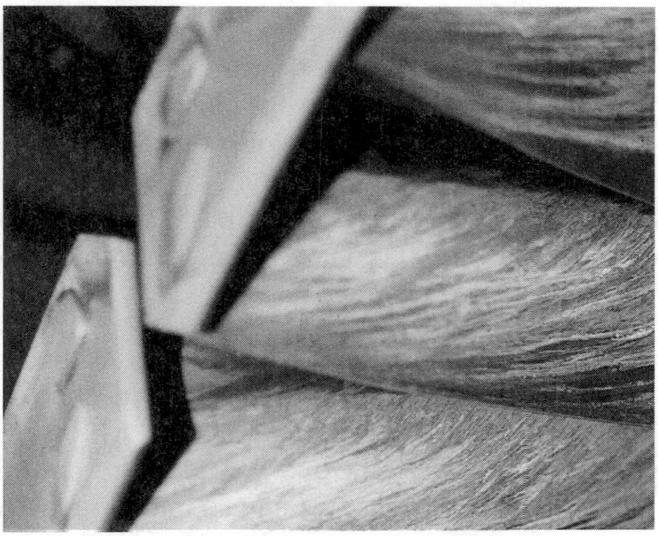

Figure 3.11.3—Water flow marks on a rotating blade row.

While blade material offers considerable resistance to washing erosion, the inner and outer rings, into which these vanes are cast or fabricated, are a softer material, whose principal characteristic is to have properties that make it suitable for the forming process. Figure 3.11.4 shows a series of stationary vanes forming the inlet

Figure 3.11.4—Severe washing erosion at the base of a fixed blade row.

stage of a low-pressure section of a fossil non-reheat unit. In this unit the vanes are positioned and held in place by cast iron inner and outer rings. The water entering this stationary blade row is from the crossover, and the cast iron forming the inner ring can be seen to be suffering from the effects of "washing." The material loss is sufficiently severe; the inner sidewall form has been destroyed, and if this damage is allowed to continue the vane could eventually detach.

Another form of this diaphragm erosion is shown in Figure 3.11.5(a), (b), and (c). In these three pictures are shown different degrees of washing erosion on the sidewall. In each case water has collected on the stationary vane and flowed out towards the outer sidewall. In (a) the material loss is minimal, but has been sufficient to destroy the inner sidewall form. In (b) there has been a heavier loss, and the position of the original sidewalls can be seen. Again the sidewall material is cast iron. In (c) there has been a deep cutting of the outer sidewall in a cast iron last stage stationary blade row. The material loss is clearly seen at the discharge edge, and while this may eventually weaken the attachment of the vane, considerably more material must be removed before structural integrity is affected.

In each of these three situations, this form of damage is unlikely to have any significant effect on the reliability of the unit, although it could reinforce the effects of nozzle passage frequency excitation. The potential effect on efficiency is minimal.

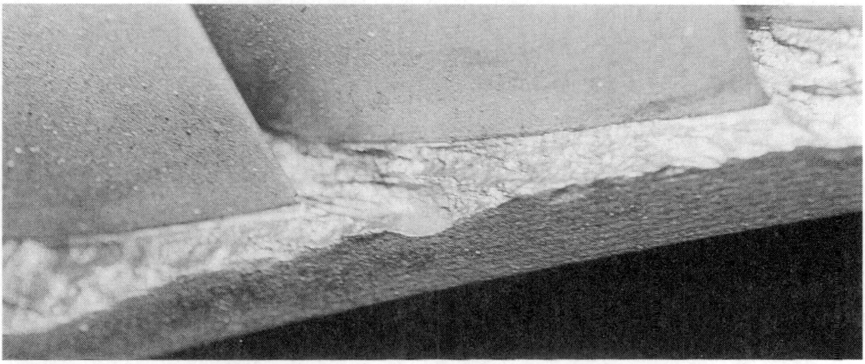

Figure 3.11.5(a)—Light washing erosion on the outer ring of a diaphragm.

Figure 3.11.5(b)—Severe washing erosion in a cast iron diaphragm. Grooves exist at the discharge edge.

Figure 3.11.5(c)—Washing erosion removing material from the diaphragm outer ring, exposing the blade vanes.

Washing erosion on the casings

The washing erosion that occurs on the casing is normally a consequence of moisture centrifuged from the rotating blades, which impacts on the casing, and then flows at high velocity over the inner surface. In the majority of instances encountered of damage to the casing, the damage is caused by a combination of these two phenomena of impact/washing, contributing to the surface material loss. Figure 3.11.6 shows the axial gap between two stationary blade rows, where the material loss to the stationary blade row can be seen on both the casing and the inlet portion. This type of damage is not uncommon, and unless it undermines the security of the stationary blade row, should not be of concern. Figure 3.11.7 shows erosion on the vertical surface of a water catcher belt, where moisture not captured by the belt has caused a material loss on the vertical surface of the diaphragm outer ring/casing.

In Figure 3.11.8 is shown the material loss pattern that has occurred on the outer ring of a diaphragm locating a stationary blade row of a nuclear unit, and in which is located a radial seal strip for sealing above the L-1 stage rotating blade. The material loss has not been severe, but to the extent the security of the sealing ring is in question. In this case the operator has the choice of removing the seal ring or making some other form of replacement. This damage was repaired by the insertion of a complete ring after machining away the affected area and sufficient material to securely attach a new insertion. The attached ring was produced from carbon steel, and a seal weld was completed for the entire circumference to ensure no leakage and washing erosion would occur. The repaired ring is shown in Figure 3.11.9.

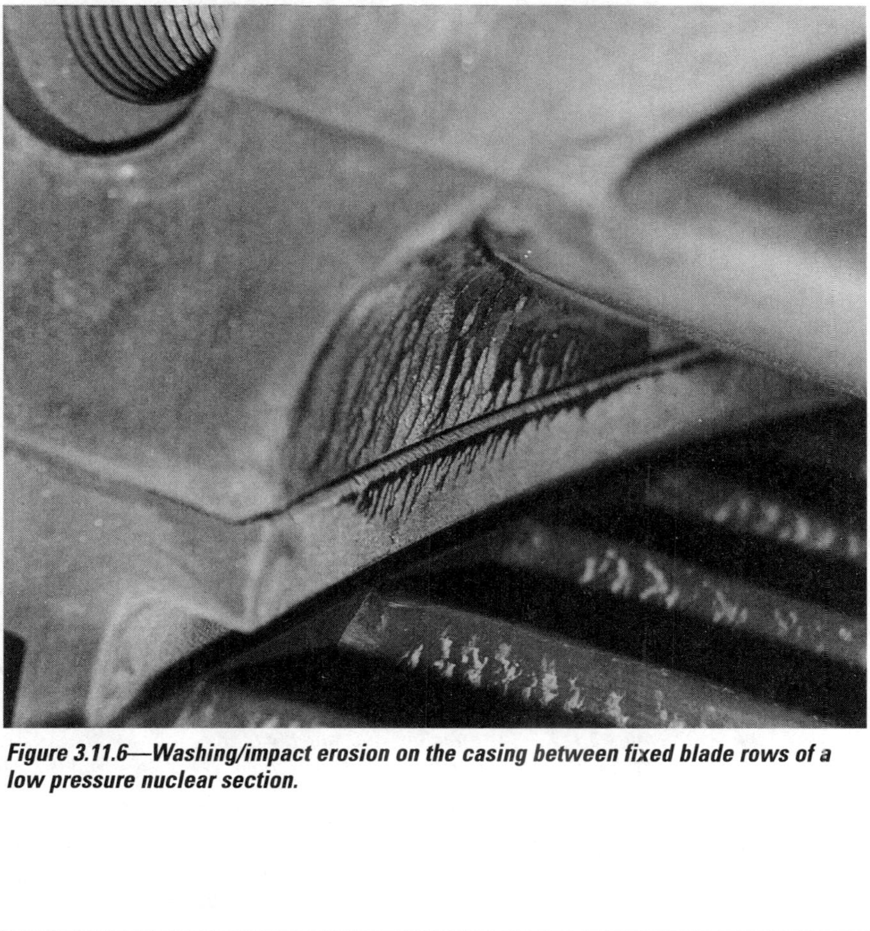

Figure 3.11.6—Washing/impact erosion on the casing between fixed blade rows of a low pressure nuclear section.

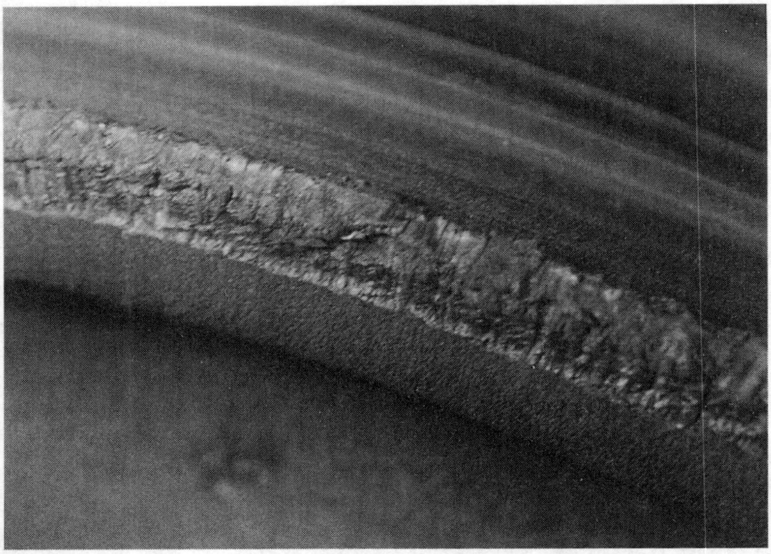

Figure 3.11.7—Washing/impact erosion above a rotating blade row.

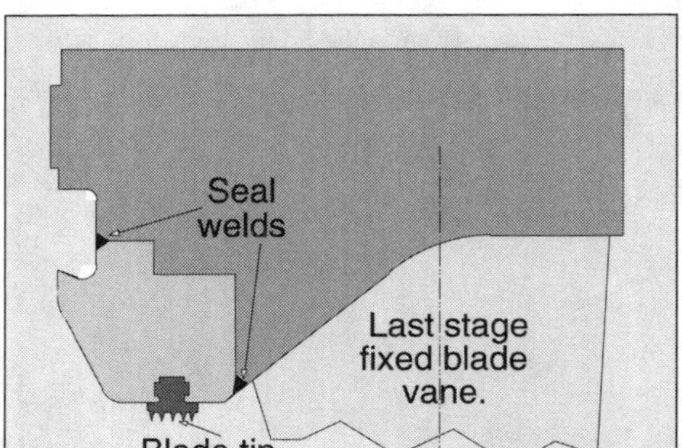

Figure 3.11.8—The attached ring, and the form of the new seal ring locating slot.

Figure 3.11.9—The diaphragm of figure 2.11.8 after refurbishment by machining away the damaged area, and attachment of a replacement ring.

WATER INGESTION INTO THE STEAM PATH

The steam turbine is an integral part of a thermal cycle, whose working fluid is steam, and/or a water/steam mixture at the lower steam conditions. Because of the potential damaging effect of water, and the design procedure of collecting and draining this moisture wherever it forms (both by direct condensation during the expansion process, and from "low collection points" at shut-down), there are many locations within the cycle where water is available in relatively large quantities.

The turbine unit is designed for the entry and exit of both steam and water at various locations. There exists the possibility of water induction, or re-entry into the turbine, under a number of cycle conditions. This re-entry can be caused by the return or injection into the steam path water taken from other parts of the unit or cycle. Those portions of the cycle from which the water is most often returned, include the main steam inlet lines, the hot and cold reheat lines for units with reheat, and feed heating extraction lines.

In the event water does re-enter the steam path, depending upon its location, quantity, and form of re-entry, the unit can sustain various forms and degrees of damage. This damage can be in the form of thermal cracking of various components from "quenching," bent rotors, and rotating blades. There are different forms of damage depending upon the extent of water re-entry and the length and design of the blade rows. This damage can include severe impact erosion and cracking of tie wires to cracked and bent blades.

There are in existence, recommendations for unit operation and the design and installation of systems to protect against this form of damage. These manufacturer recommendations should be reviewed and followed to the greatest extent possible to ensure the unit is provided with the best means of protection available.

During transient operation when loads may be rejected, valves closed, or operating in a manner to exert a controlling influence on overspeed, there is the possibility of sudden pressure decays, flow reversals, and the possibility of water flowing or being drawn back into the steam path.

Recommendations for including protective devices in the system design exist. These recommendations define the requirements and design of systems to provide maximum protection to the internal portion of the steam turbine. While these recommendations were prepared originally for the utility industry, they also provide good guidance for industrial users where the potential for such damage exists. The systems include means of detecting the presence of water. Instructions are available to help operators minimize damage when it has been ascertained that water has in fact been ingested, or has re-entered the unit.

For the industrial user these recommendations have considerable applications provided the units have or can be equipped with detection devices.

Forms of damage

When water re-enters the steam path, with the unit hot and at speed, it has the potential, and almost always does cause some form of damage. When such an ingestion incident is known to exist, it is prudent to shut down the unit, open it, and inspect. Had opening not been planned, this very action of inspection is expensive, both in terms of the labor and material involved, and the possible effect it will have on a planned maintenance schedule for a power system. When the unit is inspected, if some of the damage is minor, correction can be delayed to a future outage. However, normally some corrective action is necessary before the unit can be returned to service.

Inspection will often reveal that a portion of this damage can be corrected without the need for major replacement, but may still require extensive effort and resources. Other damage cannot be corrected using existing hardware, and the only recourse is to replace the affected components. Such a situation can either force the unit from service for extensive periods, or require it be returned to service considerably de-rated until replacement parts can be delivered. This option of de-rating can also require the unit be opened for an additional outage, which will make the unit unavailable for a second opening, and also involve the additional cost of opening.

The forms of damage introduced by water ingestion, and are significant to the steam path components are:

Impact type damage. Impact type damage is of significance on the less robust portion of the steam path, which is the actual rotating blade row, including the stage hardware. When water re-enters the steam path, as "slugs," it will move downstream in the direction of the steam expansion. This water will impact with the blades introducing high impact forces, or shock loads on them. In some instances this is sufficient to cause rupture. Long blades in low-pressure sections will often sustain cracks in the tie wires and/or coverbands. Figure 3.12.1 shows a blade pair from a row destroyed by the ingestion of water slugs. In this case the blades of small axial width were completely bent to destruction from the root section.

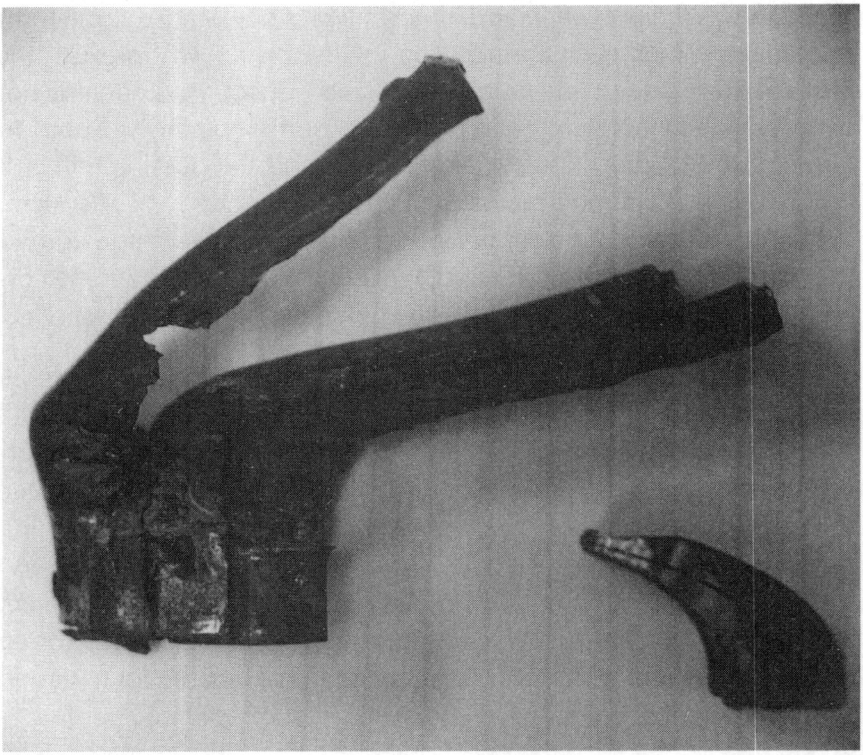

Figure 3.12.1—Bent blades. The bend produced by the ingestion of a "slug" of water from an extraction line.

Erosive damage. If water re-enters the steam path, not in the form of slugs, but as a continuous carry-over from some location—a blocked drainage or feed heating extraction line—where water is collected from possibly several stages, but is unable to be removed due to drain line blockage or a closed valve, then this water will flow back into the steam path, from which it has just been removed.

It is possible however, that this water has now been concentrated, perhaps from several stages, and at the one radial position. This concentrated water flow will—if the blade tangential velocity is sufficient—cause extensive moisture impact erosion, which can destroy the tip section of the rotating blades.

Quenching. Quenching is the phenomena by which heat is removed from a metallic piece by the sudden immersion in a cooling fluid. This is a common process in steel and other material production, and has a profound effect on the material properties. In the case of steel, as used in the steam path components, it hardens the material, but it also becomes brittle, requiring further heat treatment to regain its ductility, and be suitable for the loads it will be required to carry.

In the case of water ingestion into the steam path, the water in addition to impacting the rotating blades can also impact the inner and outer casing surfaces of the rotor, which would remove heat from them. In effect this would quench them, modifying their material properties at the surface. This same quenching can also produce cracks onto the surfaces.

- *Rotors*—In the event rotors are quenched, there are two possible forms of damage mechanisms that can be introduced, *i.e.*, bending, and cracking—both requiring remedial action because of the high rotational stresses
- *Casings*—If the casing is quenched, the forms of damage mechanisms that can be caused are basically the same as

> those affecting the rotor. Cracking is not as serious, because the stress levels are not nearly so high in the casing, however each situation should be evaluated. Bending or distortion of the casing can be significant, with it going elliptical. The horizontal joints reducing in diameter, and possibly some degree of humping, reduces the radial clearance above the blades

Steam path rubs. Often water re-introduced into the steam path will cause changes in the axial and radial alignment, by changing the actual positions of the major components, *i.e.*, the rotor and casing, as discussed above. Because the steam path components operate with small axial and radial clearances there is a possibility these spatial changes will cause "rubs." These will, in the least severe case, open the axial and radial clearances at seal points. In more severe circumstances rubs will be produced that will cause "heavy" contact between the stationary and rotating portions sufficient to generate heat that will "burn" the individual components, and can bend a rotor.

These heavy rubs can generate sufficient heat that material properties required of the various components for acceptable operation are compromised by hardening and becoming brittle. Under these circumstances of heavy rubs and excessive heat, the damage will normally be of such magnitude that from efficiency considerations alone, the alignment changes will modify clearances and lap within the unit so its performance will be significantly reduced. From structural consideration, it would be unwise to return the unit to service without a full evaluation and assessment of the risk involved.

If the rubs are not severe, it is possible some corrective action can be accomplished by slowly heating the unit before returning it to service. The rotor can be straightened sufficiently to return it to service by a slow heating and straightening process, which can often be done with the top half casing in place.

REFERENCES

1. Sanders, W.P. "Moisture Damage in the Turbine Steam Path and its Impact on Life Extension," *Turbomachinery International*, Vol. 33, No 1, January/February, 1992

2. Binnie, A.M., and M.W. Woods. *The Pressure Distribution in a Convergent-Divergent Steam Nozzle*, Proc. Inst. Mech. Engrs., 1938

3. Gyamarthy, G. Grundlagen. *einer Theorie der Nassdampf-turbine*, Juris-Verlag, Zurich, 1960

4. Stodola, A. *Steam and Gas Turbines*, 5th Edition Berlin, Springer, 1922

5. Kearton, W.J. *Steam Turbine Theory and Practice*, Published by Sir Issac Pitman & Sons, Ltd, Seventh Edition, London, 1958

6. Church, E.F. *Steam Turbines*, Published by McGraw-Hill Book Company, Inc., 1950

7. EPRI research report RP0735-01

8. EPRI research project 2818-07; report TR-102729: *Turbine Efficiency Improvement Investigation*, prepared by Stress Technology, Inc., Rochester, New York, September, 1998

9. Moore, M.J., and C.H. Sieverding. *Two Phase Flow in Turbines and Separators*, Chapter 4: "A Review of Instrumentation for Wet Steam Measurement." Published by Hemisphere Publishing Corporation, London, 1976

10. Sanders, W.P. *Water Droplet Formation and Flow Leading to Bucket Erosion in Steam Turbines*, Canadian Electrical Association Meeting, Winnipeg, March, 1967

11. Engel, O.G. *Impact of Liquid Drops,* Symposium on Erosion and Cavitation, American Society for Testing and Materials, 1961

12. Engel, O.G. "Erosion Damage to Solids caused by High Speed Collision with Rain," *Journal of Research of the National Bureau of Standards,* July 1958

13. Sanders, W.P. "A Review of Steam Turbine Blade Operating Problems," *I. Mech. E.* Toronto, February, 1981

14. Greco, S.W. Private correspondence concerning Unit #8 at Oak Creek Power Plant

15. Kirillov, K.I., P. Fadiv, and S.V. Radik. *Erosion of the Trailing Edge of Moving Blades in Axial Turbines,* Teploenergetika, 1972 19(4) 38-41

16. Dickenson. R. "How to Stop Water-Induction Damage," *Power,* February, 1979

17. General Electric Co: "Water Induction in Large Steam Turbines," *Design Recommendations,* Publication GER 25504E

Chapter 4

Operational Events Giving Rise to Steam Path Damage

INTRODUCTION

The steam path can be subjected to a number of operational events that degrade the overall quality of the unit, and so degrade performance, both in terms of efficiency and availability. Under certain circumstances these damaging mechanisms can be sufficient to force the unit from service, and degrading influences can result in considerable expense to the owner.

Not every machine is subject to each of the "damage mechanisms" that follow. But each machine can be if the operational conditions are such that the opportunity for their occurrence exists.

This chapter will examine the most common of these, and where possible provide an explanation of their characteristics, some brief explanation of the circumstances that allows them to occur and provide some guidance to their avoidance, if such avoidance is possible.

Some of these mechanisms are well understood and some allowance is made for them in the original configuration of both the cycle and the details of the turbine unit. Other mechanisms are often unanticipated, and can be particularly present in units of a prototype design when the boundaries of design are extended and experience extrapolated to provide guidance.

FOREIGN OBJECT IMPACT DAMAGE

There are many instances when the blade vanes, both stationary and rotating, are damaged by impact with solid objects that have been generated within, or gained access to the steam path. There are a number of sources for these objects, which can be of various sizes depending upon their origin. The damage they cause will depend upon their mass, their location in the steam path, and the environmental temperature that give the material different mechanical properties.

These objects have the potential to effect the operation of the unit, and must be given consideration. They include:

- *The larger objects, if generated within the steam path, may be the mass of an entire blade, which having suffered failure in the root section, detaches and rebounds within the remaining blades*

This can cause significant consequential damage, destroying a number of the blade elements within the row, some of which may also detach. Such failures, when large blades are involved, can often affect the dynamic balance of the unit, introduce dangerous level of rotor vibration, and force it from service.

Often when smaller blade elements fail, they do not affect balance, and there is no indication such a failure has occurred. The transient on any vibration-recording device may be so small it is not noticed, and the unit will continue to operate in the same manner with no evidence of failure. That a failure has occurred is often not detected until the unit is opened at the next inspection/maintenance outage.

- *The objects—whether or not they affect balance and require shutdown to take corrective action—will almost certainly cause mechanical damage if they're free in the steam path*

The exceptions to the occurrence of this impact damage are the possible failure of a piece of stage hardware, such as a coverband or tie wire. When they become loose, they can wedge at the casing inner surface, possibly at a diaphragm or steam or water extraction slot, and not re-enter the steam path. This is rare, but can occur. Another possible exception to the occurrence of damage is when a last-stage blade (or a piece of last stage hardware in any section) fails and the generated missile is carried directly into the exhaust. For a low-pressure section, the missile can be carried directly into the condenser tube.

When a piece of debris re-enters the steam path and impacts with the blades, it causes mechanical damage that introduces an "impact pit" and/or causes material rupture. The extent and form of the damage depends upon the size of the piece, the temperature of the surface with which it impacts, and the point of the blade or other surface on which it impacts.

Should the impact occur on a thin section of the profile, the deformation that results is likely to be more severe and damaging to the aerodynamic form, than if a more robust part is struck.

When objects too large to pass through the blade passages are generated, they may wedge between the stationary blade rows. Often the rotating blades cut the large piece into a number of smaller segments, which rebound in the axial gaps between the stages until they are reduced to a size small enough to pass downstream. This chopping action does considerable damage to the blade inlet or discharge edges (whichever is slicing the large particle), and normally results in the replacement of the entire blade row. The pieces passing downstream then impact on the following blade rows, causing damage to them.

- *Surface roughening and efficiency loss are the consequences of the object's impact with the vane surface*

This is not, however, the only form of loss caused by object impacts. There are other types just as damaging to unit performance. The major factors introducing losses are:

- Surface roughening associated with the production of impact craters on the vane surface

These craters appear at the point of impact, which for small objects tends to be at the inlet on the suction face. However, as the objects become larger, there is no method of generalizing location or severity. The craters tend to make the boundary layer separate from the vane surface prematurely, and in so doing create vortices that will then exist throughout the expansion. Where the boundary layer separation occurs, will depend upon the axial location of the crater.

- By impacting with the inlet nose—which can be relatively thin on many designs, particularly those of an older vintage—the nose is deformed

This modifies the inlet angle causing an incidence loss. The impacts can also be sufficiently damaging that they close down the inlet, making it more difficult for the steam to enter the blade passage.

- Impacts on the discharge edge from objects existing in the axial gap between the stationary blade discharge and rotating blade inlet will often close down the stage discharge area

This will modify pressure from the stage and modify the stage enthalpy and blade velocity ratio. It will also modify the discharge angle from the optimum. Damage to the discharge edge can modify the discharge area, the steam flow, and the pressure distribution throughout the steam path.

- Debris near to the blade tips can cause damage to any seal or leakage control devices and deform them, increasing leakage that occurs there

Free hard objects in the steam path, whatever their origin, are destructive and have the potential to degrade efficiency, and possibly cause the unit to be forced from service immediately. They may also introduce a condition that will deteriorate, possibly causing failure at a later date, or extending an outage while replacement parts are obtained.

SOURCES OF THE IMPACTING OBJECTS

To cause damage within the turbine steam path, missiles must gain access, or be present in some manner. The most likely causes for them gaining access, or being present follow:

The result of mechanical failure of some portion of the steam path. For example, when a component or component part detaches and becomes loose within the blade system, the resulting debris will be centrifuged to the casing and rebound, most likely re-entering the steam path and possibly making many damaging impacts before its dangerous effects cease. Such debris can find several final locations in the unit:

- Lodging in the casing in a feed heating extraction line or similar location
- Lodging in a diaphragm or stationary blade row
- Being chopped into smaller pieces and distributed throughout the steam path
- Being carried into the low-pressure section and transported into the condenser. In many instances, larger pieces carried into the condenser will cut or severely damage tubing. This will possibly cause other problems associated with the ingress of cooling water, which will be chemically unsuitable for use in the turbine system
- Being carried over into a feed heater, steam extraction piping, feed heater, or other extraction line

Debris carried into the unit from the boiler. While this material generally consists of weld bead (or even stick ends resulting from initial construction and post-repair problems), other pieces have been found in steam lines. The damaging effects can be minimized by the use of fine mesh screens for short periods, after initial start up and after return-to-service, when repairs have been made. There is an additional pressure drop (and efficiency degradation) associated with the use of the fine mesh screen, but the removal of weld bead is important. The fine mesh screen need only be used for about six to eight weeks.

Debris entering the unit from some external source. This is unlikely, but there are instances in which components external to the steam path proper can fail and enter the blade system.

Parts left in the unit during an outage. Good maintenance practice makes this uncommon, but often small pieces such as nuts, bolts, and small tools will be dropped into the unit during an outage when the unit is being inspected and possibly refurbished. These pieces are most often metallic, and have the potential to cause significant damage upon restart.

Of the various sources of steam path debris, the generation of a missile from steam path components themselves is the most common and most likely to cause the most severe structural damage in terms of requiring component replacement. Such missiles are the result of damage in their stage, and may cause consequential damage that may be greater than the damage suffered by the stage in which the initial failure occurred.

Formed internally

This is probably the major source of objects capable of causing serious damage, because they can be of considerable mass.

The extent of the damage depends upon the size of the failed component centrifuged into the steam path. Missiles could cause massive damage on the first row of impacted elements and then be chopped into a number of smaller objects, which would cause less extensive damage (from a structural integrity perspective) as they pass through the steam path. Figure 4.3.1 shows a control stage that has suffered a blade failure at the root upper load bearing surfaces. The blade has detached and become a missile within the unit—there is relative little damage to the other blades in the row. Adjacent blades need to be replaced, but other elements are, from a cursory examination, in an acceptable condition. Nondestructive examination confirmed these blades had not suffered severe damage.

In this stage there was a large axial space between this row and the following stationary blades inlet. Therefore, the detached blade was centrifuged into this steam chamber where it did not re-enter the main steam flow.

By comparison, it is possible for a blade of much smaller mass to fail and cause significant damage because there is no axial space for it to rest in.

Fig. 4.3.1—A failed blade in a control stage. There is minor damage to adjacent blades and the cover band.

Massive objects can cause damage sufficient to force the unit from service for extensive periods. If the rotor is thrown out of balance there can be extensive rubbing damage, which could be harder to correct than the failure of a steam path component. Figure 4.3.2 shows the stationary blade row of a stage following a massive failure. Here the vanes have suffered impact damage, causing tears adjacent to the outer diaphragm ring where objects have been present in the axial gap between the rows.

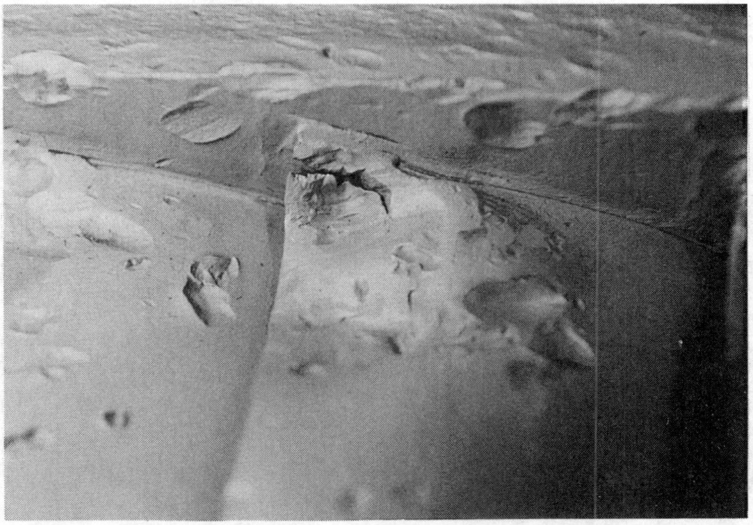

Fig. 4.3.2—Impact damage to a stationary blade row. The damage resulted from the failure of a large rotating blade which was trapped in the axial gap between the rows. A tear can be seen adjacent to the side wall.

Formed in the boiler

The boiler is a major source of small particles (*i.e.*, weld spatter) that are deposited on the tubing and eventually break loose and travel to the turbine valving system. The screens mounted around the valve inlet capture larger beads but smaller beads pass through the screen and impact the steam path. Figure 4.3.3 shows a screen removed from service—weld beads can be seen. Figure 4.3.4 shows typical damage resulting from small weld objects carried through a steam strainer screen.

External to the steam path

For missiles to be generated externally and enter the steam path is not common, but it does occur. If there is failure of any valve or screen component, there is a high probability they will gain access to the steam path, and will impact on the blade elements if they can pass through the first stage nozzles.

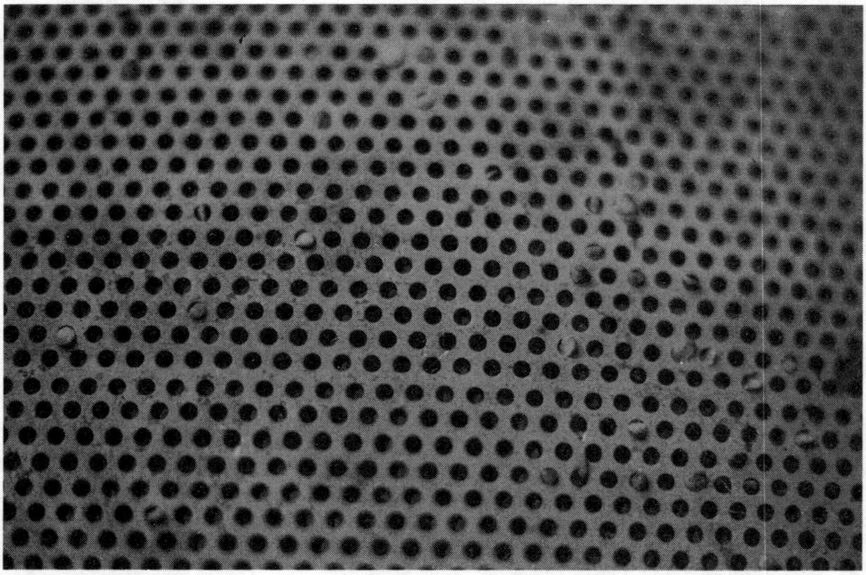

Fig. 4.3.3—The screen removed from around the valves, showing metallic debris which was too large to enter the steam path.

Figure 4.3.5 shows the result of the failure of a valve seat on the control stage stationary blade row. Here a portion of the valve seat was wedged between elements of the stationary blades, and a portion managed to travel into the axial gap between the stationary and rotating blade rows. The rotating elements are shown in Figure 4.3.6. This damage was sufficient that smaller missiles were formed, traveled downstream, and caused damage on a later stage (Fig. 4.3.7). This impact damage became progressively less severe as the distance from the control stage increased.

Fig. 4.3.4—Peening type damage on a control stage stationary blade row. This damage was caused by weld bead from the boiler superheater section and possibly other small particles.

Fig. 4.3.5—Damage to a control stage stationary blade row, caused by the failure of a valve seat which then passed into the steam path and become trapped between the fixed blade discharge and rotating blade inlet. The nozzle partitions are damaged to the extent they must be rebuilt by welding.

Fig. 4.3.6—The rotating blades from the control stage showing in figure 4.3.5. These blades have suffered massive damage, and must be replaced.

Fig. 4.3.7—Damage sustained further down the steam path by the stage showing in figures 4.3.5 and 4.3.6. Having been chopped into smaller particles, the valve seat passed down the steam path causing subsequent impact damage.

Drop ins

Problems associated with metallic pieces left in or dropped into a unit during an outage can be major ones to most users. "Drop-ins" can be of any size, limited only by their ability to bypass the components that prevent their being seen before the unit is closed up and returned to service. Figure 4.3.8 shows remains of a bolt left in or dropped into a unit during an outage and not removed. This material origin was traced from an analysis of the type of material.

Fig. 4.3.8—Pieces of a stud or bolt dropped into the steam path during a maintenance outage.

IMPACT DAMAGE CLASSIFICATION

Unfortunately, it is impossible to classify this type of damage. The question that always arises is, "Should it be classified in terms of the amount of area affected? In terms of the closure of the inlet or discharge edges? Or in terms of the most significant damage that causes surface roughening, or causes small ruptures in the surface of the blade?"

The following discussion considers various types of damage and its general form. However, in evaluating any situation, it must be considered that a form of damage acceptable at one radial location on one stage may be totally unacceptable at a different location on another. Therefore, the following descriptions of damage are discussed, but no offer is made of the most significant.

What *can* be stated is that some forms or levels of deterioration will obviously force the unit from service. Others—and those more perplexing to the engineer—ask, "When failure is not imminent, what level of damage is acceptable or tolerable?" These are levels that contain not only uncertainties difficult to judge (in terms of their effect on both efficiency and the probability of failure if corrective action is not taken), but are also dependent upon many other more complex factors:

- How critical to grid system security is the unit on which damage has occurred?
- What is the potential cost penalty of continuing to operate with the blades degraded?
- What is the cost in terms of time and material of replacing or repairing the damaged elements?
- What is the probability of the elements failing before the next planned outage?

These primary considerations are not considered exhaustive, and the operator may well have to consider other factors that override many of the economic requirements. However, if a situation could lead to a massive failure, then safety becomes a consideration and action must be taken.

Peening type damage

Peening type damage occurs when small pieces of debris that the steam transports through the steam path collide with a blade surface. The high relative velocities between the blades and the metallic objects are sufficient to cause plastic deformation (impact craters) on the blade surface. This form of damage occurs on both stationary and rotating elements, indicating that the objects have themselves achieved a high velocity within the steam.

Figure 4.4.1 shows the tip regions of a rotating blade row that has suffered extensive peening type damage. There is also some small amount of damage on the cover. This is not considered extensive, but will have reduced the effectiveness of the integral seal produced on the cover. This damage is consistent with the impact of small particles, formed in an earlier stage, that are trying to gain access to the blade row. It can be seen that this damage does not extend too far back into the passage.

The various sources of the small objects capable of causing this peening type damage include the following:

- *When many fabrication (welding) processes used in the manufacture of thermal cycle equipment are completed—equipment built in place—considerable quantities of weld bead are often left in the components and pipe work connecting them*

  Many of these beads are attached by heat fusion to metallic surfaces. Such fusion is not a metallic bond, and these objects are capable of loosening during operation.

It is therefore common for such weld beads to be left inside the boiler and other vessels and structures. Some are loose and some are attached to the plates and tubes. Indeed, it would not be possible for the manufacturer to remove this economically. During operation, this bead will experience both steam pressure and temperature expansion forces causing it to detach from the surface to which it is attached. These forces may however, not be of sufficient strength that the bead detaches immediately upon start up. Under these circumstances, the unit can operate for some considerable period before there is sufficient effect to loosen it. If the bead loses some of its adhesion, it eventually detaches. The bead is then free to be carried through components and pipe work into the turbine.

Fig. 4.4.1—Peening type damage occurring on the inlet edge of a rotating blade row. This level of damage can be tolerated from structural integrity considerations but will have an adverse influence on stage efficiency.

It is common practice for the majority of manufacturers of steam turbines to supply fine mesh screens for use during initial start up, and use after any repairs requiring "in-place" welding that cannot be cleaned at completion. These screens are used over the normal coarse screen surrounding the main inlet valves. After a period—normally some weeks or months—these fine mesh screens are removed, and any collected weld bead or other debris, is removed with it. At this time the unit goes into its normal mode of operation, with only the coarse screen in place.

This fine mesh screen cannot be tolerated in the unit indefinitely because of the excessive pressure drop associated with its use and the consequent loss in unit output and efficiency. Economics of plant operation dictate this fine mesh screen should normally be removed after about six weeks from initial start up. By that time, the majority of the weld bead and other fine material entrained in the system would have dislodged and been carried out. There is a possibility a portion of this weld bead will continue to adhere to the equipment surfaces for a greater period and then become loose after the fine mesh screen has been removed. At that time, this bead is free to enter the turbine steam path, with the possibility of causing peening type damage.

- *Components that detach within the steam path will impact with the rotating components immediately in their vicinity*

These impacts "chop" (or break down) these detached pieces into small objects of a size capable of causing damage. The generated or chopped objects are then free—and small enough—to pass through the steam path and make impact craters in subsequent stages. Figure 4.4.2 shows a stage that has suffered damage and shed a portion of its coverband that was cut into smaller pieces and caused peening-type damage on the following rows of the unit.

The rotating blades' ability to chop depends upon stage temperature. At the higher temperature, the detached components are more easily broken down into smaller objects. However, the rotating blade material is also softer at the higher temperatures, and may also suffer more severe damage from the impacts.

- *There is also the possibility that debris existing in other parts of the system eventually may be able to migrate into the turbine and be carried through the steam path*

This debris can include small parts left in a unit at maintenance outages and other foreign material that gains access through the performance of work on the unit. For these reasons, it is normally necessary for maintenance staff to be aware of any part or components dropped into the unit during repairs or inspection. It is normally less expensive to recover these parts than leave them in the unit.

Fig. 4.4.2—A high pressure stage where a portion of the cover band has detached causing impact damage both to the blades in that row and to elements downstream.

When solid particle debris that exists in the turbine—irrespective of its method of entry or generation—strikes the blades, it will, upon impact, carry sufficient force to damage the vanes. It can also impact with the coverband, tenons, and any flow-restricting device (leakage control seals) located above the blades. While this damage is usually in the form of small craters on the surface of the component, it can cause a significant frictional loss.

There are no general rules as to the repair procedure, or when they should be employed. Each case must be assessed individually, based on:

- the potential loss in efficiency
- the possibility that such damage could deteriorate to the extent the unit integrity is placed in jeopardy

Figure 4.4.3 shows peening damage on the inlet nose of a vane with water collecting grooves. Such impact damage could effectively remove the water-centrifuging capability of these blades, increasing the moisture content of the steam path beyond the design-anticipated level.

Fig. 4.4.3—Minor impact damage to a vane inlet edge, but in this case the impacts are threatening to block the moisture drainage channels.

Damage due to loose material in the steam path tends to be more acute on the rotating blades, *i.e.*, objects causing massive or heavy craters on the rotating blades will cause peening type damage on the stationary blade elements. This is probably because there is a lower velocity between the objects and the stationary blades so less damage is likely to be caused there.

Deformed vane inlet edges

Large objects in the axial gap between the stationary and rotating blade rows always carry the possibility of deforming and partially closing the vane outlets. This is shown in Figure 4.4.4, where the entire rotating blade row inlet has been damaged and partially closed. This unit suffering high levels of vibration due to some phenomena—not necessarily the failure that closed the blade inlets—is shown by the evidence of the heavy rubs on the coverband inlet side.

Fig. 4.4.4—Impact damage on a vane inlet edge. These impacts are closing down in inlet portion to the steam passage, restricting steam access to the row.

Large surface craters

Occasionally, a mechanical rupture occurring within the steam path is not of a magnitude that the unit is immediately forced from service. Indeed, many such events are not even detected, or detectable, until the unit is removed from service for a maintenance outage. These ruptures therefore go undetected, and the unit continues to operate for extensive periods of time with a damaged condition present. While such damage will cause a reduction in stage and unit efficiency, this may not be detected unless the most sophisticated testing devices are used to measure section efficiency and unit output.

Unless it escalates to an unacceptable degree by further ruptures, this situation is accepted—often in ignorance of its existence—and unless blades in other stage elements have been damaged as a con-

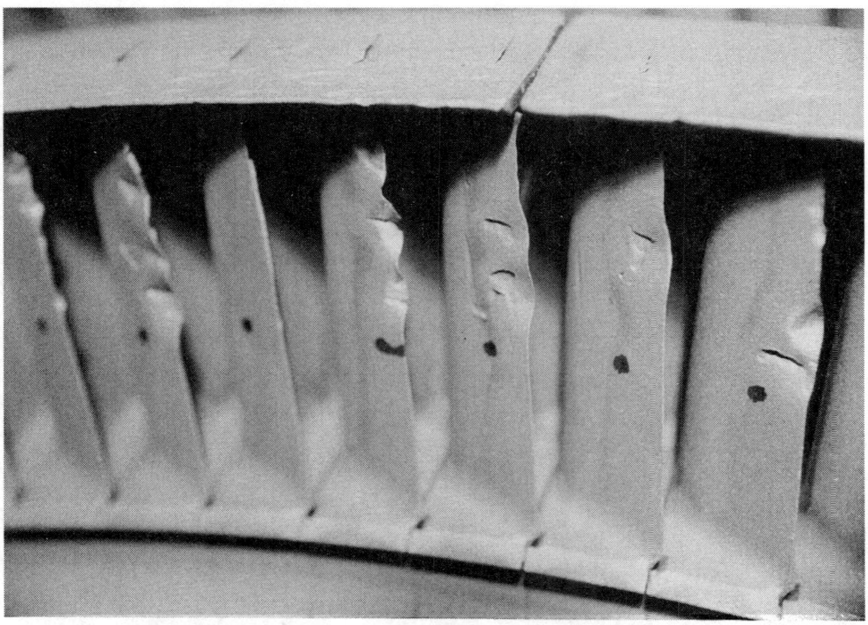

Fig. 4.4.5—Large impact craters on a rotating blade inlet edges. These craters have occurred in a region where tears can easily develop. This situation should be corrected before the unit is returned to service.

sequence of the situation, this is of no great concern to the operator. However, when this condition is discovered upon opening the unit, an evaluation must be made and the situation corrected if further damage could result.

This type of situation is shown in Figure 4.4.5. Large craters and tears exist on the inlet nose of a blade row, and many of these indents, craters, and tears could, if not corrected, eventually force the unit from service. The decision to decline corrective action and return this unit to service can only be taken recognizing the risk involved. A similar failure is shown in Figure 4.4.6. After separation, the detached main objects are chopped into smaller objects. A portion of cover-band from the L-1 stage has detached after a high-cycle fatigue failure of the tenon. Such a failure would probably not be noticed—there might be a marginal increase in the level of vibration on the bearing closest to the stage—but this might not draw attention to the failure. However, the consequence of this coverband detachment can

Fig. 4.4.6—A missing segment of cover band from the penultimate stage of a low pressure expansion.

be seen in Figure 4.4.7. There has been damage caused to the inlet edge of the following (last or exhaust) stage rotating blade. This does not appear to be significant at this time but there is always the possibility of a crack propagating at some later date.

Fig. 4.4.7—Impact damage on a last stage blade, where the cover banding of figure 4.3.6 has impacted in the erosion shield region.

When minor ruptures occur, they represent the detachment of parts of the blade system and the generation of loose material within the steam path. This material is then free to be chopped and rebounded between the components, where it normally will be cut into objects small enough to pass through the blade openings with the main steam flow. During their passage through the steam path, this debris impacts on stationary and rotating vanes, damaging or deforming them, and causing an efficiency loss—possibly producing significant mechanical damage. Figure 4.4.8 shows the damage caused to a sealing strip attached to a stationary blade row outer ring. This type of damage can represent a significant loss in unit output if it becomes excessive. Impacts on the face of the outer ring also are seen.

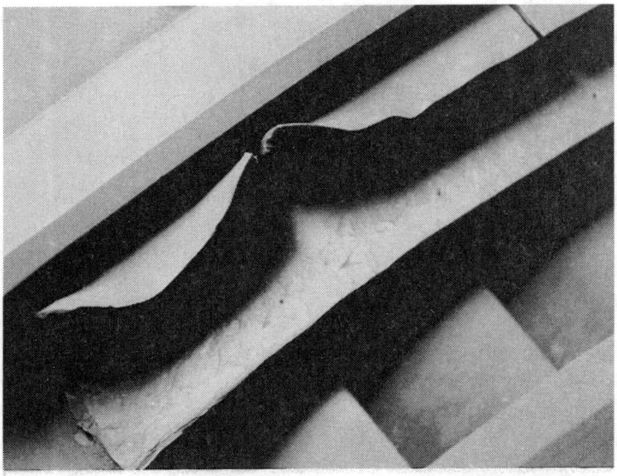

Fig. 4.4.8—A damaged radial seal strip. Because of the thin section, seal strips are particularly susceptible to impact type damage.

Massive craters and damage

Any component failure (mechanical rupture) occurring within the steam path has the potential to generate massive metallic debris. This debris will often be chopped by the rotating blades into smaller pieces, and pass through the remaining downstream portion of the steam path, probably causing minor mechanical damage as discussed previously. Due to this chopping action, and during passage through the blading system, these initially large pieces can cause severe impact damage to the components, depending principally upon their size, hardness, and relative velocity at impact.

While the rotating blades are most susceptible, the stationary blades can also suffer massive damage. This is shown in Figure 4.4.9 for a stationary blade row and in Figure 4.4.5 for a rotating blade. Another example of this form of major damage on a high-pressure stationary blade row is depicted in Figure 4.4.10, where portions of the discharge edge are missing, and large sharp-edged craters exist on the suction face. Figure 4.4.11 shows another form of this damage on rotating blades. Complete discharge edges have broken loose as the result of impacts from free material—in this case, generated within the row itself.

Fig. 4.4.9—A stationary blade row with very large craters at the discharge edge.

Material ruptures

Depending upon the point and angle of impact, and the size of the particles that impact on the vanes, tears can be formed. This is most likely to occur when the impact is on the thinner inlet nose of the rotating blade, or when debris is trapped between the discharge from the stationary and inlet to the rotating rows.

A tear is shown on the discharge edge of a stationary blade row (Fig. 4.3.2), and in a rotating blade row (Fig. 4.4.5). Objects trapped between the rows caused both tears. A similar tear, in which there are multiple impact craters on the stationary blade vane discharge edge, and a tear on one element at the outer sidewall is shown (Fig. 4.4.12). Figure 4.4.13 shows a similar situation on a rotating blade row, where a heavy impact has caused a tear to appear on the discharge edge. Impact craters can also be seen on the coverband.

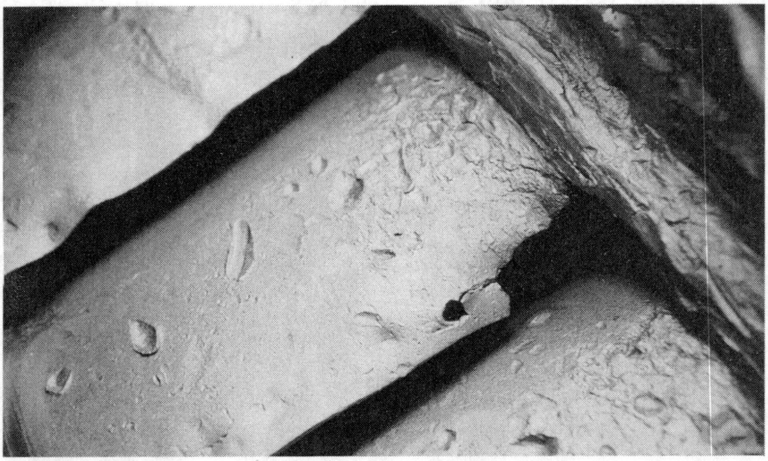

Fig. 4.4.10—Massive type craters on the discharge tail of a fixed blade row.

Fig. 4.4.11—High pressure blades having been impacted on their inlet edge, have suffered material rupture.

Fig. 4.4.12—A stationary blade with material ruptures at the discharge edge.

Fig. 4.4.13—A rotating blade row damaged by large particles trapped between the blade rows.

Elements trapped between the rows

Major damage normally occurs when an object too large to pass downstream becomes trapped between the rows. It is not wedged in a position where it will cause damage and is free to rebound and be chopped into smaller pieces before passing downstream. Aspects of this situation can be both advantageous and disadvantageous:

- While in the inter-stage position (unable to migrate downstream), the object will not damage the other blade rows. If this situation occurs and the piece can be removed, the damage will remain limited to two blade rows
- Stationary blades can normally be repaired, returned to service, and continue to operate in an "as new" condition. There could be the necessity of an extended outage to make such weld repairs
- Rotating blades suffering damage of this type most often require complete replacement. This will normally require an extended outage

Figure 4.4.13 is a rotating blade row that has suffered extensive damage. The elements must be replaced before the unit can return to service. In this case the damage extends to the coverbands and the tenon heads, which have been completely destroyed by the rubbing of objects between the coverband and casing. A similar situation is shown in Figure 4.4.14, where objects have become trapped above the coverband, causing serious gouging damage. The outlines of the tenon heads can still be seen. There is deep grooving where the debris has gouged the coverband outer surface.

Vane damage and assessment for correction

When impacts occur on a vane, and it sustains damage that deforms the profile, at what point is it necessary to replace, refurbish, or repair the affected elements? This is a major consideration whenever impact damage is found.

We know if any profile deformation has occurred, blade efficiency will deteriorate from its previous condition. Depending upon the nature of the damage, deformation of one profile can modify the shape and expansion efficiency of two passages.

Fig. 4.4.14—Grooves in a cover band as a result of debris trapped between the cover band and the casing.

In general, peening type damage can be dressed. While not restoring efficiency to the original level, it can at least minimize losses. The following line of corrective action is suggested:

For stationary blade elements. If the extent of damage is at or beyond the level of that shown in Figure 4.3.4, corrective action must be considered. However, unless there are significant tears, or situations indicating objects will detach after return to service, the economics—fuel cost and load factors against repair costs—may also influence the final decision.

For rotating blade elements. Rotating blades are more sensitive to damage. While peening type damage can be dressed to reduce efficiency losses, no decisions should be taken that will place the security of the unit in jeopardy. Apparently insignificant damage may quickly deteriorate once the unit is returned to service.

While a user can often accept efficiency deterioration and continue to operate with damage, few situations arise where a unit can be "operated at risk." To do this, the situation must be fully analyzed. It must be known that deterioration can quickly occur with the consequences fully understood.

SOLID-PARTICLE EROSION (ABRASION)

Today's fossil-fueled turbines work under higher pressures and temperature conditions and are expected to endure more frequent starts and shutdowns. This has given rise to solid-particle erosion—a damaging phenomena causing significant material loss from the various components in the stages following the initial entry or reentry of hot steam from the boiler superheater or reheat section.

Damage is introduced by the action of hard scale that forms on the water side of the boiler tubes. It is carried over into the turbine steam path, where it contacts the various surfaces of components and removes material from them. In such instances, the first one or two stages of the unit suffer a recognized pattern of material loss in various regions after the entry of steam. This damage can be severe. Material loss occurring in regions which can affect both the efficiency and structural integrity of the unit.

This damage occurs principally on the pressure faces of the stationary and rotating blade vanes, at the inlet and discharge edges. However, such damage is also found as a material loss on the coverband and tenon heads.

These damaging effects have been determined to be the consequence of hard particles of scale, which have been exfoliated from the inside water surface of the boiler tubes. This scale is carried in the form of small particles into the first stages of the unit, where they come into high velocity contact with the stationary and rotating blades. The scale is composed primarily of magnetite, which is extremely hard, and initially has sharp or jagged edges. These damaging phenomena can affect the stages of the high and reheat sections of the unit. It is believed there are two basic mechanisms contributing to this damage:

- A gouging or cutting action that occurs on ductile blade materials. This is most severe when the impact angle is from 20° to 40°. These damage incidents for oblique impacts are a function of particle velocity raised to a power "n," where "n" has been reported to have values of from 2.0 to 4.0
- A "chipping" type mechanism, in which small particles are broken from trailing edges. This is a brittle type failure found most commonly in austenitic materials

Material removal mechanisms can therefore be either abrasive or have an impact effect on the scale—or a combination of both. However, any removal mechanism must take into account material losses on the surfaces being eroded, and how they react to scaling at the high temperatures at which the steel operates.

As scale enters the steam path, the steam will transport it through the first stage nozzles. After its initial impact, a portion of the scale will be centrifuged to the casing, where it may take a number of routes through the remaining stages of the expansion, in some cases colliding with and damaging other portions of the steam path. Other portions of this scale will rebound in the axial gap between the stationary and rotating blade rows, eventually passing through the rotating blades and causing some level of damage to them. Eventually all of this scale will migrate to the blade outer flow sections. Continuous impacts with the blades, and the centrifugal force imparted to the scale, means the scale will be deposited at the casing into various niches, where it will remain.

The scale is formed in the boiler tubes as a result of continued operation at high temperatures, and is attached to the tubes. However, as temperatures change within the tubes—particularly at start up and shutdown—scale will exfoliate and become free within the tubes themselves. From this observation it is clear that units on cycling duty, or subject to a large number of cold starts, will suffer more exposure to this form of damage. They will require a greater level of monitoring and ultimately require greater levels of maintenance.

SCALE FORMATION

Boiler tubes are normally produced from a low-alloy steel (perlite), which lends itself to relatively easy oxide scale formation on the internal surfaces.

Figure 4.6.1 shows magnetite scale on the steam side of the boiler tube. During startup and shutdown, when temperatures within the boiler and reheater tubes are changing—or when temperature transients occur in the vessels due to uneven firing or other conditions—

both the tube and the scale expand and contract. Because the tube and scale have different coefficients of expansion, shear forces develop along the interface with temperature changes. These forces are of sufficient magnitude that with large temperature changes, they will cause rupture at this interface, allowing scale to detach.

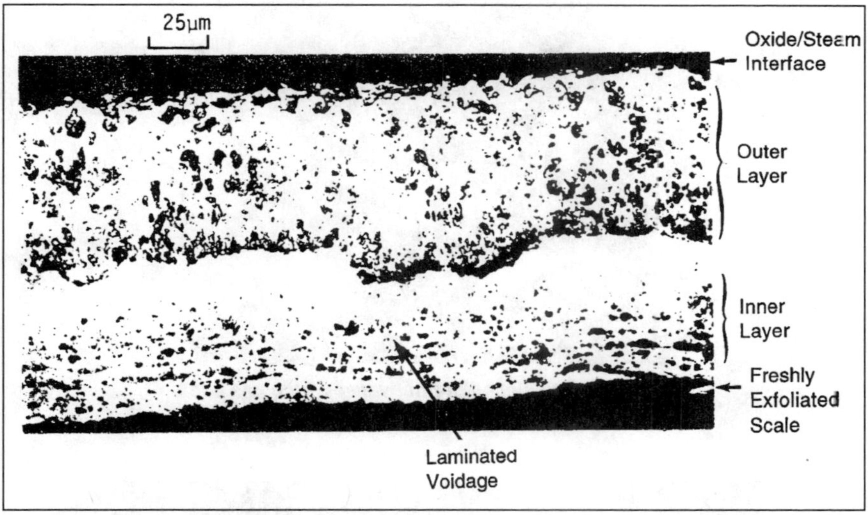

Fig. 4.6.1—Magnetite formation on the steam/tube interface.

Increased flow velocities through the boiler tubes can help loosen and detach the oxide scale. When this scale breaks loose, it is carried over into the steam turbine stages, damaging the components with which it comes in contact. Figure 4.6.2 shows the stem from a stop valve pilot that has been exposed to scale impact for two years; Figure 4.6.3 shows the valve seat from this same valve. The material loss due to impingement by the scale can be seen. After passing through the valves, the scale enters the main steam path.

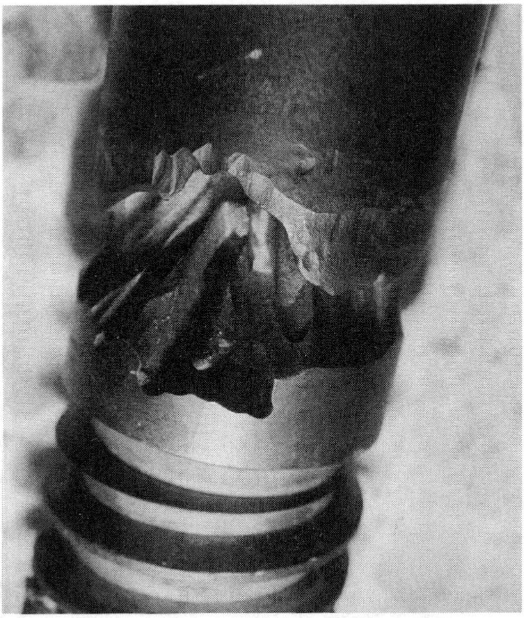

Fig. 4.6.2—The stem of a pilot valve which has been subjected to solid particle erosion.

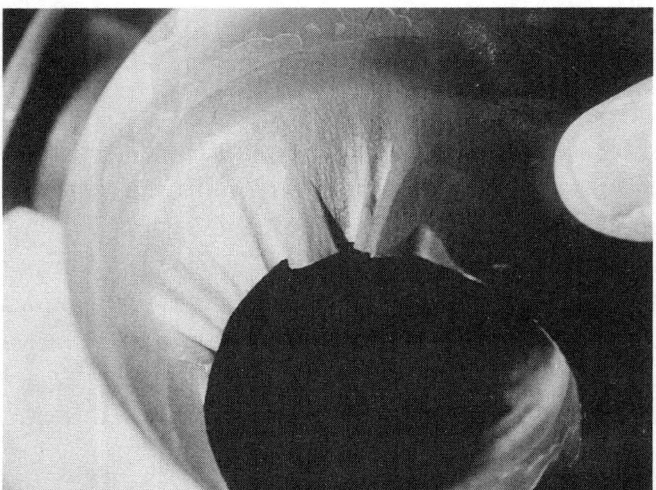

Fig. 4.6.3—The seat of an eroded pilot valve.

THE EROSION MECHANISMS

In considering the mechanics of the cutting or abrasive actions that remove material from the stationary and rotating rows, it is best to review typical stage velocity diagrams, and consider the probable track of the particles between and within the steam expansion passages.

Actual material losses depend mainly upon stage geometry and the turning angles of the vanes. These losses are therefore design sensitive. However, general explanations of the mechanics of the two types of row (stationary and rotating) can be developed, and with observation of individual units, a better understanding of the individual loss modes developed.

The stationary blade row

Consider the velocity profiles of the stationary blade row shown in Figure 4.7.1. In such a blade row, steam carrying small particles of scale enters the blade passage. Because steam particles within this passage are of smaller mass, they can be accelerated by the release of thermal energy deflected through the vane-turning angle. Both the pressure surface of vane "R" and the suction face of the adjacent vane "S" influence them. At the throat discharge, the steam leaves the row at the design angle.

However, because the transported scale particles are of greater mass, they cannot be accelerated through the complete passage turning angle while being deflected to some degree through the passage. Some of them make impact contact with the pressure face of the stationary vane "R," causing a cutting or abrasive action on its surface. That scale contacting the vane at point "j" will be predominantly impacting, while the scale contacting at "k" will be predominantly gouging. However, every piece of contacting scale will both impact and gouge. It is a matter of extent, and that extent varies from "j" to "k," with the actual damaging mechanism being dependent upon the impact angle.

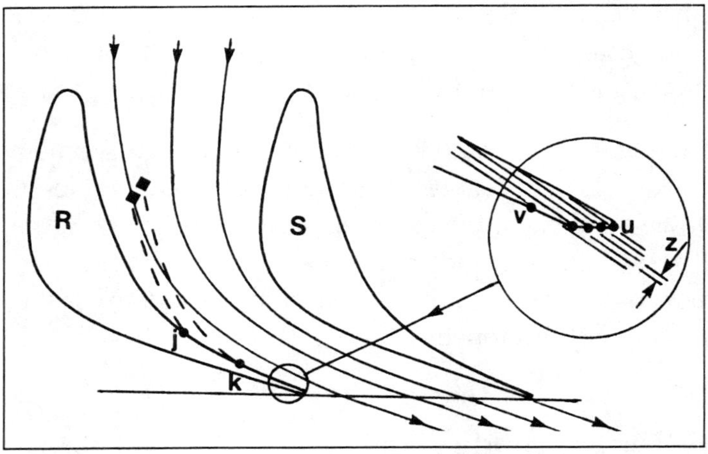

Fig. 4.7.1—The steam and scale tracks between the blades, also showing the potential wear pattern at the discharge edge.

When it contacts the pressure surface of vane "R," the scale will react on it and remove material. This material removal will continue to penetrate, removing more surface material towards the discharge edge because of the greater quantity of scale present along the surface of the vane. Shown in more detail in Figure 4.7.1 is the effect of subsequent levels of penetration "z." The amount of vane material remaining after these subsequent levels of penetration will provide only small visual changes to casual inspection up to the position where suction face material is removed. At this point, the more characteristic loss patterns occur.

The rotating blade row

The tracks of the scale in the rotating blade row are influenced by the forces exerted on them by the flowing steam and by the centrifugal effects of rotation. This means that for the rotating rows, it is necessary to consider the scale moving in the axial/tangential plane and also the radial direction. What's more, in the rotating blade rows, the mechanism of material loss is somewhat more complex, having the complication of high radial flow components introduced by the centrifugal forces, imposed on the drag forces acting on the moving scale by the main steam flow.

The axial/tangential planes. Consider velocity triangles in the axial/tangential plane for the steam and scale in a typical rotating row in Figure 4.7.2. Here the velocity triangles of both steam and scale entering and leaving the rotating blade row can be seen.

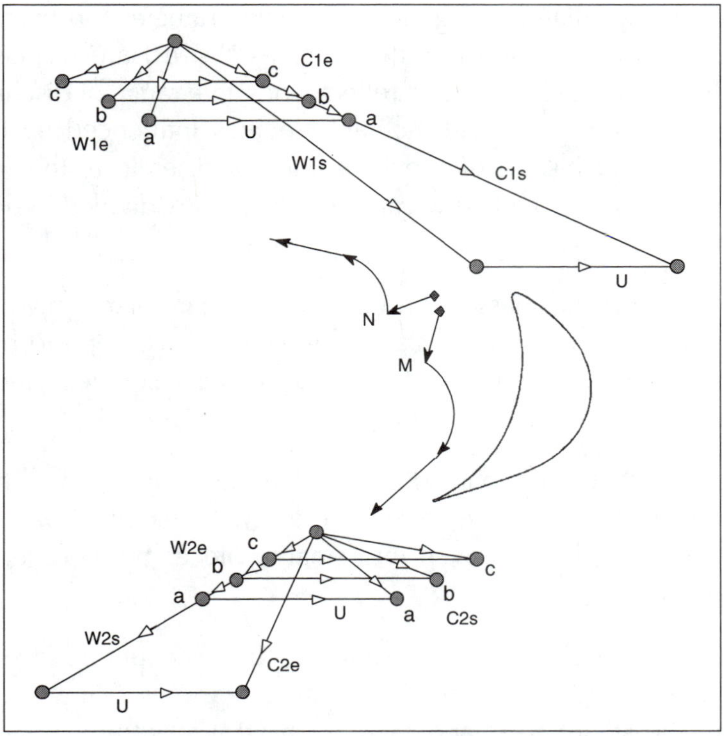

Fig. 4.7.2—The velocity triangles for steam and scale at discharge from the stationary blade row.

As shown in this vector diagram, the rotating vane, having a tangential velocity "U," is arranged to accept the inflowing steam from the stationary row with a relative velocity "W1s." This steam flows into the expansion passage between the blade vanes and discharges (with a substantially axial velocity) into the following stationary blade row. The particles of scale have essentially the same discharge angle as the steam leaving the stationary blade row. It should be recognized that there could be discharge angle differences for that portion of the scale contacting the stationary vane discharge edge. Such scale could rebound or deflect from the steam angle. The relatively larger mass of the particles of scale will leave the influence of the stationary blade row at lower velocities "C1e."

In the vector diagram (Fig. 4.7.2) velocity triangles for three different particles (a, b, and c with velocities "W1e") are displayed. These different velocity triangles reflect a possible range of discharge velocities for the scale, with actual velocities influenced by their size, shape, and history of previous impacts. Therefore, the scale entering the rotating blade row can be conveniently divided into two groups:

- Scale that has passed through the stationary blade row without collision with the stationary vanes. It will tend to be of smaller size, and have somewhat higher velocities, since it has not lost energy by impact
- Scale that has impacted on the stationary vanes. It will have lost energy, and therefore have a lower velocity, but might also have been broken into smaller pieces by the effect of impact

It is impossible to take all these factors into account in developing a detailed model. It is sufficient to assume the discharge velocities from the stationary blade row can have a spectrum of values from "W1e(a)" to "W1e(c)" as shown for the inlet velocity triangles.

It is now necessary to consider the possible trajectories of scale particles at both ends of the probable velocity spectrum. Consider first that scale entering with a velocity "W1e(a)" (the scale that has passed through the stationary blade vanes without impact, and will possibly be of smaller size). This scale will enter as shown in Figure 4.7.2—striking the rotating vane suction face at position "M." At impact, this particle will possibly fracture into smaller pieces and rebound into the main steam flow. It is then probable all or most of this scale (or smaller particles) will be carried through the steam passage to enter the following stationary blade row. Scale particles entering the rotating blade row with a velocity "W1e(c)" will strike the vane at position "N" and rebound. Because the majority of this scale has already impacted the stationary blade vanes, it will tend to be of larger size, and the majority will probably rebound into the axial gap between the stationary and rotating rows. This scale will be considered in more detail in the following section, which discusses the radial flow effect.

The scale carried through the expansion passage to discharge will leave the rotating blade row with velocities and directions as indicated as "C2e," which again can have a spectrum of values of both velocity and discharge angles. The steam will leave the row at the design predicted velocity "C2s" in a substantially axial direction.

The radial plane. Due to the centrifugal effect on the working fluid in the rotating blade passages, scale particles flowing in the passage formed between these vanes will experience considerable radial forces that will tend to make them flow to the outer radii of the row, where any coverbands will prevent them being centrifuged radially outwards onto the casing. Figure 4.7.3 shows scale paths indicating the radial flow towards the underside of the coverband, though the coverband prevents these particles from leaving the row. However, upon leaving the blade row, these particles will have a large radial flow component at its discharge edge due to their mass. This will cause a large proportion of these particles to flow onto the casing. Some of this scale will rebound from the casing and re-enter the main steam flow. A portion will be retained there in any form of hideout or cavity that exists. Often when a unit is opened, some scale can be recovered, principally from the lower-half casing. A probable scale flow pattern in the rotating blade is shown in Figures 4.7.3 and 4.7.4.Here the effects of steam leaking over the coverband are also shown.

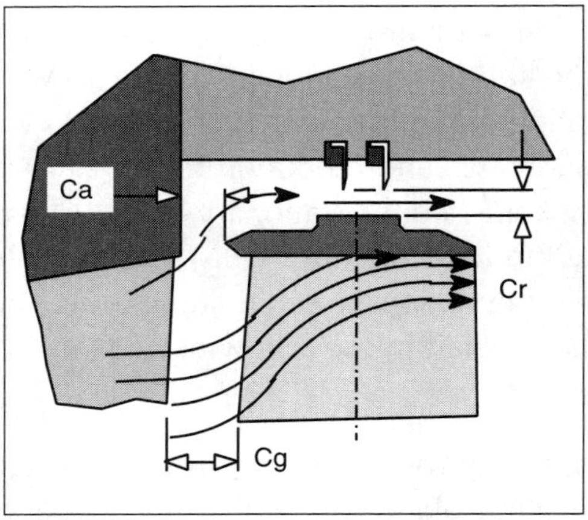

Fig. 4.7.3—The radial flow of scale particles after leaving the stationary blade row.

The amount of steam leaking over the coverband has a direct effect on the quantity of scale reaching the casing through any seals—axial or radial—that may be designed into the stage. Figure 4.7.4 shows the effects of leakage and the interference that can occur above the blade tip. This figure also shows the possible paths the scale can take. After initially passing to the casing on the rotating blade inlet side, it can impact on the casing and will swirl in the radial clearance above the blade. This scale will eventually be carried past the radial seal and interact with the steam discharging from the rotating blade row. It can possibly be influenced, to a lesser degree, by the scale centrifuged to the casing from the rotating blade. Another consideration in terms of the seal influence is that as seals rub, clearances will be opened, producing an increase in the leakage steam quantities and steam forces affecting the scale particles. Greater steam forces tend to effectively suppress scale radial flow and permit less scale centrifuged to the casing. Scale that fails to reach the casing and is retained in the steam flow will be carried into the following stage.

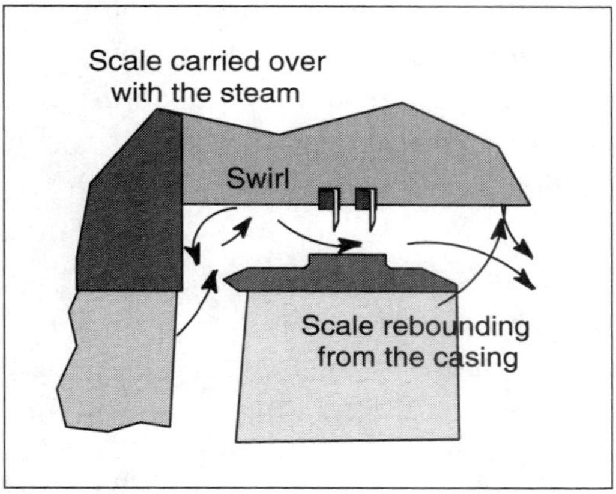

Fig. 4.7.4—Scale rebounding from the casing inner surface and moving past the seal strip.

Consider now in more detail possible paths of the two "extremes" of particles "a" and "c" (Fig. 4.7.2). The particles indicated by condition "a" with a velocity "W1e(a)" enter the steam path at a high velocity and penetrate the expansion passage to the point "M" before they impact on the suction face. The majority of these particles are transported through the blade passage. While they will be centrifuged radially outwards to the casing, a proportion of them will flow into the following stationary blade row, possibly after impacting on the casing inner surface. This centrifugal action within the blade passage is shown is Figure 4.7.3.

Of greater concern in terms of material loss and damage, is that scale (particles "c" entering with velocity "W1e(c)") rebound into the axial gap "Cg" between the stationary and rotating rows (Fig. 4.7.3). This scale, causing some level of damage to the vane with its initial impact and upon rebounding some distance back into the axial gap "Cg," will be re-entrained into the main steam flow and be free to re-enter the rotating blade row causing further damage.

Upon leaving the vane after impact at point "N," this scale will have some small radial component. Upon entering the axial gap, it will tend to flow or migrate radially outward with some velocity. Depending upon the size and direction of the scale, it will flow back and strike the stationary blade row, where it will rebound and be carried by the steam field force back into the rotating blade row. This type of impact and the consequent damage can be repeated as many times as the particles are forced by the steam back into the blade expansion passage. Figure 4.7.5 shows the possible path of a single piece of scale that enters the rotating blade row near the root section. Its multiple impacts in the flow ultimately lead to the tip section, where it will either pass over the coverband or through the blade row. It is also possible that any given piece of scale will fracture into smaller pieces developing individual paths appropriate to their size and relationship with the steam field. Similarly, scale entering at other locations will rebound and flow towards the tip section, so there is a gradual accumulation of scale towards the blade tip. It would be expected that damage would become progressively more severe towards the tip.

At unit startup, the quantity of steam and the force of the steam being admitted to the unit is relatively small. This is when the majority of scale exfoliates and flows to the turbine. However, there could be some delay in the entry of this scale to the steam path—particles may be too large to pass through the fine mesh screen, become trapped in pockets and hideouts in the valves and inlet pipes, require some time to break into pieces small enough to pass through the screen, and other scale may eventually be torn from hideouts and pass to the turbine. Therefore, while scale is carried to the turbine at start up, its entry to the turbine in a form causing damage may be delayed by various mechanisms, and the damage process could continue for some time after each return from service.

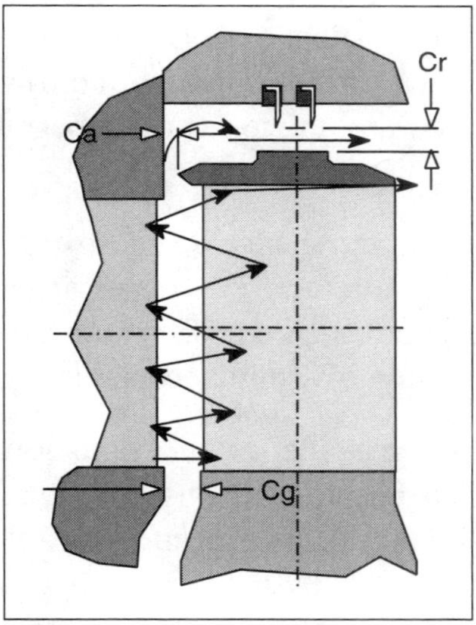

Fig. 4.7.5—Scale rebounding between the stationary and rotating blade elements and migrating out towards the tip section and seal.

MATERIAL LOSS PATTERNS DUE TO SPE

Material loss patterns that occur on stationary and rotating blade rows are now familiar to the majority of operators, and immediately recognized. Such damage is characterized by the modification it produces to the cross section on both sets of vanes and sidewalls, and the effect it has on the sealing system above the rotating blades. Removing this material therefore modifies the shape of the expansion passages, and is capable of adversely influencing the steam path expansion efficiency and the unit reliability.

These loss patterns from the stationary and rotating rows have certain similarities, but the form the damage within the two rows takes can also be quite distinct. This is due to different forces influencing the scale paths as it flows though the rows. It is of value to consider the two rows, and the forms of damage to which both are subjected.

The scale's hardness and shape can cause erosive damage before its entry to the steam path, and it is necessary for the operator to examine the valve control system to ensure the level of damage these components sustain is within acceptable levels.

Early turbine damage—after initial admission or after return of steam from the boiler reheater section—can be subject to relatively severe damage as a consequence of this scale. Such solid-particle erosion of the unit is normally a slow process, although its effects and presence can be inferred from both pressure distributions throughout the unit and state line efficiency measurements. This is because a gradual process must be monitored with test-quality instrumentation. It is often difficult, immediately after start up, to differentiate between the effects of this erosive damage and damage due to impurity deposition on the steam path expansion surfaces.

Removal of material occurs primarily at start up, and if steam is admitted to the steam path through partial arcs of a row, it is common for those segments which are first to admit steam to suffer extreme damage. This can be seen from an examination of a control stage nozzle box. Figure 4.8.1 shows a portion of a control stage nozzle plate, in which material loss is not even around the circumference. This row has suffered material loss from the stationary blade vanes on their discharge edge; the uneven nature of the erosion damage is also shown. Such loss from the discharge edge affects the discharge area and angle. If not corrected, it will cause a significant deterioration in expansion efficiency.

Fig. 4.8.1—Portion of a control stage nozzle plate showing chipping erosion at the discharge edge of the vanes.

The turbine valve system

Steam-transported scale is capable of causing damage to all metallic surfaces with which it comes in contact—including the valves. However, erosion damage is velocity sensitive, and will have a greater effect on those surfaces that are subjected to the high velocity flow of steam and scale across them.

Of those units suffering valve system damage, the damage occurs before the steam enters the steam path. However, the damage also affects performance of the unit, and it should be prevented from progressing to the greatest extent possible, even if this requires the expensive option of replacing parts. The scale that enters can erode both the stems and seats—in fact, any surface where steam reaches a higher velocity as it is accelerated across the surfaces by the pressure drop that occurs in the valve system. The actual material loss patterns are a function of particular valve geometry and the manner (or direction) in which the inlet pipes admit steam. Similar damage can be produced on the inlet valves, where steam is returned to the unit from the boiler reheater, again carrying scale with it. However, such valve damage tends to be less severe, as these intermediate valves do not exercise the throttling or control function of the main inlet system.

As scale leaves the boiler's superheater section, the working fluid transports it through the main steam leads and the valve system of the unit to the steam path of the high-pressure section. This control valve system contains a stop valve that remains fully open at all loads. It is intended as a primary control device that closes only on loss of load and an anticipated emergency overspeed condition. The second valve in this system is a control valve that controls the quantity of steam being admitted to the turbine in response to the demand for load placed on the unit by the operator. It is normal for one of the main stop valves to contain a pilot valve designed to be opened at start up against the initial steam pressure. This pilot valve admits steam sufficient to start the unit rolling and overcome the initial friction forces preventing rotation. This valve will also provide sufficient steam to achieve a small load on the generator. The pilot valve is therefore a prime candidate for erosive damage, because it is subjected to the major portion of the initial charge of steam containing the scale as it enters the unit (see Figures 4.6.2 and 4.6.3).

One preferred method of initial start up when scale is present, concentration is normally high, and loads are small (+/- 5.0%), is to admit steam through the complete admission arc, *i.e.*, control valves operate in parallel rather than sequentially, giving a more even distribution of scale around the complete inlet annulus.

The propensity of the valves to be damaged is dependent upon design geometry. Not all valves erode evenly even with similar patterns of operation. It is difficult to predict the erosion potential of a particular valve as relatively small changes in geometry affect flow patterns of the steam and therefore scale distribution, having a marked effect on damage patterns. Figure 4.6.2 shows a pilot valve stem that has been in operation for about two years, and Figure 4.6.3 shows the seat from this same valve. Both exhibit severe damage patterns and material loss.

The control stage nozzles

After steam has passed through the main stop and control valve system (including the strainers), it enters the steam chamber where it is required to turn through relatively large angles—axial and tangential—to reach the inlet to the control stage nozzle blade vanes. Because the transported scale is of relatively larger mass, it is unable to accelerate around the inlet plenum of the nozzle boxes to the same extent as the steam, and will tend to concentrate towards one region of the chamber, *i.e.*, scale will not be evenly distributed around the perimeter of the nozzle segment served by any valve.

This effect is shown diagrammatically in Figure 4.8.2. The first valve of a four-segment nozzle box admits steam "1" (with vertical steam inlets). Here the steam (shown as a dashed detail) will enter and flow in equal quantity to all circumferential locations from "u" to "v." Because scale has a considerably larger mass, it will flow, full lines, predominantly to the lower partition at "v." For this reason, erosion of the nozzle partitions towards the horizontal joint "HJ" will increase. This erosion damage will also be most severe in the nozzle segment "1," which is the first to admit steam to the unit. It contains the pilot valve and will carry the unit from about 5 to 25% load.

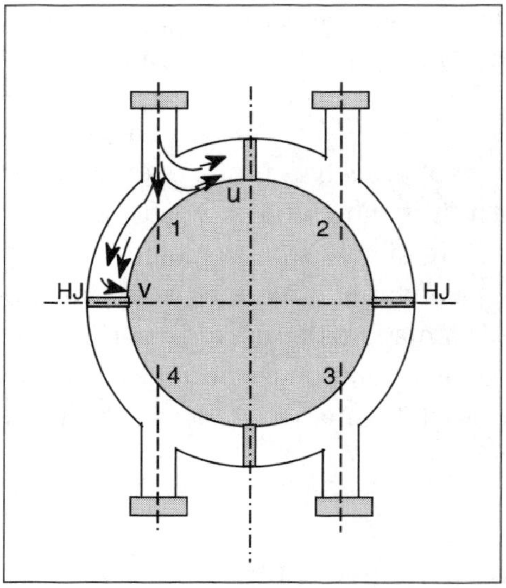

Fig. 4.8.2—Scale flowing with the steam to the first nozzle chamber.

The stationary blade rows

After passing through the control stage—stationary and rotating rows—particles will find relatively even distribution around the circumference and, therefore, the diaphragm of the second and subsequent stages. It is anticipated that any erosive damage would be relatively evenly spaced around the total circumference of the row. Experience has found this to be the case, with minor variations due to some flow inequalities possibly introduced by the quantity of steam flowing with entrained scale, through the individual passages; and the possibility that individual discharge edges may have minor differences in thickness, particularly after repair and dressing. However, for all practical purposes, the damage is relatively even.

There is one exception to these "even damage" observations, and are seen on those stages employing extended vane partitions to add mechanical strength to the diaphragm and limiting its axial deflection. Figure 4.8.3 shows a section through a stationary blade row with such extended partitions "s" and "t." If the steam and scale flow enters axially, there would be no difference in the quantity of scale entering the individual throats. However, if the scale enters at some shadow angle "Ψ" in the tangential direction, scale distribution will not be even. Figure 4.8.3 shows passages "K" will experience scale ingestion equal to one tangential pitch. Passage "L" will experience little or no scale, as it is protected by extended vane "t." Passage "J" will receive scale for passages "J" and "H." The scale that would enter "H" is deflected into "J" and provides similar protection to passage "H" as is afforded to passage "L" by partition "t."

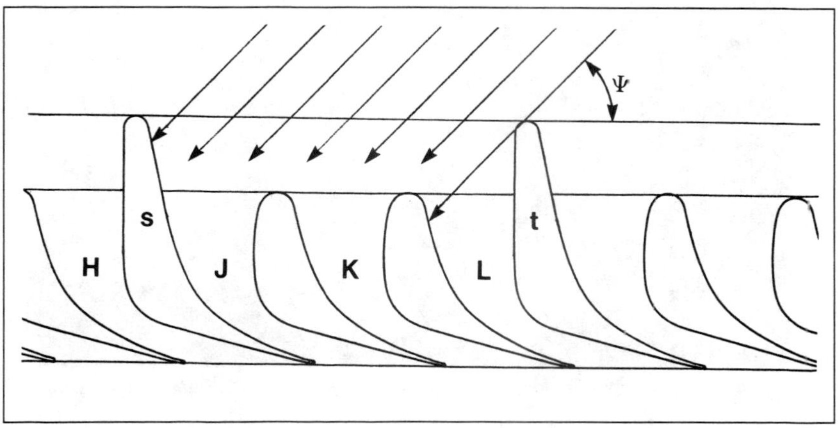

Fig. 4.8.3—The effect of "acute angle" entry of steam and scale to a stationary blade row, showing the uneven distribution of scale.

In fact, this model is based on purely theoretical, "unsubstantiated assumptions." The actual pattern depends upon the scale angle, which will vary from particle to particle, and possibly be quite random. The number of expansion passages existing between the extended vanes will also influence it. However, the principle as presented here can be observed from an examination of diaphragms with extended vanes as they are removed from service.

In the radial direction, the scale entering these diaphragms will be concentrated towards the outer sections. This is because the scale will be predominantly that which has passed through the preceding rotating blade row, and will have a radial flow component to its direction. It is normally noted that the material loss by erosion is greater towards the tip section of the stationary blade vanes (Fig. 4.8.4).

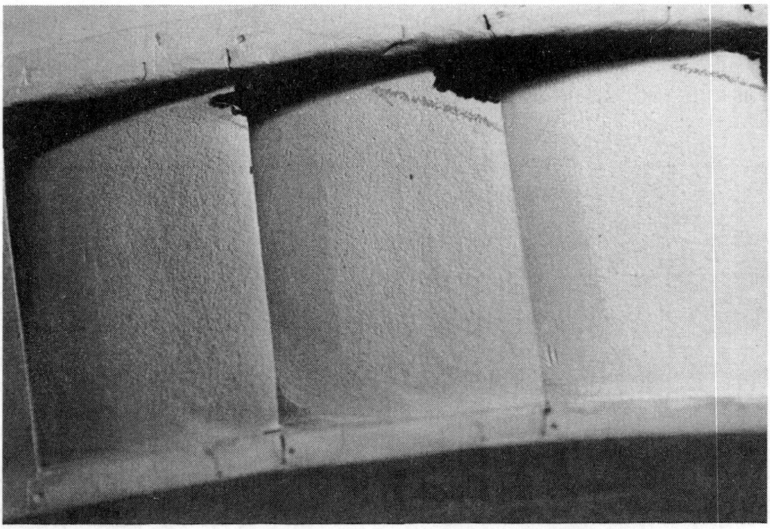

Fig. 4.8.4—The material loss in the outer regions of the stationary blade row.

From the stationary vane inlet nose. In the previous section we discussed material loss suffered by the diaphragm stationary vanes. In addition to this damage, certain stages will suffer a material loss from their inlet nose—damage that tends to be concentrated towards the outer diameter.

Scale entering the stationary blade rows does so at a relatively low velocity. However, since the steam has a low velocity at the points "C2s" (Fig. 4.7.2), it would be unable to accelerate the scale, which enters at a velocity "C2e" (joined by scale rebounding from the casing, probably in quite random directions) and accelerates to an acceptable velocity and direction. This scale will enter the stage influenced primarily by the direction in which it enters the axial gap ahead of the ring of stationary vanes. This relatively low-velocity scale will tend to concentrate at an outer diameter.

This form of erosion when it occurs is not normally severe because low velocities are involved. Such damage is shown in chapter 1, Figure 1.9.12, where the metal removal is concentrated just below the outer sidewall of the diaphragm. Here the amount of material removed is not high, but the vane is in tension at its inlet edge due to the axial loads on the inner diaphragm ring or web. These grooves can represent a "notch sensitive" region. While this form of damage is unlikely to cause failure, it should be corrected if it progresses to a depth that would weaken the axial strength, and the ability to resist axial pressures causing excessive axial distortion of the inner web.

From the stationary vane discharge edges. Material loss that occurs from the vane discharge edge is produced by one of two distinct mechanisms, or the two in combination. However, these losses are the result of scale flowing over, or impacting with the thin material at the edges. The loss mechanisms that affect each discharge edge are in reality a function of the vane material properties and the susceptibility to impact. Two failure modes are operable:

- *A gouging, or grinding action that removes material by an abrasive type process*

This process is relatively slow, and removes material at a reasonably uniform rate as a function of start ups. Such a material loss is shown in Figure 4.8.4. A high rate of material loss is shown in Figure 4.8.5, where a previously rebuilt vane has suffered extensive material loss. The filler material used to make this repair did not offer sufficient resistance to renewed solid-particle erosion. Because scale flow density is greater at the outer diameters, these regions wear at a considerably faster rate than the inner portion.

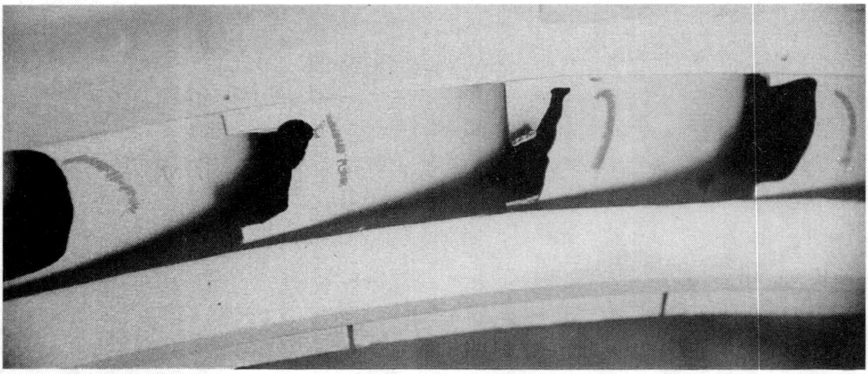

Fig. 4.8.5—Material loss from a stationary blade vane of martensitic material.

- *A brittle type fracture that occurs most commonly in vanes produced from austenitic steel*

This condition—known as "chipping"—can accelerate blade edge loss, and can occur so early in the operation of the components that subsequent erosion due to abrasive particles can erase evidence of this type of metal loss. Figure 4.8.1 illustrates partial loss by the chipping condition. This problem is best resolved by the use of a thicker (0.040") trailing edge.

It is possible this mode of failure is caused by an initial thinning of the discharge edge by abrasion, resulting in a weakening (gradual change) of the section. This produces changes to its natural frequency, introduces panel mode vibration, and/or continued operation at a valve point. This causes the loss of the small semi-circular pieces of material.

For *martensitic* steel the most commonly encountered damage mode is that of gouging or abrasion. Because martensite is sufficiently ductile, it can absorb the effect of impacts without causing rupture. While removing discharge edge material may produce a natural frequency (which would cause failure), the material's ductility is sufficient that high stresses are not introduced by the effect of impacts or erosion. Austenitic material, however, is less ductile and has a high level of stress induced in it by impacts. It will suffer failure under the effects of repeated impacts. Failure modes in either martensitic or austenitic materials are capable of being repaired, however, the weld filler material and stress relief requirements may be different.

From the stationary vane sidewalls. There is a tendency for scale entering the stationary and rotating blade rows to migrate by centrifugal action towards the outer diameters of the row, where it can cause more severe damage. This particle concentration in the outer flow regions will often cause gouging type damage to the outer sidewalls. This effect can be seen in Figure 4.8.6, where gouges have cut into the outer surface positions. Both the vane discharge edge and the sidewall have been eroded, modifying the discharge angle and increasing the discharge area.

Fig. 4.8.6—Material loss from the stationary blade row outer side walls.

This area increase due to sidewall material loss may be relatively small, but for small radial height stages it can significantly add to the total discharge area. This will also encourage steam to flow over the tip section of the rotating blades, because the outer lap design values have been destroyed.

The rotating blade rows

When a unit is examined, the most dramatic material losses can be seen on the stationary portion of the unit. These require immediate attention. Repairs can be made during a normal outage, once it is decided where repairs will be made—on-site or at a service facility remote from the plant. It is also necessary to determine the degree and extent of any stress relief operations required.

However, rotating blades also suffer material loss. While they do not suffer the same level of loss as the stationary blade rows, they can also be placed at risk because of the stresses induced by the effects of rotation. When rotating blades suffer material loss, they need to be monitored to the extent that:

- an estimation can be made of the rate of deterioration
- the point in the maintenance cycle at which replacement blades should be made available

There are three distinct areas where damage should be examined and some effort made to quantify and record the condition.

From the rotating vane inlet nose. Particles discharged from the stationary blade row enter the rotating row at a variety of angles and velocities (Fig. 4.7.2). Impacting scale will remove material from impact points on the inlet nose. As described earlier, the concentration of particulate increases towards the outer section of the rotating vane. The effects of increasing loss towards the outer section is shown in Figure 4.8.7(a), and the resulting cross sections in Figure

4.8.7(b). In terms of efficiency, the major effect of this damage is the modification made to the vane inlet angle—decreasing or eliminating the nose inlet radius and modifying the mean inlet angle.

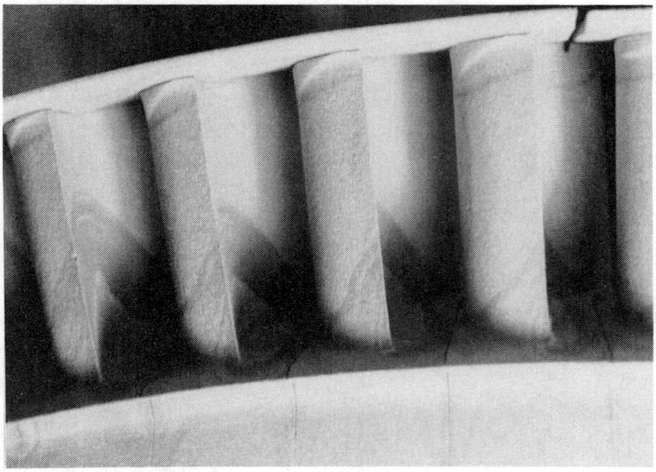

Fig. 4.8.7(a)—The material loss pattern from the rotating vane inlet nose.

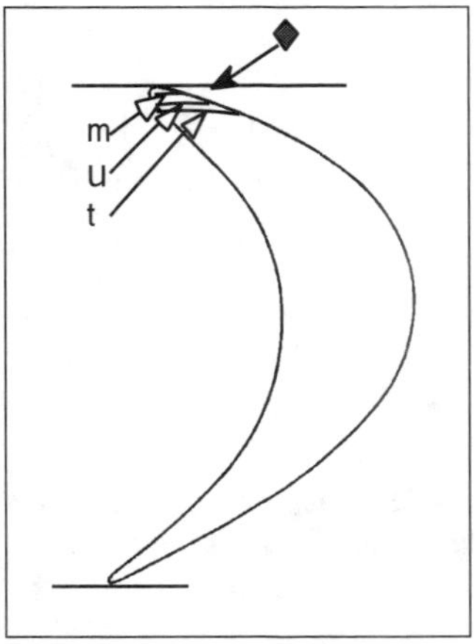

Fig. 4.8.7(b)—The variation of profile section, after erosion, as a function of radial height. "m" is the mean section, "t" the tip, and "u" between the two.

Structurally, this material loss will not significantly weaken the blade in the earlier stages of damage, but will eventually weaken the vane to the extent the blade will need to be replaced. This loss will also modify the natural frequencies of the blade, but is not considered a major concern.

Caving. When solid-particle erosion damage is found on a rotating blade row, it is advisable to examine the pressure face just below the coverband. Many high-pressure and reheat stages have integral coverbands—others have coverbands attached by riveting. The scale that is present in the rotating stage is, as previously discussed, centrifuged to the underside of the coverband where it will cause abrasive damage to the pressure face of the vane, and can cause erosive damage to the tenons in the case of tenons attaching coverbands.

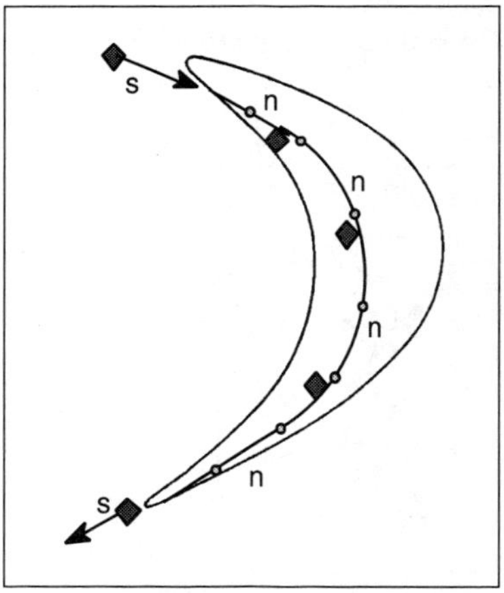

Fig. 4.8.8—Diagrammatic representation of the "cave" profile "n.....n".

This "undercutting" of the vane at the tip section just below the integral portion of the band is termed "caving." Caving represents a loss of pressure surface material. Erosion into the profile reduces its section there, and could eventually weaken the section to the extent failure will occur. Figure 4.8.8 shows a representation of the vane cross-section at the cave. It can be seen that an integral coverband will place a centrifugal bending load on the vane, and as the section is reduced, stress intensity will increase. In Figure 4.8.9, a blade removed from a high-pressure stage with an integral coverband, shows caving. This can also have some small adverse influence on stage efficiency.

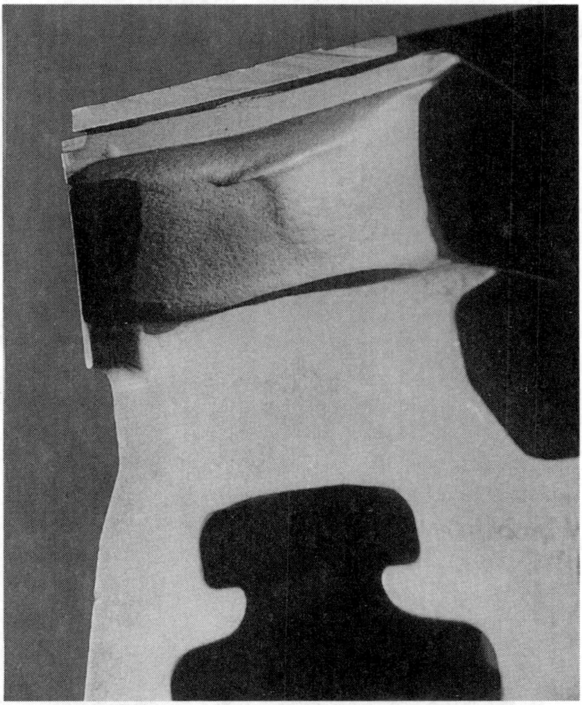

Fig. 4.8.9—A vane with caving on the pressure surface at the tip section.

From the rotating blade discharge edge. When such damage occurs, it is normally symptomatic of "caving" and represents an extension of the cave from the position of maximum loss to the thin discharge edge. Whenever material loss as shown in Figure 4.8.10 is observed, the blade should be examined for caving.

This material loss causes a stress-sensitive region at the blade tip. This is not considered significant if the stage contains a coverband attached by riveting, but in this case, the greater concern is tenon integrity at the inner surface of the coverband. However, if the rotating blade row contains an integral coverband that applies a centrifugal bending load to the vane tip section, that should be examined carefully for tears at the thin discharge edge.

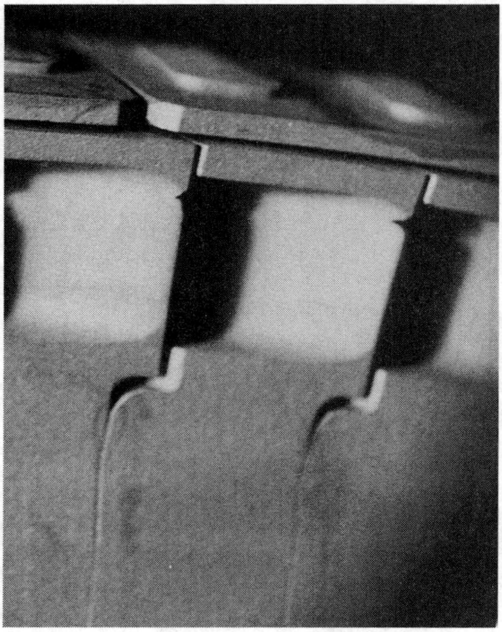

Fig. 4.8.10—The discharge edges of rotating blades showing the "caves" just extending to the vane edge.

Associated stage hardware

Because of scale present in the rows and the centrifugal effects on the scale, a portion will find its way to the casing inner surface, the coverbands, tenons, and the seals are located there. When these components are damaged, they exhibit material loss patterns at the outer radii. This damage should be observed, monitored, and if necessary, corrected before irreparable damage is sustained—damage that can only be corrected at high costs.

Three components above the rotating blade row can suffer material loss, and need to be considered:

The coverbands. Figures 4.7.3, 4.7.4, and 4.7.5 show how scale is carried out past and over the coverband by steam that leaks over the rotating blade row. Also, as discussed in an earlier section, the quantity of steam that leaks over the cover—and by implication, the quantity of scale carried with it—is dependent upon the leakage area at these seal locations. In addition, scale reaches the casing inner surface after repeated impacts between the stationary and rotating blade rows (Fig. 4.7.5). Therefore, because of the presence of scale above the blade, the potential for damage in this region exists, and the potential for damage increases as the unit seal system "rubs" or as the unit ages. Seal maintenance is therefore important in the inlet stages of a unit.

Scale has the potential to cause solid-particle erosion damage to the coverband. For damage to occur it is necessary for the scale to either impact on the coverband surface (before or after impact with the casing) or flow across surfaces at velocities sufficient to cause material loss by abrasion. The blade tips are rotating at high tangential velocities, and the scale, as seen from Figure 4.7.4, will enter the radial clearance above the coverband at considerably different velocities and angles to those of the steam. It would therefore be possible for material loss to occur on many surfaces, depending upon the cross section of the coverband. While material loss does occur on these surfaces, so far the damage has rarely proven to be significant, and any damage sustained is minor.

At coverband batch-end positions (Fig. 4.8.11), discontinuities in the tangential run of the coverband include a tangential gap ("g") that allows for coverband movement under thermal transients. In many of these designs, it becomes necessary to remove overhang material from corners to reduce the coverband curling stress to acceptable levels. At these overhang positions there is often damage as a consequence of the impact of scale. This damage is relatively minor, and would call for no corrective action.

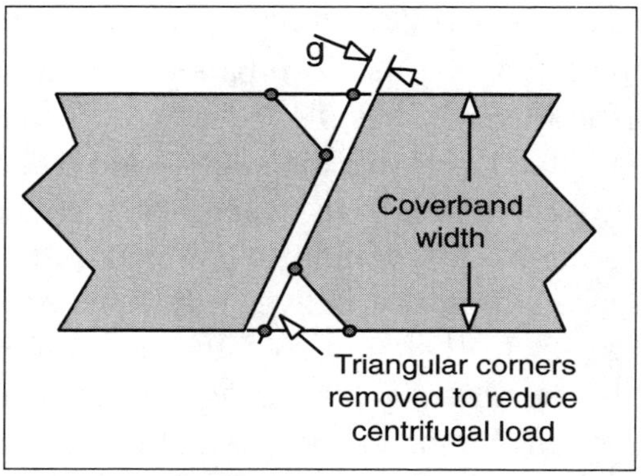

Fig. 4.8.11—Cover band end positions which can be subjected to solid particle erosion.

The tenons. Scale that is centrifuged out from the stationary blade row will be carried to the radial gap above the coverband, and will eventually pass over the blade coverband. This flow over the coverband will cause erosive material loss from tenons that are formed above the coverband outer surface. This material loss can eventually endanger the integrity of the tenon. Figure 4.8.12 shows the general arrangement of the fox-holed tenon, indicating the extent of coverband provided to protect the tenon.

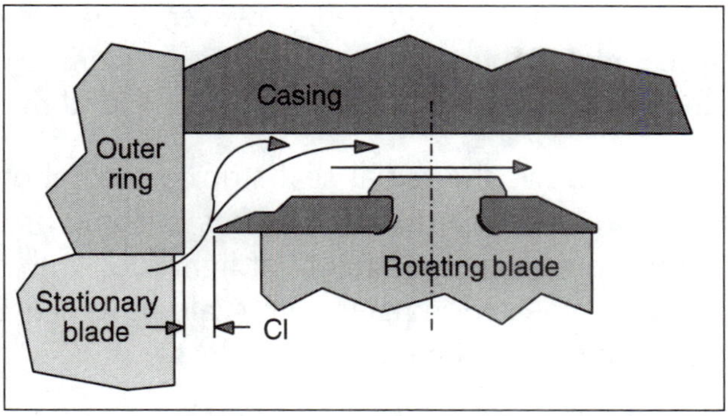

Fig. 4.8.12—Scale path over a simple tenon.

If fox-holed tenons are used (as they are on many control stages), this becomes significantly less of a problem, as the coverband material provides protection for the tenon heads shielded by the coverband itself. If the coverbands do not provide a fox-holed attachment, this should be considered as a corrective action if damage becomes significant. However, if this is done, it is necessary for the coverband to have adequate radial depth to allow a fox-holed attachment. Alternately, the coverband can be replaced and increased in radial thickness sufficient to allow the fox holes to be produced. In either case, an analysis of the stresses induced in the tenon head and coverband must be considered.

If the material loss at the tenons is occurring on a row that does not have an integral inner coverband, the coverbands can be secured by the use of under coverband welding. If applied correctly, this can strengthen the attachment and allow the rotor to be returned to service.

The radial seal strips. The radial seal strips shown in Figures 4.7.3, 4.7.4, and 4.7.5 represent a barrier to leaking steam. Since the scale moves in basically the same direction as the steam, it also represents a barrier to the flow of scale over the coverband. It would be expected these seal strips would also suffer erosive damage but, in fact, these strips are often unaffected; however, there are occasions where such seal material loss does occur. This is often in the form of chipping, and semi-circular pieces can often be found missing.

Figure 4.8.13 shows the radial seal strip formed above a high-temperature control stage. Strip material has become brittle due to extended exposure with the possibility of light rubs with the cover-band. Subsequent repeated impacts with scale have caused a combination of gouging and chipping material loss.

There are many regions within the inlet stages of the high and reheat sections where material can be lost. A number of these damaged regions will affect expansion efficiency, and certain of them can potentially cause mechanical damage that can force a unit from service. Other consequences may extend a planned outage while replacement parts are obtained. When damage is noted and materi-

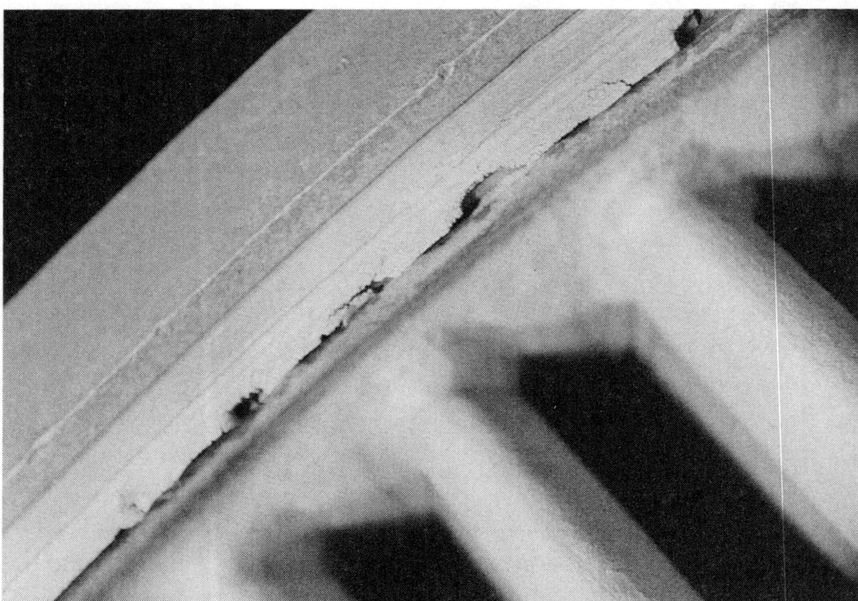

Fig. 4.8.13—The radial seal strip from above a control stage. Here the strip has embrittled from extended exposure to high temperatures, and continued impact from hard scale has caused the loss of small pieces of strip material.

al losses have the potential to compromise performance significantly, it should be monitored and decisions made to modify or replace affected parts before a serious condition occurs.

Figure 4.8.13 shows the radial seal above the control stage of a unit with a steam inlet temperature of 1,000°F. On this strip there has been a material loss due to a combination of embrittlement and continued impacts from scale exfoliated from the boiler.

PROTECTIVE MEASURES AGAINST EROSION

Erosion causes sufficient loss of output and represents such a financial penalty—that considerable effort has been expended in finding methods of preventing and/or slowing the loss rate. These methods include changes in stage geometry, the manner in which the units are operated (particularly at start up), the use of protective coatings, and changes to the boiler materials and cleaning methods. Details follow.

Stage geometry changes. These include modifying the nozzle outer wall form and the arrangement of the blades. Also, changes have been made in increasing the axial clearance between the discharge edges of the stationary blade rows and the rotating blades. This will, of course, reduce the erosion on the rotating blades but has no effect on the stationary rows.

Changes to the boiler cleaning and operation. The scale that impacts the steam path and causes solid-particle erosion is generated in the boiler. Therefore, it is possible that changes in the design, operation, and maintenance of boiler tubing can limit the damage.

Protective coatings. Steam path materials can be provided with surfaces that are hard and able to resist the abrasive effect of the scale. This coating technology is becoming a major contributor to extending steam path life. Various processes and coating materials are available and being investigated.

Procedures for turbine start up. Scale is exfoliated from boiler tubes during times of boiler tube temperature change. Therefore, if procedures for turbine start up are modified, it is possible that generated scale can be distributed around the entire nozzle admission, rather than concentrating over a small arc of admission, and so the erosion—while still present—will not be as destructive.

These various correction and preventative procedures are discussed further in chapter 8.

SPE INFLUENCE ON STAGE PERFORMANCE

Potential effects of erosive damage on the reliability of various components of the steam path have been described. Some of these effects are obvious. It can also affect efficiency of the stage, as will be reviewed next.

Area change effects

The loss of material from the stationary and rotating blades can have an effect on the efficiency of the stage by modifying the shape of the vane and the discharge area.

Vane material loss. As material is removed from the discharge edges of the stationary and rotating vanes, the throat is formed on a different portion of the adjacent vane suction surface. Since passages formed between adjacent vanes are always convergent (with the exception of certain last stage elements), it is clear the discharge area will change and modify the discharge pressure. The stage enthalpy drop changes the velocity ratio "ρ" (ratio of blade velocity to steam velocity.

Sidewall material loss. Material loss from the sidewalls most often affects the outer sidewall, and most often sidewalls of the nozzle vanes. In removing material from these locations, the steam will discharge from the nozzle at a larger diameter and—depending upon the lap—can be directed towards the casing wall to flow over the coverband. The grooves formed in the outer sidewalls will also encourage scale to flow outwards to a greater extent and will cause erosive damage to the seal system located above the blade tip.

Steam discharge angles

Material lost from the discharge edge of the various blade rows modifies the effect on the steam discharge angles. This modification causes steam to enter the following blade row at an angle for which the vanes were not designed. There will therefore be some "incidence loss" and a reduction in stage efficiency.

Increased area and local pressure

As material is lost from the vane discharge edges, the pressure distribution throughout the unit is modified. Pressures at the various stages are altered, as are stages at which steam is available for extraction, regenerative feed heating, and any other cycle requirement. This will affect the cycle performance and therefore unit heat rate.

Cooling stages

There are certain double flow designs of turbine sections which have a small pressure difference from discharge at the turbine end flow to the generator end flow on the first stages. This is done to provide protection to the rotor from overheating by encouraging a cooling steam flow from one end to the other. This steam flow is intended to carry heat out of the rotor material, this heat being generated by frictional heating "churning" of the steam in any chambers, or dead spaces existing between the two flows.

As stage discharge areas change due to erosion it is possible that this cross flow will cease and the heat generated in these dead spaces will heat the rotor, causing a change in its material properties, with the consequence the rotor will sustain some level of damage reducing its operating life.

STEAM PATH COMPONENT RUBBING

The axial and radial clearances of the steam path are normally held within "tight" values to minimize the amount of steam that leaks bypassing stages and blade rows, or escapes from the steam path at shaft end glands. Because of these tight clearances—and the inevitability of system disruptions causing shaft vibration and other transients—situations will arise that cause contact between the stationary and rotating portions of the unit while it is at operational speeds. When such contact occurs there is a frictional reaction between the two sets of components. This contact, or rubbing, causes considerable frictional heat to be generated. The steam path components are immersed in the working fluid, so when heated by friction to extremely high temperatures they are immediately quenched.

Such quenching can introduce complex changes to the structure of the materials and, if the rubs are both severe and prolonged, can deposit material from one component on another.

Heat generated by frictional interaction between the components represents the destruction of a portion of the rotational energy of the rotating components. However, steam may still expand through the blades, even if the condition causing the rub has abated and the unit has been disconnected from the system. Unfortunately, many instances arise in an operating situation in which the unit does not trip and steam continues to be admitted to the unit (continuing to drive the rotor). Unless the situation producing the rub is self-correcting, it will continue, possibly causing irreparable damage.

Conditions can exist in which sufficient frictional heat is generated such that the metal temperature of the rotating components will rise above the austenitizing temperature. This material is then quenched. Because this heating and quenching are occurring simultaneously, the change to the material structure tends to be contained to the surface of the components. However, the final depth of structural change will depend upon the heat generated (the pressure of the rub), and the period of time this condition was experienced. A transition range exists within the material, and relatively close to the surface.

Should these rubs occur between major operating components and seal system components, the level of heat generated will be relatively small because the seal system (labyrinth strips) is designed so the seal portion will wear, reducing the heat raised. Knife-edges are used on labyrinth seals so only a minimal amount of heat is generated. However, these rubs, if they are persistent, will ultimately "groove" the rotor and cause small changes to the surface material.

In many component designs, seals are produced as an integral part of the rotating or stationary blade row. These seal strips are intended to provide an axial or radial seal. Such integral seals are dimensioned so if rubs occur, they open the clearance. While this represents a deteriorating of efficiency, it is one that will not cause any significant hardening of the major component material and the damaging effect is relatively minor.

Other rubbing situations may be induced by contact between loose materials, which can be generated from a variety of sources, *e.g.*, detached portions of the steam path or other major components. Such material can simply be left in the unit during an outage, where it later becomes lodged in a position and causes rubs. Figure 4.11.1 illustrates a portion of a blade tip where tenons have failed, allowing the coverband to deflect radially outward, where it rubbed against the inner surface of the casing. This action did not cause significant damage to the casing as the stage had no radial seal.

Various forms of damage can occur as a consequence of rubs in the unit, and these can affect various components of the steam path. Such damage is discussed below.

Fig. 4.11.1—A rotating blade row with failed tenons. This tenon failure allowed the cover band to rub on the casing inner surface.

Rotor rubs

Rotor rubs are normally induced as a consequence of high levels of vibration occurring at start up or shutdown, during overspeed transients, or when a significant failure within the unit causes a dynamic unbalance. Rubs can also occur as a consequence of casing "humping." This occurs if the rotor is brought to rest and heat rises in the casing, causing the upper half to be hot relative to the lower half. If this occurs, any premature effort to start the unit will cause rubbing contact between the rotor and lower half casing. In fact, any condition causing a significant temperature difference between upper and lower halves of the casing has the potential to cause damage.

Figure 4.11.2 shows a portion of a blade row, in which the casing has "humped" because the lower-half insulation was removed, causing a rapid cooling or quenching of the lower half and an upward deflection. In this case, the rotating blades made contact with the lower half casing and were deformed.

Unfortunately, there are occasions, however infrequent, when a heavy rub occurs on the rotor, causing a permanent bend to be induced in the central spindle or body of the rotor. This situation must be corrected before the unit can be returned to service, otherwise it becomes impossible, or extremely difficult to balance it, and steam path radial alignment is destroyed. Typical situations producing such heavy rubs are:

- *water ingestion*—When a portion of the rotor is quenched, causing a material structural change, producing a permanent bend in the rotor body
- *the loss of a large portion of the rotor mass, such as failure of one or more large blades*—This loss of mass produces a dynamic unbalance, which can induce a heavy rub between the rotor and a stationary portion

- *unbalance and heavy vibration with large deflections of the rotor at start up and below the first critical speed*—If efforts are made to increase speed when this condition exists, the amplitude of vibration will increase, aggravating the situation, and potentially causing heavy rubs and permanent bends

Fig. 4.11.2—Blades having suffered a "heavy" rub have deformed tangentially.

When permanent bends are induced in the rotor, it is often impossible to adjust the balance by the addition and removal of balance weights. Under these circumstances it becomes necessary to consider means of straightening the unit (see chapter 9). It is also necessary to quantify the extent of bending to provide some indication of the degree and means of correction that will prove most effective.

Many rotors suffer light rubs between the sealing system seal strips that are carried by the stationary portions of the unit. These rubs are rarely causes for concern, and for all intents of system security are neglected. There are occasionally heavier rubs in the seal region—again, these are not significant in terms of modification of the rotor material because of the light section of the seals. With the "Hi-Lo" type seals it is common to see grooves rubbed into the castellations as a consequence of light to medium rubs.

Rubs on the rotating blades

There are situations in which blade vanes and root blocks will be subjected to heavy rubs against the stationary elements. This is particularly the case if the thrust bearing is lost and there are large axial movements of the rotor train. Figure 4.11.3 shows a blade row in which the thrust bearing has wiped, with a resulting heavy contact between the blade vane inlet edges and the stationary components of the unit. These blades were heated and quenched, causing an increase in their local hardness. It was necessary to replace them.

A similar failure—due to thrust bearing failure—is shown on the blades of Figure 4.11.4. It is viewed from the backside, where the integral coverband has been deformed. In this case, there has also been significant damage to the radial seals produced integrally with the coverband.

This form of axial rub can also occur if the unit is cooled or heated too fast, with a significant difference in the rate of axial expansion between the stationary and rotating components of the unit (see discussion of "long" and "short" rotor conditions in chapter 2). Figure 4.11.5 shows a portion of a coverband inlet edge where rubbing with the diaphragm face has occurred. In this instance, the rub will have decreased the sealing effect but have little or no effect on coverband security.

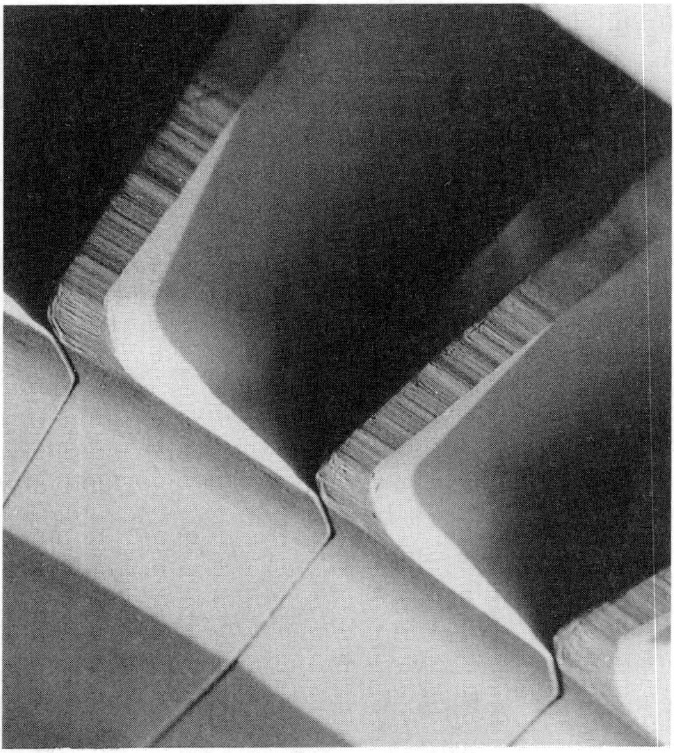

Fig. 4.11.3—Blades rubbed on their inlet edge.

This heat, generated by rubbing, is partially dissipated through the various components that come in contact, but is mostly dissipated by the steam. This is particularly so if the rub is sustained for any period of time. However, as the rubbed components are quenched, the action of quenching introduces hardness and brittleness into the material. Because the rotating blades are subject to high levels of both direct and alternating stress, the increasingly brittle nature of the blade material has a high probability of introducing cracks in the vanes to the extent they will eventually fail.

Fig. 4.11.4—Blades rubbed on their integral cover band.

Coverband rubs

On the majority of stages, coverbands are designed to run with close axial and radial clearances—particularly those in the higher pressure and temperature regions of the unit. Therefore, these portions of the blade system are relatively susceptible to damaging effects from any transient condition that affects the running clearances by relative movement between the stationary and rotating parts, whether this is temperature or vibration initiated. Figure 4.11.5

shows the consequences of an axial movement causing a rub on the coverband integral seal knife-edge. This rub has destroyed the initial clearance and made the tip leakage more severe, with a consequent loss in efficiency and output.

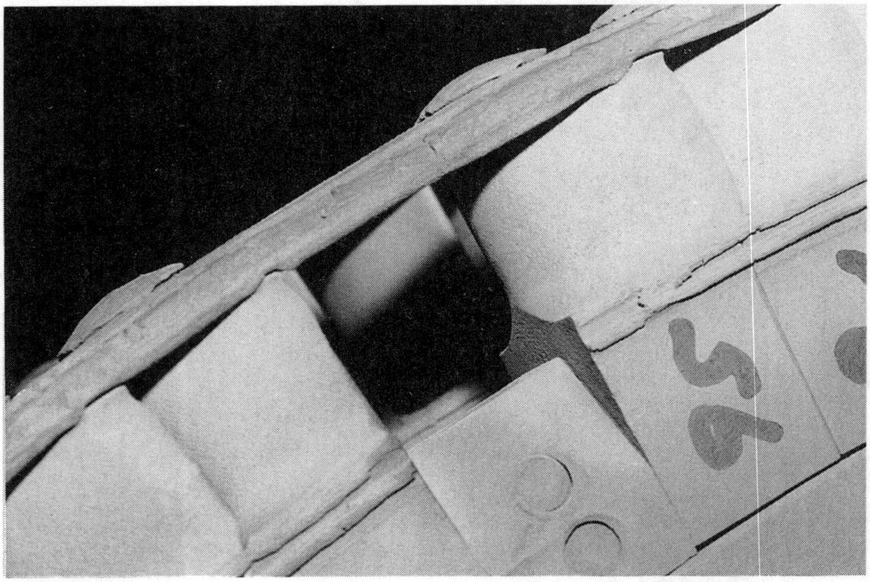

Fig. 4.11.5—A "light" rub on the inlet edge of the cover band has destroyed the axial seal.

A more severe rub in the radial direction is shown in Figure 4.11.6. Here, some system transient has caused the coverband to rub against the seal strip, which was produced from hard chromium steel. The rub produced a deep groove in the discharge edge of the band and would also have produced a significant amount of heat, to the extent the band has hardened, become brittle, and eventually failed at the edge. The exposed vane tip discharge edges can be seen. While this type of damage can be tolerated, it must be recognized that the band (or large portions of it) could eventually detach and cause impact damage further down stream.

Fig. 4.11.6—A severe rub which has modified the material properties of the cover band, causing a localized failure under the effect of operating stresses.

Figure 4.11.7 shows severe damage caused by metallic debris that has been trapped between the coverband outer surface and the casing. The debris has gouged grooves into the coverband outer surface, totally destroying it. At the time of unit inspection the coverband had not detached, but could be expected to do so within a relatively short period if not replaced. This coverband was removed, the tenons weld rebuilt, and a new coverband attached.

Diaphragm vane rubs

Any rubs that occur between the stationary and rotating blades have the potential to cause damage on both sets of vanes. This type of damage was discussed previously and shown in Figure 4.11.8. The inlet to a diaphragm shows large pieces of foreign matter, generated within the steam path, that have wedged at the inlet to the stationary portion of the steam path, with evidence of heavy rubs on this "wedged in place" debris.

Fig. 4.11.7—Heavy rubs on the cover band outer surface. Debris had been trapped in the radial space between the cover band and casing.

Fig. 4.11.8—Rubbing damage on the stationary blade vanes. Caused by foreign objects generated within the steam path.

When such a rub occurs, the form of the vane inlet is destroyed. Because of this, there is disruption of the entering flow and boundary layer separation, and considerable energy is lost in the stage.

Rubs on the diaphragm inner webs

Rubbing contact between the stationary and rotating portions of the unit can affect the diaphragm to the extent it will distort. To be damaging, the rubbing must occur on the inner web, and will need to exert high axial pressure. Greater damage will be done to the rotating blades. Corrective action for the diaphragm requires refurbishment to restore the steam path (discussed in chapter 2). There are two forms of distortion that can be experienced:

- *The diaphragm will "dish"—expanding away from the blade or wheel on which it is rubbing*

The movement will be greatest at the inner diameter adjacent to the rotor, and the form of the rub can be at any radial location on the diaphragm, depending upon the stage geometry, and the extent of any axial creep that may have occurred during the operating life of the unit. Figure 4.11.9 shows a diaphragm that has suffered a serious rub consequent upon the failure of the thrust bearing. In this case the heavy metal deposits on the inner web are rotor material. This diaphragm was able to be corrected and returned to service.

- *The diaphragm will become elliptical—the upper and lower halves "opening" or "closing" at the horizontal joint, modifying the steam path alignment*

Such a situation is shown in chapter 2 (Fig. 2.9.8). A similar distortion is shown in Figure 4.11.10.

Fig. 4.11.9—Rotor material deposited on the diaphragm vertical face, as a result of a hard rub.

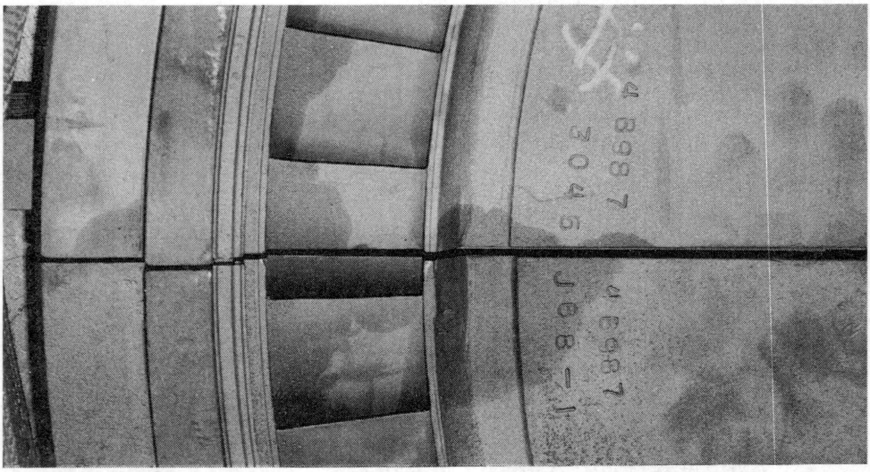

Fig. 4.11.10—The gap produced at the diaphragm horizontal joint from the heat of a rub.

Rubs on the casing inner surface

The inner surface of the casing can be rubbed under certain operating conditions, including high vibration or during excessive overspeed transients, and possibly at times when the casing has "humped."

This is not as serious as many rubs that occur in a unit, however, if this rubbing is continuous, there can be some degree of casing distortion. But again, the amount is small and in the majority of cases can be ignored. Situations can arise in which some level of correction is required, although the rotating component is normally the most seriously affected. The possible causes of these rubs are:

- *Excessive overspeed*—The rotating portion of the unit grows radially due to the effects of centrifugal load causing radial contact
- *High levels of unit vibration, causing radial contact*—If this condition occurs while the unit goes through critical speeds at start up or shut down, they are normally of short duration and damage to the casing is minimal
- *When the casing humps or distorts (becomes elliptical)*—This can consume the radial clearance above the blade tips

It is not common for axial rubs to occur on the casing inner surface, but each unit should be examined for its particular geometry—in the design stage and when a unit is opened at an outage. This examination should be undertaken during design to determine if this is likely to occur.

The extent to which such casing radial rubbing occurs depends upon design of the radial clearance above the blade and the form of any coverband. It is possible for casing distortion to influence the radial clearances, and these will need to be monitored as the unit ages. It is normal that when rubbing occurs, the rotating portion will sustain more significant damage than the casing, which while suffering material loss will not have serious damage done to it. A relatively light rub by coverbanded blades on the inner surface of a casing is shown in Figure 4.11.11. Here the steam extraction holes for feed

Fig. 4.11.11—A rub on a casing inner surface.

Fig. 4.11.12—The results of casing distortion, showing grooves rubbed into the horizontal joint by integral seals on the rotating blades.

heating can also be seen. This rub was not of sufficient severity that the casing distorted, nor was any corrective action required. Another radial rub is shown in Figure 4.11.12. Here, the casing has distorted, becoming elliptical, and the integral seals on the rotating blades have caused grooves on the inner surface of the casing.

FRETTING CORROSION

Fretting corrosion with subsequent fretting fatigue can occur when there is the potential for relative motion between two surfaces nominally in tight contact. This motion can be either repeated high-frequency contacts between the surfaces or repeated high-frequency sliding of one surface relative to the other. This striking or sliding motion induces high temperature at the impact or rub surfaces, and results in oxidation at the highest temperature contact points—a temperature that can convert the steel to a form of iron oxide (Fe_3O_4). Such damage is not common in the turbine stage, but on those occasions when it occurs, is capable of leading to serious failure. Damage is usually in the form of a pit or other mechanical discontinuity at the material surface, producing stress concentration and the initiation of a high-cycle fatigue failure.

Fretting corrosion is a mechanism that induces damage and surface material loss, but does not rely on the presence of a corrosive to produce the damage; therefore, in terms of comparison (or representation to the corrosion considered in chapter 6) is a quite distinct form of damage.

The rapid, repeated movement that induces fretting is often the form of movement that produces high-cycle fatigue. Therefore, when fretting corrosion occurs in the turbine steam path, there can also be a high-cycle fatigue failure. Fretting occurs on surfaces in constant repeated motion, not continuous motion. The latter circumstance is necessary for the heat to increase locally sufficient to cause localized oxidation.

Two mechanisms account for this oxidizing/damage mechanism—the wear-oxidation and the oxidation-wear theories. The difference is initial cold fusion between the particles tearing material from the surface and producing heat, or heat being produced by the frictional effect and introducing oxidation. The difference is relatively moot for the engineer, whose primary concern is the damage that is caused, the deep pits that are formed, and their potential to cause serious failures.

The process of "fretting" relies upon relative movement of one metallic surface to another with which it is in contact (repeated impacting or sliding), and high surface pressure contact between them. The consequence of this movement is the production of excessive surface heat and the formation of an iron oxide. These conditions can exist throughout the steam path, so the possibility of this form of damage should always be anticipated.

The construction of the turbine steam path can produce a number of locations where the potential for fretting occurs. Many of these can be eliminated by close control of manufacturing tolerances, however, they can exist as a consequence of operating conditions beyond the control of the operating engineer. Some of the regions most affected by this phenomenon follow.

Tie wire/hole interface

The most likely regions within the steam path for this phenomenon to occur are at the point where a lacing wire passes through a drilled hole in the blade vane. For fretting to occur, there needs to be a highly localized pressure contact between the wire and hole inner surface—a condition that can occur when there is misalignment between the wire and hole. Figure 4.12.1 is the diagrammatic representation of this condition, in which the middle blade suffers misalignment and the vane is "twisted" to allow penetration of the wire.

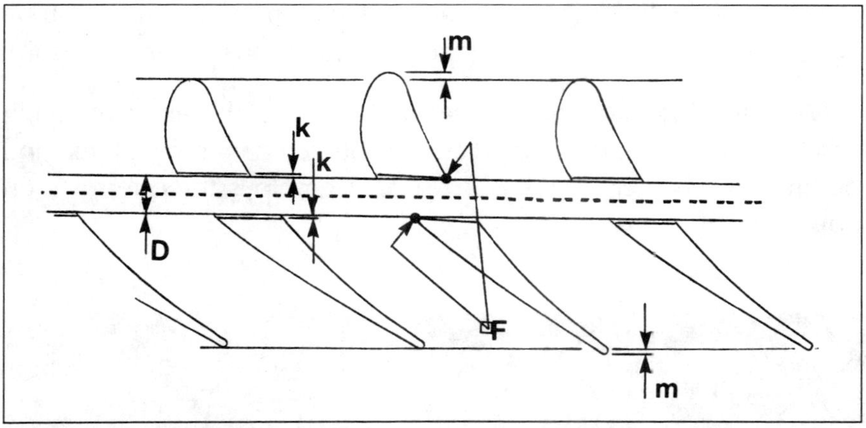

Fig. 4.12.1—Vane distortion and high pressure contact due to wire/hole misalignment.

With this condition, the normal radial clearance "k" will be consumed, requiring a twist to allow the wire to be assembled. Consequently there will be high-pressure contact at the two contact points "F" and the vane will distort from the inlet and discharge edges by amounts "m." These values of "m" will not necessarily be the same on both edges. With this form of design—relying upon centrifugal pressure to maintain wire to vane contact—there is the possibility of repeated sliding under the action of the steam vibratory forces. Such sliding is particularly evident only when small contact areas are provided, *e.g.*, occurring when the vane is twisted and contact is almost "point contact." This rubbing provides the conditions for fretting and cracks will eventually be induced.

An example of rubbing/fretting damage that had not yet caused failure is shown in Figure 4.12.2. Here, poor alignment of the vane tie wire holes has caused a major misalignment to the extent excessive wear on the vane can be seen on two blade vanes. The material loss has caused hole enlargement in the axial direction on opposite sides of the hole. The poor radial alignment of the vanes can be gauged from an examination of the pitch variation seen in this figure.

While failure had not occurred at the time of this examination, it can be noted that the wire had already been repaired by the method outlined in chapter 9 (Fig. 9.4.13), that employs three ferrules on both sides of two vanes.

Fig. 4.12.2—Blade elements with enlarged tie wire holes as a result of excessive rubbing between the hole surface and tie wire.

Fretting at coverband segment interface

In certain designs, coverbands are produced integral with the blade vane and mounted to the rotor. In such cases, they are in contact at the coverband tangential faces. To discourage the development of fretting corrosion, these surfaces must be in sufficiently hard contact that relative motion is not possible.

These coverband segments must have sufficient surface contact pressure to help ensure vibratory forces developed on the blades are not sufficient to cause fretting on them. Figure 4.12.3 shows a coverband that has developed fretting material loss under the influence of these contact forces.

Fig. 4.12.3—Fretting on the interface of an integral cover band.

A design feature now used on many rows with this integral coverband is the concept of "pre-twisting." This form of blade produces the root and coverband section with slightly different wedge angles, that are sufficient to cause a small pre-twisting and hard surface contact at the coverband segment. Many such rows are now in service and have sufficient force developed between them that movement is not possible. However, rows not pre-loaded can suffer from this form of damage.

Fretting on root load bearing surfaces

The major function of the blade root or fastening is to transfer loads from the blade to the rotor (both direct and alternating). The load transfer surfaces of these roots are subjected to loads that can produce fretting if they are not mounted to the rotor in such a way that relative movement is not possible, *i.e.*, There is a high contact pressure provided by the direct centrifugal and bending loads, and there are cyclic loads with the potential for providing constant repeated motion.

Therefore, fretting will occur if blades are mounted on the wheel or rotor in such a manner that at load and temperature this motion can occur, and the cyclic loading is of sufficient magnitude. Figure 4.12.4 shows the load-bearing surface of a root with a crack developing from fretting pits along the contact surface. Figure 4.12.5 shows the exposed surface of a root where high-cycle fatigue failure has occurred. Multiple fractures start at a line of fretting pits of the type shown in Figure 4.12.4.

Fretting at root interfaces

The cyclic loads that can produce fretting on load-bearing surfaces can also produce this same form of damage on the mating surfaces of blades mounted with their adjacent blades touching (blades with high surface contact loads). This is more common in tangential entry roots, which are packed into the rotor, and require the fitting of special closure blades to close the row.

If the blades are initially packed loose, or become loose during operation—different coefficients of linear expansion can cause this looseness to occur—then fretting can occur. Rows with repeated high-impact cyclic loads (*i.e.*, control stages) are particularly prone to this form of damage. Figure 4.12.6 shows the contact face on a tangential entry root block with fretting.

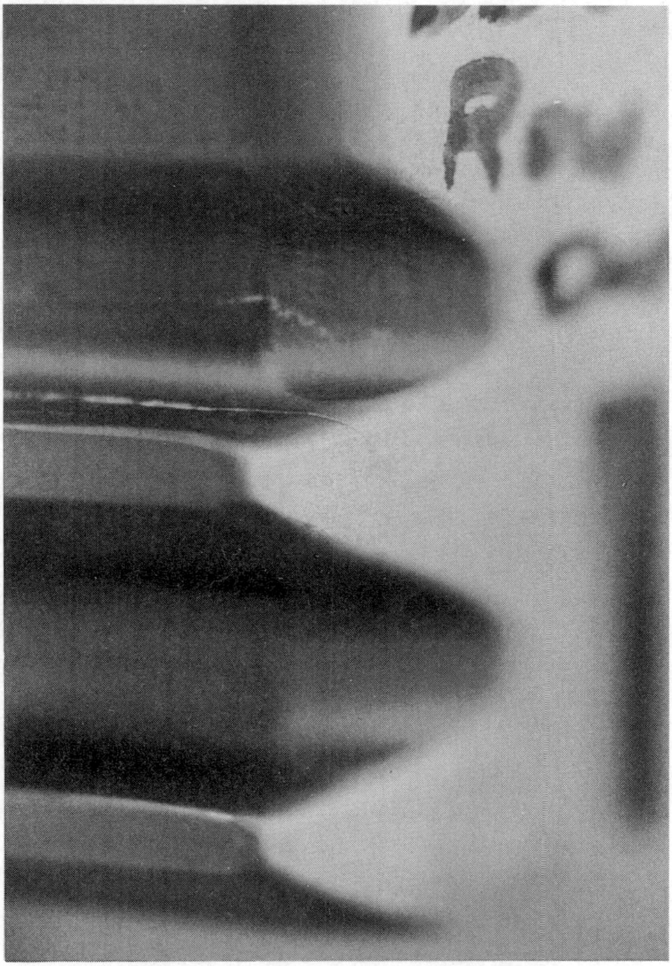

Fig 4.12.4—A fretting crack on the upper load bearing surface of a tangential entry inverted "T" root.

Fig 4.12.5—A high cycle fatigue crack, originating at a series of fretting pit, which have occurred on a root load bearing surface.

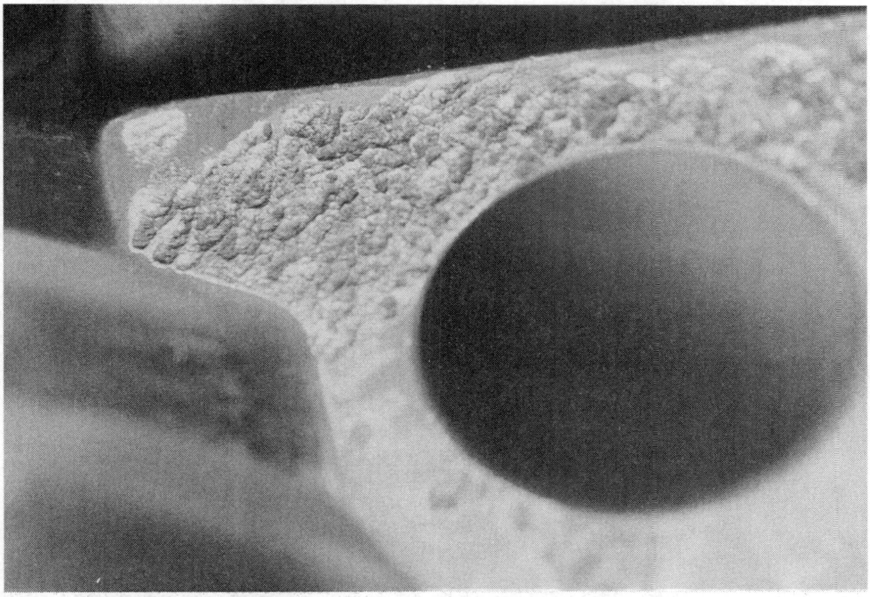

Fig. 4.12.6—Fretting damage on the verticle face of a blade root block.

REFERENCES

1. Miller, V.D.R., and A.V. Manolescu. "Maintenance Strategies and Specific Design Changes to Ontario Hydro's Lambton TGS Coal-Fired Generating Units to Mitigate the Effects of Solid Particle Erosion," EPRI Workshop on Solid-Particle Erosion in Utility Steam Turbines, San Jose, California, August 1986

2. Sanders, W.P. "A Review of Steam Turbine Blade Operating Problems," I. Mech. E., Toronto, February 1981

3. Heyman, F.J. "Notes on Solid Particle Erosion in Steam Turbines: A Brief Review and Bibliography," ASME Forum, New York, May 1974

4. Sumner, W.J., J.H. Vogan, and R.J. Lindinger. "Reducing Solid Particle Erosion Damage in Large Steam Turbines," American Power Conference, Chicago, April 1985

5. Fontana, M.G. Corrosion Engineering, Third Edition, McGraw Hill College Division, ISBN: 0070214638, 1985

Chapter 5

Steam Path Damage and Deterioration from Material Property Degradation

INTRODUCTION

Within the turbine steam path and for major components of the turbine, many forms of both static and dynamic loading introduce mechanical stress. In addition, many components that operate at high stress levels do so continuously at elevated temperatures, which while they are sensibly constant, will under certain conditions change at rates capable of causing changes within the material.

Under such conditions of load, temperature, and temperature variation, these components are, or can be subject to various forms of material property degradation and material life consumption. The design process evaluates these conditions and attempts to define the individual components so they are suitable for operation with the stresses induced by the operating conditions, enabling them to endure for the anticipated life of the unit.

However, these phenomena are capable of damage to the extent that component failure occurs. Circumstances can occur during operation where the unit is operated beyond its design limits. These circumstances represent conditions under which stresses are induced beyond those anticipated. Similarly, material properties inferior to the design specification can be introduced. While failure may not be instantaneous, there is an accumulation of damage that ultimately forces the unit from service and requires corrective action, either immediately or for a future outage.

Steam turbines are intended and designed to operate for considerable periods—possibly between six and eight years between major inspection outages. Therefore, the internal portions of the unit, including the steam path "hot sections" will not be available for damage assessment and correction between these periods. It is normally considered necessary that clearances and deflections are measured for comparison to earlier measured values recorded when the steam path became available at a previous outage (see chapter 1). These inspections and comparisons are necessary to ensure pending problems are evaluated and, if necessary, corrective action taken to avoid returning the unit to service with the risk of a forced outage.

Major material degradation phenomena that can potentially affect the steam path components include high-temperature creep and high- and low-cycle fatigue. These three phenomena will be considered in this chapter.

CONSIDERATIONS OF MATERIAL STRUCTURE

As a molten metal transforms from liquid to solid state, a random distribution of the nuclei form a *eutectic*. These nuclei then grow by the addition of other atoms of the pure metal in a closely defined and repeated matrix. The number of nuclei that form within the "melt"—and the size of the individual grains that grow and form the final solid—is influenced by the rate at which the melt cools. With a slower cooling process there is greater opportunity for the individual nuclei to initiate and crystals to form.

Because of uneven cooling within the melt, the initial formation of the nuclei is most often towards its outer boundaries, where the temperature first begins to lower by heat loss. This cooling, and the subsequent solidification of the melt continue, with individual crystals of the alloy growing until the liquid phase has been completely transformed to the solid. This formation of the nuclei is random, as are the growth and orientation of the individual crystals.

Because of this random nature, crystal alignment is such that individual crystals will not join. Instead, the solid will exist containing the crystals of the alloy, with boundaries filling the space between the individual crystals. At the boundaries of the individual metal crystals—with each crystal representing a correct spatial arrangement of the individual metallic atoms—there are discontinuities. These discontinuities are the crystal or grain boundaries and represent a structural discontinuity that can contain impurities of elements and compounds that were present, but not absorbed by the individual crystals. They may also contain trace amounts of undesirable elements unable to be removed from the original melt. This boundary material can be either enriched or depleted of one of the alloying elements, and therefore capable of influencing the total physical properties of the material.

Under development are materials for turbine application that are of a single crystal—some success has been achieved. However, the majority of materials is, and for a long time will continue to be those with a structure of individual crystals connected at their grain boundaries.

The most common melt used within the turbine materials has two distinct material characteristics—those of the pure metallic crystals and those of the material contained within the boundaries. For materials produced by modern methods (with controlled constituents and cooling rates), the characteristics of the grain boundary material can be predicted and even controlled. These boundary materials offer properties fully acceptable, and even advantageous to the total properties.

Note: Single crystal blades that have been developed are currently used in gas turbines where the operating temperature is higher. These blades can operate at higher temperatures and have a creep life greatly in excess of the more common alloy steel used for steam turbine elements.

HIGH-TEMPERATURE CREEP

Creep is a material phenomenon affecting the high-temperature regions of the steam path and those components exposed to a tensile or shear stress. More accurately, creep is defined as "high-temperature creep," because while materials under stress creep at all temperatures, it is only at higher temperatures that these deflections—and ultimately material ruptures due to stress—become significant. It is necessary to recognize that the higher the component temperature, the faster the creep deflection, and the faster the final failure will occur.

The design process is aware of, and allows for, creep deformation in those components and regions where it can occur. If the conditions exist sufficiently to produce creep in a component, then that component will suffer the effects of creep deformation. What the design process will do in such a situation is determine the material, dimensions, and other necessary characteristics so the component will deform by an amount that can be tolerated for the operating life of the unit. However, it is necessary once again to consider that the unit may operate beyond its design tolerances. Under that situation, creep rates and total deflection could occur that are beyond an acceptable limit.

For a component that is affected by creep, and for the distortion to become significant, it is necessary to have present a stress—either uni- or multi-directional—and for this stress to be applied consistently over extensive periods of operation. The "creep strain" that results can occur at varying rates, and is normally, by definition, divided into primary, secondary, and tertiary phases. This means that creep rates are not constant, but vary with time. The primary phase is rapid, but as the creep rate enters the secondary phase the deformation slows to an almost constant value before accelerating to failure in the tertiary phase. However, in most cases within the steam turbine, the presence of creep is noted well before the damage extends to the extent of mechanical rupture.

For creep deformation to be considered significant within the turbine steam path, it is not necessary for mechanical rupture to have occurred, or even for cracks to have formed. This is because at many locations within the unit there are relatively small running clearances between stationary and rotating parts. Should there be rubs due to creep deformation, then significant damage can result with major components being involved. While rubs should not occur in the primary phase of creep deformation, and deformation of a well-designed component should never enter the tertiary phase, it remains the responsibility of the design engineer to select component geometry so deformation during the secondary phase will not cause such rubbing contact to occur.

The creep rate for any component, under a defined set of operating conditions, is dependent upon two major parameters—operating stress level and operating temperature.

The operating stress level. The stress levels experienced by any component is dependent upon a number of variables, including steam pressure drops, rotational speeds, and steam flow rates. Each of these parameters are capable of being varied during operation, and therefore can modify local stress levels. The stress developed on any component or part of the unit may also vary during operation.

A consideration with the transient variation of these operating parameters is that they may cause a local temporary (elastic) deformation. While not permanent, it may cause a total instantaneous deflection or distortion during the transient, such that mechanical damage could be initiated. This failure may not have occurred if the existing creep (plastic) deformation had not been so severe at the time of the transient.

The operating temperature. Component temperature as stress is applied influences the materials' ability to resist the applied stresses, and therefore control the rate at which creep deformations occur. As the temperature of the component rises, creep rate increases. For this reason, it is necessary to control boiler superheater and reheater temperatures, and control them within design specified levels. These design- imposed restrictions, while appearing stringent, are intended to extend the life of the individual components and by implication, that of the unit.

As in the case of stress, considerations of temperature transients are important, as these can represent a temporary but possibly dangerous variation from the design specified operating mode.

While the design process recognizes, attempts to anticipate, and then allows for the possibility of transients during the normal operating life of the turbine unit, their accumulated effects can still result in component damage or (in the most severe case) failure, with the unit in a forced outage status. The design process allows for the effects of transients by determining stress levels, and applying what experience has indicated is an adequate factor of safety. However, it is ultimately the responsibility of the operating engineer to limit, as far as practical, the occurrence of transients that can reduce the operating life of the unit.

CREEP DEFORMATION

Creep is a slow plastic deformation (considered briefly in chapter 2). It occurs in any component under stress for extended periods. This stress can be at a steady level or fluctuate, but the stress must be in the same direction at all times to cause deformation. The stresses required to cause this material deformation are much lower than the values required to cause failure when applied suddenly.

The rate at which deformation under conditions of creep occur is defined as occurring in three separate phases. While not entirely distinct in any material, they can be recognized from a "creep strain/time" characteristic curve (Fig. 5.4.1). This curve is constructed with the assumption that both stress and temperature were held constant during the test period.

Note: Creep deformation is a plastic flow of the material and is not recovered when the load is removed. This differs from deflection, which is elastic and recovered as load is removed, with the component returning to its original form.

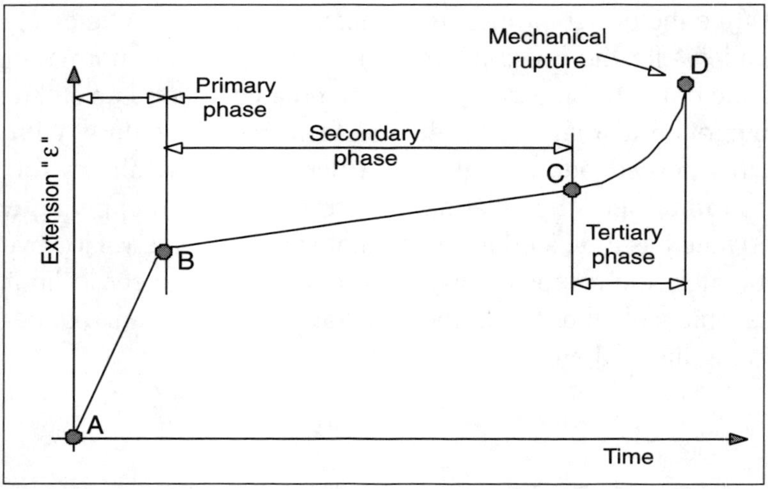

Figure 5.4.1—The phases of creep deformation to mechanical rupture.

The three separate phases to creep deformation can be explained as follows:

- *Primary phase, "A-B."* During this initial phase, relatively rapid extension occurs with application of the same constant load, normally at a slightly decreasing rate. The rate of deformation is initially rapid because the weaker locations within the boundary layers fail at relatively smaller applied loads
- *Secondary phase, "B-C."* After this initial deformation (primary phase "A-B"), the creep rate decreases to a relatively constant rate. After the initial boundary layer failures there is a continuing series of ruptures as more boundary cracks occur and fewer metallic crystals carry the load. During this time the deformation continues, but there is little or no acceleration of the rate
- *Tertiary phase, "C-D."* As the deformation enters the tertiary phase, the rate of extension accelerates until condition "D" is reached, at which time mechanical rupture occurs. At con-

dition "D" the material is no longer able to carry the applied load due to the boundary layer ruptures.

Characteristic curve families

Consider a family of creep curves for a component material under constant load but at different temperatures. A series of characteristics (Fig. 5.4.2) is obtained. In these four curves—T1 > T2 > T3 > T4—it as temperature rises so does the creep or deformation rate. The following observations can be made concerning this family of temperature characteristics:

- At the higher temperatures, the creep rate is higher in each of the three phases
- At all temperatures, the amount of time in the primary phase is about the same, but the total deformation is somewhat greater at the higher temperatures
- The time in the secondary phase is reduced as the temperature is increased, so that the creep deformation enters the tertiary phase at a lower total deformation
- The total life of the component is reduced at the higher operating temperatures

A similar family of curves can be constructed for a material held at a constant temperature and tested at various stress levels.

As the characteristic curves leave one phase and enter the next, the exact form of the curve is difficult to define. It is best described as a change in the rate of creep deflection. Because the curves are the result of experimental data, they may not be precise but provide the design engineer with sufficient information that life predictions can be made with a reasonable degree of accuracy.

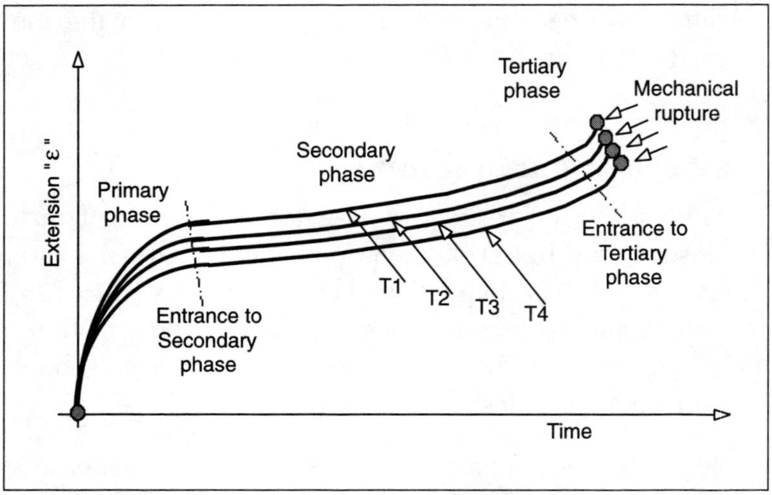

Figure 5.4.2—The effect of temperature on stress rupture characteristics at constant stress levels.

In defining the required creep properties of any material used for a steam turbine component, it is necessary to recognize that coarse grain materials creep at a lower rate than those with a fine grain structure. Unfortunately, a component that is to be subjected to creep-type loading may be produced from a fine-grain material allowing it to meet other requirements of design. To a degree, this compromises the creep characteristics of the material from which the component is constructed. These considerations require a detailed review and material evaluation. This is an evaluation the design engineer must make in the initial specification of the materials to be used for any particular application, recognizing that the creep potential of the component must be countered by other design modifications or parameters.

THE CREEP MECHANISM

Creep deformation occurs in all components subject to stress, and operating at any temperature. In practical applications, only those components at an elevated temperature and subject to high stress levels are of significance and interest to the turbine engineer. (There are threshold values of temperature and stress that vary with the material being considered.) This interest derives from the fact that those components' mechanical condition, performance, and remaining life must be observed and possibly monitored.

The term "elevated temperature" is used without definition. Its significance in the expectation and examination of creep and the temperature at which significant creep can be expected to occur is temperature that is about one half of the metal-melting temperature on the absolute temperature scale (K° = F° + 460°), *i.e.,* If an alloy steel has a known melting point of 1,120°F, then significant creep levels can be expected to occur at a temperature of:

$$0.5(1{,}120 + 460) = 790°F$$

Note: The majority of turbine steels melt between 1,300°F and 1,400°F. When they are used in the high-temperature range, it is normal to consider that creep deformation could occur in those components with an operating temperature at or above 750°F, and their design is considered for creep deformation. This provides a built-in factor of safety.

Such components are normally designed to provide a "minimum creep life" of 200,000 hours, which gives an operating life of about 30 years, at a load factor of 75%.

We know that relatively low-magnitude stresses are capable of producing creep deformation when applied for a sufficient time at an elevated temperature. Under these conditions—elevated temperature and low-magnitude stress applied continuously—components deform gradually. They begin at a high rate, which continues until the lower secondary level of deformation occurs. At this secondary condition, micro-cracks occur both at the surface and internally, and can grow until a tertiary rate of deformation is entered, which continues at ever-increasing rates until final rupture occurs. The creep rate is, in effect, the rate at which the component deforms—elongating, compressing, or even bending. It is a measure of the rate at which the component shape changes in the direction resulting from the application of the stress.

It is of interest to consider the internal mechanisms, ruptures, and crystal adjustments that occur in the material under conditions of load and temperature. It is these changes that contribute to, or cause and promote this permanent plastic deformation.

Consider the schematic (Fig. 5.5.1), representing a micrograph section through a piece of material. It is at a constant elevated tem-

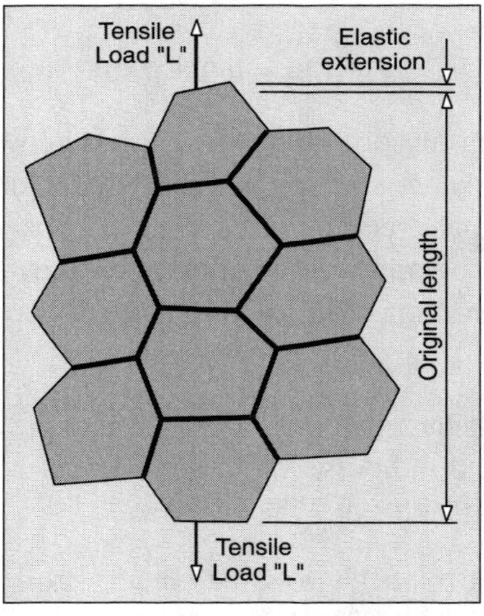

Figure 5.5.1—The schematic of material placed in tension with a small elastic extension.

perature and subjected to a tensile load "L." This load is of a size that neither rupture nor significant deformation occurs as it is applied suddenly. There is shown to be a small elastic deformation from the original length, but this is such a small amount that a major portion would be recovered if the load were removed after only a short period of application (elastic extension). However, after an extended period, further deformation will be noted—this deformation comprising any initial elastic and a more permanent plastic deformation. This deformation can be the result of two separate forms of structural change within the material occurring together, and are illustrated in Figure 5.5.2. They are explained below.

- The existence of impurities (inclusions) or micro-voids at the grain boundaries

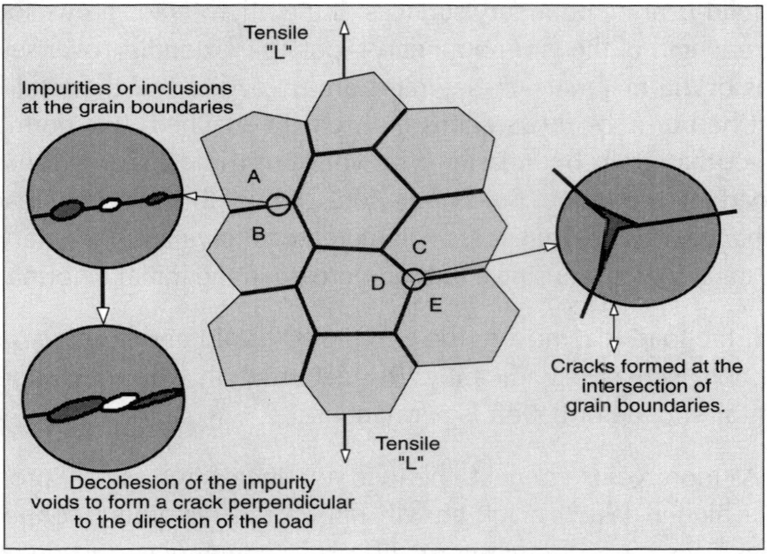

Figure 5.5.2—The effect of extended service on material structure at an elevated temperature, with the material subject to a tensile stress.

Consider the boundary that is close to perpendicular to the direction of the applied load "L," between grains "A" and "B." It indicates there are a certain number of impurities or even small voids present at this boundary.

As the load is applied, a tensile load develops across this boundary, and this face is placed in almost direct tension. The voids and impurity centers—which exist at this surface and normally have less resistance to applied load—will tend to grow along the boundary, perpendicular to the applied load. This growth will continue until adjacent voids join, and in sufficient length can be considered to form a micro-crack.

After a micro-crack forms, the remaining connected length of the grain boundary or boundaries must take up a portion of the total load. This causes an increase in the internal tensile stress across all remaining load-bearing boundary surfaces. If the micro-crack grows for the entire length of the grain boundary—possibly extending over several gains of the material—these grains are unloaded and the total load must be borne by those grains that remain attached. It is normal to expect that grain boundaries with voids and impurities will not all unload at the same time. Therefore, while the external load is unchanged the internal stress will increase, allowing further cracks to initiate and grow causing a further increase in the initial deformation.

If the load is removed, the width of the void and impurity crack will not decrease significantly. The deformation will not disappear. This (plastic) deformation is permanent.

As more voids connect, the load and local stresses rise, promoting a higher (faster) crack growth rate in the remaining connected grains. This form of crack formation is inter-granular, and de-cohesion of the voids will exist at the grain boundaries.

- Wedge cracks formed at grain boundary junctions, as shown for the common junction of the boundaries of grains "C," "D," and "E"

These "wedge cracks" extend along the grain boundaries, again unloading the affected grains and placing the load on the remaining portion of the material still connected.

Like those micro-cracks formed by void de-cohesion, this type of crack is inter-granular. However, it should be expected that those boundaries most closely inclined to the perpendicular (to the applied load) will suffer the greatest stress effect, and it should be expected that any micro-cracks that form will preferentially develop along those boundaries such as those existing between grains "C" and "E" in Figure 5.5.2.

As these various voids coalesce and greater loads are placed on the remaining material, this remaining material between the micro-cracks will undergo considerable plastic deformation before final rupture occurs.

The most common cause for creep deformation (considered in the previous sections) is when a tensile load is applied. However, it is not uncommon for extension to occur under other forms of load, *e.g.*, compressive, shear, bending, or torsion. Under these types of load, plastic deformation will occur if the conditions of stress and temperature are such that they promote the creep phenomena.

Grain and boundary layer relative strength

When a load is applied to a piece of material, and conditions of temperature and stress are such that creep deformation occurs—the deformation being caused by the formation of micro-cracks and the deformation this allows—there is a question of the relative strength of the pure grains of the material, and the material contained within the boundary layers.

At lower temperatures, grain boundary material tends to be stronger than the materials of the eutectic grains. Therefore, at lower temperatures, ruptures occurring will tend to be trans-granular, while at higher temperatures they will be inter-granular. The temperature at which failures move from the trans-granular to the inter-granular form is called the "equi-cohesive transition temperature." This temperature is dependent upon a number of variables, such as the stress applied and the amount of time the component has been exposed to the temperature at which it is operating.

Figure 5.5.3 shows a diagrammatic representation of the variation of strength of the grains of pure material and the grain boundaries.

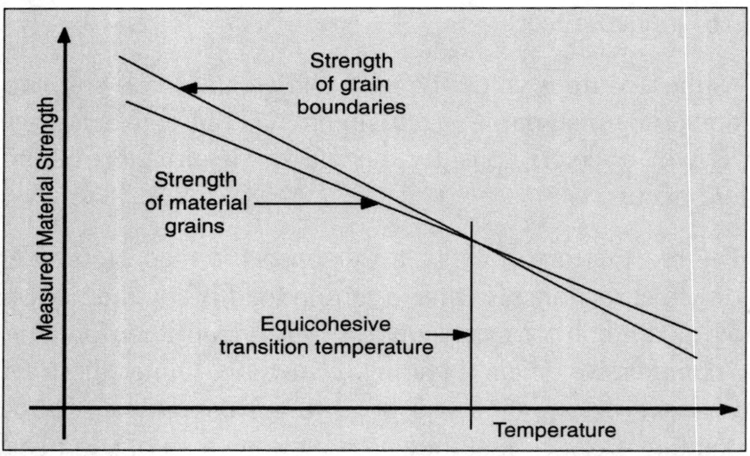

Figure 5.5.3—The relative strength of the individual metal crystals and the grain boundaries as a function of temperature.

Slip planes

"Slip" is a form of plastic deformation that occurs under the action of shear loads, causing a permanent displacement in the direction of the load. When slip occurs in a specimen subjected to creep deformation, slip planes tend to be along the grain boundaries under the load intensity. Such slip would be visible by microscopic examination or using scanning electron microscope methods.

If the material section shown in Figure 5.5.1 has a shear load applied (Fig. 5.5.4), a plastic deformation will occur through some angle (as shown), giving a total deflection "Ψ." Again, this deformation occurs because of the presence of the two parameters of stress and temperature previously discussed. Such deformation will be plastic and therefore permanent.

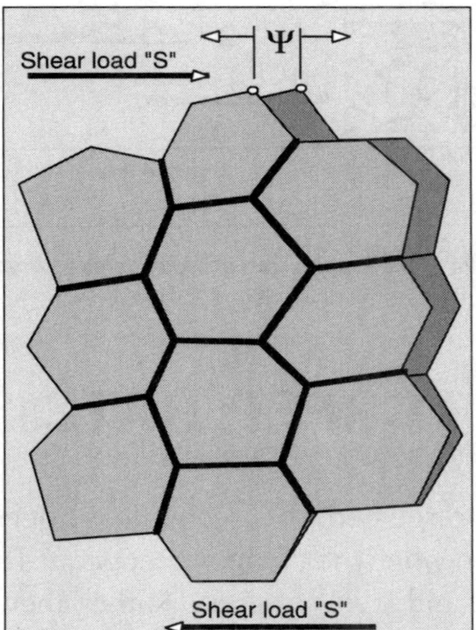

Figure 5.5.4—Material deformation when a shear stress is applied.

Creep progression

Figure 5.5.5 shows a schematic of creep damage growth. This growth damage from individual voids or cavities connect to form these cracks and continue to connect until the micro-cracks join, forming major macro size ruptures that eventually produce complete failure.

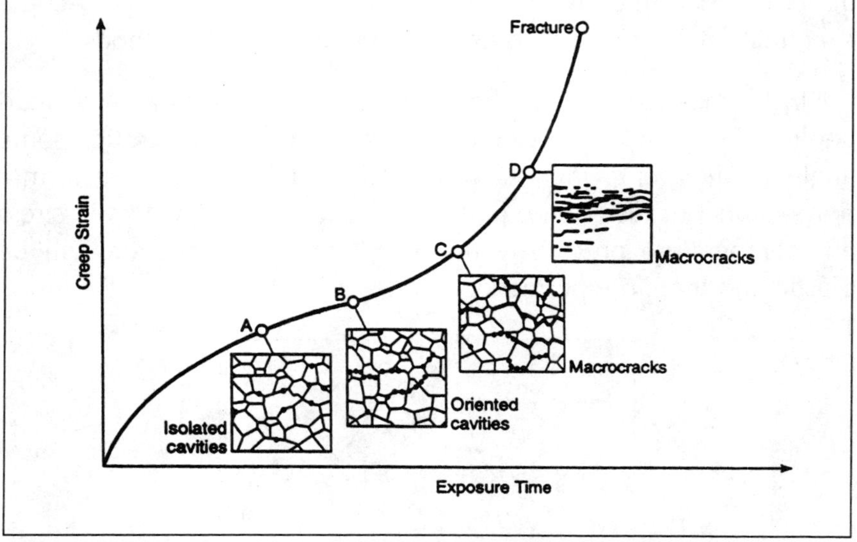

Figure 5.5.5—Schematic of the progression of creep damage from isolated cavities to macro-cracking.

CREEP RATE

As previously discussed, the creep rate is not constant under any applied load, even when the component operates at a relatively constant temperature and levels of stress. Rather, there is an initial high rate of primary deformation that slows through a secondary phase, and is maintained relatively constant for considerable periods, before entering a tertiary phase where deformation is again rapid to failure.

It is interesting to consider the material characteristics that produce these changes in creep rate. Unfortunately, no precise explanation can be found, but if we consider the crystallographic and property changes occurring within a component subjected to the conditions promoting (or inducing) creep, some explanation for creep rate can be established in general terms.

When load is applied to a component at elevated temperatures, there are degrees of "work hardening"—a process by which the hardness and mechanical strength of a material increases as a consequence of plastic deformation. This phenomenon occurs below the re-crystallization temperature and reduces the creep by increasing the material's ability to oppose deformation. When temperatures are below the re-crystallization range where turbine components operate, it is still possible for re-crystallization—a process by which large grains are refined to form a number of the smaller crystals—to occur slowly. In turbine steels this is a form of "cold working" that changes the crystal structure at the localized temperature, and tends to increase the creep rate.

A further material characteristic change that can influence the creep rate is precipitate over-aging. This phenomenon produces an increase in certain material mechanical properties in the direction the load is applied. This increases the creep rate.

How these various phenomena interact in a material to modify the creep rate depends upon time and operating environments. During short time applications of load, work hardening is the predominant characteristic. This means the initial rate of creep is high, and deformation rapid, in the direction the load is applied. However, as the time-of-load application increases, those phenomena that tend to lower creep rate begin to become effective. A balance is reached between them and the work hardening effect to the extent the secondary phase of creep rate is relatively constant. As time increases, there will be an accumulation of damage—formation of surface and internal cracks—and the creep rate will accelerate into the tertiary phase, where the rate of material deformation and damage accumulation increases to final rupture.

Prediction of creep rates

The methodology available for predicting creep rates is not exact. However, there is sufficient experimental data available that certain empirical relationships have been formulated, sufficient that a reasonable level of accuracy in rate prediction can be achieved.

There are three basic equations defining the instantaneous, primary, and secondary strains:

Instantaneous $$\varepsilon = \varepsilon_o + C_1 t^{1/3} + \dot{\varepsilon} t$$

Primary $$\varepsilon = \varepsilon_o + C_2 \left[1 - \exp(-C_3 t \right] + \dot{\varepsilon} t$$

Secondary $$\varepsilon = \varepsilon_o + \frac{\dot{\varepsilon}}{C_4} \ln\{1 + C_5 [1 - \exp(-C_4 t)]\}$$

where:

εo is the instantaneous strain
t is time in hours
and
C1, C2, C3, C4 and C5 are empirical constants dependent upon the mechanical and chemical properties of the material

In the tertiary stage of creep the strain is best defined by equations:

where:

$$\varepsilon_t \propto \exp[C_6 (t - t_3)]$$

$$\varepsilon_t \propto \frac{1}{(1 - D)^{C_7}}$$

t is the time in hours
t3 is the time in hours to enter the tertiary range
D is a constant that varies with time

Note: The constants defining these deformation rates are material dependent, which can be defined.

CREEP IN STEAM PATH COMPONENTS

Since many components of the turbine steam path are subjected to continuous loads and elevated temperatures (elevated as defined earlier), then it should be expected that some level of creep deformation will occur during their operating life. This section will consider those components susceptible to this form of damage.

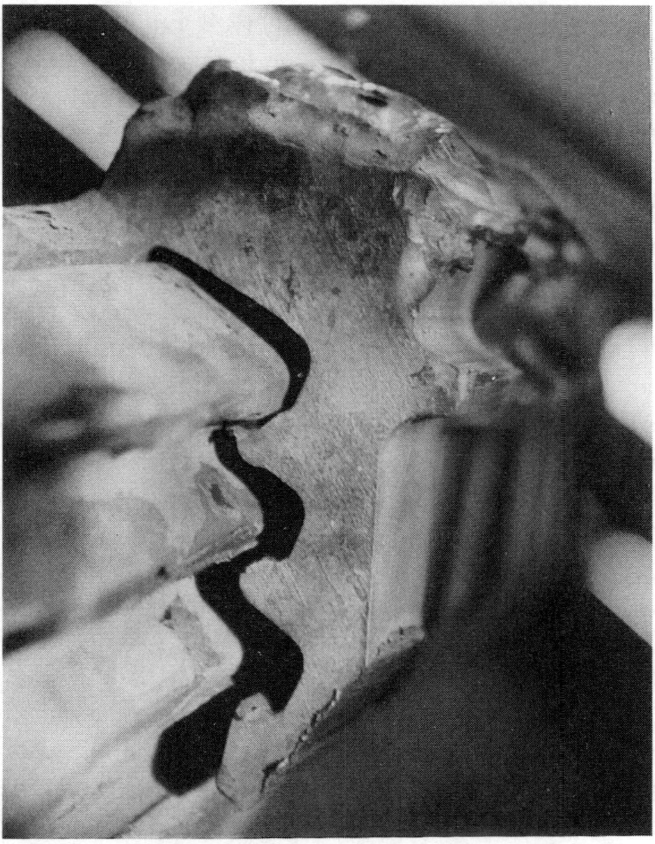

Figure 5.7.1—A control stage blade root having suffered creep damage.

Rotating blades

The rotating blades are subject to continuous loads during operation. In the early stages of the high and reheat sections of the unit, these loads are applied at temperatures sufficiently high that creep deformation can and will occur. The extent of this deformation and the need to replace blades as a consequence is dependent upon a number of factors, *e.g.*, the magnitude of the load, and the time for which the blades have been in operation. Figure 5.7.1 shows a blade from the control stage of a high-pressure section with initial steam temperature of 1,100°F, and which has been in service for 160,000 hours. This blade has lasted remarkably well, but under the influence of stress and temperature over an extended time, has ultimately crept until the load bearing surfaces on the root ligaments have started to move outward past the transfer surfaces on the rotor. This action has sheared the side grips. Shown as Figure 5.7.2 is the form of the creep deformation of a similar blade root, indicating the initial and final operating condition, with the "lift" due to creep indicated.

Another factor that can induce damage in any component, but especially on rotating blades, is failure to hold certain design specified dimensional tolerances. Figure 5.7.3(a) shows the closing blade from the first rotating blade row of a reheat section, with a reheat temperature of 1,000°F. In this unit a "false start" was made in producing the root pinhole. The consequence of this is shown in Figure 5.7.3(b)—It did not have sufficient surface to transfer its load. Therefore, under the influence of centrifugal loading, the pin distorted or crept, allowing the blade to migrate radially outward.

In general, high-temperature rotating blades are manufactured from an AISI 410 or 403 martensitic type stainless materials. There are, however, some high-temperature/high pressure applications where either an AISI 422 or an austenitic type material can be employed. A concern with austenitic materials is they have a different coefficient of expansion. It is necessary to account for this difference during the design phase to ensure that under thermal transient conditions the unit

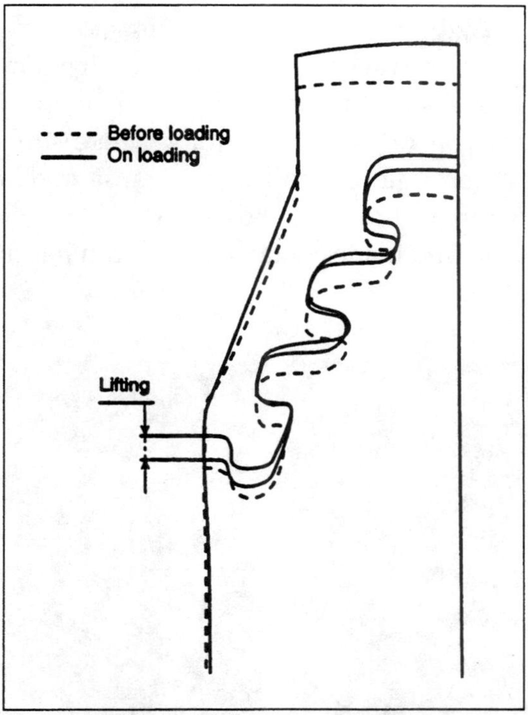

Figure 5.7.2—Creep deformation in a straddle type root of the form shown in figure 5.7.1

is able to accommodate these differences, and their possible effect on unequal load sharing between the load bearing surfaces.

Note: When replacing rotating blades for a stage in the creep range, it is preferable to determine any creep deformation that has occurred, as producing blades to the original dimensions may not allow equal load sharing between multiple load transfer surfaces.

Rotors

Rotors—like other components subject to high temperature and high pressure—are subject to creep deformation. This damage occurs most often in regions of the rotor where stresses are high. These are most often regions where load is transferred from the blade root to the rotor rim. Figure 5.7.1 shows a wheel rim where the blade root has deformed and blades detached. This rim has also deformed due to creep.

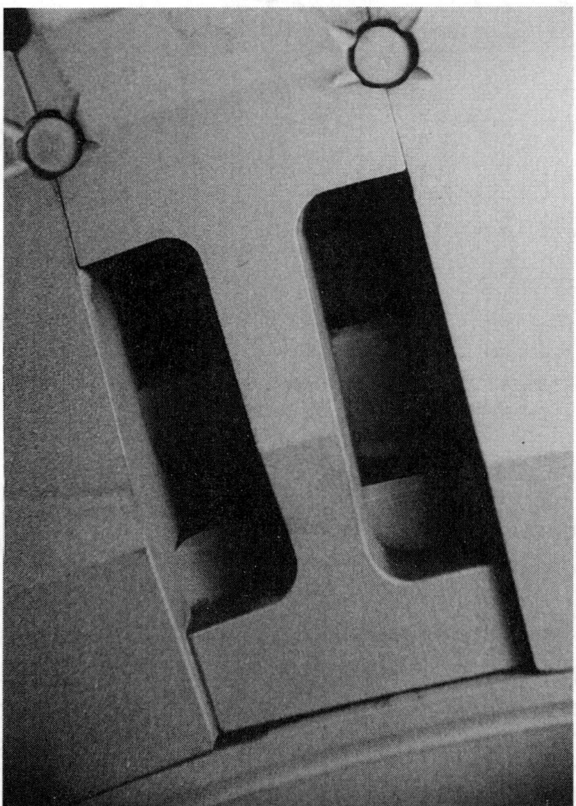

Figure 5.7.3(a)—Root creep see figure 5.7.3(b).

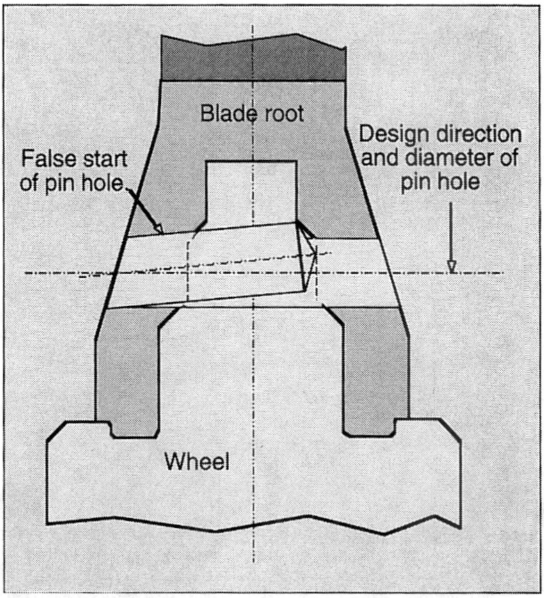

Figure 5.7.3(b)—Showing the false start on the pin hole at manufacture, permitting blade root creep.

High-temperature rotors are of a complex form and the highest temperature and stresses may not occur at the same location. However, there is normally sufficient coincidence between these two contributing factors that creep has a high probability of occurring, if only at a relatively low rate.

The rotor can also suffer damage as a consequence of combined "creep fatigue." See *HCF Failure Surface Appearance.*

Diaphragms

Diaphragms are produced in three parts—an outer ring, a blade annulus consisting of a number of vane elements, and an inner web. Each of these three components of the diaphragm is subjected to the pressure differential of the stationary blade row. The outer ring is supported against the locating slot in the casing, and does not distort under the effect of pressure. However, both the vanes—produced from a martensitic steel of the same general mechanical and chemical properties as the rotating blades—and the inner web are subject to pressure differential, and will eventually take up an elastic and, if at sufficiently high temperatures, plastic or creep deformation.

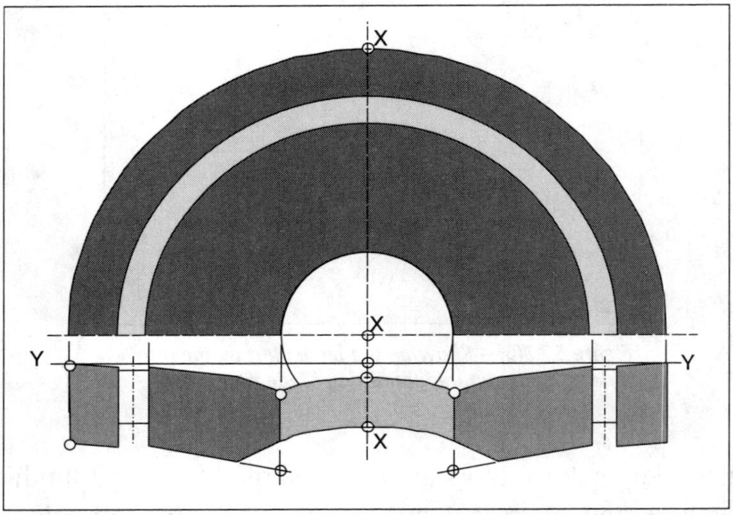

Figure 5.7.4—Diaphragm deflection, being a maximum at the inner diameter on the horizontal joint.

As steam is initially admitted to the unit, stresses are set up in the diaphragm. These initial stresses induced in the various components of the diaphragm are of little significance in themselves, their magnitude being in the order of 5,000-10,000 psi. However, the axial loads, while not high, result in a down-stream deflection of the diaphragm that is a maximum in the inner web inner diameter adjacent to the rotor. Figure 5.7.4 shows the deflection of a diaphragm at the inner diameter. The maximum stresses that occur in a diaphragm due to the pressure differential it experiences, occur in the element along the axis "XX," which is at 90° to the "YY" axis or horizontal joint. It can also be seen from this diagram that the maximum deflection does not occur on the axis of maximum stress, but rather on the horizontal joint which, during operation, receives less support from adjacent material than does the vertical center line "XX."

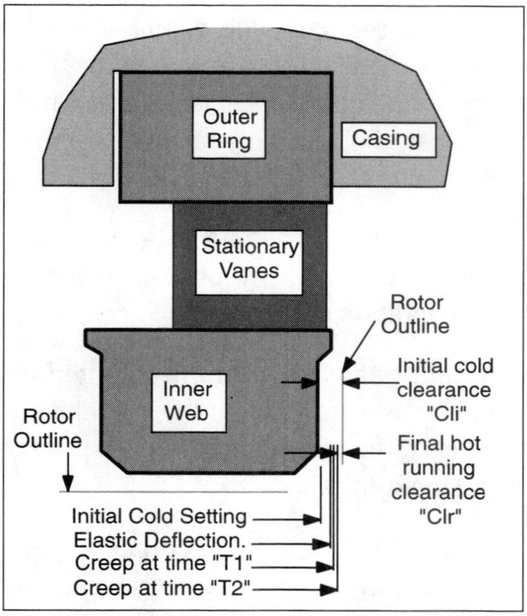

Figure 5.7.5—The stages of deflection of a diaphragm with time exposure.

The outer rings and webs in the high-temperature regions are produced from forged, high-strength alloy steel. The major concern with the manufacture of diaphragms is the ability of these components to resist for extended periods at high temperatures, the axial pressures developed across them. The outer ring is located in the casing to form a steam-tight joint that will not deflect to any significant degree. However, both the vanes and inner web will deflect downstream under the influence of the axial pressure gradient. It is therefore necessary to consider the effect of time on operating deflection and the influence this will have on axial clearances within the steam path.

Diaphragms have, compared to the rotating components, relatively low levels of stress induced in them. These components are unlikely to fail in a catastrophic manner. However, when steam is admitted to the unit, there is a pressure differential stress induced in the vanes and inner web that will cause it to deflect downstream. When a unit goes into service, each diaphragm will experience an initial elastic deflection followed by a time and temperature dependent plastic deflection. Figure 5.7.5 shows a cross section of a diaphragm in its operating position relative to the rotor. It can be seen that:

- there is an initial cold clearance between the inner web, at its inner diameter, and the rotor "Cli," when the rotor is in the stationary condition

- when the unit starts and steam is admitted, there will be an initial elastic deflection, under the influence of the stationary row pressure differential, consuming a portion of this clearance reducing it to "Clo"

- after a time "T" there will have been creep deformation, which is an additive to the elastic deformation. After this time period, the creep deflection will have increased, reducing the clearance to "Clr"

As operation continues, the initial "hot running clearance" is reduced by a deformation of the creep deformation, and a condition could eventually exist that a "rub" will occur between the diaphragm inner web and the rotor. This rub is normally at the diaphragm inner diameter, or at any location where "tight" running axial clearances might exist. Such a rub will generate a considerable amount of frictional heat. These heated regions are immediately quenched by the surrounding steam and introduce a hardened zone on both the diaphragm and rotor. While the condition of the diaphragm may not be immediately serious, the rotor will be burnt to the extent it will become particularly hard, and brittle, and should not be operated without corrective actions being taken.

This rub can often occur during a "start up" or "shut down" situation, when the rotor could be subjected to "short" and "long" conditions (see chapter 2) modifying the normal operating clearances.

Note: When the unit has stopped, and the steam shut off, the elastic deflection will be reduced to zero. However, the plastic deflection is permanent, and will not be recovered. It is recommended to monitor critical clearances as a part of the normal maintenance record in stages operating in the temperature region, where creep could be present.

In an effort to increase resistance to creep deflection, and to reduce the extent of initial plastic deformation, some manufacturers produce their steam path stationary vanes from different profiles. The vanes used in those stages having a predicted high creep deformation are evaluated for the effects of pressure, temperature, and steam momentum loads. If these loads are excessive, it is common to use extended axial width vanes as shown in Figure 5.7.6. Such vanes are used on a portion of the total in each stage, and extend the axial depth of the steam path from "W" by an amount "E," to "Wu."

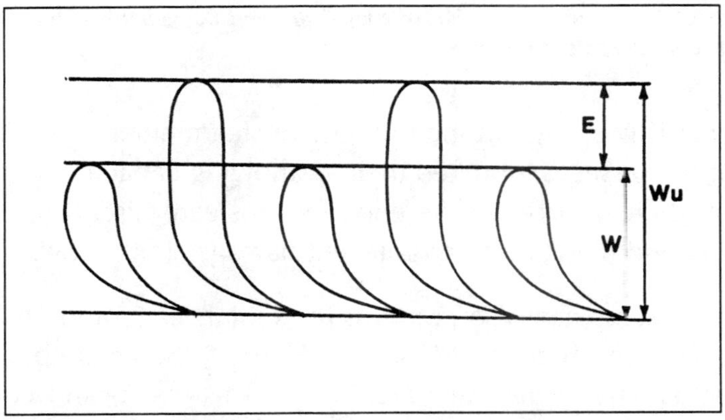

Figure 5.7.6—Extended section vanes, for axial strength, within a steam path.

The extended portion of these vanes is selected to preserve, to the greatest extent possible, the aerodynamic form of the expansion passage, and has little or no effect on the flow distribution of the steam. This vane extension has little or no influence on the performance of the stage, because the velocity of the steam entering from the previous rotating row is low and practically at 90° to the tangential direction. However, if the steam enters at some angle "δ," as shown in Figure 5.7.7, this could change the amount of steam entering each expansion passage, and cause minor changes to the stage flow characteristics, having an effect on the "swallowing capacity" of the individual passages.

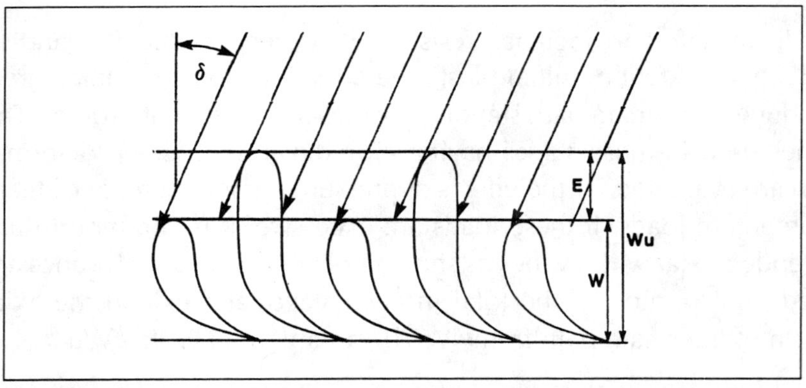

Figure 5.7.7—The effect of "axial stiffeners" on steam flow distribution in the event steam does not enter axially.

Note: This change in the quantity of steam entering the individual nozzle passages modified their swallowing capacity. As a consequence, there will be a difference in the steam discharge velocity, pressure, and direction from adjacent passages in the row.

Diaphragms can also creep in the radial plane and affect their horizontal joint. This effect is seen in Figure 5.7.8 where both diameter increases and decrease are shown. This effect, and corrective actions are discussed in chapter 7. In Figure 5.7.9 is shown the clamped halves of a diaphragm, before correction, where the halves have moved in, opening a gap at the horizontal joint.

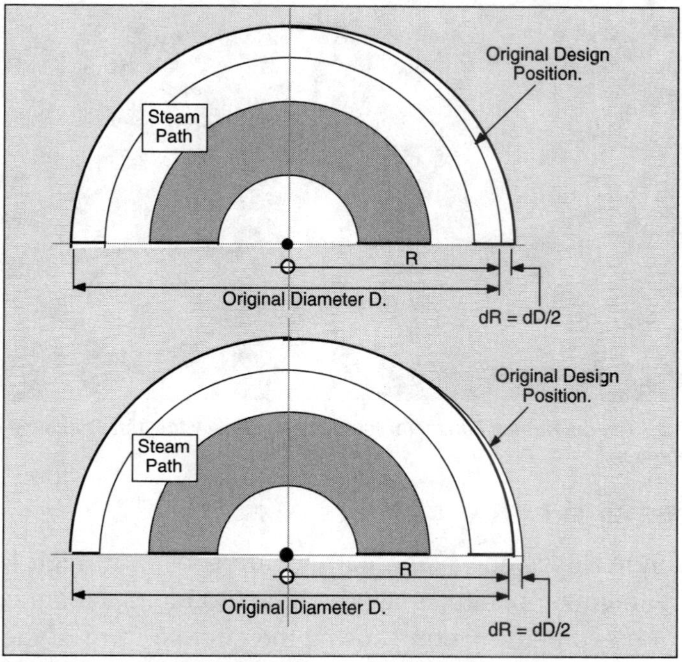

Figure 5.7.8—Diaphragm radial creep deformation. In (a) the diaphragm has crept inwards reducing the diameter by an amount "-dR", and in (b) the diaphragm has opened by a similar amount "+dR".

Note: When considering the axial or radial distortion of diaphragms it must be considered that creep is not the only mechanism introducing this condition. It is also possible that residual stresses locked into the component due to a welding or other process could cause this condition to occur. It is also necessary to recognize while this condition can be corrected, in cases of creep where distortion is removed, voids that exist between the crystals of the material will not be removed and deflection will occur again at a relatively fast rate.

Figure 5.7.9—The diaphragm horizontal joint gap caused by the halves moving in due to creep deflection.

Casings and blade carriers

Turbine casings and blade carriers are subject to high levels of stress and high temperatures. Under these conditions there can be a degree of creep deformation causing the circular form of the casing to be modified to an elliptical form, with the horizontal joint either increasing or reducing. The possible consequences of this deformation are dependent on details of the design. In certain circumstances such deformation can cause serious damage and even introduce difficulties in removing the studs and lifting the top half casing or blade carriers, because of binding between the casing and horizontal joint studs. In other situations the difficulties involved are minor, will not influence steam path alignment, and can therefore be tolerated.

Note: Cylinders that are hotter on the inside surface than the outsides tend to "move-in" at the horizontal joint. This can make the removal of horizontal joint bolts difficult, and has been known to make unit disassembly difficult, even resulting in the destruction of a large number of studs. In the worst cases, even after the top half casing has been removed, the lower half will move in and "grab" the rotor, making its removal difficult without causing damage to the rotating blades and shaft end seals.

Figure 5.7.10—The inner surface of a casing where creep deformation has moved it in to the extent heavy rubs with the rotating blade integral seals have grooved the casing.

Figure 5.7.10 shows the horizontal joint of a casing that has "moved in" at the horizontal joint, producing heavy rubs between the casing inner surface and the integral seals produced on the blade tips. In this design the stationary blade rows are carried in the inner casing, and so any distortion it suffers will cause a misalignment between the stationary and rotating blade row, reducing and ultimately destroying the "lap." This type of stationary blade row deformation can also produce a tangential variation of the steam force, possibly introducing stimuli into the rotating blades, causing some form of vibration.

Note: As in the case of diaphragms, the distortion present in casings can be due to phenomena other than creep, and the comments regarding their correction and future deformation are also valid for casings.

Threaded components

Many studs and bolts are always in a state of high tension, and operate at high temperatures. If the components' screwed ends are contained within a thick horizontal flange it is unlikely they will extend by any significant degree. Similarly, a thickheaded nut will not allow significant deformation. However, the shank can and will extend with time. This deformation should be monitored as the life of these components should be considered finite and should be changed whenever there is an indication that their life has expired.

HIGH-CYCLE FATIGUE

High-cycle fatigue is a mechanism capable of causing mechanical rupture in a component as a consequence of the application of repeated cyclic loads. These are loads whose maximum values can be considerably lower than those required to cause failure by a non-alternating application, when applied without shock, and in the same direction to the same component.

This mechanism of failure is perhaps the most common observed in failed components of the steam turbine. The high frequency of this type of failure is a consequence of the many sources of cyclic loading generated within the steam flow. These cyclic loads, while of low magnitude, are of high frequency in a component that is rotating at high speed, and therefore can accumulate many stress cycles within a relatively short period. In addition, there can be present many regions of stress concentration as a result of damage caused by some other initiating mechanism. Components normally able to sustain the initial form of damage and be unaffected would fail as a consequence of the application of a low-magnitude alternating load.

The design process establishes the dimensional and material requirements for each component of the steam path, and selects them so stress levels (direct and alternating) induced by operation will not exceed values that can be tolerated by the components. It is often relatively minor damage that will initiate damage, and then high cyclic loading will drive the components to failure.

High-cycle fatigue is most destructive in regions where high direct stress levels exist. The most common location to discover high stress levels is where there is stress concentration. There are three basic causes of stress concentration:

Section changes. Design considerations often require section or form changes. These changes can introduce regions where small fillet radii or other form changes are required. This causes load to be applied in a non-uniform manner, which increase stress levels locally.

Material deficiencies. As materials are produced for turbine application, they are carefully examined to ensure faults, (*e.g.*, inclusions) do not exist in regions where they represent a possible compromise to the material quality. However, it is possible for such faults to go undetected. It is also possible for faults to be introduced particularly during any manufacturing process involving the use of heat. These faults (*e.g.*, hard spots) can go undetected despite the best efforts of the manufacturer to eliminate them.

Machining or assembly marks. There are many manufacturing and assembly processes where surface marks, scratches, and gouges can be introduced. Some of these may even be hidden at completion of the process, making them more difficult or impossible to detect. These can act as stress concentration centers, capable of causing damage.

THE HIGH-CYCLE PHENOMENA

The alternating stresses applied to a component vary between a maximum and minimum, and vary from a mean value. Therefore, if the mean stress is denoted by "σm," and the alternating stress has a range of "2σa," then the maximum stress to which the component is subjected is "σm+σa," and the minimum "σm-σa." This varying stress effect is shown in Figure 5.9.1. In fact in practice there are three forms (or stress ranges) in which this stress is applied:

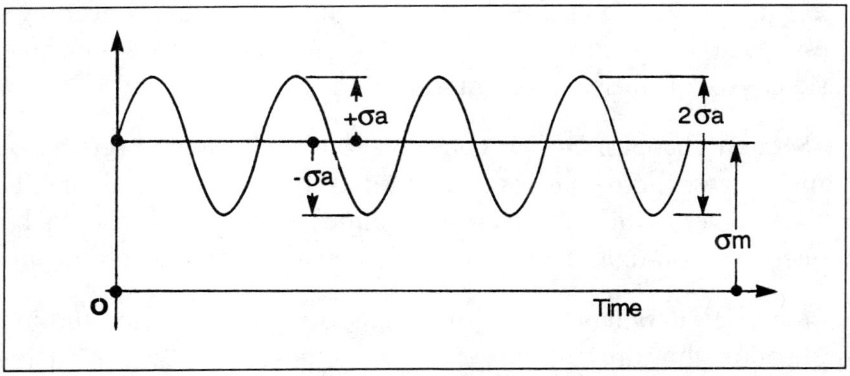

Figure 5.9.1—An alternating load of "2σa" applied to a constant load of "σm".

- The alternating stress changes from a maximum to minimum around a zero mean stress. This is shown in Figure 5.9.2(a). Therefore, the direction of the resultant stress varies during each cycle
- The alternating stress varies about a mean stress. It's lower in magnitude than the mean stress and the total stress is at all times in the same direction. This is shown as Figure 5.9.2(b)
- The alternating stress varies from a maximum to a minimum, the minimum being equal to zero. This is shown in Figure 5.9.2(c)

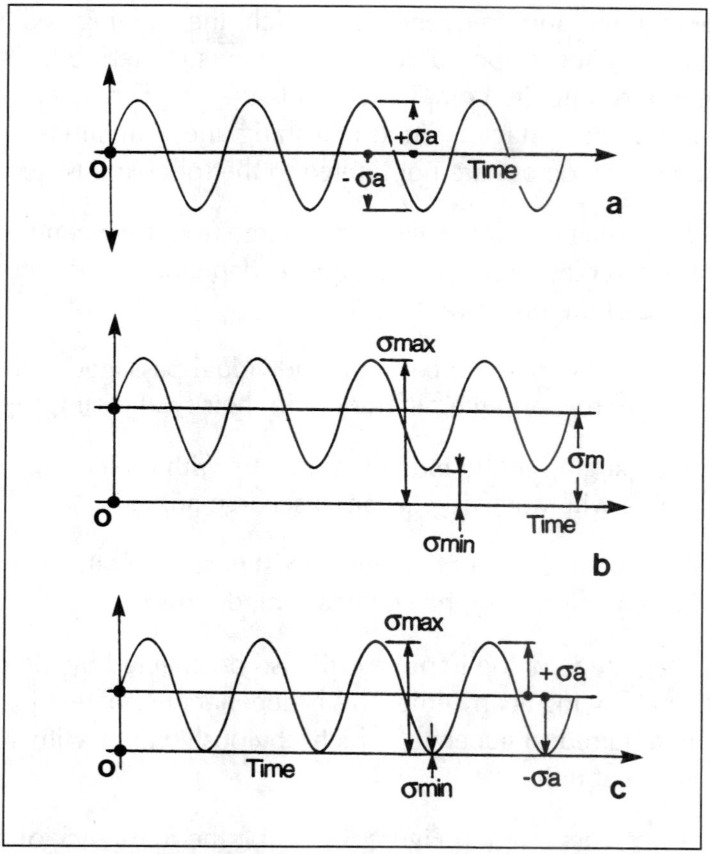

Figure 5.9.2—The three forms of alternating load application.

In defining these three forms—direction and magnitude of alternating stress level and its variation, as depicted in Figure 5.9.2—it has been assumed the stress is applied in a *sinusoidal* form and maximum and minimum values of the applied alternating stress are always of the same magnitude. These basic assumptions are not necessarily true. Within any steam path, there can be considerable variation from them.

The manner and frequency in which the varying loads are applied is dependent upon a number of factors related to the steam path geometry and its compliance with the standards set by the designer. Important factors in establishing the magnitude of the steam force include but are not limited to the following issues:

- The strength of the wake at discharge from between two stationary blade vanes. This can be dependent upon compliance of the blade vane
- The inlet conformance of the individual passages in the stationary row and the differences in their swallowing capacity
- The gauging of the individual throats within the steam path, which will modify the steam discharge angle
- Any inconsistencies that might exist in steam path form at the horizontal joint of the stationary blade row

The magnitude of the impulse will also be affected by the quantity of steam flowing at any time. This is dependent upon the load the turbine is required to generate, which obviously varies with system demand on that unit.

It has been assumed in Figure 5.9.2 that the frequency of application of the stress remains unchanged. Within the turbine steam path of a unit designed for constant frequency power generating this is a valid assumption. However, in other turbines designed to operate at varying speeds, this cannot be assumed. Because the accumulation of damage is frequency sensitive, such units can accumulate different amounts of damage at different times. The most damaging situation to arise in the steam turbine is when the frequency at which the impulses developed within the steam path is coincident with the natural frequency of the components these impulses are interacting with. Under these circumstances the impacted component can vibrate to failure with a relatively short period of time.

From these considerations it is clear that an analysis of the steam path alternating stress levels can be complex. A normal procedure in the design phase is to select and dimension the various components to fall within certain limitations. Then, should damage be sustained in operation from some form of vibratory loading, at that time a detailed analysis can be made to identify the circumstances of failure.

ROTATING BLADE VIBRATORY STRESSES

The blades of a turbine row are usually the most susceptible to high-cycle fatigue, and have induced in them stresses caused by the vibratory stimuli developed within the flowing steam during operation. The magnitude of these stresses vary from stage to stage, and are influenced by the frequency and magnitude of the vibratory load sensed. The physical characteristics of the blade—size, vane form, and the material from which they are produced (physical properties and damping characteristics)—will also influence the total effect of the stimuli on the blade. Various sources of vibratory stimuli in the steam path are given in Table 5.10.1.

Section	Harmonic	Typical Sources
High Pressure Section	• High per rev (40x)	• Nozzle tolerance limits • Upstream wake degeneration • Structural turbulence
Intermediate Pressure Section	• Nozzle passing frequency (NPF) • 2 x NPF • 3 x NPF	• Nozzle wakes • Diaphragm harmonics • Diaphragm harmonics
Low Pressure Section	• One per rev • 2 per rev • Multiple/rev • Medium/rev • High/rev	• Relative displacement nozzles to blades • Diaphragm joints • Structural supports in flow path • Diaphragm harmonics Aeroelastic disturbances • Nozzle turbulence harmonics • Upstream wake degeneration Structural turbulence

Table 5.10.1—Steam Path Exciting Forces

These various modes of vibration can cause the blade to vibrate in the axial, tangential, and torsion modes. There are various harmonics of these modes, and the magnitude of any induced stress is a maximum when the frequency of the impressed force, or stimulus, is in phase with one of the natural frequencies of the blade itself. This coincidence causes an increase in the vibration amplitude, and the strain and stress this induces.

The forces acting on the blade and contributing to vibratory loading are the steam bending forces developed as the result of the steam flow across the vane. Many of these loads are predictable and the design process avoids introducing these into the blading at a level where they can cause failure or damage. While the magnitude of these steam forces may be low, if their frequency of application is close to a natural blade frequency, they can soon induce a failure into the rotating blade elements.

Nozzle impulse effects

A common cause of damage or failure in the turbine rotating blade system is the effects of nozzle impulses. A nozzle impulse results from fluid flow distortions. Distortion results from uneven flow patterns between the stationary blade vanes. As the rotating blade vane traverses the nozzle pitch, it is subjected to a variation of steam pressure impulses, and therefore the forces developed across its tangential and axial length. A typical impulse/force diagram for three nozzle pitches (four vanes) is shown as Figure 5.10.1, where the steam force "F" on the blade at any instant is varying dependent upon its position in the tangential direction across the blade outlet.

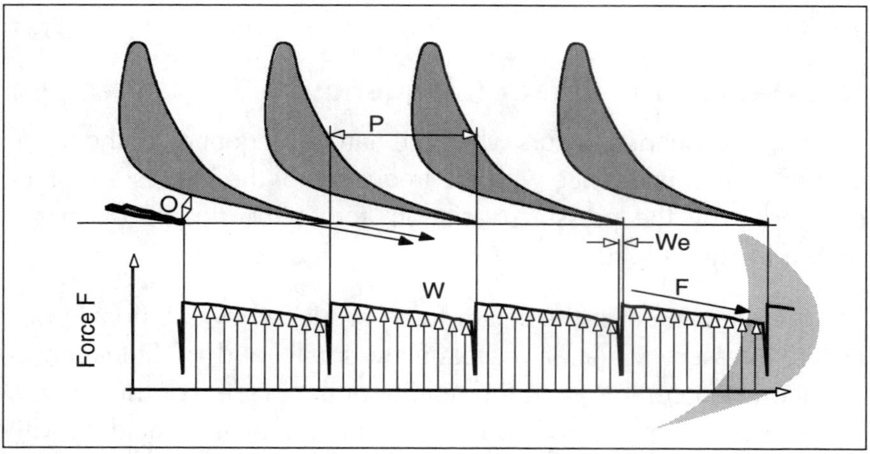

Figure 5.10.1—The variation of steam force "f" across the pitch "P" of the fixed blade row.

Also, shown in Figure 5.10.1 is the effect of a viscous trailing edge wake. The magnitude of the force produced by this wake effect is dependent upon several factors, including the flow passage shape, and the extent to which the boundary layer separates from the nozzle trailing edge. Shown in the lower portion of the diagram is a schematic of the steam force "F" at discharge from the blade row, drawn above some arbitrary position. The actual steam pressure decreases across the nozzle passage, and at the discharge edge, where "wakes" ("We") form, there is a small reduction in pressure due to this effect. This steam force can be seen to be acting on the vane. Because it is a varying force, the entire blade is subject to an alternating force, which induces a cyclic stress of varying magnitude across each nozzle passage.

For rotating blades, this nozzle impulse effect is possibly the most common cause of failure. For most designs the normal magnitude of the combined tensile and bending stress are not high enough to cause failure. However, the effects of this varying nozzle force can be magnified to produce dangerous levels of alternating stress.

Nozzle passing effect (frequency)

Blade resonance occurs when the natural frequency of the blade is at or near coincidence with the frequency of the impressed forces developed by the pressure variation across the nozzle discharge shown in Figure 5.10.1.

The magnitude of the steam force produced by the nozzle discharge pressure variation is relatively small, and of little consequence when compared to the rigidity of the blade. For this reason, if the natural frequency of the blade is not at or near coincident with the nozzle passing frequency, the blade dynamic stresses will not be influenced to a considerable extent.

To quantitatively establish this effect, an amplification factor to be applied to the steam bending force can be determined.

Consider the blade pair from a rotating row shown as Figure 5.10.2. It is subject to nozzle impulses. Consider each to act as a cantilever under the influence of these impressed forces. Also assume the blade has internal damping due to its molecular structure.

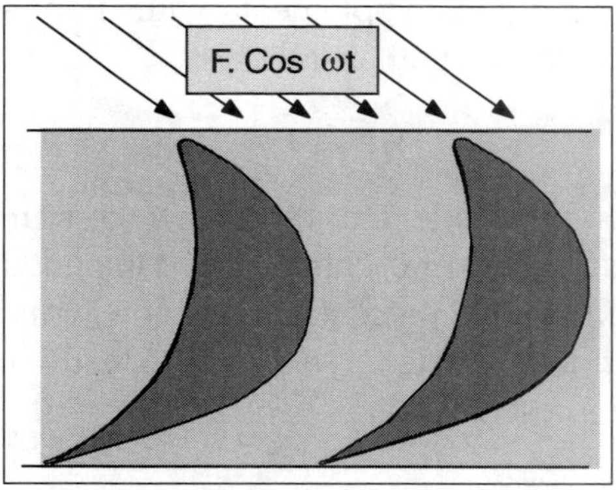

Figure 5.10.2—The steam force developed on the rotating blade row.

The equation of motion of the blade is given by:

$$M\frac{d^2y}{dt^2} + Fd\frac{dy}{dt} + Ky = F\cos\omega t$$

where:

K = Blade stiffness

M = W/g

Fd = Internal damping force
(Numerically = Magnitude of damping force when the velocity is unity)

Y = Displacement of the blade from equilibrium in time "t"

Velocity coefficient is Φv, and Φv = Fd. dy/dt. This can also be written in the form:

$$\frac{d^2y}{dt^2} + a\frac{dy}{dt} + by = c$$

The complete solution of this equation consists of a decaying time-transient portion (the complementary function) and a sustained portion (the particular integral) as follows:

$$y = y_s + y_t$$

In this equation, the first term represents the transient vibration, which can be assumed to be completely dampened by molecular friction. The second term represents the forced vibration and is maintained by the periodic nozzle force. Once a steady state is attained, this vibration is represented by:

$$y = \frac{c}{\sqrt{\left[b - \omega^2\right] + a^2\omega^2}} \cdot \cos\left[\omega t - \lambda\right]$$

Where $"a" = \frac{Fd}{M}$, $"b" = \frac{K}{M}$, and $"c" = \frac{F}{M}$

"y" has a maximum value when cos (ωt - 1) = 0, π etc. "c" is negative because initially the response "y" lags the driving force, c = F/M. Then:

$$y_{max} = \frac{c}{\sqrt{\left[b - \omega^2\right] + a^2\omega^2}} = \frac{c/b}{\sqrt{\left[1 - \left[\frac{\omega}{b}\right]^2\right]^2 + \frac{a^2\omega^2}{b^2}}}$$

Now, $c/b = F/K = \Delta$

Now, c/b = F/K = Δ (where "Δ" is the deflection produced by the static force equivalent.

"Fn" is the nozzle passing frequency
"Fr" is the blade natural frequency

where:

"Ad" is the dynamic amplification factor

$$Ad = \frac{1}{\sqrt{\left[1 - \left[\frac{\omega}{b}\right]^2\right]^2 + \frac{a^2\omega^2}{b^2}}}$$

Which reduces to:

$$Ad = \frac{1}{\sqrt{\left[1 - \left[\frac{Fn}{Fr}\right]^2\right]^2 + Cd}}$$

$$= \left[\frac{a}{2\pi f_R}\right]^2$$

where: **"Cd" is the damping coefficient**

Figure 5.10.2 shows the steam flow from stationary blades into a rotating blade row. Here the force onto the blades is shown as "F.cos ωt," produced at a frequency of "Fn" cps. Depending upon the natural frequency of the rotating blades there can be considerable amplification in the magnitude of vibration. Figure 5.10.3 shows the amplification factor "Ad" shown in the equation immediately above. This is shown as a function of the ratio blade natural frequency at operating speed "Fr" to the nozzle passing frequency "Fn." These curves show the influence of various damping coefficients "Cd" on the amplification factor, and therefore, the dynamic stresses developed in the blade.

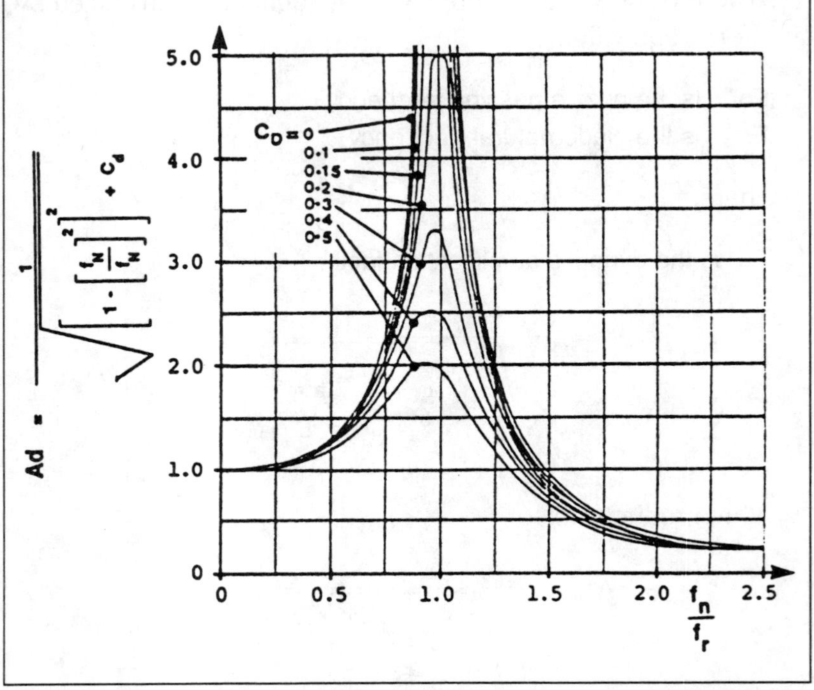

Figure 5.10.3—The dynamic amplification factor "Ad", as a function of the ratio of "nozzle passing frequency" to blade "natural frequency" for a range of damping coefficients "Cd".

Combined stress

In evaluating total stresses on a blade it is necessary to be able to determine the effects of the combined static (centrifugal), and vibratory (steam bending) stresses. A convenient method for doing this is to use the "Gerber's Parabola" or Gerber's Diagram. Gerber made an analysis of the fatigue results of various metals, and determined the effects of the combined stresses that could be represented by Equation 5.10.6. This expression is shown on curve Figure 5.10.4(a), with the limiting and actual stresses as shown.

where:

$$\sigma' = Fv\left[1 - \left[\frac{f}{T}\right]^2\right] \quad \mathbf{5.10.6}$$

σ = The alternating (limiting steam bending) stress
Fv = The limiting range of alternating stress, = "T/2"
f = The mean direct stress = "fc" + "fb/2"
T = Tensile ultimate strength of the material

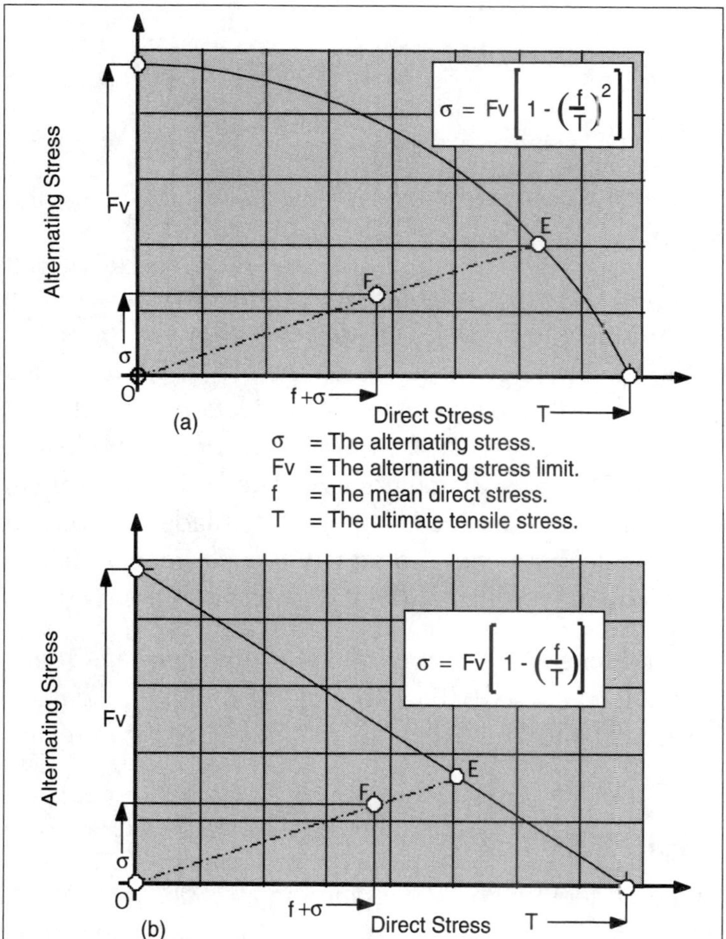

Figure 5.10.4—In (a) is shown the "Gerbers Parabola" and in (b) the Modified Goodman Diagram.

This expression indicates that a parabolic relationship exists between the combined "direct" and alternating stress "σ" and the limiting stresses "f" and "Lv" that the blade can tolerate. It was subsequently found that certain materials, including nickel and chromium alloy steels, do not accord with this relationship, but rather enjoy a linear relationship as defined by the modified Gerber or Goodman equation 5.10.7. This Goodman straight-line curve is shown on Figure 5.10.4(b).

$$\sigma' = Fv\left[1 - \frac{f}{T}\right] \qquad \textbf{5.10.7}$$

This expression is termed the "modified Goodman linear relationship."

Both the Gerber and Goodman expressions can be represented as shown in Figure 5.10.4. This diagram allows a representation to be applied to the known stresses on the blade, and can be adjusted to account for the dynamic amplification factor "Ad" of the alternating stresses, as determined from the frequency of the loads applied to the stage. There are difficulties associated with determining exactly "all" of the vibratory loads applied to a blade row, but predictive methods exist. These can also be left to experimental determination for a blade family profile, and the elements position within the unit.

To determine the "factor of safety" for the combined stresses, the ratio OE/OF could be used as an indication of the blade acceptability, based on either the Gerber or modified Goodman relationship.

Example 5.10.1

An alloy steel blade for operation in a 3,600-rpm unit has a predicted natural frequency of 5,360 cps when corrected for the stage operating temperature. These blades are to be used in a row for which the designer has a choice of stationary blades elements to give the correct discharge area and angle.

The available nozzle partition selection will be selected from:

Zn = 96 vanes - "A"
Zn = 84 vanes - "B"
or
Zn = 80 vanes - "C"

The predicted direct stress "f" is 13,350 psi; the limiting steam bending stress "σ" (without amplification) is 4,640 psi. At stage temperature the UTS (T) is 87,250 psi and the limiting fatigue stress "F" is 47,750 psi. The blade material has a damping coefficient "Cd" of 0.018. It's a matter of which stationary blade arrangement appears the most satisfactory.

Solution:

Nozzle passing frequency "Fn" with 96 partitions
= 60 x 96 = 5,760 cps - "A"
Nozzle passing frequency "Fn" with 84 partitions
= 60 x 84 = 5,040 cps - "B"
Nozzle passing frequency "Fn" with 80 partitions
= 60 x 80 = 4,800 cps - "C"

The dynamic amplification factor, (applying Eq. 5.10.5) is:

For 96 partitions:(Fn/Fr)= 1.0746, therefore Ad = 4.88 - "A"
For 84 partitions:(Fn/Fr)= 0.9403, therefore Ad = 5.64 - "B"
For 80 partitions:(Fn/Fr)= 0.8955, therefore Ad = 2.15 - "C"

For each possible stationary vane application the direct stress is 22,350 + (Ad x σ")/2

For 96 partitions: 13,350 + 4.88 x 4640 = 24,672 psi - "A"
For 84 partitions: 13,350 + 5.64 x 4640 = 26,435 psi - "B"
For 80 partitions: 13,350 + 2.15 x 4640 = 18,338 psi - "C"

For each possible stationary vane application the alternating stress is Ad.σ"

For 96 partitions: = 4.88 x 4,640 = 22,643 psi Position - "A"
For 84 partitions: = 5.64 x 4,640 = 26,170 psi Position - "B"
For 80 partitions: = 2.15 x 4,640 = 9,976 psi Position - "C"

It can be seen from the Goodman Diagram, Figure 5.10.5 plotted for the three possible nozzle vane arrangements. The amplification factor "Ad" will dominate the choice for this stage. "B" with 84 stationary vanes is close to the Goodman limiting line, while "A" with 96 vanes, is just a little safer. Again, this represents a marginal design with a poor safety factor. However, "C," with 80 vanes, is the most promising with the coarser pitched nozzle partitions or stationary blade elements. This alternative "C" is therefore a safe design, and represents an acceptable choice.

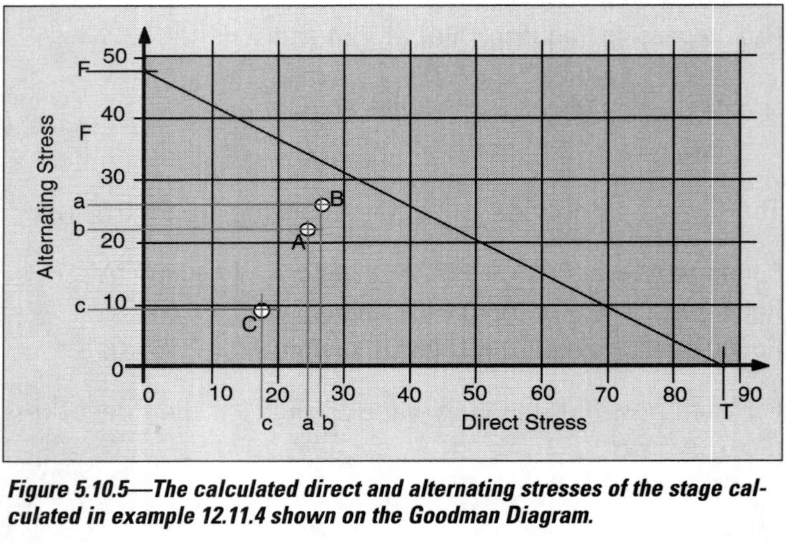

Figure 5.10.5—The calculated direct and alternating stresses of the stage calculated in example 12.11.4 shown on the Goodman Diagram.

The control stage on a unit can be subjected to high dynamic loading as a consequence of partial admission. These dynamic loads result from the rotating blades moving into and out of arcs of admission and then into dead bands where there are no steam forces. Where the nozzle passing effect represented a condition of a "ripple" of steam force at admission to the rotating blade row, the effect of partial admission is total removal and reapplication of the steam force. This effect places heavy loading on the control stage, and must be designed to withstand this condition, which is capable of causing forced vibration irrespective of the natural frequency of the blade.

Partial admission effect

A simplistic diagram of the steam force, or loading diagram for the single revolution of a blade, is shown in Figure 5.10.6(a). This figure shows the transient loading applied to the blade and the resulting deflection due to the steam jet effect.

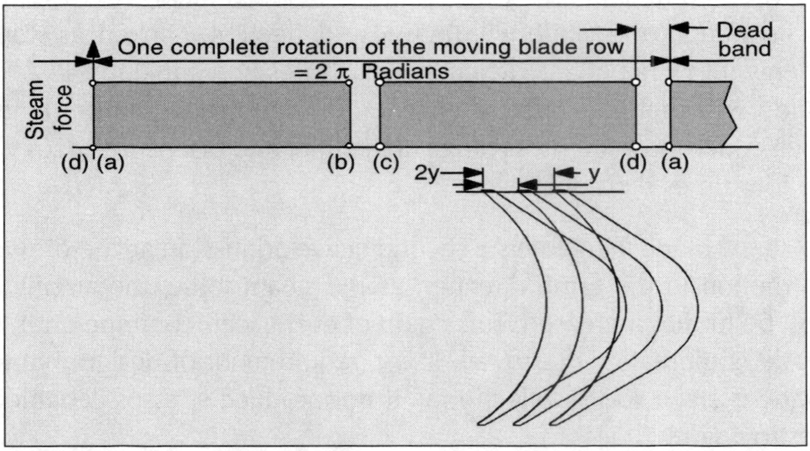

Figure 5.10.6(a)—A simplistic diagram of the steam force diagram for one revolution "2π" of the moving blade, showing "active arcs" and "dead bands".

In this figure, steam acts upon the rotating blade as it enters an active arc at "a." Here the steam jet deflects the vane, introducing a stress in the vane proportional to the deflection. At the instant the force is applied (position "a") the kinetic energy of the steam jet causes the blade to deflect, thus storing strain energy. This deflection of the blade continues until the stored strain energy equals the kinetic energy of the steam jet. At this condition the blade will have deflected by an amount "y." This kinetic energy will then dissipate itself in the blade by deflecting it by a further amount "y." At completion of this energy conversion, the total deflection of the blade will be "+2y," at which condition the stress due to this deflection "2y" is equal to "σm." The blade bending stress will be twice that at the equilibrium position. (This explanation assumes there is no internal molecular damping from the blade material.)

There will also be a small magnitude stress "+/-σa" due to the nozzle impulse effect.

If the blade being considered then rotates and moves out of the active or admission arc at "b," with the blade at the lower extremity of its motion, the blade will then lose deflection, convert its stored energy into kinetic, and deflect back to a position equal to "-2y." The blade will therefore vibrate at twice its normal amplitude. Under such conditions, the stress range at the inlet and discharge edges will be four times the static value.

If the blade then enters a second active admission arc "c-d," with its motion in the same direction as the steam force, the amplitude will be further increased. This chain of events can continue until the blade is ultimately destroyed. It is a requirement of design that the blade is sized so its deflection will not produce stresses capable of destroying it.

Actual stress levels can now be determined with considerable degree of accuracy by finite element methods.

The actual steam force diagram

The theoretical steam force diagram (Fig. 5.10.1) represents forces as applied in a simple analysis. In fact, there are factors that modify the form of this diagram.

First it is not possible for the force to be applied instantaneously, and a finite time "dti" is required at inlet for the blade to sense fully the axial effect of change in the steam pressure. Similarly, the force cannot be removed from the blades instantaneously, and requires a time "dto" to be removed completely. Therefore, the axial steam force at positions "a," "b," "c," and "d" [Fig. 5.10.6 (b)] will modify in form.

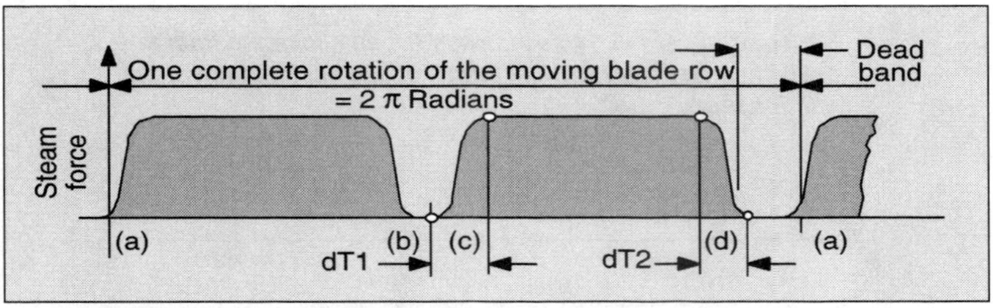

Figure 5.10.6(b)—A modification of the simple steam force diagram to reflect the effect of "entry into" and "exit from" the fixed blade row active arcs.

A second factor that modifies the force diagram at entry to an active arc is the effect of the steam flow over only one surface of a blade passage. This is termed the "dip" effect. Under the circumstances shown in Figure 5.10.7, the steam begins to enter the passage between the vanes. Before the blade senses the application of load, there is a venturi effect on the vane suction surface that will pull or deflect the vane towards the stationary blade row, tending to unload it. This effect is shown as a negative force "-dF1," which exists just prior to the application of full load to the blade.

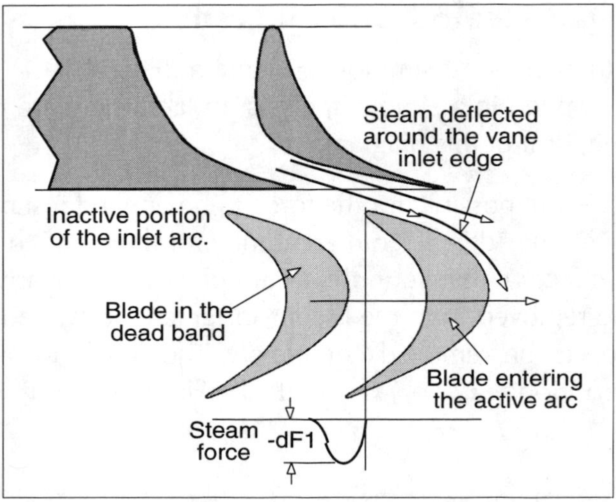

Figure 5.10.7—The effects of "entry to" an active arc from a dead band, and the resultant forces on the suction face of the rotating blade vane.

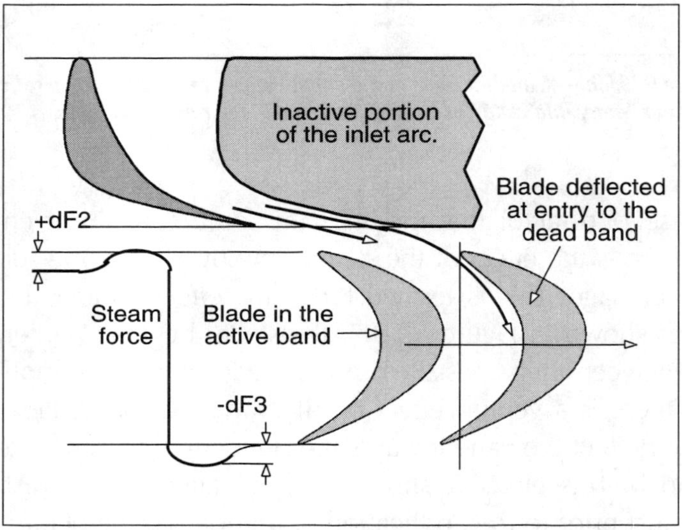

Figure 5.10.8—The effect of enter to an inactive arc from the active, and the effects of the steam load applied to the rotating blades.

Similarly, as the blade moves into the inactive arc (dead band), and the steam enters the blade passage, for an instant there is a force produced on the pressure surface of the vane at exit from an active arc (Fig. 5.10.8). There is no counterbalancing force on the suction surface because that vane is now in the inactive arc, and has no steam present. Therefore, there is a small unbalanced force "+dF2" acting on the blade. This is termed the "overshoot" effect. In addition, for the next short interval of time, there is a small negative pressure on the blade due to the flow through the passage. This effect is shown as "-dF3" in Figure 5.10.8.

Rotational stiffening effect

Large blades have natural frequencies that are lower than the short blades. However, they accumulate a considerable number of cycles, and will fail by, or be driven to, failure by some primary damage site within relatively short periods of time. Also these blades are "tuned"—*i.e.,* their "manufactured frequency" is adjusted to fall within a "band of acceptability" before installation.

For these large blades, their centrifugal force has no effect on their mass, but it has a considerable stiffening effect. Therefore, as speed of rotation increases, the natural frequencies of the blades increase. The Campbell equation for centrifugal stiffening is developed in the following manner.

Consider a particle of mass "m," with an elastic support so stiffness "Rs" is required to produce unit deflection. Its natural frequency "fo" is expressed by:

$$fo = \frac{1}{2\pi}\sqrt{\frac{Rs}{m}}$$

If this same particle is supported in another manner with an elastic stiffness "Rc," its new frequency "fc" will be given by:

$$fc = \frac{1}{2\pi}\sqrt{\frac{Rc}{m}}$$

If both these stiffening effects are applied simultaneously, the frequency will then be "fr," and can be found from:

$$fr = \frac{1}{2\pi}\sqrt{\frac{Rs + Rc}{m}}$$

Then if "Rs" is the stiffness furnished by elastic supports, and "Rc" is the stiffness due to centrifugal effects, and assuming the latter is proportional to "N," the speed of revolution is rps.

$$Rc = \beta\left[4\pi^2 mN^2\right]$$

Using the equation for "fo," and eliminating "Rs" gives:

$$f_r = \sqrt{f_o^2 + \beta N^2}$$

In these expressions "β" is an arbitrary constant, and the value of "β" usually lies between 2.0 and 3.0 for the lower modes.

It is clear from this expression that as the speed of rotation increases, the natural frequency of the blade will increase. Therefore, it is necessary to be able to represent this change of frequency in a manner that enables the effect on any blade to be predicted under any variation of speed. This is what the Campbell Diagram achieves.

The "β" factor. Most last stage blades have a tip/root diameter ratio that does not vary significantly from one manufacturer to another. Also, most last stages are designed for a substantial 50% reaction at their mean diameter, and a variation of reaction along the length of the blade to account for the radial flow or vortex effect of the steam. This means the blade inlet and discharge angles will not be significantly different from one manufacturer's element to another. There will also be minor section differences that reflect the research each manufacturer contributes to development of specific designs. Such differences are not expected to cause significant effects on the "β" value.

fo	<———— 127 ————>							
fo2	<———— 16129 ————>							
N2	<———— 3600 ————>							
β	1.8	2.0	2.2	2.4	2.6	2.8	3.0	3.2	3.4
βN2	6,840	7,200	7,920	8,640	9,360	10,080	10,800	11,520	12,240
fr	150.4	152.7	155.1	157.4	159.6	161.9	164.1	166.1	168.4

Table 5.10.1— Operating Frequencies as a Function of Fundamental Frequencies "fo," N = 60 cps

A sensitivity analysis is made for a blade for different values of "β" from 2.0 to 3.0 in Table 5.10.1. The variation of blade frequency at 60 Hz, for a blade, having a standstill frequency of 127 cps is shown in Figure 5.10.9.

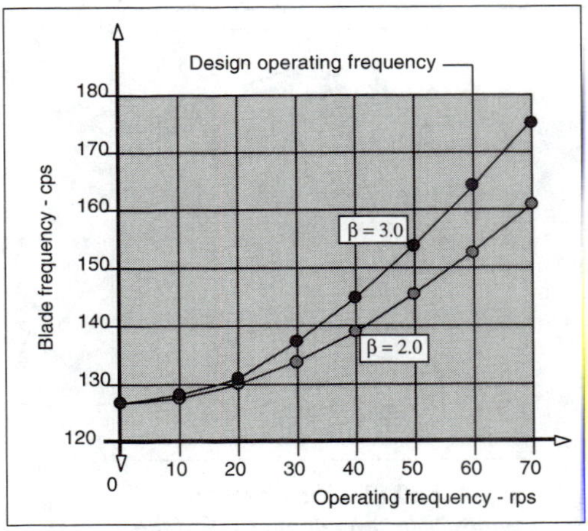

Figure 5.10.9—The rotational stiffening effect on a blade for "β" values from 2.0 to 3.0.

Even if the blade section geometry were to be so distorted in manufacture that its "β" value was off design by 10%, it can be seen from Figure 5.10.9 this would not alter the operating frequency too much. In fact, a 10% variation represents more than twice the value that could be expected. As an example consider a blade produced with a 5% error along its entire length, having an effective mean depth error of 5%.

The blade frequency can be found from:

$$fo = Ky \cdot \sqrt{\frac{Ie}{Ae}}$$

where:

fo	=	Blade frequency in cps
Ie	=	The effective section modulus along length
Ae	=	The effective section area along length
Ky	=	A constant

The effective width and depth are shown in Figure 5.10.10.

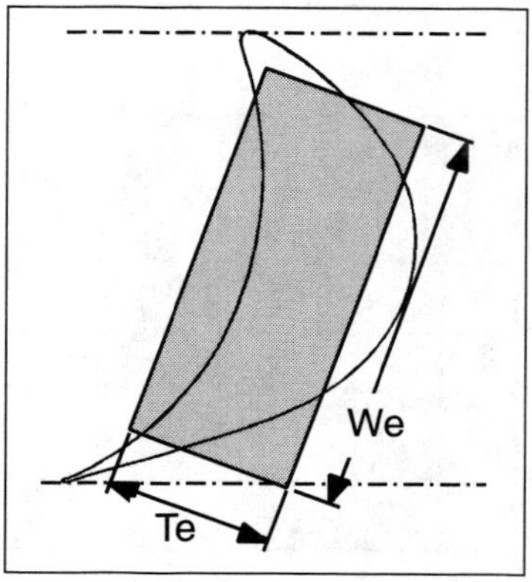

Figure 5.10.10—The effective width "We" and depth "Te" of the vane profile.

$$fo_{error} = Ky \cdot \sqrt{\frac{We^3 \cdot Te^3}{We \cdot Te}}$$

It can be seen that a 5% variation in the "β" factor will have a marginal effect on operating frequency "fr." However, it is of value to evaluate the effect of "β" variation on a typical last stage blade, and determine how at its extreme values, it could place a blade into a "risk" category. As an example, consider a blade with a predicted standstill frequency of 127 cps, for operation at 60 Hz, with a 2-pole generator. The divergence of the centrifugal stiffening curve can be shown on the Campbell Diagram, Figure 5.10.11. The values of "fr" are shown calculated in Table 5.10.2.

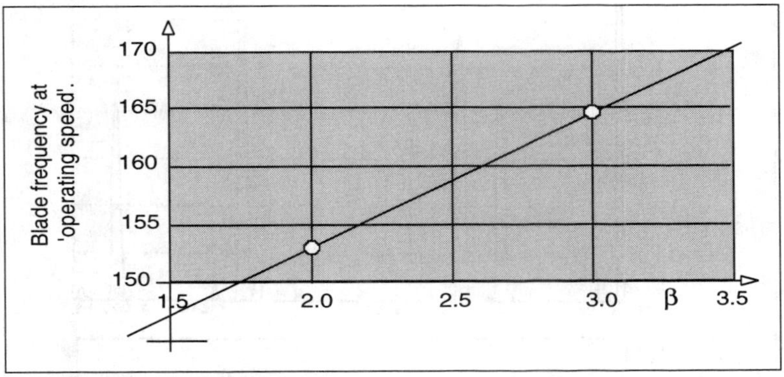

Figure 5.10.11—The variation of the value of the operating speed fundamental frequency for a blade with a standstill frequency of 127 cps, as a function of "β".

The vibration signature of two "identical" L-1 blades are shown in Figures 5.10.12(a) and (b). From these two figures it can be seen there is relatively close agreement at the lower values—the fundamental and first two harmonics—but a degree of divergence at the higher values. This is typical for normal longer blades, and the manufacturer will under certain circumstances adjust these elements, if there is a need for such tuning.

N	5	10	15	20	30	40	50	60	70
N^2	25	100	225	400	900	1600	2500	3600	4900
β	<———				2.0				———>
βN^2	50	200	450	800	1800	3200	5000	7200	9800
fr	127.2	127.8	128.6	130.1	133.9	139.0	145.4	152.7	161.0
β	<———				2.0				———>
βN^2	75	300	675	1200	2700	4800	7500	10800	14700
fr	127.3	128.2	129.6	131.6	137.2	144.7	153.7	164.1	175.6

Table 5.10.2 —Blade Frequency as a Function of Rotor Speed "β" = 2.0 and 3.0

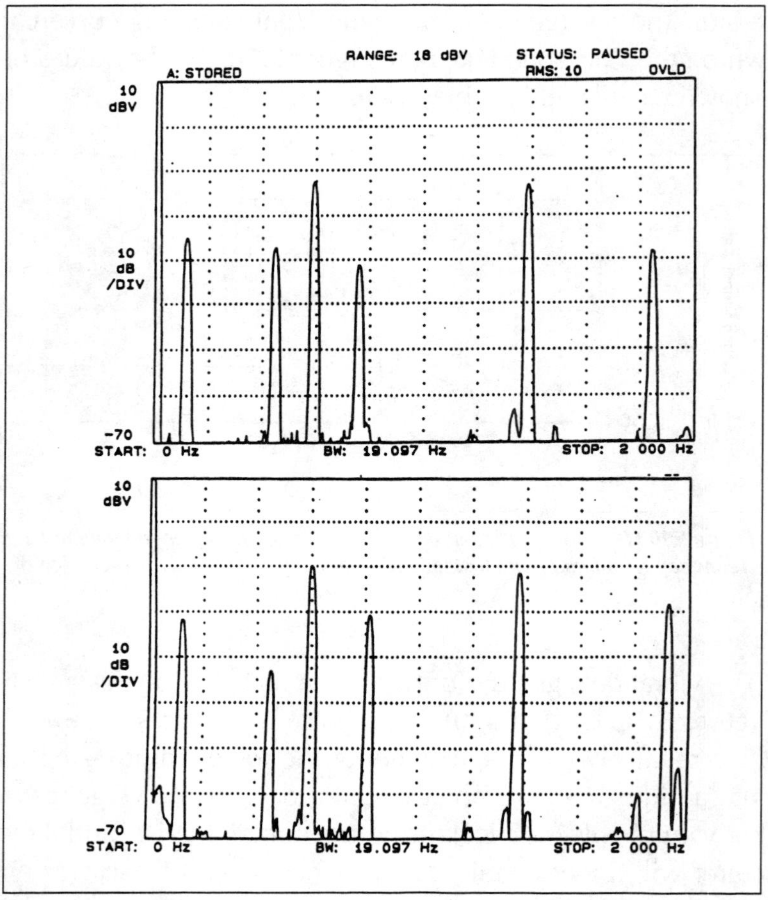

Figure 5.10.12—The vibration signatures of two nominally "identical" blades for an L-1 stage. The fundamental and lower harmonics are shown.

The tolerance band. In an earlier section, the manufacturers' dimensional tolerance band for a blade vane was discussed without any effort being made to quantify the value of such tolerances, or what effect dimensional conformance by a blade supplier would have at either extreme of the tolerance spectrum.

Blade frequency is a function of blade material properties and the physical dimensions of the blade itself. For modern units designed for high output, materials used by all manufacturers for their blades are similar. The chemical composition and physical properties are sufficiently close, that a family of blade materials exists whose "Young's Modulus E" is not significantly different from one material and supplier to another. In terms of blade tangential frequency, the most significant mechanical property is "Young's Modulus," and for torsional modes the "Shear Modulus."

Modern machining techniques allow blades to be manufactured within fine tolerance bands, achieving only marginal variation of section properties along the length of the blade. This will mean that only minor variation of "β" values are likely to occur from one element to another.

It is necessary to attempt to assemble the existing information on blade frequency, and assess realistically what frequency spread can be accepted. Table 5.10.3 lists typical frequency bands over the first 5/rev points required to give a "safe" frequency split at operating speed. The "Δf" value is typical of the assigned values in any design situation.

This table gives the extreme ranges for "safe frequency," based on the assumptions that $2.0 \leq \beta \leq 3.0$. Therefore, any blade produced with a standstill "frequency range fo" between those calculated in Table 5.10.3 (safe range) should be within a limit that, at operating speeds, operation will not be coincident with any revolution stimulus, or within its tolerance band as defined in the lower portion of the table.

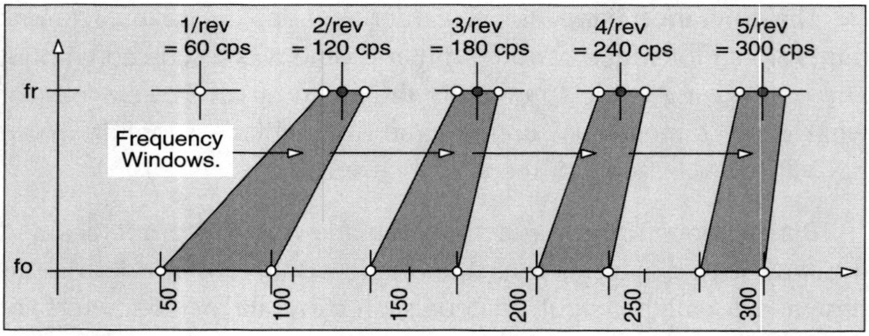

Figure 5.10.13—The safe ranges or "windows" of a blade at standstill frequency "fo", and at operating speed frequency "fr" for the lower 5/rev. stimuli.

Figure 5.10.13 shows the "safe ranges" of frequency as determined in Table 5.10.3, and the predicted maximum frequency split for different limits on "β," as shown in Table 5.10.4 determined from these values.

N / rev	1		2		3		4		5	
Frequency "f"	60		120		180		240		300	
Range +/-Δf	3		6		8		10		12	
f-min/max	57	63	114	126	172	188	230	250	288	312
fo^2	3249	3969	1296	15876	29584	35344	52900	62400	82944	97344
N rpm	<—				3600					—>
β	<—				2.0					—>
βN^2	<—				7200					—>
fo (β = 2)			76.1	93.1	149.6	167.8	213.8	235.2	275.2	300.2
β	<—				3.0					—>
βN^2	<—				10800					—>
fo (β = 3)			46.9	71.2	137.1	157.6	205.2	227.4	268.6	294.2

Table 5.10.3—Fundamental Frequency and Tolerance Band.

Frequency Split	1 - 2		2 - 3		3 - 4		4 - 5	
Frequency Range	63	114	126	149	188	230	250	288
Limiting "β" value	2.0	3.0	2.0	3.0	2.0	3.0	2.0	3.0
fo limit	0 - 46.9		93.1 - 137.1		167.8 - 205.2		235-2 - 268.6	

Table 5.10.4—Acceptable Values of Operating Frequencies

It is probable that many last stage blades from various manufacturers will not lie in this identified safe range, as constructed from the data in Table 5.5.3. However, these safe ranges are predicated on the extreme values of "β" from 2.0 to 3.0. An actual blade may well have a "β" value that is greater than 2.0, less than 3.0, and its value is unlikely to change by more than +/-5%. This will obviously increase the safe range for that blade. Therefore, when reviewing the standstill frequency of any blade, it is not possible to judge its performance potential from standstill frequencies unless the "β" value is known also. It is possible, however, to judge that if the standstill frequency falls into one of the safe ranges of Table 5.10.3 and Figure 5.10.13, the blade will operate safely at 60 Hz.

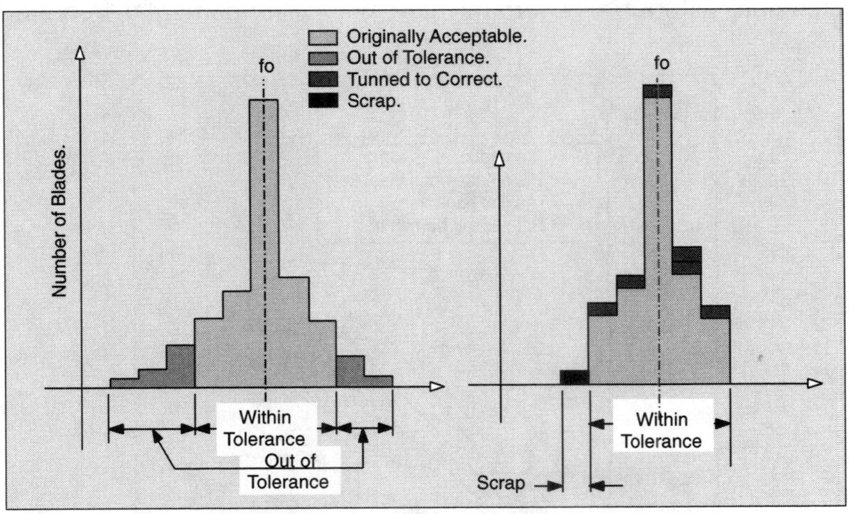

Figure 5.10.14—In (a) is shown "as manufactured" frequencies for a population of long blades, and in (b) the frequencies of this same group after individual tuning.

A manufacturer producing blades with frequencies when measured, will have values that produce a normal "Gaussian Distribution" curve [Fig. 5.10.14 (a)]. The base or total spread is a function of the tolerances the manufacturer is able to hold for these elements. Any blade outside the manufacturer's tolerance band has the potential to have a frequency beyond the safe range of that assigned by the designer. The out-of-tolerance blades will not necessarily be frequency rejected however, their rejection depends upon the manufacturers' tolerances. Such blades, if they violate frequency requirements, will either be reworked or rejected as scrap. If the distribution as manufactured is examined, and the "+f" and "-f" elements are reworked, the Gaussian curve will then be modified as shown in Figure 5.10.14(b).

If the fundamental frequency at standstill is between the 2 and 3/rev, the first harmonic frequency should also be checked in this manner. Torsional vibrations are only rarely of significance, and then only if the row has no coverband. However, free standing or blade elements without a coverband and with a fundamental frequency

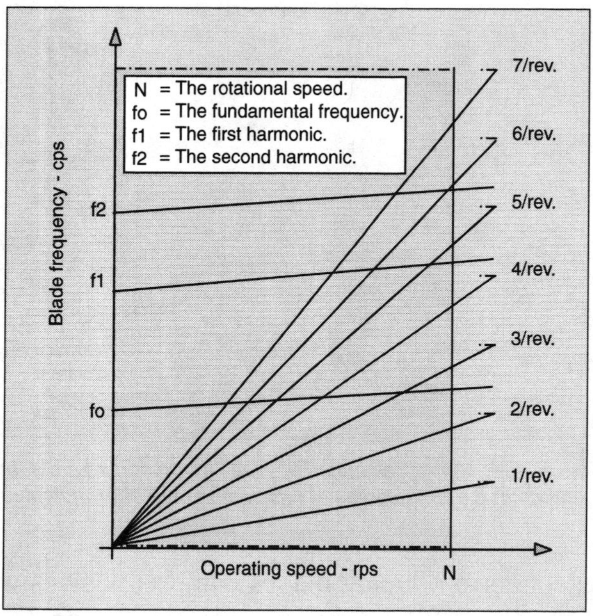

Figure 5.10.15—The "Campbell Diagram", showing the relationship between blade natural frequencies, and the rotational speed of the rotor. The Campbell Diagram indicates the interference between the blade frequency and possible exciting stimuli.

below 137 cps should be checked for the first torsional mode. Measuring the force required to produce unit angular deflection in the blade, and comparing this with a blade of known torsional frequency can do this.

The variation of blade natural frequency as a function of rotational speed is shown in Figure 5.10.15. This form of curve is called the "Campbell Diagram," when the possible per-rev (/rev) stimuli are also plotted. This curve is named after Wilfred Campbell, who undertook much of the early work on turbine vibration. He was first to plot a diagram showing the variation of frequency against rotational speed. For a blade to operate successfully, it must avoid coincidence with a per-rev stimulus, as shown in this diagram.

If the "Campbell Relationship" (see equation on page 559) is considered to hold true for all blades, irrespective of size, then it can be seen that centrifugal stiffening will cause a considerable percentage increase as the natural frequency becomes smaller.

$$fo = \frac{k_1^2}{2\pi h^2}\sqrt{\frac{EI}{m}}$$

For unit length m = A . ρ

$$fo = \frac{k_1^2}{2\pi h^2}\sqrt{\frac{EI}{\rho A}}$$

For steel ρ = 0.283 lb/cu. in., and E = 30,000,000 psi

$$fo = \frac{1.875^2}{2\pi h^2}\sqrt{\frac{32.2 \times 12 \times 30 \times 10E6I}{0.283A}}$$

Therefore, h (short blade)

$$= 500 = \frac{113250}{h}\left[\frac{I}{A}\right]^4$$

For typical short blade "h"

$$"h" = \frac{15 \times (l)}{(A)}$$

For typical ratios of (l /A) short blades can be defined as those below about 5.0″ vane length

The percentage change in natural frequency of a blade with its natural or fundamental frequency up to 1,200 cps is shown in Figure 5.10.16. This figure is drawn for a "β" value of 3.0. It is obvious the lower the natural frequency, the greater the percentage effect on the natural frequency caused by rotation.

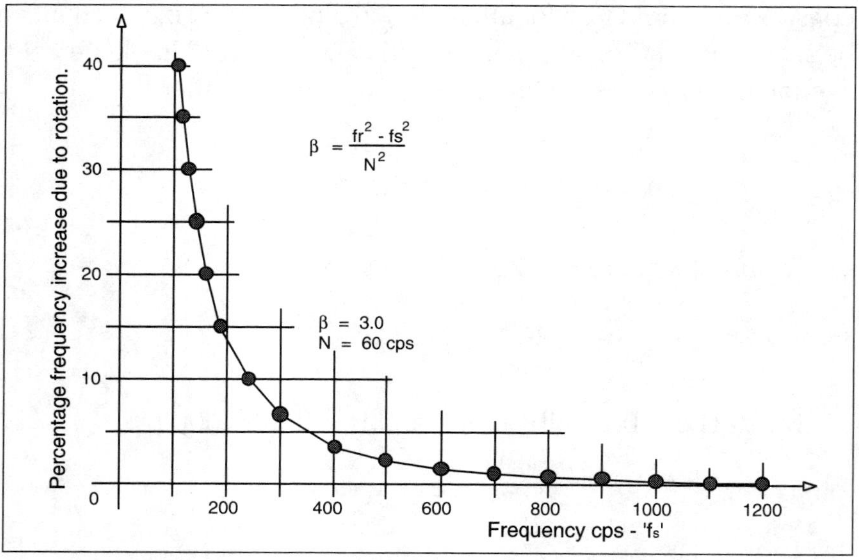

Figure 5.10.16—The percentage change in blade natural frequency as a function of blade natural frequency.

The Campbell Diagram

The Campbell Diagram is a pictorial representation of the variation of blade fundamental and harmonic frequencies as a function of the speed of rotation. The equation on page 542 was developed to show the stiffening effect of rotation on blade frequency. This equation often called the "Campbell Equation" can be used to determine the value of the blade frequency at speeds above the standstill value, and demonstrate the potential for interference with an exciting stimulus at operating speeds. This diagram is therefore used to determine the adequacy of a blade in avoiding operation at various conditions, where it could be operating at risk.

Figure 5.10.17(a) shows the increase in blade frequency as a function of the rotational speed from "0" to "Nr" rps. If this frequency increase is added to the standstill frequency "fo," then a typical Campbell curve for the fundamental frequency can be seen in Figure 5.10.17(b) for a long blade with a 2-3/rev split on its fundamental tangential mode. A similar analysis for the harmonics will demonstrate increases in the harmonic frequencies. The first harmonic (2nd frequency) is normally the first tangential mode, and the second harmonic (3rd mode) is normally the first torsional. It is assumed on this diagram that the third harmonic (probably the 2nd tangential mode) occurs above the 7/rev and is not shown on this diagram. In the case of this blade, it has a fundamental "standstill frequency" of "fo0." This occurs at an operating speed of "No." As the rotor speed increases, the blade fundamental frequency increases, until at operating speed "Nr," the blade has a fundamental frequency of "fr0." The first harmonic frequency is shown as "fo1" at standstill "No." This increases to a frequency of "fr1" at the operating speed of "Nr." Similarly, the 2nd harmonic increases from "fo2" to "fr2" at operating speed. In Figure 5.10.17(b), the frequency at operating speed is between the 2 and 3/rev for the fundamental, between the 4 and 5/rev for the first harmonic, and the 6 and 7/rev for the second harmonic.

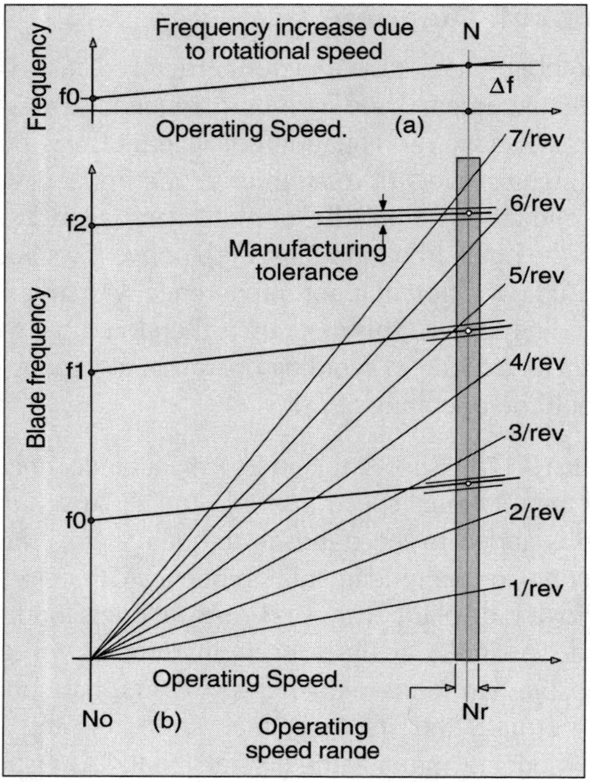

Figure 5.10.17—The Campbell Diagram for a blade with a 2-3 split on it fundamental, a 4-5 split on the first harmonic, and a 6-7 split on the second harmonic.

The blade manufacturer normally determines blade natural frequencies at a "standstill" condition within test rigs (*i.e.*, when the speed of rotation is "zero"). This information is used to predict the frequency at operating speed. Also, there are computer methods allowing the standstill frequency to be determined by finite element methods, and then using these same techniques to predict the frequency increase with rotational speed.

Manufacturer's frequency tolerances. Different manufacturers employ somewhat different philosophies to arrive at the frequency tolerance applied to their blade elements. Some simply state a frequency band or tolerance of a stipulated number of +/- cycles per second; others will employ a +/- percentage for variation from the stimuli position. Fortunately, both have similar results, despite marginal differences in their magnitude depending upon the tolerances specified. In either case, these tolerances can be represented on the Campbell Diagram.

The tolerance band specification selected by the manufacturer generally depends upon the first tangential mode predicted by calculation. On blades expected to operate below the 2/rev locus, the manufacturer normally elects to use a +/- cycle tolerance. Those elements with a fundamental frequency above the 2/rev locus tend towards a percentage distribution from the per rev stimuli.

An examination of a portion of the Campbell Diagram (Fig. 5.10.18) shows between the 2 and 3/rev locus lines on which is superimposed the influences of both the lower frequency tolerance from the 3/rev locus, and the upper frequency tolerance of the 2/rev locus. It can be seen that the blade has a small vertical range in which it can be manufactured. The blade can then be placed into operation without the fear of being excited by either 2 or 3/rev stimuli. This diagram also shows that the operating speed tolerance normally imposed by the manufacturer is safe for the unit. This permitted speed variation "+/-ΔN" defines a window "abcd" into which the blade operating frequency must fall to be "safe" during operation.

Depending upon the value of "fr," the manufacturer will be concerned that the blade has the ability to move into the frequency tolerance "a" of the upper /rev locus as the speed reduces, or into the frequency tolerance "c" of the lower /rev if the speed increases significantly beyond the rated operating frequency. The window "abcd" represents the safe zone for this blade during operation, and the manufacturer will produce and tune his elements to ensure they operate within it.

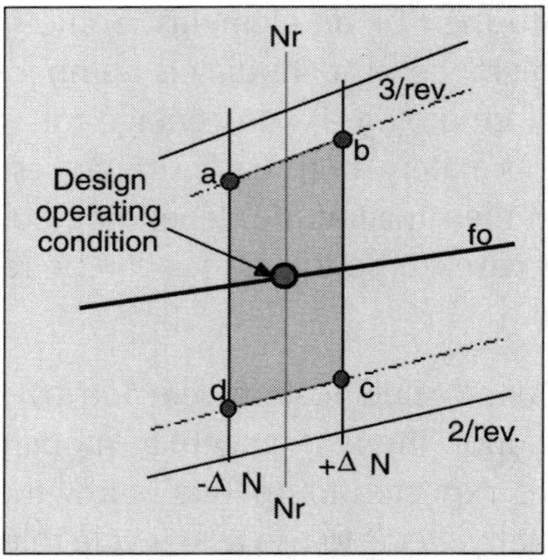

Figure 5.10.18—The "frequency window" between the 2-3/rev. lines on a portion of the Campbell Diagram. The possible range of operating speeds is shown as +/-ΔN.

There is a pinch point, as shown in detail "c," where the lower window limit "d" approaches the 2/rev line, which can have a tolerance placed upon it. Some manufacturers specify their operating limits as a percentage of the /rev lines as shown in this detail.

Frequency distribution and tuning. A turbine blade profile is produced within an envelope of tolerances. The designer specifies this envelope can be achieved with the existing manufacturing facilities. Within this envelope the blade will have aerodynamic and mechanical characteristics and properties that will ensure the blade operates at a level of efficiency and mechanical integrity consistent with design expectations.

The natural frequencies of the long last-stage blades are dependent upon section geometries and the rate at which these change along the profile from root to tip. Therefore, because of element-to-element variation, it cannot be expected that each blade element will have the same natural frequency. The manufacturing process produces a series of blades with marginally different profiles and marginal differences in section area, section modulus, and mass distribution along their length. This produces a distribution of measured frequencies for any row of blade elements.

A typical distribution of the fundamental frequency of a row of long blades is shown in Figure 5.10.14. Certain elements have, because of inconsistencies in the basic manufacturing process, a frequency that places them outside an acceptable range, *i.e.*, at operating speed they will not have a frequency falling within the "safe window" of Figure 5.10.18. If these elements can be corrected without violating other design requirements, they can be corrected and used. However, it is possible that such elements would have been rejected because of other forms of noncompliance with design requirements.

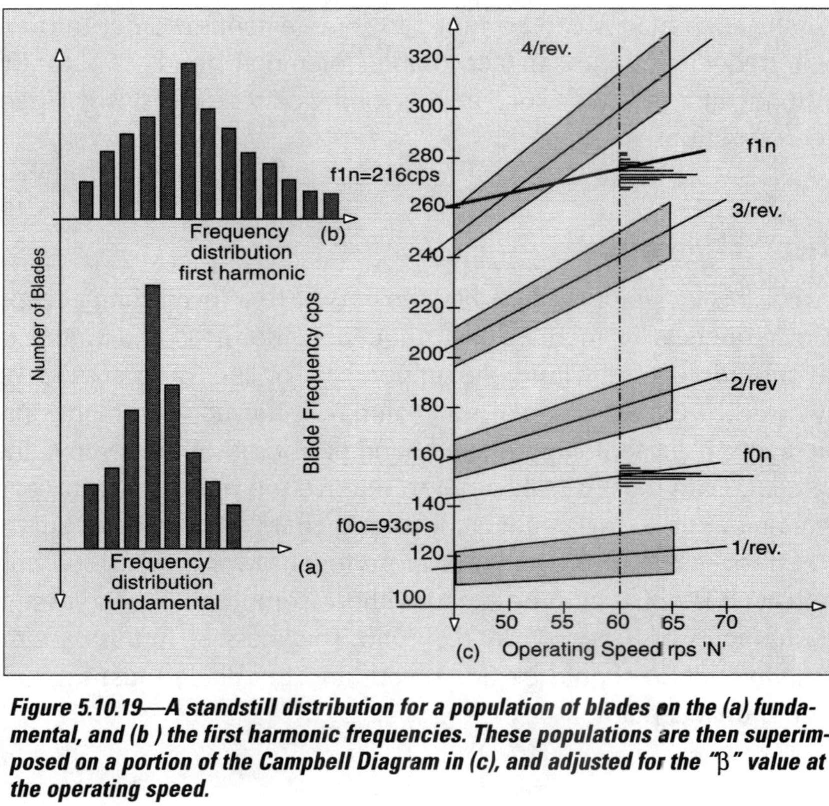

Figure 5.10.19—A standstill distribution for a population of blades on the (a) fundamental, and (b) the first harmonic frequencies. These populations are then superimposed on a portion of the Campbell Diagram in (c), and adjusted for the "β" value at the operating speed.

Figure 5.10.19 shows a distribution diagram of the fundamental and first harmonic frequencies of a long blade, and where these frequencies occur on the Campbell Diagram at operating speed. These frequencies have been corrected for the effect of centrifugal stiffening. Normally the manufacturer will measure the standstill frequency of all large blades to ensure their acceptability. Figure 5.10.20 shows a frequency test being conducted on a low-pressure stage blade element that has been repaired by the attachment of a new inlet nose erosion shield.

In any sample of long blades whose frequency has been measured, there are likely to exist elements of blades whose frequency places them in an unacceptable frequency band—the "-f" group and the "+f" group. These blade elements are expensive to produce, and the manufacturer will attempt to salvage them if at all possible. Usually such blades can be tuned to some extent, in order to bring their frequencies into an acceptable tolerance band. This tuning must not alter the blade such that its form violates other design-specified conditions.

The "-f" group

For this group of blades, the elements have a lower than acceptable frequency. To increase this frequency, the measured values of the profile section within the upper half of the vane should be reviewed. If excess material, or material at the upper dimensional tolerances is present, lapping and hand polishing should remove the material. Care must be taken not to remove too much material, and to maintain the correct radius forms and change of radius of curvature. If the blade is already at or close to its lower dimensional tolerances, it is unlikely to be tuned without compromising the design requirements of "Chord C" and "Profile Thickness T." In this case no rectifying action should be attempted, and the blade must be considered unacceptable.

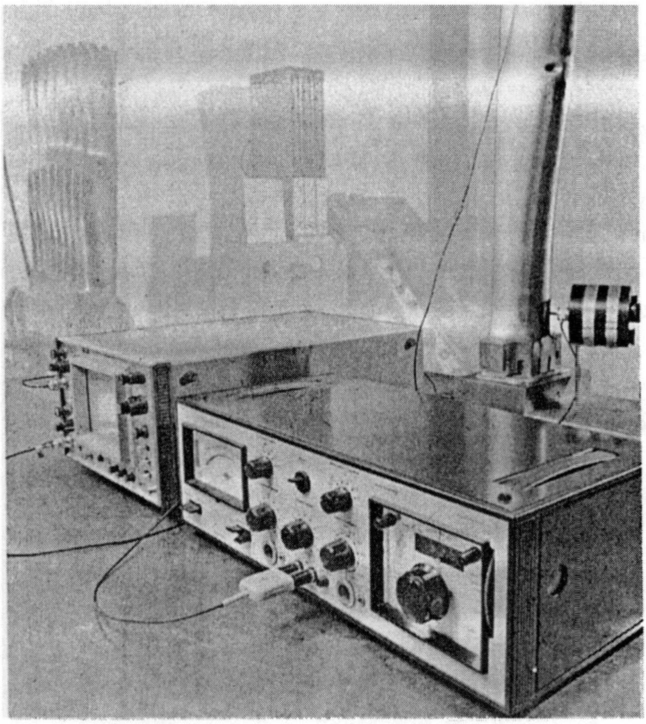

Figure 5.10.20—A frequency test on a group of three blades. These blades are connected by an inner and outer tie wire. To help ensure accurate values, the blades are weld attached to a firm steel base.

The "+f" group

These blades have a natural frequency above the upper frequency limit. In this case, it is possible to lower the frequency by hand polishing or lapping the lower third or quarter of the vane. Care must be taken in the following areas:

- Do not remove any material from the discharge edge, unless excess material is beyond twice the design tolerance
- Do not remove any material from the root platform-to-vane blending radius (Fig. 5.10.21) on either the pressure or suction

face, unless there is excessive amounts present, sufficient to distort the vane form

- Material on the profile thickness "T" (Fig. 5.10.22) should be removed at a ratio of about two to one from the pressure face provided this could be done without violating the profile envelope of manufacturing tolerances. If the root vane section and lower profiles do not have extra material on them close to the upper tolerance band, it may be that such blades cannot be corrected. It would be unacceptable to thin it too much near the root because this area is the most sensitive to blade profile direct stresses, both centrifugal and bending
- No material should be removed in the area of the lacing wire holes

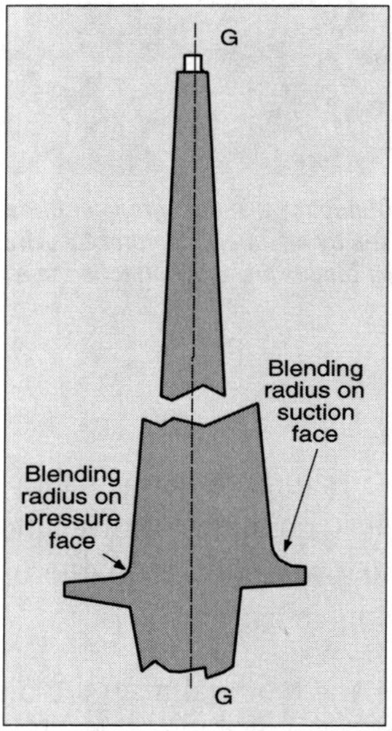

Figure 5.10.21—The root blending radius, and the positions where material should not be removed without careful evaluation.

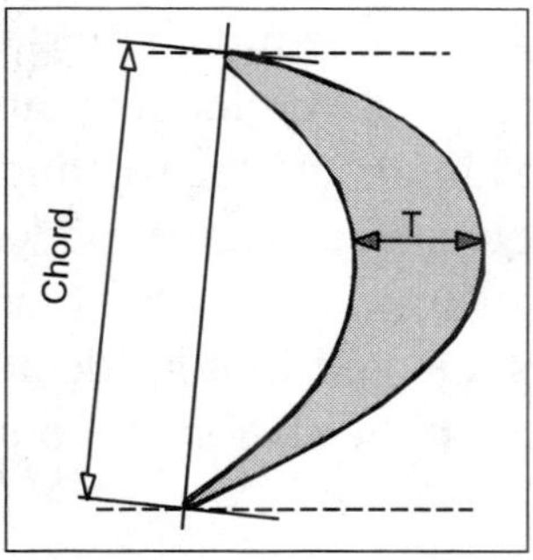

Figure 5.10.22—The basic characteristics of the profile "C" and "T".

MATERIAL PROPERTIES

Factors that define the ability of materials to resist high-cycle fatigue loading are determined in precisely the same manner as are all other mechanical properties of material—*i.e.*, by mechanical testing. Due to the importance of material fatigue resistance in modern design practice, a considerable amount of data on those materials most commonly used has accumulated. This data accumulation applies to turbine component materials, and is considered proprietary by the turbine manufacturers who have developed specific materials suited to their design practice and application. Much of this information is available in the public domain concerning mechanically and chemically similar materials—*i.e.*, It is readily available to the turbine user when considering the nature of a failure, the suitability or substitution of a material, or when making some other evaluation or investigating.

Accompanying creep in a material there is a general softening. However the changes are normally insufficient for this to be an indicator that creep is occurring in any particular component or region.

Those components of the steam path most likely to be subjected to high levels of impulse or fatigue loading, and therefore susceptible to high-cycle fatigue failure, are the rotating components and the various components such as the rotating blades and the stage hardware that is attached to them. These are the most susceptible to damage, because they are subjected to considerably higher direct and alternating stress levels than the stationary components.

Form of material data availability

Basic data required by the design engineer includes relationships that exist between the alternating stress levels to which the component is exposed, and the number of cycles required to induce failure. This relationship has been developed for the most common materials used for steam turbine components that are subjected to high frequency cyclic loading. However, when examining and applying this data to design analysis and failure investigations, it is essential to recognize there are factors that modify or influence these values.

During the design phase, the designer establishes the basic parameters required of the steam path components. The design process then examines the forms various components can take to meet these. After a preliminary selection from among the possible choices, the designer determines the stress levels that will be induced, and selects and dimensions them so they will operate with an adequate factor of safety.

A relationship exists between the parameters of limiting alternating stress and the number of cycles required to induce failure in each candidate material for each component. A typical relationship is shown as Figure 5.11.1 for an alloy steel material. This curve shows the number of cycles to failure as a function of the applied stress level. From this curve it can be seen that as stress levels are reduced, a condition is reached (and below where) failure will not occur. This stress level is known as the "fatigue or endurance limit." This endurance level stress will be dependent upon the temperature at which the component is operating.

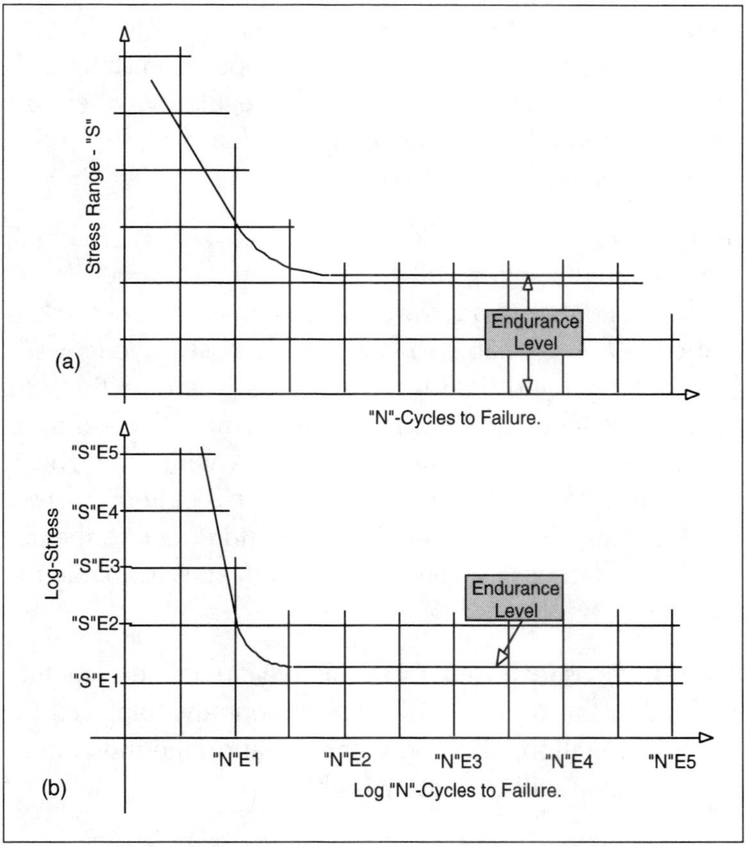

Figure 5.11.1—The stress "S" to number of cycles "N" to failure. Curve (a) represents the experimental data for a typical turbine steel and (b) represents the more common form of this data Log "s" -Log "N".

The "S-N" curve (endurance limit)

The simple "S-N" (stress-cycles to failure) curve of Figure 5.11.1(a) depicts the basic relationship between the mean stress required to induce failure, and the number of cycles required for failure to occur. However, a more convenient manner of plotting these values is to plot "log σa" to a base of "log N." This curve also defines the endurance limit.

While fatigue type failures can initiate from a smooth surface in the steam turbine because of its mode of operation, fatigue failure normally initiates at some small surface irregularity, where a small pit or other abnormal stress-concentrating defect exists, and where alternating stress becomes a maximum.

The initiating point will develop as a surface micro-crack that with repeated load cycling will grow in depth as a function of time. This crack will continue to grow until the remaining un-cracked portion of the surface is unable to carry the imposed cyclic load (plus any direct loading) and final rupture occurs. The direction of crack growth is influenced by the stress pattern, and there may be an apparent change in the direction of the crack with time. This directional change is caused by the change in stress pattern as the crack modifies the stress, in terms of magnitude and direction the component is subjected to, as less load-bearing surface is available to carry the total load.

When surfaces exposed by a high-cycle fatigue fracture are placed together the surfaces will join without any gap, *i.e.*, there is no plastic deformation, and upon magnification the individual crystals of the material will be clearly visible.

Factors affecting fatigue strength

The fatigue strength is defined by experimental data and shown on the "S-N" curves for any material. It represents the material strength under ideal conditions, or at least laboratory conditions where the environment is controlled. In actual practice stress patterns are complex, and the environment is often difficult to control or maintain at constant and acceptable conditions. There are a number of factors that can influence or modify the actual ability of a material to resist fatigue type damage. The majority of these influences reduce a materials tolerance to stress. The following issues address the principal modifying influences that decrease the fatigue resistance.

Surface marks. Test data is normally determined from the use of test pieces prepared with smooth surfaces. These surfaces are free of damage so test results are not contaminated by the presence of modifying factors. In the steam path components it is almost inevitable that surface irregularities will exist from the action of corrosion, solid particle impact, or erosion. These craters, indentations, and other irregularities will act to concentrate the local stress and can therefore act as an initiating site.

Stress concentration. The geometries of the steam path's various components are complex. They contain cross-section changes and cross-section modulus changes sufficient to introduce a number of locations where there will be stress concentration. This concentration occurs as a consequence of the geometry. While the design engineer attempts to minimize the potential damaging effects, it is not possible to eliminate them entirely. The design process, however, will reduce their damaging effects to the greatest extent possible.

Temperature. The mechanical properties of materials are temperature sensitive. Therefore, as the temperature of a component changes so will its ability to resist the effects of cyclic loading. There are two aspects of temperature changes influencing the damage caused by cyclic loading:

- The change either to increase or decrease temperature will modify the mechanical strength of the material and decrease or increase its ability to resist the loads induced by vibration
- The natural frequencies of the various components is a function of the "Young's Modulus" of the material. Therefore, as temperatures change, so will natural frequencies. This tem-

perature change can raise or lower the fundamentals and harmonics of the various components, which can increase or decrease the margin between exciting forces and the natural frequencies

Heat treatment. A consequence of heat treatment is the heat-affected zone ("HAZ"). Many portions of the steam path are subject to heat treatment (by brazing or welding processes) as part of the initial manufacturing processes. Also, heat treatment can be undertaken as part of many refurbishing techniques applied to individual components.

The HAZ has a tendency to be brittle, and therefore more subject to the stresses induced by cyclic loading.

Decarburization. Decarburizing is a process of carbon being removed from the surface of a material by some form of localized heating. Such a situation can occur, for example, if hot weld bead is dropped onto a surface, or a naked flame touches a surface during a manufacturing or refurbishment process. A decarburized surface is brittle and will not withstand the cyclic loading as readily as unaffected material.

Damping. If a component is receiving energy impulses at or close to one of its natural frequencies, and no damping is present in the system, theoretically there would be a build-up of energy in that component until vibration amplitudes reached such proportions causing the component to fail. This failure occurs because the stresses induced would exceed the proportional limit of the blade material at that temperature. Fortunately, during operation, the vibration amplitude and the magnitude of the stresses induced in the various components are damped to a degree. Principal among these damping phenomena are the following:

- *Internal friction of the material*—Internal friction causes an inter-granular hysteresis loss—the molecular friction between the grains of the material. The size of the hysteresis loop, and loss, is independent of the frequency of vibration, but is a function of amplitude, material, and method of loading (tension, shear, or a combination of both)

Both the crystal structure and the grain boundary materials influence the extent of internal damping, as well as the amount of plastic deformation they can sustain.

- *Fluid action*—In this case, energy losses are due to viscous shear of a gas or liquid against the vibrating surface of the system (damping forces are proportional to velocity). For turbulent flow losses, damping is proportional to the square of velocity
- *Ferromagnetic action*—Some materials (*e.g.*, chrome alloy steels) have a randomly orientated magnetic field within their metallographic structure. Any traversing of these fields by adjacent particles due to motion of the component during vibration causes internal energy dissipation. It manifests itself as heat in the blade material
- *Dry friction*—If any relative motion occurs between the various parts of the stage hardware—blade root to wheel, vane to tie wire or tenon to coverband—frictional losses occur. These losses are a function of the normal force causing the vibratory motion, and the coefficient of friction between the two rubbing surfaces

Residual tensile, bending, or shear stress. It is possible that during the manufacturing phase of the various steam path components, residual stresses could be inadvertently induced into them (or as a consequence of a shrink or interference fits, or by some other manufacturing process). The ultimate consequence of these processes could be that a stress is induced into a component, which will be increased in magnitude as the unit goes into service. These resultant stresses—operational plus residual—could then become sufficient to initiate failure (or drive to failure) of a component that under conditions of correct compliance with the design specification would not fail.

Other conditions exist that can reduce the severity of fatigue. Some of these can be induced as part of a conscious effort to increase the fatigue resistance of the components, while others, like those reducing life, would be as a consequence of the manufacturing process. Among these influences are the following.

Shot peening. A process used in many applications to increase the fatigue resistance of a component is to shot peen it in sensitive regions. This process introduces a compressive stress in the outer surfaces, thereby requiring an initial tensile stress to overcome the compressive effects before any damaging tensile effects are applied.

The process of "shot peening" places the surface of a component in compressive stress by the repeated impact from selected solid particles. This stress increases the ability of a material to resist cyclic loading. This process is applied to certain areas of the steam path components where the possibility of cyclic loading and high-cycle fatigue damage are present.

Note: The shot peened layer is very thin. Therefore, handling after this process is complete must be undertaken with considerable care or there is the possibility of minor damage, eliminating the positive effects of the peening.

Residual compressive stress. Just as residual tensile, bending, or shear stresses introduced by the manufacturing process can reduce the fatigue life of a component, so residual compressive stresses can increase it.

While such compressive stresses will reduce the tendency to fatigue damage in one location, it must be considered that these stresses, like other forms of residual effects, will be counterbalanced by tensile or opposite effects in other locations. Therefore, the complexities introduced by residual stress should always be considered as negative, and an unpredictable aspect of manufacturing. They are to be avoided to the greatest extent possible.

FATIGUE STRESSES AND THEIR REPRESENTATION

The design process requires methods that quickly and accurately provide guidance to the engineer concerning how components react when subjected to the various forms of stress to which they will be exposed in operation. There are two forms possible:

- *Direct stresses*—the sum of all those unidirectional stresses to which a component is exposed, *i.e.*, they do not change from one direction to another while the unit is operating at constant load, speed, and employing steam of constant conditions. To this direct sum must be added the influence of alternating stress that, in some instances, is in the same direction as the constant direct stresses, and therefore adds to its overall magnitude

- *Alternating stresses*—Those whose magnitude and direction change. Normally the design process keeps these at relatively low magnitude and, unless there is some form of amplification, will not impose undue loads on the components

Combined direct and alternating stresses

Modern computational methods using finite element analysis (FEA) allow the actual combined effects of direct and alternating stress to be determined with considerable accuracy. However, the process of completing these evaluations is complex, time consuming, and therefore expensive. Certainly in the early phases of design, these processes of individual FEA for each component would be impractical. The engineer requires methods to quickly review the combined effects of these stresses and obtain a fast and accurate evaluation of acceptability.

The diagrams of Gerber and Goodman. The most effective methods are those of the Gerber or Goodman Diagrams. These methods, and the equations of Gerber and Goodman for various groups of material, were discussed and demonstrated earlier and applied to rotating blades.

The effects of operational modification (on the Goodman Diagram)

Consider the Gerber or Goodman Diagram of Figure 5.12.1. Here the effects of combined stress are shown as point "E" with a fatigue limiting stress of "J" and a direct limiting stress "T" at the stage environmental temperatures. For this component, the design evaluation process has determined the design alternating stress to be "σa," the maximum design direct stress as "σf" (the sum of the alternating stress "σa"), and the direct stress "σd." The stresses induced in this component are well within the material limitations, and if the unit continues to operate as specified (and no damage occurs), this stage will not experience any form of failure. The factor of safety can be taken to be the ratio "OG/OE."

Consider potential effects of certain operational phenomena that could modify the stress levels.

Unanticipated stress concentration. Should any form of damage occur that causes stress concentration at some location on the surface or other position within the component, direct stress levels will increase to "σc" at the location of the produced discontinuity or damage. The consequence of this increase is to raise the direct stress from a value "σf" to "σf*" locally, where "σf* = σc + σa." This localized stress increase will increase the combined stress to a condition "H," which obviously reduces the factor of safety to "OG*/OH", Figure 5.12.2.

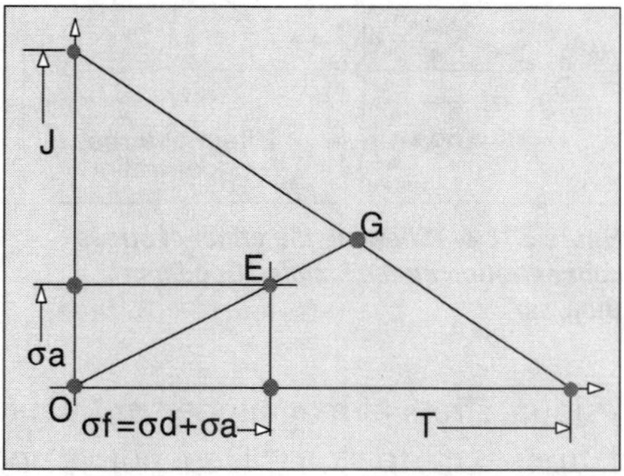

Figure 5.12.1—The "Goodman Diagram", showing the combined effect of direct and alternating stress.

Should this damage concentration occur where the direct stress is not a maximum—and localized stress does not increase beyond the value of "σf"—then there is no change in the overall maximum or limiting stress levels.

In many instances of surface damage, alternating stresses would also have been modified (normally increased), as damage would have modified and normally reduced the bending modulus of the component, causing some level of increase in the alternating stress.

The modified form of the combined stress shown in Figure 5.12.1 is shown as Figure 5.12.2. However, this diagram has assumed there has been no increase in the alternating stress levels.

Residual stresses. In the event a component is supplied with a residual stress of unknown magnitude, it will certainly modify the direct stress (and possibly the alternating stress), depending upon the location, form, and nature of the residual effect.

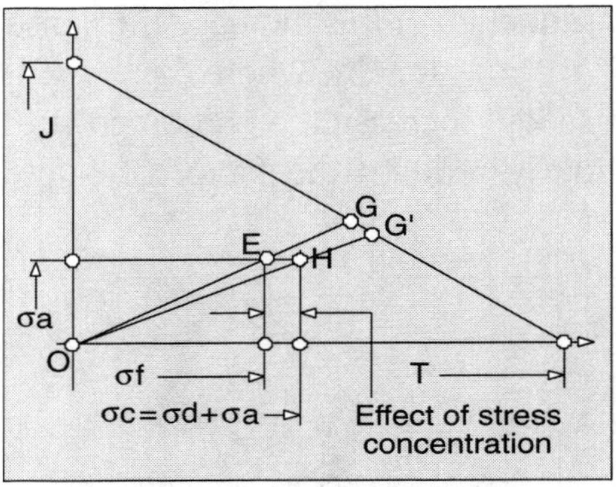

Figure 5.12.2—Showing the effect of stress concentration on the simple "Goodman Diagram".

Consider a residual stress of magnitude "σr," which has modified the direct effect from "σf" to "σfr." Depending upon the type of stress, this can represent an increase or decrease in the maximum direct stress by an amount +/-δσd. Similarly, the residual can modify the maximum level of alternating stress by an amount +/-δσa. The presence of a residual stress provides an envelope within which the final combined stress can fall.

Therefore, the range of stress—or possible envelope—can be large as shown in Figure 5.12.3, depending upon the magnitude of the residual effects. In this figure the range of stress and its variation from the calculated combined value of "E" is shown as the window "abcd. " The magnitude of the residual can have a significant effect on the margin of safety of the component. Unfortunately, in most instances, the residual will tend to increase stress levels, both direct and alternating. The exception to this is the case of shot peening, where a compressive surface stress is induced to minimize the effect of high surface tensile stresses. Shot peening will have little effect on the direct stresses because the depth of stress modification is small.

Environmental temperature change. Changes in the steam environment within the steam path will have an effect on the form of the

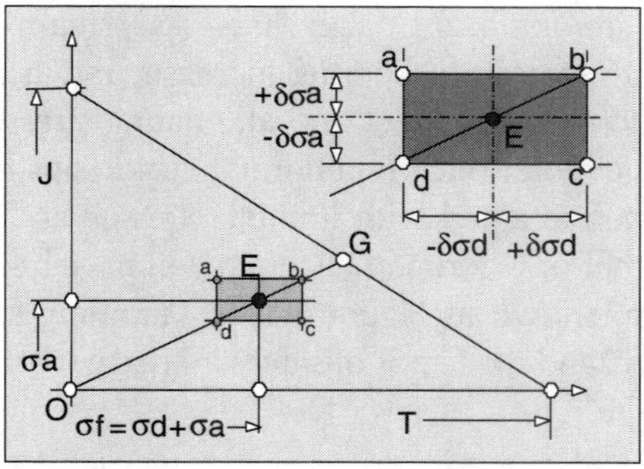

Figure 5.12.3—The effects of residual stress, either tensile or compressive, provide a "window" of total stress on the "Goodman Diagram".

Gerber or Goodman Diagram. Consider first the effect of changing the steam inlet pressure. This modification will make some small change in the pressure drop across the individual components, which will have been designed to operate at some constant pressure ratio. However, this change effect is small and can normally be considered negligible.

Note: In a situation of having to remove a row of blades for any reason, steam temperature will increase at all stages downstream of the removed row. This will affect metal temperature and reduce the material properties to the extent the factor of safety is reduced.

However, a change in the steam temperature will have a direct effect on the material properties, and as a consequence modify the factor of safety to the combined stress. Consider the effect of temperature changes as shown in Figure 5.12.4. If there has been no change in the actual stress levels associated with this temperature change, the Goodman Diagram will remain unchanged. However, changes in the operating temperature will modify the diagram. Consider:

- *Steam temperature decrease*—As steam temperature decreases material properties increase, and its ability to resist the effect of both direct and alternating stresses increase as well. Therefore, once equilibrium conditions of temperature have been attained with limiting stresses at "T-F" and "J-F," the Gerber or Goodman Diagram will have been modified to the form shown as Figure 5.12.4—Limiting stresses of "Jd" and "Td" and the factor of safety increased from "OG/OE" to "OG*/OE"

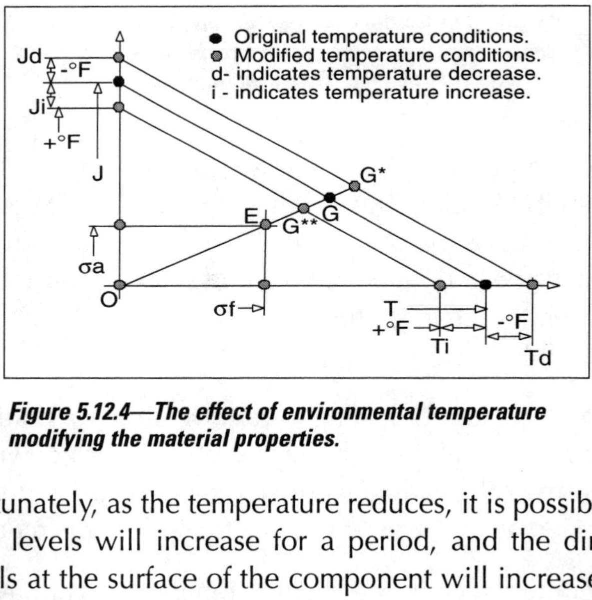

Figure 5.12.4—The effect of environmental temperature modifying the material properties.

Unfortunately, as the temperature reduces, it is possible the thermal stress levels will increase for a period, and the direct tensile stress levels at the surface of the component will increase to significant levels for a short period. This increase, while of short duration, could cause high stresses to occur, and their damaging effects are cumulative.

- *Steam temperature increase*—In the event of a temperature increase, the ability of a material to resist direct and alternating stresses is reduced. Therefore, the Gerber or Goodman Diagram shown in Figure 5.12.4 will be modified by a reduction of the limiting direct and alternating stresses to "Ji" and

"Ti." This transient effect is to induce a compressive stress in the outer fibers of the components. While this tends to be less destructive of component integrity as a temperature decrease, the damaging effects are present and once again cumulative. In this situation the factor of safety will have reduced from "OG/OE" to "OG**/OE"

Component frequency variation. If some form of mechanical damage or other condition exists within the steam path (*e.g.*, material temperature change, as previously discussed), it can modify the natural frequency of a component. This effect must be considered in terms of its possible influence on the combined stresses.

We have previously seen indications of amplification of the maximum amplitude of vibration—and therefore, stress—as a function of the ratio of the frequency of the implied force and the natural frequency of a rotating blade. This same amplification effect is true for any steam path component that is free to vibrate and is receiving load impulses. There is a need to consider how this frequency change can amplify the stress levels of any component, and how this will be reflected on the Goodman Diagram. This effect can be seen on Figure 5.12.5.

In the event the component frequency changes—moves closer to coincidence with the applied impulse forces—then the amplification factor "Ad" will intensify, increasing the magnitude of the alternating stress. The direct stress is the sum of the direct and alternating stresses ($\sigma f = \sigma d + Ad.\sigma a$) and the total alternating stress is "$Ad.\sigma a$." The effect of this stress increase, when "Ad" is large, is shown in Figure 5.11.5. Here the operating stress is represented by point "K," which gives a factor of safety well below 1.0. The component can be expected to fail.

Component and unit speed variation. One of the most significant high-frequency applied forces generated within the steam turbine are those due to the nozzle passing effect. These forces have the potential to influence the stress amplification factor "Ad" in any component experiencing these impulse effects. The nozzle passing frequency is a direct function of the number of stationary blades or discharge apertures in the stationary blade row and the rotor speed. Therefore, for rotating components, if the rotational speed varies, then the value of "Ad" will also vary, as does the magnitude of the applied alternating stress. In these circumstances the applied stress is modified, and if "Ad" has increased, the factor of safety has been reduced, as shown in Figure 5.12.5.

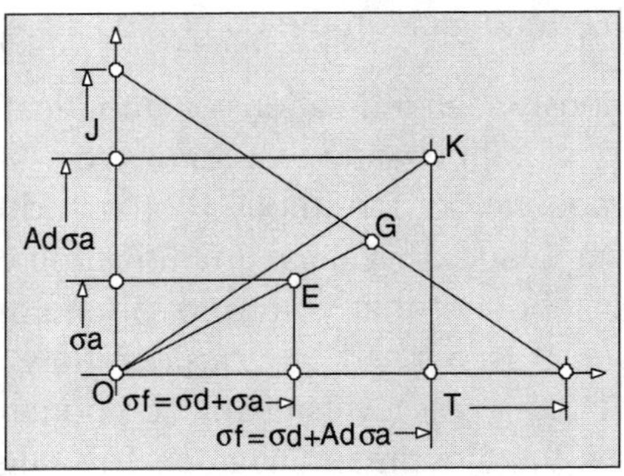

Figure 5.12.5—Variation in the natural frequency of the component can modify the stress amplification factor "Ad".

CRACK GROWTH

The rate at which a crack [initiated by high-cycle fatigue (HCF) or any other mechanism] will grow by HCF is dependent upon a number of factors:

- *Material temperature.* The temperature at which the material is operating will have an effect on the rate of growth, being faster at the higher temperatures

- *The level of component applied stress.* The higher the direct stress applied to the component the faster the crack will grow
- *The frequency of the applied load.* The growth of any crack is a consequence of stress reversal on the crystals of the material. Therefore, the higher the rate of stress reversal is applied, the faster the crack will grow
- *The magnitude of the applied force.* The higher the magnitude of the alternating load, the greater the stress imposed on the failure surface and the crack tip
- *Crack tip shape.* The form of the crack tip affects the crack intensity factor, which in turn affects the rate at which a crack will propagate through the material

The remaining life of any component found cracked can now be determined with reasonable accuracy by computational methods. These remaining life prediction methods can be used to establish the need to continue operating or replacing damaged components. These methods will also permit a unit to be returned to service with confidence that replacement parts can be made available supporting the next planned outage.

Crack form

Cracks can be either intergranular or transgranular. The form and progression of the crack is dependent upon the levels of stress and temperature at which the component is operating. For high stresses at high temperatures the cracks are most often transgranular; similarly, when the cracks form and grow over long time periods the cracks are transgranular. However, for medium range stresses (well below the UTS), the cracks are intergranular.

Cracks can often start at the low stress range as transgranular, but as stress intensity increases with reducing area, or increased stress concentration, they will become intergranular.

The prediction of crack growth

The rate at which high-cycle fatigue cracks grow is dependent upon several factors:

The frequency of the applied load. The frequency at which the alternating load is applied will affect the accumulation of damage. Also, it is possible that as the crack grows, the natural frequency of the component in which the crack is growing will modify with the crack size. This can also move the natural frequency into and out of coincidence with the applied stress.

The magnitude of the alternating stress. The magnitude of the applied stress will modify the load at the crack tip, causing an acceleration or retardation in the rate it progresses.

The magnitude of any direct stress. If the component is subject to a direct stress and an alternating stress—while this will not modify the alternating stress range—it will increase the maximum stress to which the component is subjected. The time to failure will therefore modify.

The material properties. Different materials have different levels of resistance to mechanical rupture due to applied high-cycle loads. Twelve percent chrome steel has a high resistance, and represents one of the reasons this material is most often chosen for application within the turbine steam path.

The material temperature. The environmental temperature at which the component is operating is fundamental to the mechanical strength and resistance of materials to mechanical stress. Therefore, the higher the temperature of the component the faster the crack will progress.

HCF FAILURE SURFACE APPEARANCE

When high-cycle fatigue is the predominant damage mechanism, a fracture surface will exhibit a crystalline nature. However, most often when a failed component has been exposed to normal operating transients, its surface will appear to originate at some initiation point, where some form of discontinuity in either form or material properties exists, and is sufficient to cause some level of stress concentration. At this point of stress concentration, a crack will initiate after being subjected to a cyclic load for a sufficient period.

In practice it is unlikely the cyclic load will continue without some form of interruption, *i.e.*, Some transient condition (*e.g.*, a load or temperature change) will occur, a shut-down and start-up cycle will be undertaken, or some other transient will occur that will raise stress levels to which the component is subjected. The effect of these stress up-ramps is to increase the level of stress at the crack tip to the extent there will be micro-tearing, causing some small level of tensile failure. This tensile failure at the crack tip will appear as the familiar "beach marks" characteristic of HCF failures from an operating component.

These "beach marks" (or conchoidal striations) are symptomatic of load and stress changes in an operating component. Figure 5.14.1 shows a portion of a blade root where several failure initiation sites exist. These initiation sites were located at "fretting fatigue" pits that caused highly localized stresses.

The "beach marks" on this rupture surface are clear. However, between these marks crystalline rings indicate the growth of the crack out from the initiation sites. Had the transients of load, temperature, and (possibly) speed that drove the HCF not occurred, it is probable the final failure would have been delayed, though it still would eventually have forced the unit from service.

Figure 5.14.1—A high cycle fatigue fracture surface on a blade root.

Note: There is no clear definition of when a failure is high-cycle fatigue and when it is low cycle. The differences will be discussed later. It generally has been considered that:

Low cycle (LCF) 1 < cycles < 10,000 High cycle (HCF)

However, this is considered a convenient definition not based in scientific fact.

In addition to the vibration of individual blades there is also vibration in the blade groups. This is particularly true if the groups are discrete—*i.e.,* batching is not continuous and connected through 360 degrees. Figure 5.14.2 shows a blade row where failure has tended to be at the end elements due to a natural frequency of the group that introduces high levels of alternating stress in the end elements.

Figure 5.14.2—A last stage rotating blade row where the end elements of banded groups have suffered failure.

CREEP FATIGUE

Creep fatigue is a combination of two separate mechanisms—A component subject to high temperatures, and stress (exposed to creep) is also subjected to thermal transients as a consequence of its normal operation.

Creep fatigue is most often found in components of thick sections in which the quantity of material affected is such that heat will be unable to soak into it to the extent excessive differential thermal expansion can be avoided. In casings exposed to temperature differentials across their thickness, these phenomena can be particularly damaging. Also, in rotors having large regions operating at high temperatures, it is possible for these phenomena to occur, particularly at fillet radii where stress concentration can be high, and at section changes in the wheel to rotor portion of impulse design units.

TEMPER EMBRITTLEMENT

Temper embrittlement is a phenomenon affecting the hotter stages in the high pressure and reheat sections of turbines. It occurs as a consequence of the migration of impurities such as sulfur, phosphorus, tin, and antimony to the grain boundaries of the material. The inclusion of molybdenum in the steel decreases the susceptibility to embrittlement.

These phenomena occur as the result of long-time exposure "soaking" at high temperatures. As a result, material toughness is reduced. The soaking temperature is normally considered to be in the range of 750° to 1,050°F for low-alloy steels used in the manufacture of steam turbines. This causes a reduction in the impact value of these materials if they are operated at the normal temperatures experienced in the high pressure regions of the high pressure and reheat sections of the units.

This situation can present a serious condition in rotors where embrittlement can lead to failures. Detecting this condition is best determined by undertaking material tests from samples removed from convenient locations in the rotors (*e.g.*, the bores) when they become available for examination.

While it is now possible to control steel production to limit the amount of impurities or trace elements that can induce this condition, there are large numbers of units in operation produced before the refinements in steel production became available. These units require careful monitoring as they age to prevent catastrophic failure.

LOW-CYCLE THERMAL FATIGUE

Low-cycle thermal fatigue is a form of damage that often introduces the need for expensive replacement or repair. These phenomena are most often associated with changes occurring in the temperature of the steam, which in turn modifies the temperature of the turbine components. However, any large repeated change in stress or strain can initiate or drive such a condition to failure irrespective of operating temperature changes.

Turbine components—stationary and rotating—experience uneven thermal expansion and contraction as a consequence of a number of operating phenomena, over which the operator at best may have limited control. These high-rate environmental steam condition changes can therefore cause high-magnitude internal strains to develop in the component materials. The components can either be surrounded by steam that is changing temperature (e.g., a blade, rotor, or diaphragm web) or have one surface exposed to a high-variable temperature and another exposed to a relatively constant, but lower one.

Damage that occurs as a consequence of high-strain changes is termed "low cycle" (LCF) because the number of cycles required to introduce material rupture is significantly lower than the number required to produce "high-cycle" failure (HCF). Therefore, the principal differences in cause and effect between high- and low-cycle fatigue include the following:

- For high-cycle fatigue to occur, a large number of cycles are required. Low-cycle fatigue requires significantly fewer fluctuations of the stress cycle
- The stress and strain level in low-cycle fatigue is high, whereas the high-cycle levels of stress can be well below the yield point of the material
- When low-cycle fatigue occurs there is material deformation at the fracture surface. High-cycle exhibits no detectable deformation

There is no clear delineation in the number of cycles required to define a transition from "high-" to "low-" cycle damage. A total of 10,000 cycles has been suggested as the limit, but a more realistic definition might be based on the appearance of the fracture surface and the presence (or absence) of mechanical deformation of the material.

Another major difference between incidents of high- and low-cycle fatigue is the majority of the time-to-failure spent in HCF is in the initiation of the crack, whereas with LCF, the majority of the time-to-failure is spent in the propagation. Such cracks will normally form and become visible early in the life of the cyclic action.

A consideration in this matter is that LCF cracks will be visible early in the life of the component, where an HCF crack may be well advanced in the incubation phase. Failure can occur early in component life after nondestructive examination (by conventional dye and magnetic means) has indicated the component is uncracked and is returned to service.

Because turbines are exposed to steam temperature variations, they can and often are subject to low-cycle damage. Such damage can occur in both stationary and rotating components. It is important to consider those factors that have a causal effect in producing excessive temperature variations occurring beyond boiler control, and are in fact a function of the manner in which the units are operated. Such temperature variation is clearly a function of the temperature of the steam delivered from the superheater or reheater, and the extent to which this temperature is controlled by the attemperation system.

As the load demand on a particular unit changes so can the temperature at which the steam is delivered, and the variation of temperature throughout the steam path. The manufacturers of steam turbines will often provide guidance as to the acceptable rate of temperature change, and often provide means of determining the rate at which component life is reduced as a function of temperature ramp rates.

Unfortunately, as a unit ages it is often slowly relegated in the order of generating preference. The factors that influence this include:

- The heat rate of the steam cycles into which the unit is built. They would be higher than the newer units in the system
- The internal efficiency of the turbine, and the extent this efficiency is maintained at outages. As unit load factors decrease, the need for maintenance expenditures becomes more difficult to justify. Therefore, turbine steam path efficiency will tend to deteriorate
- Known reliability of the unit and its ability to respond to changes in load demand
- The cost of the fuel used in the power cycle, and poorer heat rate. These make unit use more difficult to justify
- The location of the unit relative to the load demand center

Each of these considerations—and possibly others—will be factored into the utility dispatch system, defining the order in which units should be started and the priority assigned to them to pick up, drop, or change load to meet system demand changes.

THERMAL TRANSIENTS

A thermal transient occurs each time there is a change—increase or decrease—in the temperature of the steam surrounding the various components within the steam path. Depending upon the extent or degree of change, various levels of stress can be induced in the components due to internal changes required in the material at the new temperature, *i.e.*, components expand or contract to meet the temperature changes.

The increase or decrease in steam temperature will produce an adjustment of the temperature of the components. The rate at which heat can soak into or be rejected from them will influence their internal temperature distribution. For this reason, an important characteristic of turbine materials is its thermal conductivity—its ability to conduct heat from the surface to the interior portions upon heating, and from its interior to the surface on cooling. It is also important that once heat has reached the material surfaces it can be radiated into the steam or can accept heat from a hotter steam surrounding the metal surfaces.

This characteristic of the component and its ability to accept or reject heat is influenced considerably by the surface condition of the components. This includes deposits that may have been built up, their composition, and their thickness. The ability of these deposits to transfer heat from the steam to the metal, and vice versa, can influence the stresses induced in the components.

Causes of transients

There are a number of situations that give rise to temperature transients within the unit. They include both predictable and unpredictable effects. Predictable effects are a function of operation, load demand, and other factors within the control of the system engineers. That does not guarantee that the transients are kept within acceptable ranges. Naturally, a system will do all possible to limit temperature changes to levels and rates that are within design specification; unfortunately, this is not always possible and system requirements can occasionally override prudence, so the unit is subject to conditions that contribute to damage accumulation. While such situations are rare, it must be recognized that they can and do occur.

The more common situation is one where steam temperature changes and produces thermal stress changes in the various components, including the following.

Unit start up. The process of returning a unit to service after a shutdown is dependent upon the length of time the unit has been out of service and the amount of cooling that has occurred during this time. The shutdown period will determine the optimum or maximum rate at which the unit is brought to speed and both the rate at which temperature and load is increased. There are definitions of "hot," "warm," and "cold" starts that define these rates.

Unit shutdown. When a unit is removed from service for any reason, engineering definitions should control the rate at which steam temperatures (initial and reheat) are reduced to keep within acceptable levels the rate at which the temperature of the components is reduced. These definitions can also define the period during which steam temperature and its level are to be maintained on the unit. This is particularly important if the unit is not to be opened for inspection or maintenance, but is to be returned to service after a short period (*e.g.*, removing a unit from service for a weekend).

The extent to which units are to be shut down will depend upon load demand, which is itself dependent upon environmental conditions of temperature and humidity. Residential load is relatively constant from day to day (Fig. 5.18.1), but there are definite variations between weekends and weekdays. High industrial load is present during weekdays only. Also, there are considerable differences between summer and winter peaks, and spring and fall, when demands for heating and cooling are not present to the same degree.

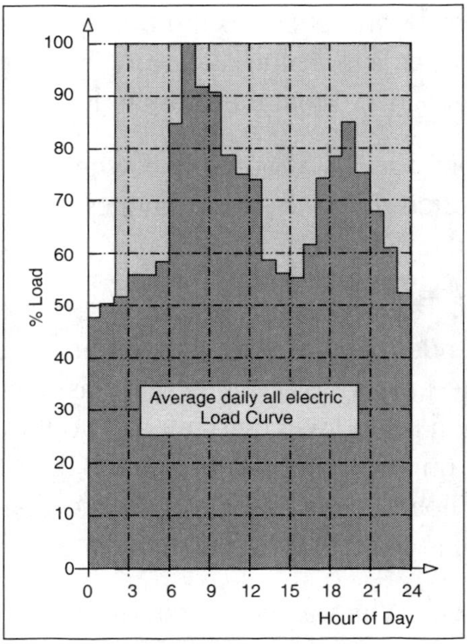

Figure 5.18.1—A typical Winter's day residential load profile.

Load changes. Changing the outputs of one or more units accommodates changes in load demand. These changes depend upon the time of day and the day of the week. Figure 5.18.2 shows the variation in residential demand of a typical North American utility. During a normal day, there is a surge in demand during morning and evening peaks when households are preparing meals. This load variation is about 40 percentage points. While industrial loads will be highest between these peaks, there is still a significant difference between the hours of nightfall and day. This causes significant demand on the load carried by various units. While it is not practical to shut down a unit for one-day variations, it does require that certain units are taken off-line during the night and then restarted each weekday to meet peak demands.

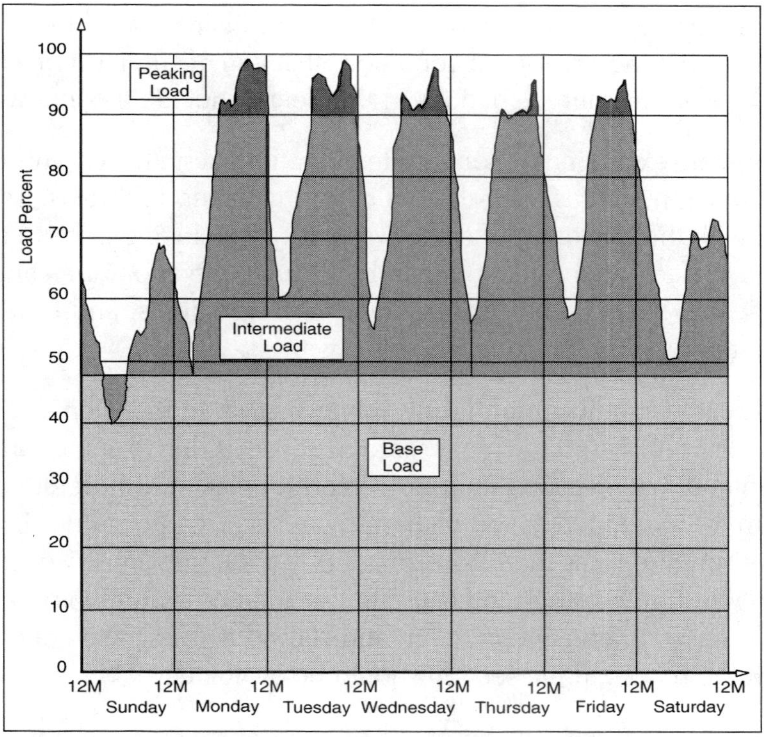

Figure 5.18.2—A typical weekly load profile of system demands.

Valve and load rejection testing. It is not uncommon for units to be subjected to certain types of testing, sufficient to ensure that safety and protective devices are operative. These tests can include valve testing, overspeed checks and adjustments, load rejection tests, and certain other tests specific to a particular installation, all of which will cause a redistribution of temperatures throughout the steam path. If these tests are planned and the extent of temperature variation is known (if it occurs), their results can be allowed for in the accumulation of data known to the operating staff.

Unpredictable situations can cause excessive thermal gradients across a large number of components. This can introduce extensive damage. While a single incident is unlikely to cause failure, it can contribute damage, which is cumulative and can shorten the life of the unit. The most common conditions that can occur in service (and produce temperature variations in the steam) include the following:

Boiler excursions. Boiler attemperation systems control discharge temperatures from the boiler and reheater. Problems associated with the operation of these systems can include excessive temperature fluctuations within the turbine. Such operations are capable of causing heating and cooling beyond design limits, and have the potential to induce high levels of stress.

Excessive windage on faulty valve closure. In the event there is a closure of the turbine inlet or intercept valves due to some form of mechanical or electrical fault in the control system, it is possible for steam to be trapped in the high-pressure section of the unit. Under these conditions, as the rotor continues to rotate it will cause excessive "windage heating" and cause the steam temperature to rise. This can be a very serious condition, and should it occur, the condition of all the blades in the section should be evaluated.

Load rejections. In the event of a system disruption and load rejection (causing the unit load to be rejected), there is a sequence of events that tends to raise the temperature of the steam in the turbine unit. There are two sets of circumstances to consider:

- *The unit not going to emergency overspeed.* In this circumstance, the control valves close but the main stop valves will remain open, carrying station auxiliaries. The control valves will continue to admit steam to the unit, slowly reducing it to normal operating speed when the unit can be reconnected to the system, assuming the system is operative and can accept the load. The unit is brought back to generate power for the system. Even though the unit is continuing to receive steam there will be a maldistribution of temperature throughout the steam path and some elements will be subjected to excessive temperature gradients

- *The unit going to emergency overspeed.* Should the unit go to emergency overspeed then the control and stop valves will close and steam will be present in the unit causing some windage heating. Steam drains from the unit will release the pressure and discharge the steam out of the sections, but there will still be heating causing thermal transients in the turbine components

Severity of transients

It is difficult to quantify the severity of transients. We know that the stress induced depends upon a number of factors related to the material strength, its ability to conduct heat, the temperature differential, and the rate at which the temperature change occurs. Means of calculating the stresses induced are considered elsewhere in this chapter.

Temperature distribution

Under normal operating conditions a temperature distribution develops throughout each component reflecting the temperature of the steam surrounding that component (the steam from which it is drawing heat). This distribution remains constant, on the assumption the temperature of the steam does not change. As the steam temperature changes—due to some modification of the unit load, boiler conditions, system disturbance, or adjustment—the component temperature will modify.

Consider the major components influenced by high levels of stress and temperature changes.

The casing. As a pressure-containing vessel, the casing will have a pressure differential developed across it. Associated with this pressure differential is a temperature gradient. In terms of the temperature gradient there are two forms of casing to consider:

- *Those casings where the steam is surrounding both sides.* Both are subject to changes in inlet steam conditions, or other conditions causing changes. In this casing design there is a change of temperature on both sides; heat will soak into, or flow out of both inner and outer surfaces. In this instance, the thermal gradient is developed from both sides of the casing and compressive or tensile stresses developed on both sides

- *Those casings on which the outer surface is surrounded by air, and on which the temperature is ambient or influenced by ambient conditions.* Usually insulation surrounds the casing outer surface, and will not allow immediate heating, cooling, or heat transfer through the casing wall. The outer surface is not surrounded or influenced by changes in the steam temperature

Under these circumstances, the inner surface will experience temperature changes at the same rate as the steam temperature changes, but the outer surface will be influenced more by the insulation, as atmospheric temperature can be considered to be constant.

A low-cycle fatigue crack initiated at a point of high-stress concentration can be seen in Figure 5.18.3. This crack has formed at a small fillet radius behind the nozzle plate and grown into the body of the casing.

Figure 5.18.3—A low cycle (thermal) fatigue crack in a high pressure casing. This casing has undergone 166,000 hours of operation.

The rotor. The steam path portion of the rotor is in a totally enclosed steam environment, except for the shaft ends that project through the casing to the bearings, which are at or near atmospheric temperature. However, for considerations of rotor temperature gradients this "end temperature effect" can be ignored. Its effect would be relatively small with the temperature gradient designed into the shaft end seal arrangement.

Figure 5.18.4 shows a plot of high-pressure/reheat rotor temperatures under normal operating conditions at full load (a), at the end of a shutdown (b), and at the end of a start up reaching full load conditions (c). It can be seen that the highest temperatures (at full load) exist at the surface and the temperature has penetrated the diameter of the rotor so the thermal gradient is axial rather than radial. A gradual decline in this temperature along the axial length of the shaft is produced by the expansion of the steam and its release of thermal energy in the blade rows.

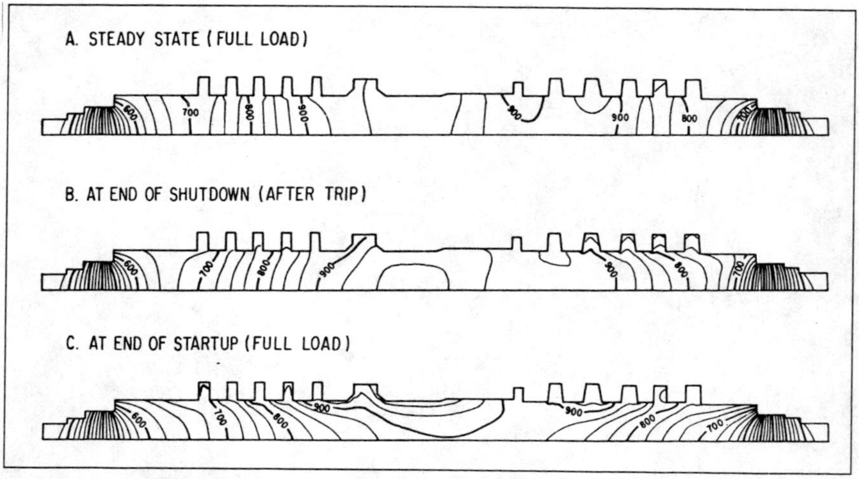

Figure 5.18.4—Temperature profiles on a combined high/reheat rotor under various conditions of operation.

In this unit the steam is admitted at the center span to the high-pressure section. It flows to the "left." After reheating, the steam is again returned to the center span position where it expends to the "right" through the intermediate (reheat) section.

It can be seen from these three figures that there are considerable temperature gradients, with the center position (b) being the hottest and in (c), the outer section being hottest. Under conditions shown in figures (b) and (c), stresses will be induced in the rotor body.

The discs and wheels. The discs and wheels are also exposed to temperature changes as steam conditions change. Shrunk-on discs tend to be used in the low-pressure/low-temperature stages, and are not therefore subject to the same level of temperatures and temperature gradients. For this reason they are of little concern in most applications.

However, the integral wheels of the diaphragm and wheel impulse designs are subject to these temperature effects. In a pure impulse stage there is no pressure drop (and no temperature drop) across the rotating blade row (and therefore across the wheel). In practice there are modifying considerations, however.

- *Reaction in the rotating blade*—The majority of impulse stages are designed to have a small degree of pressure drop across them. Therefore, a small temperature drop is associated with this pressure drop

- *Vortex effect*—The flowing steam has a radial pressure gradient between the stationary and rotating blade rows. There will be a temperature gradient in the radial direction between the stationary and rotating components

- *Diaphragm leakage*—The hotter steam leaking into the gap between the diaphragm discharge faces on the wheel, if it is contained within a diaphragm that has leakage under the seal system, will influence the stationary blade row. A large proportion of this hotter steam will flow through the pressure balance holes and tend to hold the inner portion of the wheel at a temperature slightly hotter than the outer portion. There will likely be temperature gradients in the wheels in both the axial and radial directions

- *Blade tip leakage*—Steam will leak across the blade tip in both impulse and reaction designs. This steam—like that which leaks under the diaphragm—is a throttling expansion (no work being done), and the energy released is partially converted into temperature. Therefore, the steam entering and mixing with the steam discharge from the rotating blade row will have its temperature raised by a small amount by the mixing effects. It is unlikely however, this leakage will influence the temperature on the wheel discharge side

- *The rotating blades*—The rotating blades (both reaction and impulse) will have temperature gradients across them from inlet to discharge and from pressure to suction face. As a consequence, the blades will be subject to thermal stresses as there are changes in steam temperature. Because blades in the higher temperature ranges tend to be of relatively small thickness, this temperature gradient will disappear relatively quickly. There can be high magnitude stresses induced for short periods when there are sudden temperature excursions in the steam

DETERMINATION OF THERMAL STRESSES

A *hysteresis loop* is the best way to consider those stresses and resulting strains that components endure during a cycle considered to be in the nature of low frequency/high strain. Consider a component subjected to an alternating tensile/compressive load cycle in which the total stress variation is "$\Delta\sigma$." This load can be considered to introduce forms of applied stress within the component, ranging from "$-\Delta\sigma$" (the maximum compressive effect, producing a compressive strain of "$-\Delta\varepsilon$") and at the other extreme, "$+\Delta\sigma$" (the maximum tensile effect, producing a material strain of "$+\Delta\varepsilon$"). Normally these loads are of equal magnitude in the tensile and compressive direction (magnitude "$+/-\Delta\sigma/2$"). The hysteresis loop is shown in Figure 5.19.1.

If this stress/strain curve is examined in more detail, the following facts can be derived:

- Total stress range "$\Delta\sigma$" can be divided into components "$+\Delta\sigma$" (tensile) and "$-\Delta\sigma$" (compressive). These two stresses are of equal magnitude but opposite direction

- Similarly, these stresses produce maximum strains "+Δε" and "-Δε," again of equal magnitude but opposite direction. The "+" sign represents an effect from tensile stress and "-" represents an effect from a compressive stress
- When the component is first put into service (having experienced no cyclic loading), its condition is represented by point "O" of Figure 5.19.1. Then, as an initial tensile load is applied, a tensile stress is induced that produces a linear elastic strain "Δεie" up to condition "A"

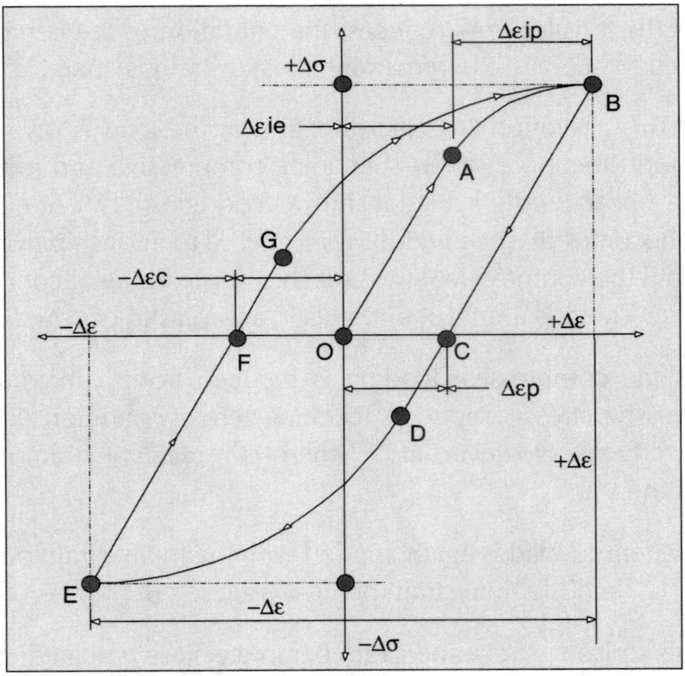

Figure 5.19.1—The stress/strain cycle for a component undergoing strain in the elastic/plastic range.

After condition "A," the extension with increasing load will continue, but will no longer follow the linear relationship. It will instead form as from "A" to "B." This strain "$\Delta\varepsilon ip$" is plastic and therefore irrecoverable, up to a total strain of "$+\Delta\varepsilon$." Therefore:

$$+\Delta\varepsilon = \Delta\varepsilon ie + \Delta\varepsilon ip$$

At condition "B," the load has reached its maximum tensile value. Now as the load is reversed, the tensile stress reduces. As the tensile load reduces to zero, the entire tensile component of stress is removed and the strain reduces, but does not reach zero. A plastic strain component remains.

As the tensile stress reduces, the condition of the component is reduced to "C" and the remaining plastic strain is "$\Delta\varepsilon p$."

At this condition "C," the direction of the load is reversed. The load—and therefore stress—becomes compressive and initially follows a linear (elastic) relationship to condition "D." After "D" the strain becomes plastic and follows a curved relationship from "D" to "E," until the compressive stress reaches a maximum at condition "E" of "$-\Delta\sigma$." There is a total strain, elastic and plastic of "$-\Delta\varepsilon$."

As the compressive loading is reduced from a maximum, the compressive stress is reduced, reaching zero at condition "F" and the strain reduces. However, at "F" there is a residual strain of "$-\Delta\varepsilon c$" remaining.

The tensile load is again applied with the initial strain being elastic to "G" and the final strain being plastic to "B."

This series of stress and strain parameters are repeated each time the components undergo a set of cyclic loading, and the accumulation of plastic strain continues to deform the component.

If compressive and tensile stresses are of the same magnitude, the diagram Figure 5.19.1 will be symmetrical about the point of origin

"O." In a turbine component the actual stress levels are a function of a number of factors and are particularly sensitive to temperature ramp and cooling rates.

The total strain range of any component undergoing the stress and strain cycle shown as Figure 5.19.1 is the sum of "+Δε" and "-Δε," and is shown as the strain range from "B" to "E" in this figure. This strain consists of two elements—the elastic "Δεie" and the plastic "Δεip" portions. The elastic portion is represented by the ranges "O-A," "B-D" and "E-G." Since the range "O-A" is present only during the initial strain cycle it can be neglected. The plastic strain ranges are from "A-B," "D-E," and "G-B," and again, the range "A-B" can be neglected. There is a relationship between the total and elastic strains for any material. This relationship is given by:

$$+\Delta\varepsilon = B.\Delta\varepsilon\ \Delta\varepsilon p^{\gamma}$$

Where:

Δε is the total strain range
Δεp is the plastic strain range

and

B and γ are constants determined for the material

This relationship is only approximate and applicable primarily to cycles of intermediate strain ranges. A relationship for a typical material between the plastic strain range "Δεip" and the total strain range "Δε" is shown in Figure 5.19.2.

Note: Due to strain hardening, the hysteresis loop gradually increases and moves in the direction of increasing stress for the same strain range.

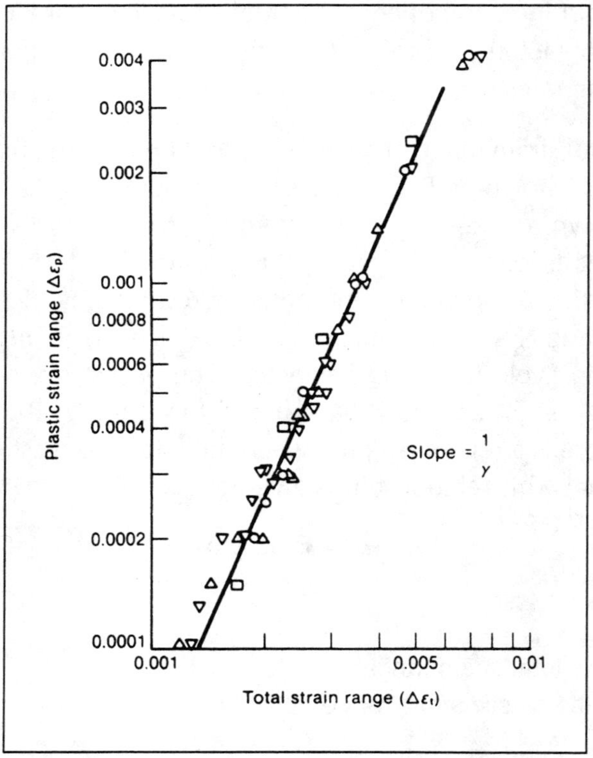

Figure 5.19.2—Correlation between total and plastic strain as defined by equation 5.19.2

COMPONENTS OPERATING AT HIGH TEMPERATURE

Many turbine components in the high and reheat sections operate at high temperatures and are subject to cyclic loading, both high frequency/low stress and low frequency/high stress. These components are simultaneously subject to creep and fatigue.

The steam turbine consists of components that are subject to thermal transients during start up and shut down, as well as load changes and any system disturbance causing steam temperature changes at rates that exceed the design specified limits. This is anticipated during the design phase, and normally there is margin in the design to accommodate a certain amount of this and the resulting damage. Unfortunately, those components that operate within the creep range for any particular material are also those components most seriously affected by thermal transients. The extent of stress induced and the consumption of component life are dependent upon a number of factors. These include the following elements:

The material from which the components are manufactured. The martensitic and austenitic steels from which steam turbine components are produced have the ability to accept and reject heat at a rate that, while inducing stress, is not as severe as other materials with superior mechanical properties that could be used.

Austenitic steels have a higher coefficient of thermal expansion than martensitic materials. However, the rate of change experienced by these two basic steels are relatively the same; therefore, once stable thermal conditions have been reached, any transient will see a relatively common expansion at any interface. But, as can occur in some designs, problems can be introduced where an austenitic material is weld-attached to a martensitic because of stresses induced during the transient period. Similarly, since scale formed on many components acts to impede the transfer of heat from the steam into the steel of the component, a large variation in the thickness or composition of scale formation on a component can introduce some differences in the rate of temperature change, and therefore stresses within any single component.

The rate of temperature change. The rate at which the temperature of a component changes will influence the stress induced in it. The rate of component heat acceptance or rejection is termed its *emissivity*. This characteristic is a measure of the material's ability to accept and reject heat through its surface. "Total emissivity" must also consider the rate at which heat is transferred through any scale deposited on the surface. This implies that the interface of the scale/steel may not be constant over the entire surface during any transient, or the rate at which constant conditions are reached will vary from position to position on the component surface.

Once the heat condition in the outer layers of the material surface changes, there is the requirement of thermal conductivity of the material and its ability to dissipate the thermal change throughout the material. These two material properties of emissivity and thermal conductivity will therefore influence the stresses and the resulting strains induced in the component.

The number of cycles the unit experiences. As discussed previously, temperature excursions (causing either temperature increases or decreases) will induce alternately tensile and compressive stress in a component. We also know the rate at which these temperature changes occur influence the magnitude, and each cycle will increase the total plastic strain to which the components are subjected. It is also clear that these strains and damaging effects are cumulative. Therefore, each thermal transient will cause some level of damage if the degree of temperature change is sufficient to introduce a plastic strain into the component.

For these reasons, operators should limit the rate of temperature change and the number of cycles as far as possible. It is recognized however, that this is an ideal means of operation, and financial considerations of fuel economics and load demand do not normally permit this, particularly on older units within a system.

Those components subjected to significant thermal gradients are also normally in the creep range. Therefore, stresses induced by thermal change add to the creep stresses. For those components in the secondary range of creep (range "B-C" of Fig. 5.4.1), any temporary change in the stress level can cause a reduction of the remaining creep life and accelerate its progression into the tertiary range.

Components are normally designed to operate in the creep range and have an anticipated life of 200,000 hours before either requiring replacement or entering the tertiary range of deformation. Imposing a thermal stress and plastic strain on these components can reduce their useful life considerably. Figure 5.20.1 shows the horizontal joint of a high-pressure casing, exhibiting circumferential cracks that result from combined creep and thermal fatigue.

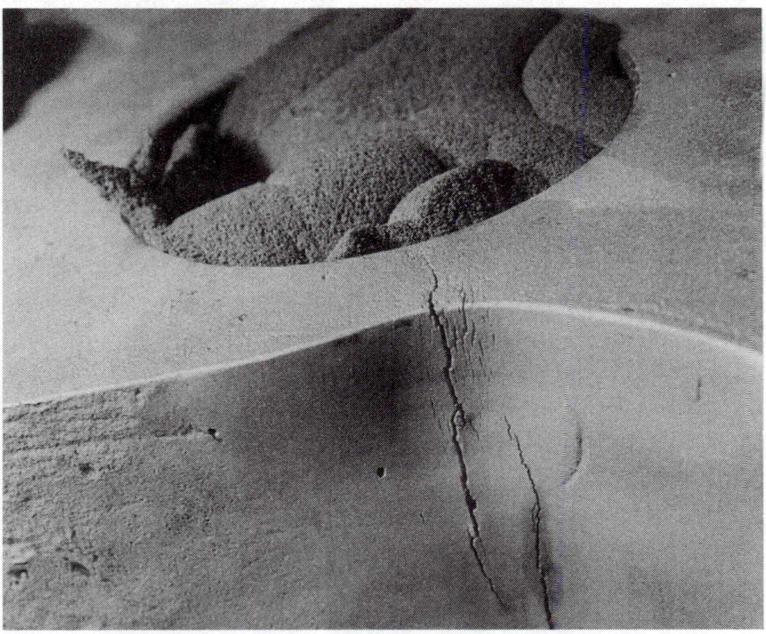

Fig 5.20.1—A crack initiating at the horizontal joint of a high pressure casing, and running in the circumferential direction.

REFERENCES

1. Neubauer, B., and U. Wedel. *Restlife Estimation of Creeping by Means of Replicas,* American Society of Mechanical Engineers, *Advances in Life Prediction Methods,* New York, NY, 1983

   McCloskey, T., B. Dooley, and W. McNaughton. EPRI Publication, *Turbine Steam Path Damage: Theory and Practice,* 1998

2. Viswanathan, R. *Damage Mechanisms and Life Assessment of High-temperature Components,* Published by ASM International, Metal Park, Ohio 44073, 1989

3. Ayano, S., S. Suzue, M. Muramatsu, and M. Yamada. *High-temperature Blade Deterioration and Life Assessment Techniques,* EPRI Steam Turbine Blade Seminar, Los Angeles, California, March 1987

4. Viswanathan, R., R.I. Jaffee, and T. McCloskey. "Metallurgical Factor Affecting the Reliability of Fossil Steam Turbine Rotors," ASME *Journal of Engineering, Gas Turbines and Power,* Vol. 107, 1985

5. Rieger, N.F. *Blade and Rotor Reliability Improvement,* STI/EPRI Seminar, Rochester, New York, June 1991

6. Swift, H.W. "Mechanical Properties in Relation to Design," *Journal of Instrument Engineers,* August 1940

7. Campbell, W. "The Protection of Steam Turbine Disc Wheels from Axial Vibration," ASME *Transcript,* Vol. 46,1924

   Weaver, F.L, and M.A. Prohl. "High Frequency Vibration of Steam Turbine Buckets," ASME *Transcript,* 80, No. 1

Kubiak, J.A., O. Argonne, A. Carnero, and J. Aguirre. *Analysis of the Last Stage Blade Group,* Instituto de Investigaciones Electricas, Cunervaca, Mexico

8. Broek, D. *Elementary Engineering Fracture Mechanics,* Published by Martinus Nijhoff, 1982

9. Rollason, E.C. *Metallurgy for Engineers,* published by Edward Arnold Ltd., London, 1956

Samuels, L.E. *Metals Engineering, A Technical Guide,* published by ASM International Metals, Park, Ohio, 1988

10. Ross, P.N., and L.G. Hauser. "Some Future Dimensions of Electric Power Generation, circa 1970-1990," *The Westinghouse Engineer,* January 1991

11. "Energy Marketing: Rate Structures Need Basic Rethinking." *Electrical World,* June 15th, 1974

12. Walker, D.N., S.L. Adams, and R.J. Placek. *Torsional Vibration and Axial Fatigue of Turbine-Generator Shafts,* IEEE-PES Summer Meeting, Paper No. F79-756-8, 1979

13. Skelton, R.P. *The Prediction of Crack Growth Rates From Total Endurance in High Strain Fatigue,* Fatigue Engineering Material Structure, Vol. 2, 1979

Chapter 6

Steam Path Damage and Deterioration from the Deposition of Contaminants

INTRODUCTION

The thermal-power cycle is a complex arrangement of individual pieces of equipment. Many act to control the expansion of steam; others come into contact with other forms of water, which may contain undesirable contaminants. These contact locations have the potential to contaminate the working fluid. Other units operate on steam removed directly from the earth; with no effort made to clean the steam to the extent the contaminants transported by the geothermal steam are removed before entering the steam path.

Contaminants carried by the parent steam into the turbine can be either chemically aggressive or non-aggressive. Aggressive compounds considerably influence the extent to which the steam path deteriorates (in terms of its structural integrity). Both can influence the efficiency with which the steam expands through the unit.

There are two major concerns with steam borne contaminants:

- reduction in efficiency due to surface deterioration of the steam path elements, and any resulting frictional losses that occur
- reduction in structural integrity resulting from corrosive action on the various components of the unit

There also can be considerations of the possible change in rotating blade natural frequencies if the deposited matter becomes of significant mass.

The extent to which surface frictional losses influence efficiency deterioration is a function of the Reynolds Number associated with the flow across the surface being considered. These losses are in turn dependent upon the steam's physical properties and flow velocities at the stage locations. They are also influenced by the dimensional characteristics of the row. Because the velocity of the steam flowing through an expansion passage is changing (due to both stage reaction and friction), the actual steam velocity—which establishes the Reynolds Number—varies throughout the flow passage. This variability is overcome by using data derived by experimentation—data normally developed by the manufacturers and applicable to their families of profiles.

While the best data (in terms of the actual Reynolds Number) will be determined by the manufacturer for operation with the same profiles at similar steam conditions, sufficient accuracy can be obtained by calculating Reynolds Numbers based on discharge velocities, which is the procedure normally used by the manufacturer.

In the high-pressure and reheat sections of a unit, deposition patterns are substantially more regular and deposits more evenly spread than those occurring in the low-pressure section/latter-stage blades, where a considerable amount of deposition occurs from the water phase.

A number of deposition patterns can be observed in any blade row, and it is remarkable that the patterns on any row are so similar from blade to blade. This is an important consideration when attempting to determine a frictional loss for any row, because if there was considerable variation from blade to blade it would be particularly difficult, if not impossible, to define a mean effective surface roughness—a major characteristic for establishing losses that occur. In fact, the judgment of mean surface roughness in a unit removed from service is judgmental on the part of the engineer making the assessment, as variation in depositions from inlet to discharge, from root to tip, and on the pressure and suction faces must be considered.

The deposition of compounds that occur on the blades is of two basic types:

- compounds insoluble in water, which can only be removed by mechanical means
- soluble compounds that can be removed by immersion or washing in water

The second characteristic can be advantageous under certain circumstances.

A more insidious consequence of chemical deposition on steam path elements is that some of these compounds are chemically aggressive and have the potential to come out of solution and deposit on the internal surfaces of the unit as steam conditions reduce on expanding through the various blade rows. Many of these compounds have the potential to migrate and accumulate in "hideouts," where under suitable environmental conditions they become chemically aggressive, causing reactions that eventually lead to component deterioration and ultimately failure.

Steam conditions at any stage are a function of boiler delivery conditions and the load the unit is generating. Therefore, as there are electrical load changes, conditions on certain blade rows will modify, moving the steam conditions at any location into either the superheated or saturated regions. Under such operating conditions the deposition, collection, and concentration mechanisms in place will provide the environment for various forms of corrosion.

This chapter will examine these two phenomena induced by deposition, and consider the effects on efficiency degradation and the potential this deposition provides to promote mechanical failure.

SOURCE OF STEAM PATH IMPURITIES

Operating engineers need to address operating problems associated with chemical ingress and contamination on a continuing basis. These compounds can gain access to the unit through various sources and mechanisms. Some occur as a result of mechanical damage within other components of the steam power cycle—most notably the condenser. Other sources are water treatment plants, where careful control of the treating process is required.

Oxygen can also gain access at many points in the cycle where sub-atmospheric conditions exist. Units are normally equipped to remove any air that enters the system before it can cause significant corrosive damage, or support a chemical reaction. However, a certain amount can, and often does, gain access to the steam path.

External sources—Carried in at system leaks

A common source of contaminants in the steam path is infiltration by untreated (raw) water into the system. This commonly occurs within the condenser, which is used to produce an operating vacuum. Cooling water may be taken from lakes, rivers, or the sea (in coastal installations). There are also river installations that are tidal.

It is especially necessary to consider the effects of installations where saline water (seawater) exists, because through access to the condenser, saline can produce a significant ingress of contaminants. Seawater contains high levels of many substances, most notably salt (NaCl), which can be highly corrosive if it gains access to the steam path.

System internal—Generated by water treatment systems

Modern power generating systems require water steam in the boiler to be demineralized and as pure as possible. This requirement introduces the need for water treatment—a process that will cause the introduction of excess chemicals into the unit steam path if not controlled within close tolerances. Certain of these treatment substances have the capability of forming chemically undesirable products.

Contaminated attemperating spray water

Water used to control the inlet temperature from the superheater or reheater portion of the boiler must be taken from a reliable and clean source (normally the feedwater).

Contaminated exhaust hood spray water

Water is used in the low-pressure hoods to control steam and blade temperature at low loads when steam will reenter the last stage blades with the potential to cause "trailing edge erosion" (see chapter 3). This spray water should be taken from a clean, reliable source. Often water taken from the condenser hot well contains impurities and dissolved oxygen that can initiate and also promote corrosion. In certain nuclear units this water can be taken from an auxiliary source.

Use of sodium sulfide as an oxygen scavenger

Sodium sulfide is often used as an oxygen scavenger in high-pressure boilers. Decomposition of the sodium sulfide can produce hydrogen sulfide H_2S in the early moisture region of the unit.

Use of cleaning material at outages

Cleaning fluids may contain unacceptable levels of caustics, chlorides, or sulfur when used for removing deposits and other cleaning activities during a unit outage. This may be acceptable for the cleaning process, but if components are not rinsed thoroughly of cleaning fluid and such residues remain, they have the potential to collect and produce corrosive compounds throughout the unit.

Manufacturing and repair lubricants

Steam turbine manufacturers employ sophisticated materials and production processes, many requiring the use of cutting lubricants and fluids containing compounds of sulfur, chlorine, phosphorus, and magnesium. These substances, even in trace quantities, should be removed from component surfaces before acceptance and before they are assembled and/or shipped to site.

Fluids are required to both lubricate and cool the metal cutting process. Such fluids can be either oil- or water-based compounds. The most suitable are dependent upon the material being cut, the cutting rate, and the depths of cut employed. Very few components of the steam turbine can be machined without the use of a cutting/cooling fluid. Therefore, residuals of these compounds are always potentially present on the surface of various components. This is a condition that can exist when the unit is new or when it is returned to service after repair, but such impurities can also be introduced when replacement parts are installed at a maintenance outage.

Shipping and storage preservatives

When components are shipped to site, many of them require protection during transportation, including steps to prevent atmospheric corrosion. If these components are to be stored at site during installation, and possibly placed in long-term inventory, they will require corrosion protection.

Materials used for such protection must be suited to the anticipated life, and may be of a nature that if not cleaned completely before installation, could introduce chemically undesirable compounds into the system or even attack the component during storage. Figure 6.2.1 shows blades removed from inventory for installation that have suffered corrosive attack.

Dissolved from system equipment

Another source of contaminants carried into the steam path are those produced within other pieces of equipment comprising the power cycle. Most notable of these are iron and copper oxides produced as byproducts of dissolving metals from tubes and equipment surfaces. This includes boiler tubing and regenerative feedwater heaters, where high-temperature, high-pressure water comes into contact on a continuing basis as the unit produces power.

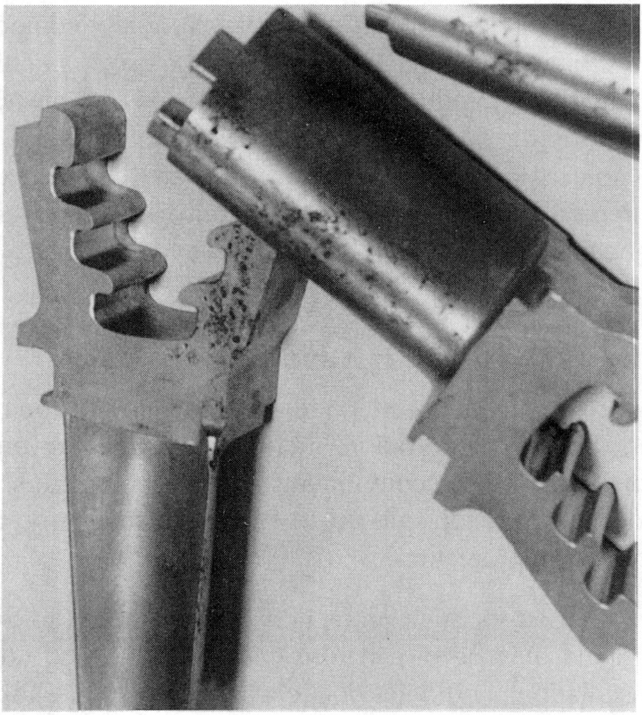

Figure 6.2.1—Corrosive damage formed on new blade elements while they were in storage.

Geothermal

The steam used in units supplied for geothermal application tends to be present at pressures relatively low—normally lower than the inlet conditions used for the nuclear cycle. This steam is also contaminated by minerals and compounds dissolved in it from the earth.

Table 6.2.1 shows the mix of gases, water, and impurities found at different sites. However, the contaminants can vary considerably from site to site, both in constituents and concentration. For direct-cycle units, these contaminants are carried into the steam path, where many are deposited on the internal surfaces, causing levels of surface roughness and corrosion considerably greater than normally experienced in fossil or nuclear cycles.

Table 6.2.1—Geothermal Steam-Water, Gas and Impurity Content

Location		***Geysers U.S.A.***		***Matsukawa Japan***		***Kakkado Japan***		***Cerro Prieto Mexico***		***TIWI Phillipines***	
Steam Quantity %	H_2O	98.05 - 99.51		99.40 - 99.80		99.85 - 99.78		00.08 - 99.78		98.43 - 99.16	
	Gas	0.39 - 1.95		0.20 - 0.60		0.03 - 0.15		0.22 - 0.92		0.83 - 1.94	
Gases Content Vol.%	H_2S	1.69 - 2.99		12.9 - 17.7		9 - 26		20.9		0.8. - 2.57	
	CO_2	63.5 - 69.3		79.3 - 85.2		65 - 87		79.1		96.64 - 98.63	
	Balance	37.69 - 34.89		1.7 - 4.2		0.5 - 15		-		0.37 - 4.4	
		Condensate	Hot water	Condensate	Hot water	Condensate	Hot water	Condensate	Hot water	Condensate	Hot water
Condensate and hot water (ppm)	PH	6.5	6.7-7.3	4.4-4.9	4.4-5.3	6.0-6.4	8.8-9.1	-	5.8-6.6	4.2-5.9	7.5
	K	-	-	0.3-3.0	120-300	0.1-0.28	55-67	-	631-2031	0.07-0.37	830
	Na	-	-	0.8-5.0	152-360	0.27-0.72	460-501	-	4406-7764	0.05-3.0	4400
	Mg	<1	-	< 2.1	4.7-14.5	< 0.1	0.01-0.02	-	-	0.02-0.20	0.19
	Ca	<1	-	< 1.6	12.8-28.8	Tr.	5.8-9.0	-	259-359	0.02-0.20	85.0
	Fe	< 0.1	-	0.45-17.8	235-630	0.01-0.05	0.06-0.09	-	-	< 0.31	0.31
	Al	-	-	< 2.0	6.5-28.6	0.01-0.04	0.14-0.64	-	-	Tr.-0.20	0.14
	Si	0.71	-	0.83-4.7	273-566	0.22-0.52	256-364	-	179-1458	0.02-1.72	345
	Cl	39	-	3.3-6.7	4.3-13.8	1.9-3.4	625-685	-	928-14934	0.91-173	8970
	SO_4	229	-	2.0-40	1188-1856	0.24-0.83	64-82	-	4.9-62	13.2-53.7	46.5
	CO_2	-	-	12.4-51	4.3-6.5	14-40	Tr.	-	42-494	106-108	-
	H_2S	5-10	30-205	10-52.2	1.7-3.3	24-58	2.1-3.4		21-165	6.0-8.9	-

Balance is: $(N_2 + H_2 + CH_3)$

In terms of deposition of impurities, geothermal units provide an entirely discrete set of considerations. This is because there can be deposits in these units completely unlike those found in units used for cycles of the more conventional fossil and nuclear fueled designs (Fig. 6.2.2).

Figure 6.2.2—The deposits on a geothermal stationary blade row. These deposits, while not corrosive, had a very significant effect on strage efficiency. The effect of a stay bar holding one stationary vane in place, and the flow divertion, and therefore deposition pattern can also be seen.

Geothermal units are arranged to be either direct cycle or indirect cycle. In direct-cycle units, naturally occurring steam removed from the earth is passed through the steam path, often with separators interposed between the steam supply and the turbine. The separators remove suspended moisture and solid particulate from the steam. There is little or no effort made to remove suspended and dissolved chemical impurities.

Indirect-geothermal cycles employ a heat exchanger vessel between the geothermal steam and the working fluid. In these units, the working fluid should have a purity comparable to that of fossil and nuclear units. However, these units can be subject to the intrusion of contaminants or impurities as the result of tube failures in the heat exchanger vessels that dependent upon the corrodents carried in, can be relatively common. In indirect cycles, the heat exchange vessels can themselves suffer heavy deposition and corrosive damage.

Geothermal steam tends to be saturated, containing at best only a few degrees of superheat. Therefore, steam path components are almost all in the saturated region, with the probability of continuous washing. Water present on steam path surfaces may itself contain impurities that are deposited and detrimental to the total performance of the unit. However, it can have some beneficial effect in terms of efficiency, as it will tend to remove some level of the deposited compounds.

Water quality

After initial unit start up and commissioning, the majority of the contaminants found in the steam path are those carried in by the steam. Therefore, control of steam quality is essential.

Note: During initial start up and commissioning, any contaminants due to manufacture, storage preservatives, and other "once-only" circumstances are removed almost entirely. However, when a component has been removed for rebuild or repair, it is a good idea to examine and possibly clean it before its return to service.

Control of steam quality is obviously synonymous with control of boiler water/steam. The recommendations outlined in Table 6.2.2 represent the upper limits considered necessary to provide steam of sufficient purity that cleaning at maintenance outages should preclude the incidence of major deposits, scaling or corrosive action. Operation beyond these upper limits should be avoided but if exceeded, immediate corrective action should be implemented.

Contaminant	Normal Operation	Limiting Condition 2 Weeks	Limiting Condition 24 Hours
Dissolved Oxygen	< 10	10 - 30	30 - 100
Sodium	< 5	5 - 10	10 - 20
Chlorides	< 5	5 - 10	10 - 20
Silica	< 10	10 - 20	20 - 50
Copper	< 2		
Iron	< 20		
Sodium Phosphate	2.3 - 2.7		
Sulfides and Sulfates:	Less than detectable, should be analyzed at least once a week.		

Table 6.2.2—Steam Purity Recommendations, in parts per billion (ppb)

Figure 6.2.3—The result of a large sea water ingress at the condenser, and deposition of salt (NaCl) on a rotating blade row.

Unfortunately, it is possible that a unit may operate for years below the recommended levels, and then a single system disruption of several hours duration can cause a contaminant ingress sufficient to produce a deposition rate highly detrimental to the continued reliable operation of the unit.

Figure 6.2.3 shows a unit using seawater for cooling in the condenser. This unit operated reliably until a severe leak of several hours duration occurred. The consequent deposition of contaminants warranted the unit be shut down and completely cleaned (including the removal of several rows of rotating blades to allow cleaning of the blade roots and the rotor portion of the root fastening). This was an expensive outage, but not nearly as expensive (in terms of time and material) as the possible need to reopen the unit within a short period of return to service and correct blade damage (and possibly rotor corrosion), with the consequent purchase of replacement blades.

THE COMPOSITION OF DEPOSITS

The composition and concentration of the compounds deposited on the steam path component surfaces are dependent upon their origin. Irrespective of the level of care taken by plant operators to limit their presence, impurities will gain access from various sources and be present in the steam. There will be a persistent deposition, possibly slow, throughout the operating life of the unit. To best categorize these life-long compounds, consider those that gain access through leakage into the cycle, from some secondary process of heat exchange or system maintenance, and those present as a consequence of some form of treatment undertaken on the working fluid.

Probably the most frequently found corrosive agents affecting the blade path—and potentially the most damaging—are sodium chloride and sodium hydroxide. The introduction of caustic contaminants into the water/steam cycle is generally attributed to the ingress of common salt (NaCl) into the system, and the dissociation of this compound into sodium (Na) and chlorine (Cl) ions, which can then recombine with ions of hydroxide (OH) and hydrogen (H) to form caustic sodium hydroxide (NaOH) and hydrochloric acid (HCl).

Since these compounds are normally controlled at extremely low levels, it is necessary to consider the mechanisms by which their deposition and concentration is increased to a level initiating some form of corrosion damage.

Once it gains access, sodium chloride readily dissolves in water. Sodium hydroxide, once formed, can be transported by the steam in the boiler feedwater through the feed heating train to the boiler. In the superheater, residual water is evaporated, leaving molecules of sodium hydroxide in equilibrium with the steam molecules. These sodium hydroxide molecules in their equilibrium condition then enter the steam turbine and flow through the steam path with the expanding steam. As temperature and pressure reduce, and the steam passes into the saturated region, the concentration will fall, i.e., the molecules of sodium hydroxide will capture water molecules to dilute the concentration and maintain an equilibrium condition.

Note: To get into the water/steam system the sodium chloride is more likely to be dissolved in water—it will be about 90% ionized. In the superheated steam the residual sodium chloride and the sodium hydroxide will exist as ionized pairs ($Na+Cl^-$ and $Na+OH^-$) in a weak association, and will be precipitated onto the surfaces of the blade path. As this action takes a finite time, some ion pairs will be carried down into the wet region and be absorbed into the water. If not deposited on the low-pressure section surfaces, the absorbed sodium chloride and sodium hydroxide will be transported into the condenser hot well.

The melting point of sodium hydroxide is a little above 600°F. At temperatures below this, these ions will come out of solution and deposit on the steam path elements, as the water droplets containing them come into contact with the steam path components. These sodium hydroxide molecules, once deposited with the water, can flow or wash into, collect, and be retained within various crevices or "hideouts" formed in the steam path. These collection regions are often not easily accessible for cleaning, or removal by the washing action of the water phase of the working fluid. There is a tendency for these deposits to collect, concentrate and remain in these hideouts.

When pressure and temperature within the steam path change, the deposited molecules will attempt to reach equilibrium conditions. This may not always be readily attainable, and may take extended periods of operation at the new conditions to achieve it. This is because of a time delay before the metal temperatures of the major components can achieve a new operating condition. Depending on their location within the steam path, the solution of sodium hydroxide in these hideouts will take varying amounts of time to achieve a new equilibrium condition. These time delays may leave a solution of sodium hydroxide in a particularly active state for extended periods each time a pressure and temperature change occurs within the steam path.

This increased activity is also possible when a unit is operating at off-design steam conditions, and changes in inlet steam temperature cause a fluctuation or variation in stage temperatures. Under these conditions, components and regions of the unit that should be in a "no sodium hydroxide region" are moved into an "active sodium hydroxide region" and can, as a consequence, suffer corrosive attack.

In areas of high temperature and pressure (superheated), stress corrosion due to sodium hydroxide activity cannot occur too readily.

This is due principally to the fact that steam and component temperatures are above the melting point of sodium hydroxide. The higher density of the steam in these regions, and its ability to prevent sodium hydroxide molecules from depositing in quantities sufficient to cause any corrosive action, also plays a role.

The number of failures and instances of corrosive damage found in the low-pressure regions of the steam turbine (due to sodium hydroxide) indicates components in this region are vulnerable, if corrosive agents can adhere to surfaces and find hideouts into which they can migrate, collect, and concentrate.

In assessing possible changes in concentration of corrodents influenced by steam conditions at various locations, the following factors should be considered:

- At full load operation, low-pressure stages vulnerable to corrosive attack may be operating in the saturated region

As load is reduced, there is a greater degree of throttling on the control stage or at the control valves on throttle-controlled units. This action modifies the expansion line, moving it to the right on the Mollier Diagram, with the susceptible stages moving from the saturated to the superheated region. The expansion in these stages now occurs in the superheat region, and some of the rows are exposed to superheated steam.

This condition change is shown on the Mollier Diagram (Fig. 6.3.1). In this unit, the expansion lines are shown for full load condition "A,B…E," and at part load "A, Bm…Em." The effect of partial arc admission or throttling is at the portion of the unit moving the high-pressure section exhaust from "B" to "Bm." The final exhaust from the low-pressure section is maintained at "Pxd," but the dryness fraction has changed from "E" to "Em," which is substantially drier.

The stages in the low-pressure section where the transition from superheated to saturated conditions are shown as stage "L-N-K." The effect of throttling is to move these stages to conditions "Lm-Nm-Km." This will modify the dryness fraction from "Xid" to "Xdp" at full load and to "superheated to Xde" at part load. The implication of this condition change is that throttling, and even the degree of superheat on a stage can vary as the load changes. There is the potential for concentration of any corrodent, and while operating in the part-load condition, corrosive action can initiate and continue.

- Failure to maintain steam temperatures from the boiler superheater and/or reheater at design conditions

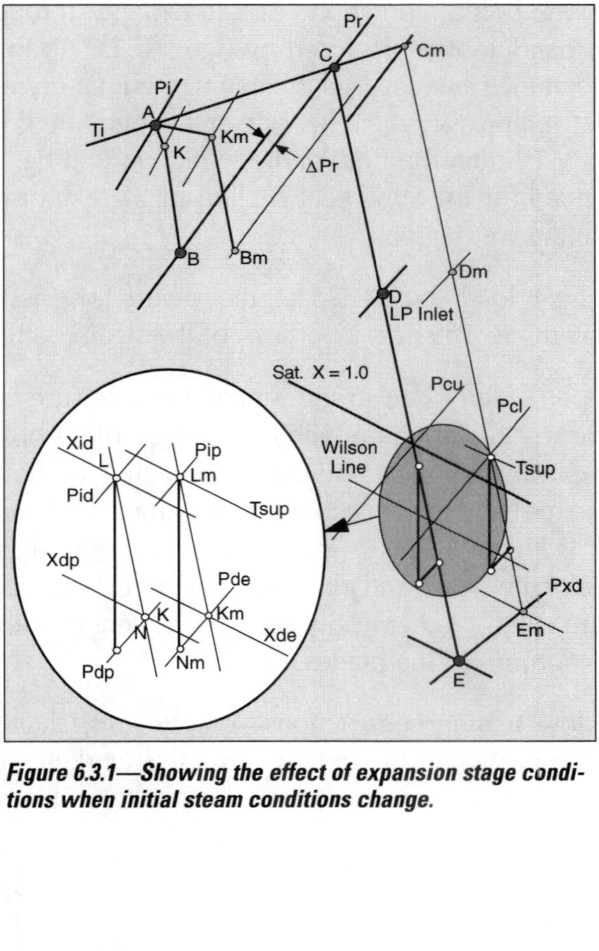

Figure 6.3.1—Showing the effect of expansion stage conditions when initial steam conditions change.

The amount of initial and reheat temperature reduction sensed by any unit depends upon the type of boiler and the load at which the unit is operating.

In general, lowering the boiler delivery temperatures while maintaining pressure at or close to design values lowers steam conditions throughout the low-pressure section. This temperature reduction can possibly lower early stages of this section into the wet or active sodium hydroxide region, at which condition any sodium hydroxide residue can become chemically active and cause corrosion.

Consider the unit expansion shown in Figure 6.3.2. Steam is admitted to the unit at condition "A" and expands through the unit with steam conditions represented by line "A...D." With the reheat temperature at the same level as the initial, the final dryness fraction will be "Xd." If, however, the reheat steam temperature is lowered by an amount "ΔT," then the final dryness fraction will be "Xdm" and stages designed for the superheat section of the expansion will be operating in the wet region.

- At light load (below 25%) the steam flow is principally through the outer flow portions of the steam path

This is particularly true in the longer, low-pressure blades. This effect causes re-circulation through the lower portions of the rotating blade row, causing windage and frictional heating of the latter stage blades at the root and lower sections of the vane. This effect is shown in Chapter 3, Figure 3.9.7. If water sprays are not operative, or the temperature of the steam and metal has not increased to a significant degree, there will be a chemically active corrodent near the root and lower vane portions of the blades.

- At flow sealing constrictions—diaphragms, shaft ends, and spill strip packings—the non-expansive working of the steam can cause local heating above the saturation temperature

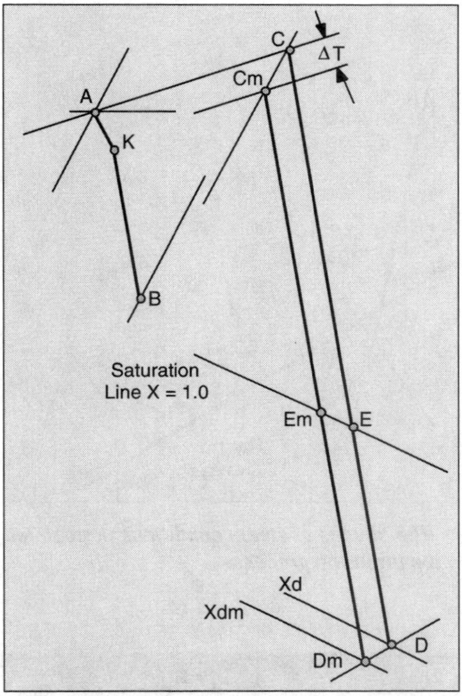

Figure 6.3.2—The effect on exhaust steam conditions of failing to maintain reheat steam temperature.

Figure 6.3.3 shows the steam conditions as they exist at the shaft-end sealing positions or at an intermediate or reheat section (condition "A" being the exhaust steam conditions at exhaust from the final stage in that section). As the steam at condition "A" leaks through the labyrinth seal strips, successive pressure drops occur and cause the steam to eventually move into a superheated range. It is possible for contaminants carried into the sealing glands to become chemically aggressive and cause corrosion at some axial location where the steam conditions support corrosion.

Figure 6.3.4 shows the shaft end position of a small industrial turbine where there has been corrosion in the gland region.

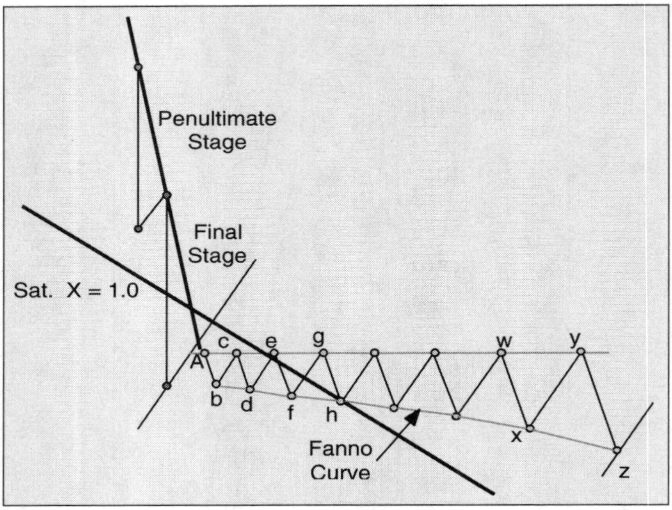

Figure 6.3.3—The change of steam conditions through labyrinth seals due to the throttling process.

Figure 6.3.4—Corrosion pitting at the shaft end seals at the high pressure end of a small industrial turbine.

Oxygen dissolved in the steam can assist or promote oxidation of the material of the steam path. For this reason, it is necessary to limit the quantity of oxygen present at any time. Accumulated oxygen is often removed from the steam/water system by the use of sodium sulfide (Na_2SO_3)—an undesirable, but relatively harmless substance at controlled concentrations. However, if sodium sulfide is used in, or finds access to high-pressure boilers, the corrosive compound hydrogen sulfide (H_2S) is formed in the system. Hydrogen sulfide is often found in geothermal applications. It is particularly unpleasant, causing corrosive pitting and leading to corrosion fatigue if the pits formed are in a region of high stress.

With regard to turbine components, many rotor and disc stress corrosion failures have been attributed to hydrogen sulfide. Unlike sodium hydroxide, hydrogen sulfide need not be present in high concentrations. In fact, it has been shown that sulfide cracking will occur for materials above a Brinell Hardness of 235 with aqueous hydrogen sulfide concentrations down to 1 ppm within 5,000 hours. Even lower concentrations can apparently lead to cracking of high hardness alloys. As discussed previously, the source of hydrogen sulfide in steam turbines is the use of sodium sulfide as an oxygen scavenger in water treatment (or its containment in geothermal steam). Therefore, the potential for long time low-level concentrations of hydrogen sulfide exists with this method of treatment. The use of sodium sulfide has in general been abandoned at most plants. Most now tend to employ an all-volatile water treatment by ammonia and hydrazine.

Classification of deposits

A 1976 study indicates compounds from various sources are deposited throughout the steam path. Obviously, compounds are more likely to accumulate in regions where local steam conditions encourage, or do not inhibit their formation and accumulation. The referenced study classified deposits into eight groups from analysis of a number of units, including oxides, silicates, sulfates, phosphates, carbonates, aluminates, chlorides, and uncompounded elements. These substances were found to exist in various forms and were distributed in different parts or pressure ranges of the steam path, depending upon their ability to precipitate from the steam, deposit, and adhere to the steam path components. The majority of these elements and compounds were chemically inert and, although capable of reducing steam path efficiency, had little potentially adverse effect upon the mechanical properties of the construction materials.

Other compounds were chemically aggressive, or capable of forming aggressive compounds. It was determined these compounds and elements were introduced into, or formed in the unit because of four principle factors:

- Poor condenser integrity. Any inward leakage of raw cooling water will add impurities to the steam, capable of causing contamination of the steam path components. This is particularly so if the cooling source is brackish or saltwater
- Insufficient chemical analysis throughout the steam/water circuit, particularly in the steam portion of the cycle
- Improper mixing of feedwater treatment chemicals
- Improper operation of demineralizers, producing sodium hydroxide and sodium sulfate

Iron and copper oxides and silicate were the most commonly detected impurities. Iron oxide is the result of corrosion of the boiler internals, while copper oxide is produced from the corrosion of feedwater heater tubes. Other substances, used as solvents and demineralizer resins, help form organic compounds. An important observation is once deposits have started to form, the first three mils (0.003") can decrease blade efficiency by about 1%. It is unfortunate that relatively small amounts of deposits can have a significant effect on state line efficiency. This is particularly so in the higher pressure, high-temperature stages, where the higher Reynolds Numbers cause losses to occur with relatively small levels of surface deterioration.

Because of these degrading effects, there is a continuing incentive for operators to reduce the formation and deposition of compounds. This is necessary both for efficiency and, in the case of chemically aggressive substances, preserving the mechanical integrity of the components and maintaining unit availability levels. Corrosive compounds continue to have a significantly adverse effect on unit availability, and operators are taking greater care to control these levels of impurities that are either corrosive, or can form corrosive compounds.

Boiler design, chemistry, and plant operating techniques are improving considerably. This is having a marked effect upon the extent of carry-over into the steam path, and therefore on steam path efficiency loss.

Silica concentration in boiler water is particularly difficult to control, and the severity of the silica problem has demanded such close attention that silica has almost become synonymous with turbine deposits. In some instances this has lead to the impression that the sole problem is silica. However, most analysis of deposits shows an abundance of metallic oxides and other compounds, which are just as damaging to efficiency, and can be much more significant in terms of forming corrosive compounds.

Deposits that form on steam path components are comprised of soluble and insoluble compounds. Deposits on those portions of the unit operating in the superheat region are mainly soluble (approximately 90%) while those in the saturated region are principally insoluble. This is almost certainly because the soluble deposits that accumulate in the wet region are continually washed by the water deposited from the working fluid, and are therefore dissolved and removed.

Since water is able to dissolve and remove much of the soluble compounds, it is possible to restore steam path efficiency to near-initial condition by washing. This washing process requires the unit be operated at light load for some period, employing low-quality steam that soon converts to a saturated condition. Water formed by condensation is deposited on the components of the steam path, thus washing and removing a large proportion of the soluble compounds accumulated on the surfaces.

Unfortunately, for advanced steam condition units, such a washing process should only be regarded as a temporary measure. This is because the insoluble deposits left behind on the steam path surfaces act as a foundation upon which the soluble components can be redeposited once the unit returns to normal service. Under normal circumstances the efficiency will have deteriorated to its "pre-wash" condition within a week or two of continued normal operation.

Figure 6.3.5 shows a time-efficiency curve for a normal steam-condition, fossil-fueled utility unit, and the effect of "water washing" with low-quality steam to remove the water-soluble compounds. It can be seen that after start up, initial efficiency deteriorates—initially at a fast rate—but this efficiency loss is almost completely recovered and efficiency restored to near-original condition by water washing. It is also clear this improvement is maintained for relatively short periods after return to normal service, as the deposits form again within a period of a few weeks.

There are, however, geothermal installations (and possibly other installations using a lower-quality steam) where the steam contains significantly higher levels of impurities than found in a normal utility unit, and where such washing has been found to be quite effective. In such installations, washing should be considered a way to retain acceptable efficiency levels.

The most effective manner of removing accumulated compounds is to open the unit, remove the rotor and diaphragms or stationary blade carriers, and mechanically clean the steam path elements with grit or some other suitable substance that removes all deposits. This is explained later in this chapter. Such a cleaning operation should become an integral part of any extended outage, whether the rotor remains on-site or is taken to a service shop. Portable blasting equipment can be moved on-site to considerably reduce the cost of this cleaning process. Since units are likely to be opened for normal maintenance outages only on a four-or-more-year cycle, the importance of station chemistry and water purity control between these outages cannot be overemphasized.

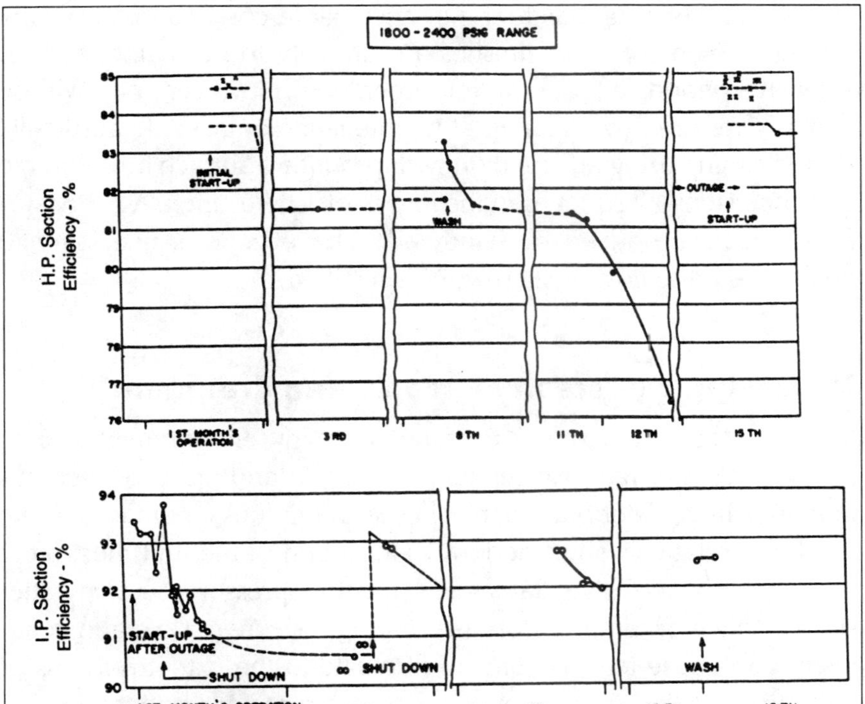

Figure 6.3.5—Time-efficiency curves for both high and intermediate pressure sections, showing the effects of efficiency on water washing and mechanical damage.

It is unlikely the original surface finish of the blade can be maintained for extended periods over the entire blade vane surface. Those portions damaged by surface deterioration will have a significant effect upon unit state line efficiency. The optimum surface finish required for any component for minimum energy losses can be determined from a knowledge of the steam conditions and the stage Reynolds Number for flow across the element. It has also been shown that polishing the steam path components in the direction of steam flow can reduce energy losses significantly.

The surface finish produced on new steam path components (stationary and rotating) is selected from design-determined requirements and what manufacturing at acceptable costs can realistically produce. Even so, these finishes are unlikely to be maintained for extensive periods after going into operation. However, they will be partially restored by cleaning at maintenance outages. Realistically, blast cleaning using any medium will produce a surface finish on the vane and sidewall surfaces in the region of 125√μ- inch. A better surface can be produced by handwork. However, it is necessary to determine if the cost of such work is justified.

Modifying the steam/corrosive environment

While corrosive action can start in many environments, some promote a faster rate of chemical aggression and produce a condition in which deterioration reaches a dangerous or unacceptable level in a relatively short period. That portion of the unit most often subjected to corrosive damage is the low-pressure section latter stages. This is where moisture is present, and where several mechanisms can cause temperatures to fluctuate to the extent portions of the unit are either in the wet or superheated regions of operation.

The low-pressure section of the unit is that portion in which water forms by condensation—the portion that exists below the "Wilson Line." This water will capture available corrosive ions, and then wash or carry them into hideouts or crevices where they will accumulate. Eventually, under suitable environmental conditions, they will become actively corrosive and attack the metals of the steam path.

This effect is demonstrated in Figures 6.3.1 and 6.3.2. Here, at rated steam conditions and full load, steam path conditions are shown as "A,B,C,D,E," with condition "K" representing the condition of the steam at exit from the first low-pressure section stage.

At full load—and after expansion through the high-pressure section, reheating, and further expansion through the reheat and stages of the low-pressure section—steam will flow across the saturation line, where the two-phase mixture will become supersaturated. Water will then form at the Wilson Line—the point, or condition, at which heat transfer from the vapor to liquid phase occurs. From this water-formation condition until the steam reaches the condenser, there is free water in the working fluid. At this point [for many fossil designs is the penultimate (L-1) stage of the unit], corrosive ions are captured and transported by the water. These corrosive ions can be carried into regions of the stage where they will accumulate and, under the correct environment, concentrate. If this unit continues to operate at full load these ions will not be able to affect the material of the row, as they will be dilute and not react with the steel. Unfortunately, two conditions can occur, both capable of making these collected ions chemically aggressive:

- steam path condition changes
- unit shutdown

Steam path condition changes

Any mechanism of operation that changes steam conditions within the unit will modify the environment at the location where the corrosive ions exist. Consider those factors leading to such changes:

Unit load changes. As the demand for output from a unit is reduced to match system demands, the quantity of steam admitted to the unit will be reduced, causing a shift to the right on the Mollier Diagram (Fig. 6.3.1). This will move some blade rows of the steam path from a saturated to superheated condition.

Initial steam condition changes. If the initial steam conditions, pressure, or temperature are adjusted to meet lower load demands, the expansion line will again adjust, causing a change in wetness at the stage positions.

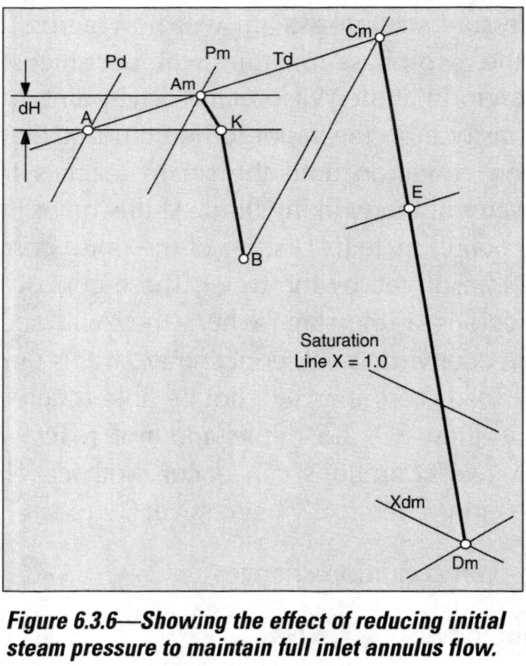

Figure 6.3.6—Showing the effect of reducing initial steam pressure to maintain full inlet annulus flow.

Consider Figure 6.3.6. To meet reduced load demand and allow the control stage nozzle to remain fully open (not introducing a partial admission loss), the initial pressure is reduced from "Pd" to "Pm." This is shown as occurring at the same initial temperature "Td." With this change in steam inlet pressure, the reheat/low-pressure expansion line is moved to the right on the Mollier Diagram. The consequence is that the low-pressure section enters the saturated region later in the expansion, and the stages moved into the superheat region will now begin to drive moisture off from the corrosive ions. This will recombine into molecules, be deposited, and become active, introducing the potential for corrosion.

Reheat temperature changes. The effects of reheat temperature changes were considered in Figure 6.3.2. Here, the effects of not maintaining the temperature modify the expansion line.

The most common adjustment is reducing the reheat temperature. Operators will occasionally do so to reduce the temperature on the first reheat stage rotating blades, which are a highly stressed component. In this situation, the amount of reheat/low-pressure expansion occurring in the superheated region is reduced, and more low-pressure stages are moved into the saturated region.

This reheat temperature reduction will move stages into the moisture region, reducing their potential to be affected by corrosion. However, it will also make some stages that have moved from the superheated into the moisture region candidates to collect corrosive ions, which will then migrate into hideouts and concentrate. If the reheat temperature is restored to its design value, these saturated stages now become superheated and any aggregation of corrosive ions could become active.

Steam path efficiency changes. If the efficiency of any portion of the steam path expansion changes, this will affect the slope (or form) of the expansion line.

Consider Figure 6.3.7—A reduction in the entire low-pressure efficiency section after leaving the reheat section causes a reduction in the available energy from this section of "dH." Here the low-pressure expansion line has moved to the right on the Mollier Diagram and will have moved stages from the saturated to superheated condition, making them more susceptible to corrosive action if corrosive ions were present on the steam path components.

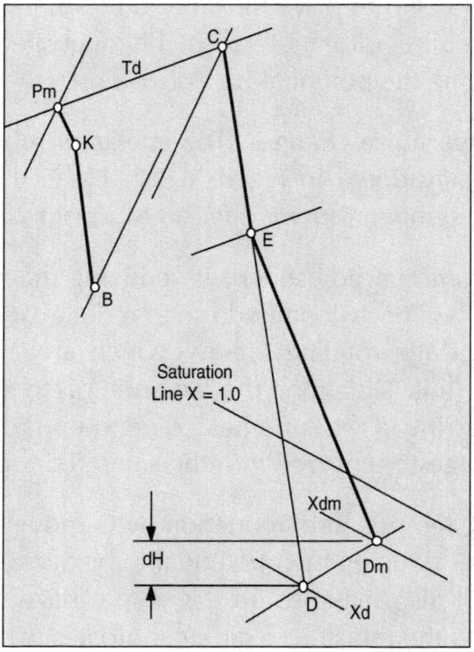

Figure 6.3.7—The effect of efficiency reduction on section "E-D" is to reduce the moisture content in the latter stages, and expose the expansion to a lower moisture region expansion.

Unit shutdown

When a unit is shut down, it is placed on turning gear until the internal temperature of the components has been reduced to a level that is safe to be allowed to come to rest. The principle concern with allowing a rotor to come to rest prematurely is that hot gases within the steam path will rise, causing the casing to hump upwards, possibly introducing permanent damage, or damage difficult and expensive to correct. There is also concern the rotor will take a minor, temporary bend, introducing a need for straightening by some suitable method before the unit can be restarted.

While a unit may cool slowly and retain safe conditions after shutdown, it is entirely possible certain components will attain a considerably higher temperature than they experience from the expanding steam. There are also locations in the unit where stresses exist even when the unit is not operating. Among these locations are the shrink-fits on wheels, at contact points where tie wires pass through blade vanes with interference, and at coverband locations. Therefore, with a localized temperature rise and the presence of corrosive agents, corrosion can be initiated. This damage is promoted by the shutdown condition, and will progress when the unit is returned to service.

THE REMOVAL OF CHEMICAL DEPOSITS FROM THE STEAM PATH

Deposits on the steam path are detrimental to its total performance. Blast cleaning will help remove such deposits and restore efficiency. However, it is unlikely any cleaning process will remove any large amounts of the chemically aggressive compounds that have migrated into the hideouts, and exist in various locations throughout the steam path. If it is suspected such compounds exist, their removal should be considered if their presence is anticipated forcing the unit from service within an unacceptable time frame.

Deposits on both the vanes and sidewalls have the potential to cause a deterioration of stage efficiency. This efficiency loss occurs as a consequence of inducing flow separation and turbulence in the flowing steam. These deposits can also contain corrosive ions that under the correct conditions can cause corrosion. Therefore, when a unit is opened for inspection and the steam path becomes available, it is a normal procedure to clean these components. This is normally done with some blast cleaning process, although hand techniques can be used most effectively. They tend to be labor intensive and more expensive.

Despite the amount of time the turbine supplier's manufacturing department devotes to preparing the steam path components to achieve design surface requirements, there will always be some form of build-up on the original surfaces—and deterioration that will introduce energy losses when the unit goes into service. This cannot be avoided.

However, the maintenance engineer has a responsibility to re-establish the original conditions of the steam path considered optimum from a performance/time/cost perspective, to the greatest extent possible and consistent with acceptable costs. Normally, re-establishing original conditions is not possible without excessive and prohibitive costs. Such actions could cause some change in the profile of the original components, which would be unacceptable. There is fortunately a suitable compromise—to blast clean the components.

Corrosive attack considerations

We know that units with recognized corrosive compounds have been returned to service. This is unwise, and will ultimately cause costs to be incurred that could be far greater than the cost of initial removal when these compounds are found. Unfortunately, the cost of removal may be high, requiring some degree of disassembly, and under the most severe conditions, the replacement of parts. The temptation to ignore this situation is often difficult to overcome.

STEAM PATH CLEANING METHODS

The removal of deposits is a normal process employed when a unit is removed from service. This cleaning is undertaken for one of two major requirements:

- Removal of efficiency-deteriorating compounds
- Nondestructive examination of components. To attempt to conduct a number of tests on components covered in deposits would not yield meaningful results

A number of considerations will have to be evaluated in selecting the most appropriate cleaning process. For instance, for a unit removed from service after only a short period of operation (*e.g.*, for a warranty inspection), the cleaning process, while removing deposits, may cause the quality of the surface to deteriorate. This is unfortunate, but may be required to allow the nondestructive examination. A too-aggressive cleaning method can cause surface deterioration to the extent micro-cracks could be hidden by microscopic deformation of surface material.

While these are minor considerations—particularly for a mature unit—the cleaning process selected should consider them. Available processes that should be considered are as follows:

Blast cleaning of the steam path

Blast cleaning of steam path components is a normal maintenance procedure, and a normal activity at every outage. If an operator intends to make efficiency assessments at an outage, the condition of the steam path components impacting on expansion efficiency should be assessed before any attempt is made to remove deposits. While the act of removing deposits may appear to be a relatively simple operation, there is always the possibility such cleaning can cause some further level of deterioration if it is not controlled and undertaken with care. Therefore, to effectively clean the components, certain parameters of the cleaning process must be controlled.

Blasting media. The media used must be of such a quality that it will remove the deposits without removing steam path surface material (metal) from the components. Common materials used for this blasting follow:

- *Grit (silicon carbide).* This must be specified as to size, shape, and age. The extent to which this grit can be reused or recycled should also be established. The grit can be used effectively in a water jet. This cleans the surface by removal of deposits and does not create the dust common to normal grit blasting. It may however, be more convenient to undertake this cleaning in the open, or in the plant basement. It is not normally possible to recover much, or any of the grit for reuse

- *Aluminum oxide.* This material removes deposits and leaves the surface in a clean condition. There is however, a small possibility this could obscure small cracks by "hiding"—peening over the edge—of any fissure

- *Walnut shell.* This is an effective material, and leaves the surface in an excellent condition in terms of both finish and ability to be nondestructively tested. This is more expensive in terms of the cleaning material, and the time required to effectively clean the surface

- *Glass bead.* This is a proven media and works well. Normal exposure times for bead are longer than with grit, but any possible impact of glass bead on the metal surface is not as severe and does not cause damage as easily. (This is better used after grit blasting and NDT/repair.)

- *Zirconium oxide.* This is a superior method of cleaning. The blasting medium is an abrasive sand taken from mines in Florida. The sand particles are rounded, and remove surface deposits while causing a minimum of surface roughening
- *Carbon dioxide pellets.* A process capable of cleaning steam path components with less impact damage. At this writing, the process has not been proven suitable for steam turbine components, but it is used widely in the nuclear industry and could possibly be developed for use on steam path components

Air pressure. The air pressure used to blast the components is critical. This pressure must be controlled, as certain blasting media can deform the surface if the air pressure becomes too high, even for short periods. The air pressure should be calibrated by the use of Almen targets. This is normally in the range of 30-50 psi.

Nozzle size and shape. The size and shape of the spray nozzle can influence the impact intensity. With certain media, excessive blasting or excessive exposure time can cause damage.

Distance of the nozzle from the target surface. This is difficult to control because the steam path does not present a flat surface to the particle jet. However, the most critical regions—those most likely to suffer mechanic deformation—must be considered when moving the jet over the component surface.

The angle of spray jet inclination. The spray angle, like the spray distance, is difficult to define and control in any particular application. In fact, the angle at which the cleaning media is shot onto the surface is not constant for the contoured surfaces of the vanes and sidewalls. However, some controlling restrictions should be placed on this together with the distance the nozzle should be from the components being cleaned.

Blast cleaning is a routine procedure, but if not controlled it can cause damage to the vanes of the blade system. Figure 6.4.1 shows the discharge edge of a stationary blade vane that has been blast-cleaned without control of the process. In this instance the thin discharge edges had degraded and thinned to some degree as a consequence of solid-particle erosion. These outlet edges have been bent over under the effect of the cleaning media (200μ grit).

Figure 6.4.1—Vane discharge edge damage caused by poor control of the grit blasting process.

This damage could have been avoided by an initial examination of the blade surfaces and general condition, and then by blasting to an approved procedure. Where blade vanes have thin discharge edges, it is best to re-calibrate the blasting procedure on the blades so they will not be damaged. When these blade elements are returned to service, they will require changing at some future outage, and the replacement blades could be ordered at the present outage for change in four to five years. The fact they are damaged to the extent shown in Figure 6.4.1 requires corrective action of some type to be taken at the present outage.

Hand cleaning steam path components

There are situations in which hand cleaning of the surface is required or justified. Normally if hand cleaning is employed, it involves chemical agents and is undertaken on individual components, using a solvent on a cloth or brush. There are advantages to this when the deposit amount is small and the original surface is still present on the steam path components. It is important that the chosen solvent does not contain sulfur, chlorine, sodium, phosphorus, or any constituent that could form aggressive compounds under the action of steam and/or heat.

Note: Solvent or chemical cleaning can be hazardous to both worker and components. Extreme care is required in this operation.

Occasionally, the operator will elect to remove deposited scale by hand methods. It is normally not recommended with use of any metallic or hard device such as a file. Often the convex side can be cleaned using a fine emery cloth drawn across the surface from the two edges. However, it is difficult to clean the concave surface in this manner, due to the surface curvature and the difficulty in gaining access to the entire expansion passage. Abrasive blasting is preferred.

On-line water cleaning

On-line water washing is a contentious matter, considered by many to be ineffective. The rationale is that many deposits can consist of both soluble and insoluble compounds. The soluble compounds will be removed by washing, but the insoluble will not, unless the water removes them by mechanical means such as impact forces.

An example is the oxides typically formed by iron and copper. These will be difficult to remove by water washing, and unless the complete scale is removed, the removal of portions will leave small step changes in surface condition causing flow separation, particularly for stages with large Reynolds Numbers.

However, certain units do appear to benefit from the effects of a water wash. This is particularly the case for low-steam condition units, and it has been demonstrated to be effective on geothermal units operating on saturated steam.

What, then, should operators consider when deciding whether to implement "water washing" procedures? Certain information is required:

- What level of improvement is achieved? The output before and after should be determined with as great an accuracy as possible
- How long is any improvement maintained? This again should be determined by repeated output tests, run frequently after washing

There is no simple answer to these questions. Certainly on turbines operated on high-temperature, high-pressure steam, there is considerable doubt as to the ability to sustain *any* improvement, but if there are improvements, then such procedures can be justified.

DEPOSITION PATTERNS

Deposition patterns are difficult to predict on any component of the steam path, as they appear to depend to a considerable degree on the type of contaminants in the system, unit load factor, and a number of other difficult to define characteristics of the unit and the mode of operation.

There are however certain inherent differences between those stages in the superheated and those in the saturated regions.

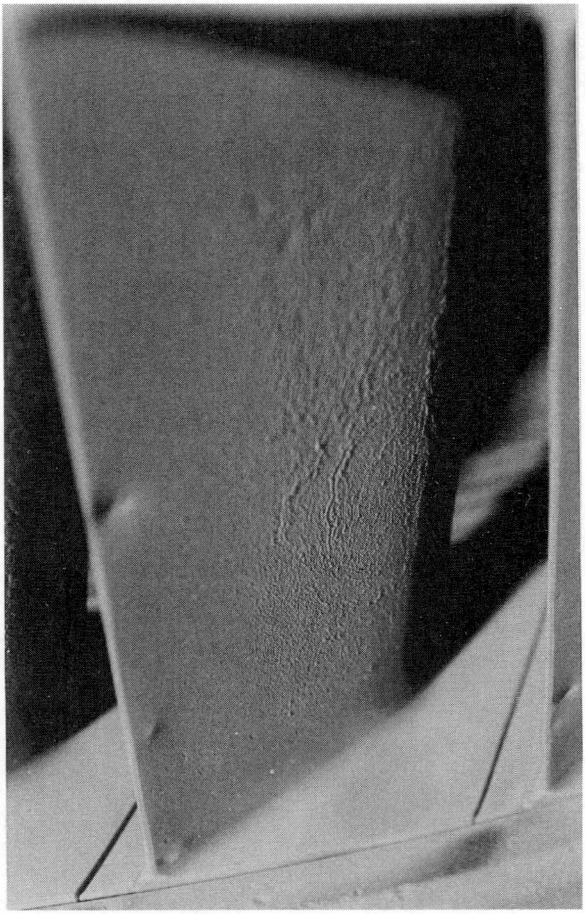

Figure 6.5.1—A high pressure reaction blade with deposits on the suction face.

Figure 6.5.1 shows a high-pressure blade from a reaction unit. The vane has a small turning angle. The deposition pattern is relatively constant across the suction face. There is a deeper deposit in the region of the throat, which can be due to a sudden reduction of the pressure at this location, as well as influence of the flow separation that might occur due to profile curvature.

The deposition that occurs on superheated surfaces is inclined to be more evenly distributed across the chord of the vanes, with deposits on both suction and pressure surfaces.

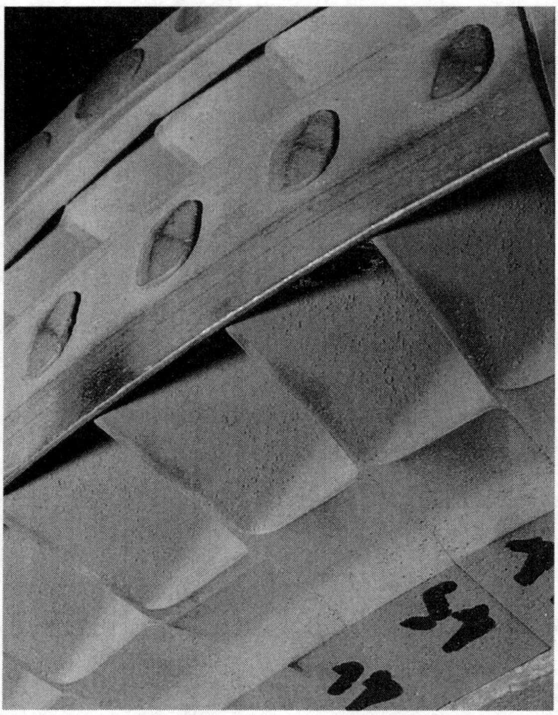

Figure 6.5.2—A high pressure impulse blade with deposits on the suction face.

Figure 6.5.2 shows blades from a high-pressure section that had been in service for more than six years after the previous outage and cleaning. Deposits are light and evenly distributed across the suction surface. As would be expected in the majority of stages, deposits on the pressure face tend to be thicker than on the suction face. However, the suction face tends to have thick deposits following the position of flow separation. (Fig. 6.2.2 shows a geothermal application.) This difference results from the steam and transported contaminants being forced into more intimate contact on the pressure side, tending to adhere in greater quantities. Figure 6.5.3 shows heavy deposits on the inlet nose of a unit, stationary blades, and Figure 6.5.4 show the suction face from this same stage. In this case, boiler control has been lost and heavy oxidation occurred, causing significant deposits on both the stationary and rotating blades.

Figure 6.5.3—Heavy deposits on the inlet nose of a stationary blade row.

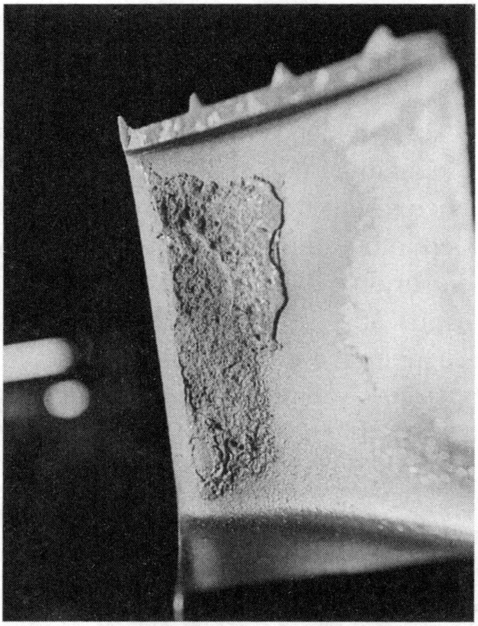

Figure 6.5.4—Heavy deposits on the pressure face of a stationary blade row.

As scale forms on the vanes and sidewalls and become thicker, it is likely that portions will become loose and detach. This is particularly the case under the influence of steam temperature changes, when differences in coefficients of expansion between base steel and the scale cause rupture at the surface interface. Scale that breaks the bond to the base metal will flake off, leaving a smoother surface below, as shown in Figure 6.5.5. This flaking and scale loss does nothing to improve the efficiency of expansion since the small step in the cleared region is more likely to cause flow separation than any surface roughness that has been eliminated.

In the radial direction, there can be a difference in the amount and rate of deposition. This deposition rate is dependent to a degree upon unit load patterns, vane turning angles, and turning angle variation along the vane. One can also see a change in the type of

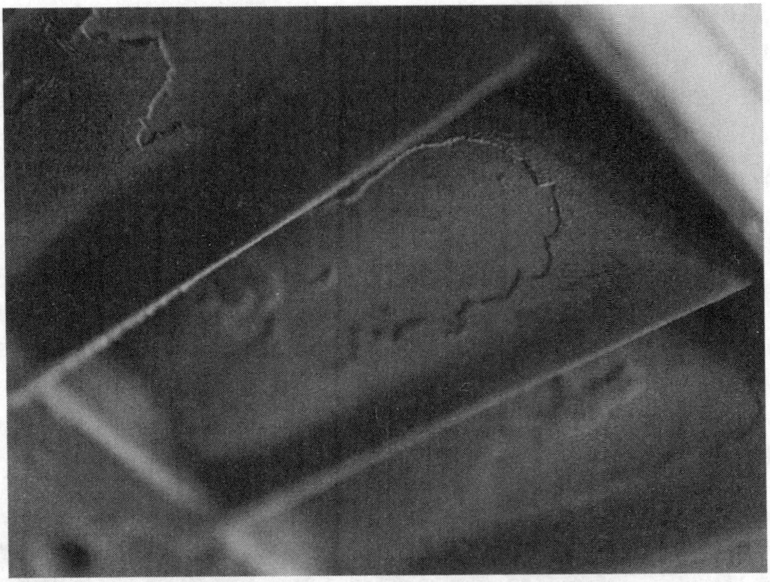

Figure 6.5.5—Deposits on the pressure face of a rotating blade, showing the effects of scale flaking.

Figure 6.5.6—Heavy deposits formed under the cover band of a rotating blade row.

deposits along the radial length of the rotating blade vane. This is probably due to the greater density of some compounds and their being centrifuged towards the coverband to a greater extent. Figure 6.5.2 shows a characteristic change, with a greater copper oxide deposit towards the outer regions of the blade.

Figure 6.5.6 shows a portion of a stage in which deposits on the vane after 5+ years of operation are relatively minor and at the anticipated level. However, there are heavy deposits under the coverband. These deposits are obviously compounds that have been centrifuged out, and have been deposited and collected on the coverband underside. This can represent a considerable efficiency loss, particularly on small radial-height stages.

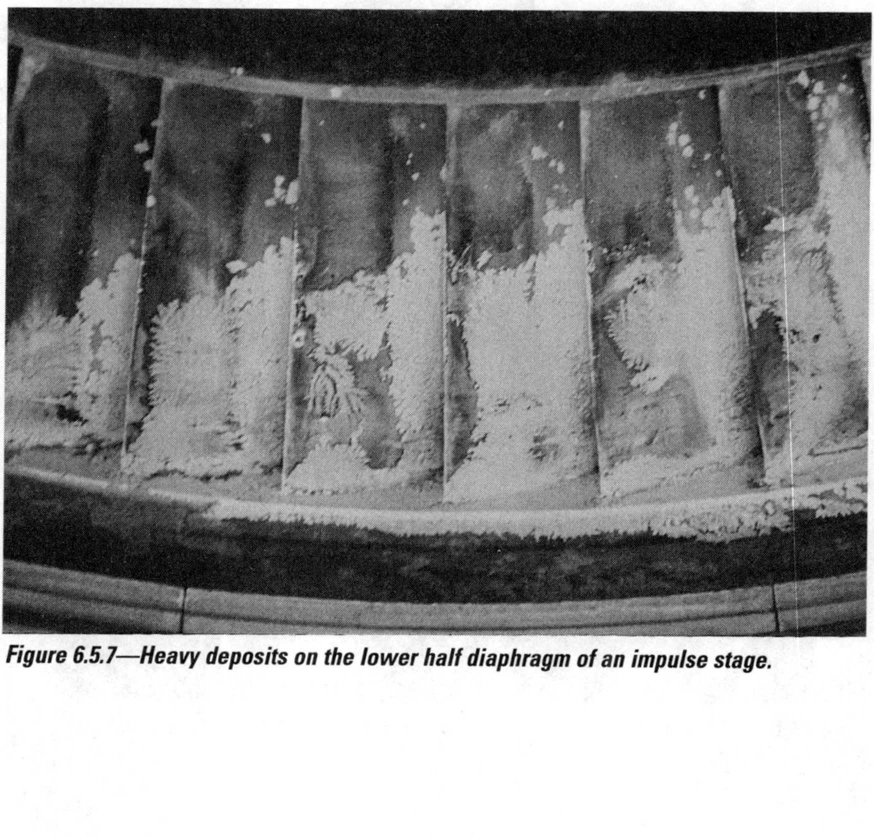

Figure 6.5.7—Heavy deposits on the lower half diaphragm of an impulse stage.

Because they lack centrifugal effects, stationary blade rows tend to have somewhat different deposition and collection patterns than the rotating rows. However, the composition of the deposits is essentially the same—they are only a function of the environmental steam conditions at the stage. Figure 6.5.7 shows a portion of a stationary vane with heavy deposits on its outer flow sections. (This was the lower diaphragm half.)

In the stationary blades of the saturated region, deposits often occur from the water film flowing across the vane in a general radial direction. This is shown in Figure 6.5.8, where deposits are in a radial/axial direction, deposited in the direction of the flow of water film.

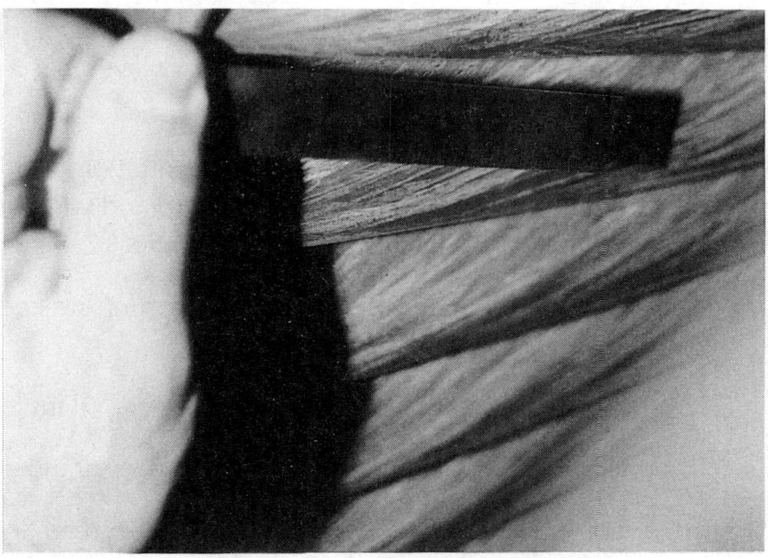

Figure 6.5.8—Deposits left on a low pressure blade from the moisture film flowing across its surface.

Defining surface finish

The surface finish defined for any component or piece of steam path hardware is selected by the designer to allow the element to be mounted, and to operate within the unit in the local steam environment. This operation must be such that the performance potential of that component is not compromised in any way.

The definition of surface requirements is the responsibility of design engineering. The specification decision is often made in concert with the manufacturing engineer, who will be responsible for selecting and defining the processes used to manufacture the piece. There are five basic considerations in selecting and specifying a surface texture (finish and direction if applicable):

- To minimize frictional surface losses from the flow of steam across the component. The surface finish can also contribute to the premature separation of the boundary layer that would increase the fluid friction losses
- To minimize the possibility of stress concentration occurring at a position where stresses levels are high, and where such concentration could lead to premature failure
- To facilitate the non-destructive examination of the component from the surface
- To allow the surface to form one face of a steam tight joint
- To provide an acceptable cosmetic surface appearance

These surface finish requirements—including the direction in which the surface has its main, or "lay" direction—are occasionally in contradiction. However, since the responsibility for defining both the degree and direction are a prime responsibility of the design engineering function, the designer is expected to evaluate and select between contradicting requirements.

Naturally, the processes used to manufacture it will influence the surface finish selected for any component. The design function will select among alternative manufacturing techniques to ensure the most demanding requirement of the individual components are met, consistent with economy of manufacture.

The production of a surface finer than the minimum requirement established by the design engineer represents an additional cost that may be difficult to justify. However, such a decision is often made to improve a surface beyond the minimum requirements, conscious of the additional manufacturing costs, but considered justified in terms of the overall appearance of the product. Such a surface finish cannot be considered to be an improvement to product quality, as it represents work beyond the minimum justified from technical considerations.

For the blades that deflect the steam and produce a diverging passage intended to expand it in a controlled manner, the most demanding requirement (in terms of surface finish) is normally that related to surface frictional loss. Surface texture requirements, from a frictional loss perspective, are associated with the stage Reynolds Number "Re," which the designer determines relative to the steam velocities, blade geometry, and the local steam environment.

Once these values of "Re" are established from the calculated characteristics at the stage, the steam flow characteristics can determine the boundary layer conditions. This includes thickness and height of the surface irregularities required to minimize the possibility of a rough surface causing flow separation and, therefore, turbulence.

For these reasons, specified manufacturing surface requirements are important to establish unit expansion efficiency, even though it is known this surface will subsequently deteriorate due to mechanical damage and impurity deposition.

At those locations on the blade where stage hardware is attached, the major considerations are those related to stress. The reason is poor surface finish can be a significant cause of failure at these locations. Where holes are produced in the blade, the surface finish must be defined and followed closely. At tenon positions, it is important that the surface be defined, and in both cases, requirements concerning surface condition must be carefully defined.

Metal surface finish specifications

Various definitions are used to specify surface finish. Each of these "texture definitions" addresses surface irregularities and the height of the irregularities above a mean or nominal surface.

Any "smooth" surface examined microscopically will be found to comprise irregularities introduced by the uneven cutting action of the tool or other manufacturing process (Fig. 6.5.9). This figure defines typical irregularities or discrepancies between an actual surface, and the mean or nominal profile.

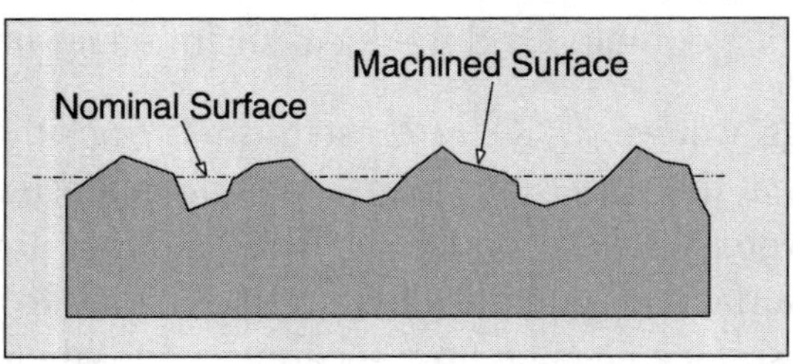

Figure 6.5.9—A nominal "flat" machined surface.

These irregularities are a consequence of either the metal removal process (and the condition of the cutting edge of the machine tool) or the surface formation from a casting or forging process. The level of conformance of the measured surface to the nominal surface is affected by diverse and complex factors:

- For metal cutting—the quality and maintenance of the cutting edge or edges of the tool piece, the means of support of the material, the rate at which the excess material is removed, and the suitability of the machine tool for that particular operation
- For a cast or forged process—it's a wealth of factors. A major influence is the quality or form of the master used to produce the surface

Figure 6.5.10 shows a portion of a machined surface, providing the most frequently used definitions of surface texture.

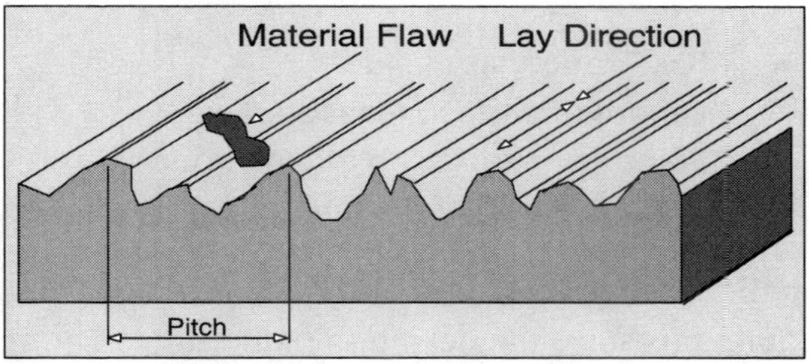

Figure 6.5.10—Definitions quantifying a surface.

Polishing produces the final surface for many steam path components (mainly stationary and rotating blades). However, the amount of material removed is small, as the polishing process does not remove material to the extent the surface contour is destroyed.

Surface roughness can be established using one of several definitions. Each definition provides a measure of the variation of the height of the irregularity from peak to peak, and compares these with the valleys between them. In order to provide a subjective measure and a means of specifying surface roughness, the following definitions of roughness have been developed, and are in use in various portions of the turbine manufacturing industry:

Root mean square (RMS) value. The root mean square value of roughness is an older definition, now tending toward disuse, but still applicable, and used by many sections of the industry. By this definition the surface roughness is determined as the square root of the sum of the mean values of deviation from the mean surface squared. This value can be calculated from:

$$Yrms = \sqrt{\frac{1}{L}\int_{x=0}^{x=L} Y^2 \, dx}$$

where:

Yrms	=	RMS roughness value
L	=	Length of surface measured
y	=	Point deviation from mean surface
x	=	Point position of the surface being gauged

Arithmetic average (AA). The arithmetic average value of roughness or texture, measures the deviation of the surface from the mean, and can be found from:

$$Yaa = \frac{1}{L}\int_{x=0}^{x=L} y \, . \, dx$$

This value is now used extensively, is easy to apply and measuring instruments are available for gauging surface directly.

Centerline average (CLA). The centerline average value of roughness is essentially the same as the arithmetic average, except the centerline is established so the area of the valleys and peaks above and below the centerline are equal (Fig. 6.5.11). In this case the surface roughness can be determined from:

$$Ycla = \frac{(U1 + U2 + U3 + \ldots + Un) + (L12 + L2 + L3 + \ldots Ln)}{n}$$

where:

U1.... Un	=	The upper peak values
L1..... Ln	=	The lower peak values
N	=	The total number of upper and lower peak values

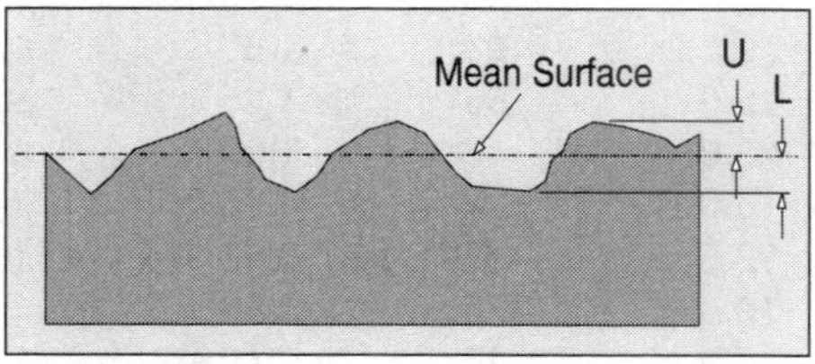

Figure 6.5.11—The surface, showing upper "U" and lower "L" variation from the mean.

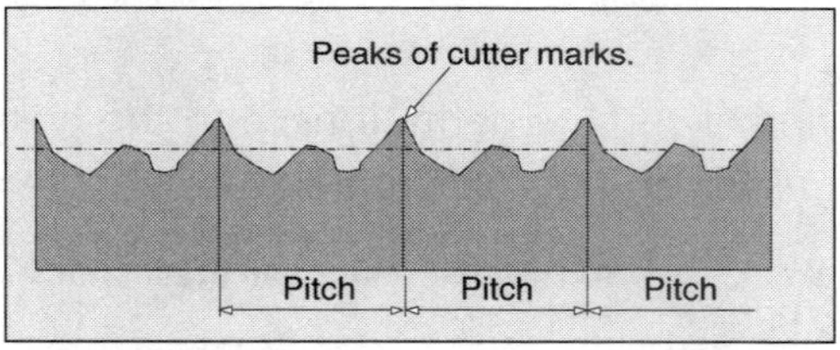

Figure 6.5.12—Showing "waviness" with repeated surface irregularities.

"Roughness" refers to irregularities between adjacent peaks produced by the metal forming action (Fig. 6.5.12). "Waviness" refers to a repeating characteristic, causing a repetitive irregularity at pitch "P". The pitch of the waviness is normally caused by the feed rate of the cutting surface. For a well-maintained machine tool mounted correctly, the waviness should be minimal.

Comparisons of surface specification. Drawings and other specifying media define surface finish requirements in either imperial or metric units. Table 6.5.1 provides equivalent value for these two scales.

Imperial System µ-in.	Metric System µ-m.	Imperial System µ-in.	Metric System µ-m.	Imperial System µ-in.	Metric System µ-m.
2000	50.0	100	2.50	6.0	0.160
1600	40.0	80	2.00	5.0	0.125
1300	32.0	63	1.60	4.0	0.100
1000	25.0	50	1.25	3.0	0.080
800	20.0	40	1.00	2.5	0.063
600	16.0	32	0.80	2.0	0.050
500	12.5	25	0.63	1.6	0.040
400	10.0	20	0.50	1.3	0.032
320	8.0	16	0.40	1.0	0.025
250	6.3	13	0.32	0.8	0.020
200	5.0	10	0.25	0.6	0.016
160	4.0	8	0.20	0.5	0.012

Table 6.5.1—Equivalence of Imperial and Metric Surface Textures

The ISO series of surface definitions, and the possible manufacturing methods used for their production, are given in Table 6.5.2.

Roughness Value µ m.	µ in.	ISO Number	Common Production Methods
50.0	2000	N 12	
25.0	1000	N 11	Flame cutting, sand casting
12.5	500	N 10	Sawing
6.3	250	N 9	Planing, shaping
3.2	125	N 8	Drilling, milling, investment casting
1.6	63	N 7	Broaching, reaming, boring
0.8	32	N 6	Turning
0.4	16	N 5	Grinding, honing
0.2	8	N 4	Polishing, lapping
0.1	4	N 3	Super finishing
0.05	2	N 2	
0.025	1	N 1	

Table 6.5.2—ISO Definitions of Surface Texture, and Methods of Producing It

STEAM PATH EFFICIENCY DETERIORATION

In its new or "mint condition," a turbine steam path will, if manufactured to specifications, attain maximum state-line efficiency without either inducing excessive stress, or introducing excessive manufacturing costs to the extent the economics of production cannot be justified, *i.e.*, the designer has established a balance between mechanical integrity and efficiency, and the optimum parts have been specified and assembled.

However, with time, this original efficiency will deteriorate. With technology currently available, it is not possible to eliminate such deterioration entirely. But it is possible to minimize its wasteful effect by paying attention to detail during design, assembly, manufacture, operation, and maintenance.

A considerable amount of skill and expertise are required to design, manufacture, assemble stationary and rotating blade elements, and to assure the quality required to achieve the high efficiency demanded of modern operation. Important factors are the shape of the vanes and the form of the passage between both stationary and rotating blades. In addition, the degree and direction of the surface finish produced on the vanes must follow design specifications. Therefore, consider carefully any action detracting from the initial manufactured condition (by modifying the surface form) or causing a modification of the profile and passage shape sufficient to produce excessive turbulence or premature separation of the boundary layer. This will contribute to inefficient utilization of the energy available from the steam.

One of the principal causes of deterioration in steam path efficiency is the accumulation of salts, oxides, and silica compounds that are deposited on the steam path components and foul the steam passages. It has been stated that the effects of scale or deposits are often more persistent and severe than all other factors that contribute to efficiency degradation. There are instances where losses in the range of 5.0% have been reported, and occasionally even more severe cases are observed.

Mean effective surface finish

In the original "as-manufactured" condition, blade vanes and sidewalls' surface finish meets design specification. It is also normal for the sidewalls and vanes to have nominally the same finish. In the case of cast diaphragms, it is possible for sidewalls to have a somewhat rougher finish. However, for cast elements of large radial height, such a roughened sidewall can be accepted to a degree. For smaller radial height elements, inner and outer sidewalls can represent a significant portion of the total expansion passage frictional area, and therefore if too rough, the total loss. For these reasons, surfaces on the shorter radial height passages should have some level of dressing to achieve design specification.

The difficulty is to know at what point the sidewalls become significant in the overall determination of loss. It is also necessary to consider the chord width of the vane and the sidewall axial length. In making an assessment of the mean effective surface finish the engineer must make a judgmental evaluation.

It is necessary to be able to quantify the frictional loss occurring within blade passages due to the deposition of various compounds. These deposits roughen the overall surface, and thus induce losses beyond those experienced by the blade rows in their "as-manufactured" condition. There are considerations relating to the deposits in the various passages, and it becomes necessary to consider some factors that allow a "mean effective surface" to be determined.

Experimental investigation shows that losses can be determined in terms of the stage Reynolds Number (Re) and its relation to the surface finish. Therefore, the velocities, surface finish, and their variation throughout the stage are critical to establishing losses. Those factors that should be considered include:

- The only stage velocities that can be predicted with ease, and with a reasonable degree of certainty are those mean velocities at inlet to, and discharge from the expansion passages
- Inlet and discharge velocities are mean values and will vary over the inlet and discharge annulus. They are slower at the expansion passage sidewalls formed by the suction and pressure faces of adjacent vanes, and the inner and outer sidewalls
- The surface finish can vary from pressure to suction face, and also as a function of surface chord position
- The surface finish can vary along the radial height of the vane, on both the pressure and suction face
- The surface finish of the inner and outer sidewalls can be different, as a consequence of deposition patterns, from the vane surfaces and each other

In estimating this "mean effective surface finish," consider the vane profiles shown in Figure 6.6.1. These are drawn for stationary and rotating elements with a large turning angle (impulse or low reaction type). For a reaction stage, stationary and rotating vane profiles are shown in Figure 6.6.2. The following factors need to be considered:

The stationary blade row (impulse)

In establishing this "mean effective surface finish" and the variation of velocity throughout the passage, the following must be considered:

- The stationary blade row has a low entrance velocity at "a," which is the steam velocity from the control stage chamber, or the carry-over velocity from the previous stage rotating blade row. This initial velocity increases through the passage to a maximum at "b." The gap "b-g" represents the throat
- The major portion of the velocity increase occurs over a relatively short distance from "f-f" to the throat "b-g." Therefore, this portion of the surface is critical to reducing the frictional loss
- Heavy deposits from "a-f" on both the pressure and suction face can cause flow disruption and boundary layer separation if they become too deep and rough
- The discharge tail "g-b" is a region where the steam will tend to separate from the vane surface, depending upon the curvature in this region. It can have a varying degree of influence on the overall "mean effective surface"

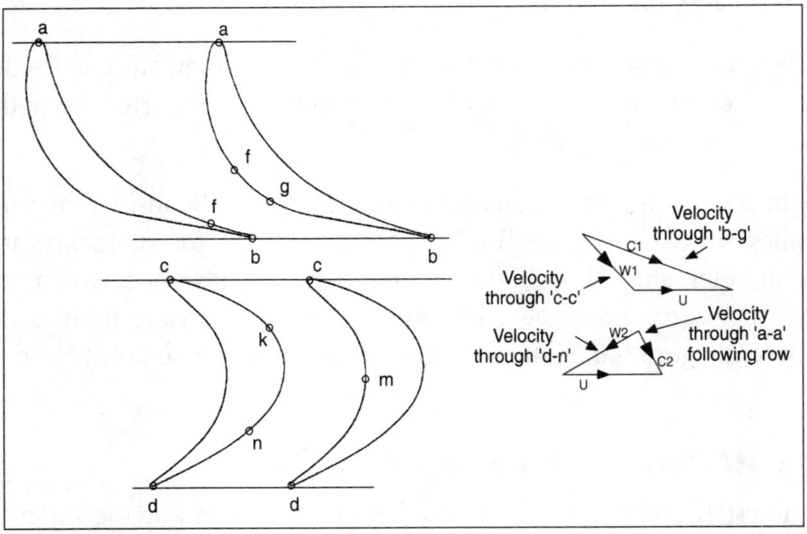

Figure 6.6.1—The low reaction (impulse) stage, and the flow influencing regions on the stationary and rotating blade vanes.

The rotating blade row (impulse)

This form of profile has a large turning angle, forcing the flowing steam, with suspended deposits, into closer contact with the pressure surface. Therefore, the following factors should be considered:

- At entry to the blade row, the velocity at "c-c" is a maximum, and the highest achieved in the blade passage
- The steam expands to position "d-d," where the kinetic energy of the steam has been removed and the velocity is a minimum in the blade row
- The surface finish will tend to be rougher on the pressure face than on the suction
- The vane suction surface region "c-k" has a high velocity across it and will cause flow separation if this becomes too rough with heavy deposits
- The suction surface "n-d" is similar to the stationary portion "g-b" on the stationary vanes but the velocities are much lower

Stationary and rotating rows (50% reaction)

Vane profiles for the stationary and rotating blade rows can be, or are, almost identical in the 50% reaction row. In addition, the turning angles of the blade rows are smaller. In general, the deposition on these reaction vane surfaces should be anticipated to be less severe than for the low reaction or impulse considered earlier for the same steam and contamination conditions.

In addition, the maximum velocities in each of these rows will be lower, because the enthalpy expended across each row is a substantially smaller proportion of the total energy available to the turbine. However, the velocity distribution in the stage is somewhat similar to the impulse, in that the velocity is a minimum at entry to the stationary row; a maximum at discharge from it, a maximum at entry to the rotating blade row, and a minimum at discharge from it. It is also recognized that in the reaction stage, the rotating blade row will produce a velocity increase, but the majority of this velocity increase will be removed from the steam by the rotating blade row.

While there are considerable similarities in the overall considerations of velocity and mean effective surface finish, there are minor differences between the stationary and rotating blade rows that need to be recognized:

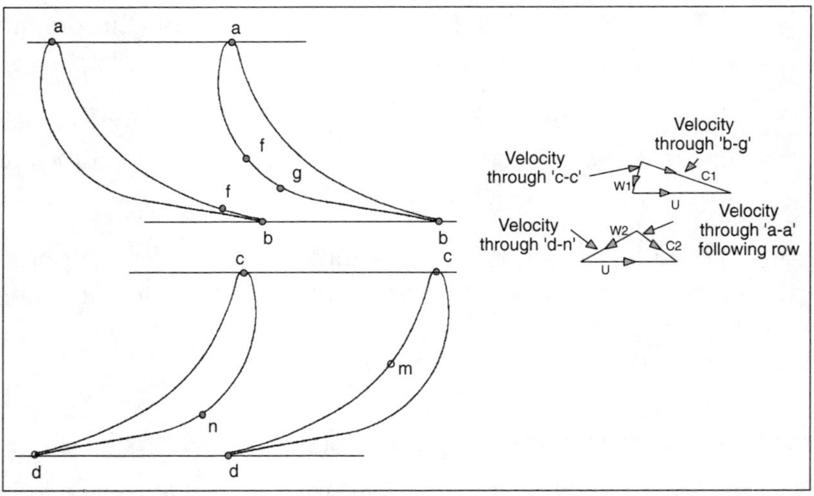

Figure 6.6.2—The 50% reaction stage, and the flow influencing regions on the stationary and rotating blade vanes.

- The inlet to the reaction rotating blade row does not have a portion such as "c-k" with the possibility of flow separation, as the inlet to the reaction row with its close-to-90 degree inlet allows a greater surface control on the steam
- The turning angles of the vanes are smaller, and so the transported contaminants will not be forced into contact with the

pressure faces on either blade row with the same degree of acceleration

It is obvious the assessment of surface finish is difficult, due to the complex nature of the deposition patterns and a number of other factors, *e.g.*, access to measure variation in a meaningful manner. With experience, the operating or performance engineer can make an assessment allowing surface finish losses to be determined and unit losses to be calculated with a reasonable degree of accuracy and with values conforming with measured values of unit deterioration.

Establishing surface finish requirements

The design engineer is responsible to determine the surface finish that is to be produced on the components of the steam path. Under normal circumstances evaluations are required only on portions of the blade vanes. In fact, the controlling influence for each vane is the manufacturing process available to produce them. It is normal for designers to have prepared standards presented in a form from which manufacturing engineers can define the production detail.

As an example of optimum finish requirements, consider a blade with a small ratio of radial height to mean diameter (h/Dm) to the extent the variation of steam parameters and properties can be considered constant along the radial height. The following are the stage parameters:

Stage available enthalpy:	24.3 BTU/#
Design stage efficiency:	97.8%
Stage Reynolds Number:	4 x10E7
Stage steam flow lb/hr.:	3,120.000 #/hr
Specified original surface finish:	32√ μ- inches

Using the curve in Figure 6.6.3, it can be determined that the surface finish correction factor "χρ" is 0.658. Therefore, the additional loss kW(sf) due to surface finish is:

$$\text{kW(sf)} = \frac{24.3 \times (1 - 0.978) \times 3{,}120{,}000 \times 0.658}{3412.14} = 321.7\ \text{kW}$$

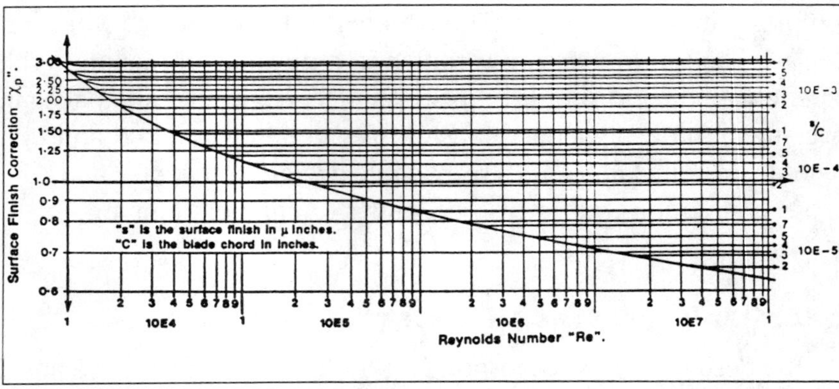

Figure 6.6.3—The loss factor, as a function of Reynolds Number for the ratio of "surface finish to blade vane chord".

As the surface finish deteriorates due to build-up, losses can be determined for each condition. Table 6.6.1 shows the calculated values of kilowatt loss with various values of surfaces finish. Figure 6.6.4 indicates the losses for this stage as a function of effective surface finish. Figure 6.6.5 shows the effect of the efficiency deterioration on the expansion line calculated earlier (where the steam condition locus in the stage moves from "A-B" to "A-X," causing a reduction in the useful enthalpy of "δH" or a total lost enthalpy of "dH").

Surface finish (μ-inches)	32	48	64	80	125	200	300
χp	0.658	0.677	0.700	0.716	0.764	0.846	0.946
Power loss-kilowatt	321.7	331.0	342.2	350.1	373.5	413.6	462.5

Table 6.6.1—Kilowatt Loss as a Function of Effective Surface Finish

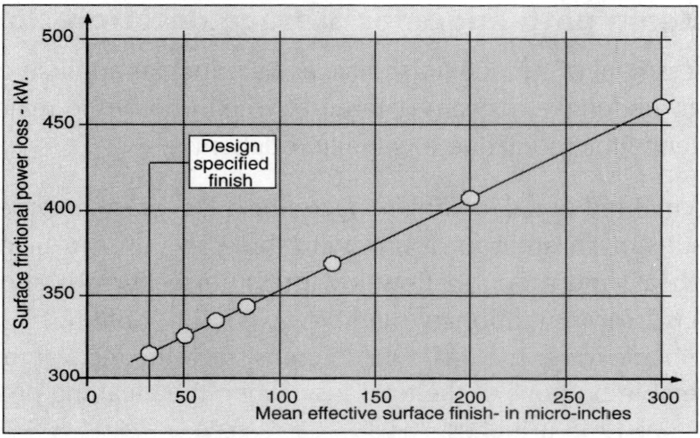

Figure 6.6.4—The stage losses as a function of vane surface finish.

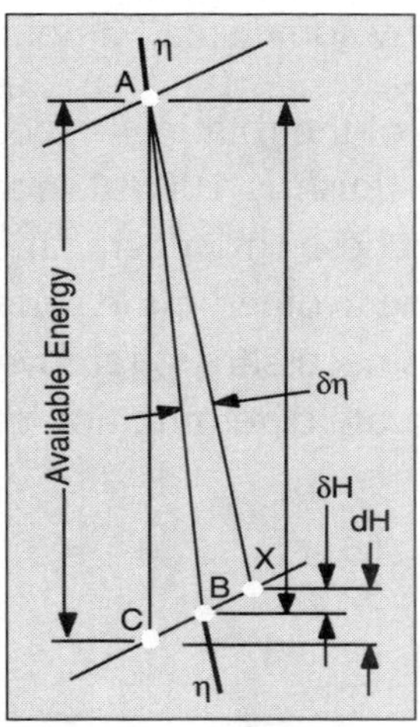

Figure 6.6.5—Showing the effect of state line efficiency deterioration of available energy.

The steam path audit for surface deterioration

Assessment of surface finish has, as its principal application, the ability to undertake a steam path audit when one of the major loss groups encountered is due to surface roughening.

The method of calculating surface finish loss was shown earlier. The results of an audit for a unit with three sections—a high-pressure, reheat, and a double-flow low-pressure section—is shown in Table 6.6.2 for the stationary blade rows, and in Table 6.6.3 for the rotating blade rows. The final two columns show the recoverable and irrecoverable portions of the total loss, when the cleaning produces a surface of 125√ μ-inches.

Note: In Tables 6.6.2 and 6.6.3 it can be seen there are rows where the estimated surface finish, as removed from service, is finer than that obtained after cleaning. This means the cleaning action will increase surface losses by some small amount.

Example: In stage 4 of the rotating blades as removed from service, the mean effective finish is 100√ μ-inches; after cleaning, it is degraded to 125√ μ-inches. This degradation introduces an additional loss that cannot be avoided if the surface is to be nondestructively examined and it is necessary to remove any scale from the surface preventing an accurate determination of the blade condition.

(1)	(2)	(3)	(4)	(5)	(6)	(7)	(8)	(9)	(10)	(11)
High Pressure Section										
Stage	Flow Lb/Hr.	Re	Surf. Fin (Manuf)	kW (Manuf)	Surf. Fin (Service)	kW (Service)	Surf. Fin (Clean)	kW (Clean)	kW Recov.	kW Irrecov.
1	557000	2.28.10E7	48	222	250	282	125	254	28	32
2	557300	1.27	48	247	250	268	125	250	18	3
3	557300	2.08	48	232	125	261	125	261	0	29
4	557300	1.93	48	232	100	267	125	250	0	18
5	557300	1.82	48	230	125	267	125	267	0	37
6	557300	1.42	48	164	125	191	125	190	0	26
7	557300	1.05	48	174	200	201	125	187	14	13
8	557300	8.97.10E6	48	182	200	210	125	196	14	14
9	436993	8.55	48	148	125	160	125	160	0	12
10	436993	7.14	48	153	125	165	125	165	0	12
Reheat Section										
11	436993	5.76.10E6	48	165	250	215	125	186	21	215
12	436993	3.78	48	175	250	205	125	186	11	106
13	436993	2.89	48	185	175	205	125	191	14	6
14	436993	2.64	48	194	250	224	125	199	5	5
15	436993	2.97	48	201	125	207	125	207	6	6
16	436993	2.32	48	213	200	239	125	220	19	7
17	436993	1.94	48	225	200	252	125	232	20	7
18	389559	1.84	48	217	200	243	125	224	19	7
19	389559	1.56	48	234	200	262	125	241	21	7
20	389559	1.36	48	247	200	276	125	254	22	7
21	389559	1.10	48	262	200	293	125	270	23	8
Low Pressure Turbine End										
L-0	175536	2.69.10E5	64	293	600	302	125	293	9	0
L-1	184366	2.69	64	253	400	253	125	253	0	0
L-2	184366	4.64	64	213	275	213	125	213	0	0
L-3	184366	8.14	64	188	200	188	125	188	0	0
Low Pressure Generator End										
L-3	184366	8.14	64	188	200	188	125	188	0	0
L-2	184366	4.64	64	213	300	213	125	213	0	0
L-1	184366	2.69	64	253	500	253	125	253	0	0
L-0	175536	2.69	64	293	300	293	125	293	0	0

- *In column (4) is shown the surface finish of the vanes, inner and outer sidewalls, as manufactured.*
- *In column (5) is shown the frictional surface loss with this manufactured surface.*
- *Column (6) defines the mean effective surface finish as the unit was removed from service.*
- *Column (7) shows the surface frictional loss with the measured service finish.*
- *Column (8) For each blade row it is assumed that the surfaces are cleaned and a surface of 125√ μ-inches was obtained*
- *In column (9) the surface frictional loss with the 125√ μ-inches finish is determined*
- *Column (10) shows the gain in output with cleaning to the 125√ μ-inches finish.*
- *Column (11) shows the irrecoverable portion of the surface loss, between the as manufactured condition and the 125√ μ-inches to which it is cleaned.*

Table 6.6.2—Surface Finish Losses: Stationary Blade Rows, Data from the audit of a 72,000kW Unit.

(1)	(2)	(3)	(4)	(5)	(6)	(7)	(8)	(9)	(10)	(11)
High Pressure Section										
Stage	Flow Lb/Hr.	Re	Surf. Fin (Manuf)	kW (Manuf)	Surf. Fin (Service)	kW (Service)	Surf. Fin (Clean)	kW (Clean)	kW Recov.	kW Irrecov.
1	557000	4.3610E6	48	223	200	377	125	309	68	36
2	557300	2.24	48	186	500	239	125	205	33	19
3	557300	2.10	48	194	250	245	125	211	34	17
4	557300	1.96	48	207	200	248	125	225	23	18
5	557300	1.76	48	218	100	217	125	217	0	5
6	557300	1.64	48	220	1000	370	125	236	135	17
7	557300	1.54	48	228	125	242	125	242	0	14
8	557300	1.30	48	241	600	362	125	253	109	12
9	436993	1.22	48	191	125	198	125	198	0	7
10	453993	1.00	48	208	250	245	125	213	32	5
Reheat Section										
11	436993	6.6710E5	48	320	500	322	125	320	3	0
12	436993	8.86	48	237	350	335	125	263	73	25
13	436993	7.04	48	254	250	330	125	280	50	26
14	436993	6.33	48	271	200	308	125	293	15	21
15	436993	5.68	48	284	125	305	125	303	0	189
16	436993	5.45	48	284	125	305	125	303	0	189
17	436993	4.52	48	290	100	290	125	290	0	0
18	389559	4.57	48	258	100	258	125	258	0	0
19	389559	3.77	48	258	125	278	125	278	0	20
20	389559	3.98	48	261	125	284	125	284	0	259
21	389559	3.41	48	281	250	2929	125	289	231	84
Low Pressure Turbine End										
L-0	175536	6.17.10E4	64	1620	600	2088	125	1620	486	0
L-1	184366	6.52	64	1662	1500	2831	125	2631	199	970
L-2	184366	1.08.10E5	64	1569	750	2087	125	1569	518	0
L-3	184366	1.40	64	1462	500	1888	125	1462	425	0
Low Pressure Generator End										
L-3	184366	1.40.10E5	64	1462	500	1888	125	1462	425	0
L-2	184366	1.08	64	1569	750	2087	125	1569	518	0
L-1	184366	6.52.10E4	64	1662	1500	2831	125	2631	199	970
L-0	175536	6.17	64	1620	600	2088	125	1620	468	0

Table 6.6.3—Surface Finish Losses: Rotating Blade Rows, Data from the audit of a 72,000kW Unit.

STEAM PATH CORROSION

The blade system is known to be susceptible to corrosive damage. The ingredients for failure are present in the form of high stresses, aggressive compounds carried into the unit with the steam, and the damaging effects accelerated by the high temperatures existing at many locations.

Corrodents are introduced into the steam/water system from a number of sources, as has been discussed. A map of potential impurity ingress sites (Fig. 6.7.1) indicates where ingress can occur (I); the transport paths (T); the deposition locations (D), and the regions most susceptible to corrosive activity and, consequently, sustaining damage (C).

From this it is determined corrodents that gain access can be transported throughout the cycle, have the ability to deposit and accumulate during operation, and ultimately result in damage or failure that could force the unit from an available status for extensive periods. It can also be seen from Figure 6.7.1 that a number of areas (pieces of equipment) become susceptible to this corrosive damage. However, the concern here is with the steam turbine.

The corrosive agents of concern contain elements or compounds containing sulfur, sodium, phosphorus, or chloride. They are quite often introduced into the system from any number of "ingress" locations—by the malfunction of water purification equipment or by chemical cleaning and preservation treatments carried out in the water/steam cycle. This cannot be considered an exhaustive list of their means of introduction, and on many occasions there is ingress from other sources.

Because the turbine is part of a complex thermodynamic cycle its working fluid must be maintained at very high levels of purity, and because the turbine interacts with other equipment that contains fluids

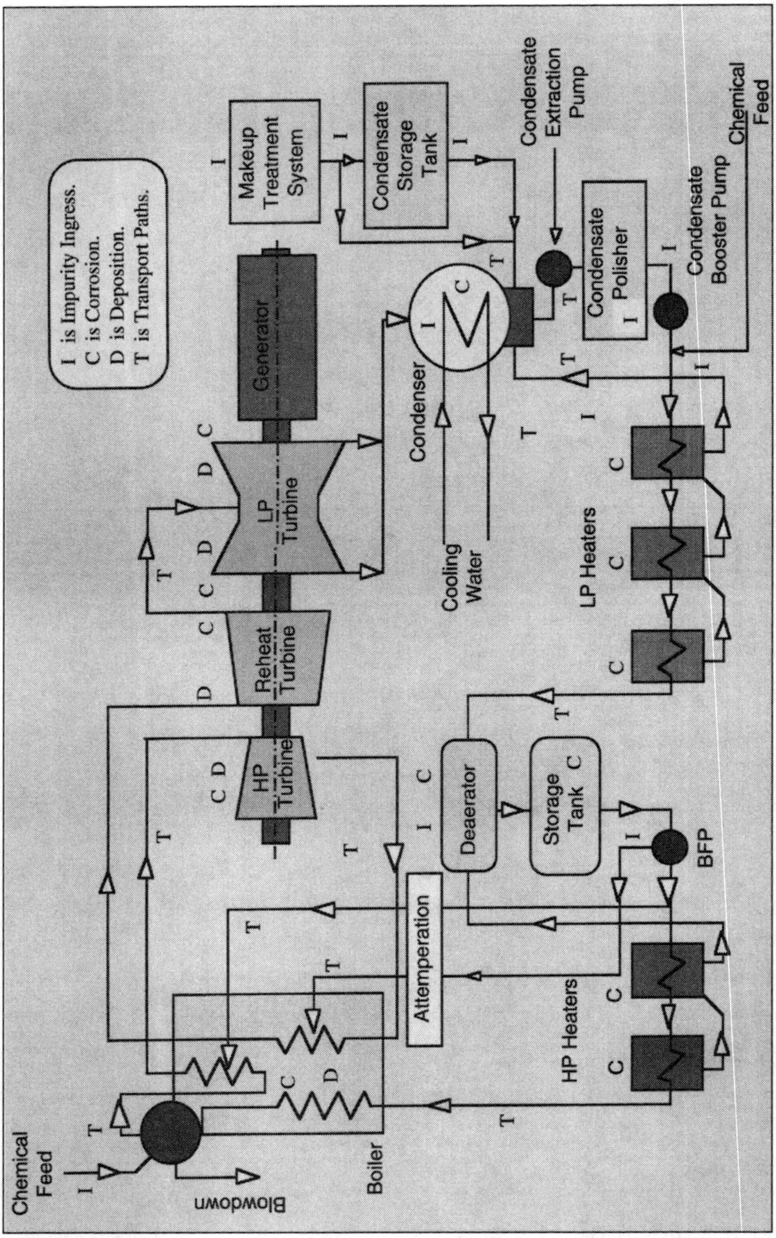

Figure 6.7.1—The fossil cycle showing the points of Ingress (I), Deposition (D), Transportation (T) and the most likely points of severe Corrosion (C).

operating at much higher level of impurities, isolation of one fluid system from another is vital. There is always the potential for the ingress of corrosive ions from these other portions of the total system into the turbine. While other portions of the system are susceptible to these impurities, it is perhaps the steam turbine that is the most sensitive and can suffer the most serious damage in the event corrosive action occurs. The need for cleanliness—including maintaining water purity by polishing and isolation—is of utmost importance in maintaining system reliability.

Only a portion of the total number of ions that gain access to the turbine from these various sources are corrosive, or have the potential to produce corrosive compounds under pressure and temperature. However, if the local steam conditions are favorable, and corrosion action occurs, this combination has the potential to cause dramatic and expensive outages, and many stages of the turbine can be considered to be at risk. The stages most seriously "at risk" are located in the region of the saturation line. This is because these stages are where moisture forms, and many of these corrosive ions come out of solution to be immediately captured by the water molecules in that region, and then transported by the suspended moisture through the remainder of the steam path (Fig. 6.7.1).

These aggressive compounds, once captured and transported by the moisture, will have a portion of their total population deposited on the surfaces of the steam path components. They will flow with the moisture films formed, and carried into "hideouts" where they can accumulate. Then, with changing steam conditions occurring throughout the unit, and depending upon a number of operating requirements, they will concentrate and become chemically aggressive under these modified conditions.

Stringent water purity control requirements are well understood, and operating staff normally has documented instruction or procedures for corrective action to modify and control quality. Unfortunately, serious leaks can occur and impurities enter the system very quickly, often before corrective action can be taken. Even very dilute solutions added to the power cycle (difficult to detect) can concentrate in hideouts during operation to the extent problems are introduced.

The hideouts

The occurrence of hideouts is a major factor in providing conditions for several forms of corrosion. Hideouts occur in the assembled portions of the steam path, and are a natural occurrence of assembly, shrink-fits, and in certain circumstances, the accumulation of errors in the manufactured parts. The most significant of these follow:

Coverband tenon holes. Coverbands are attached to the blade vane outer section by riveting a steel strip in place, using tenons that have been produced integral with the vane. A number of cavities are produced into which corrosive ions can be washed. Figure 6.7.2 shows cross sections of normal and "fox-holed" tenons. Normally, a small circumferential clearance "Cl" is left around the tenon. Because the tenon is deformed to produce a complete (or almost complete) fill of the coverband hole, cavities exist into which migration will occur. This is a relatively common location for corrosion to occur, and unfortunately is difficult to detect in the early stages when pits can form and cracks will initiate.

The peening process will also introduce a residual stress in the riveted attachment, which can promote corrosion.

Tie wire holes. Within continuous tie wire holes, wires can be loose, and are often attached to the vane by a braze attachment. In many designs, failure to align the wire and hole can result in residual stress (both tensile and compressive in the vane and wire). In the loose assembly (Fig. 6.7.3), small gaps from the wire and the hole will collect the corrosive material. As shown in (a), the design clearance is radial at "k." But in operation under centrifugal action, as shown in (b), the gap is zero at the top and "2k" at the bottom of the hole. This provides a convenient hideout.

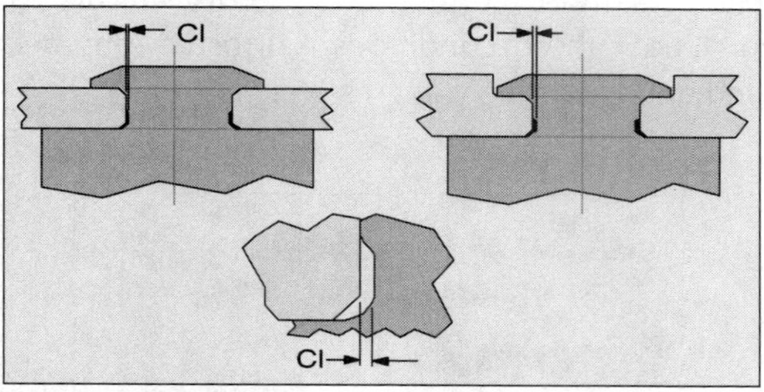

Figure 6.7.2—Hideouts produced by the riveting process, attaching the cover band ot the blade outer surface. (In (a) is the normal tenon attachment, and in (b) the foxholed design.

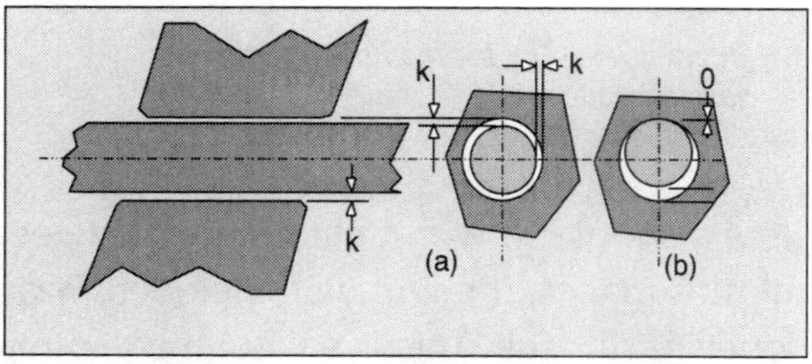

Figure 6.7.3—The loose tie wire, showing possible hideouts.

When the wire is brazed, it is common for braze to break loose at thick braze concentrations (Fig. 6.7.4). This situation provides small gaps that can act as hideouts.

Shrink-fits. Possibly the most critical corrosive action endangering the steam path occurs at the shrink-fit between a central spindle and the discs. Here a shrink-fit is established to locate the disc in the correct axial location and transmit force from the blade to the central spindle. At this shrink-fit, the central spindle is in compression, however, the shrunk-on disc has a tangential tensile stress at the shrink interface. In operation, this tensile stress is relieved somewhat, but as the unit speed reduces from operational to standstill, the stress level increases. It is also normal for these discs to dry as the unit is shut down. Under these conditions, corrosive ions become more concentrated and corrosively aggressive.

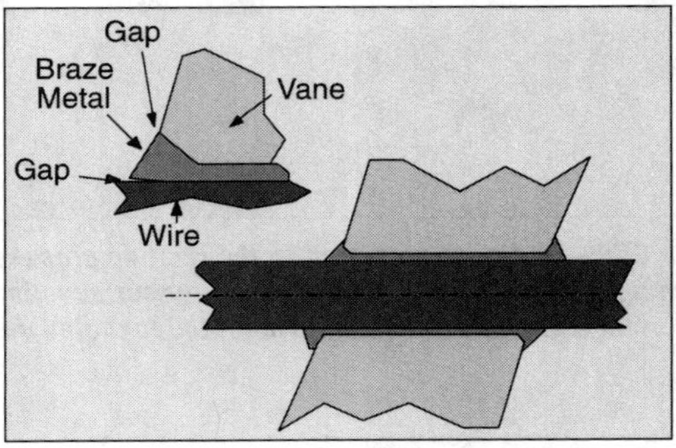

Figure 6.7.4—The brazed tie wire, showing possible hideouts.

Because it is a shrink-fit, with tight contact between the spindle and disc shrink-fit surfaces, it might be concluded that the ingress of corrosive products is difficult. This is so, but the gap requirement for the introduction of corrodents into this space is very small, and therefore ingress does occur. In addition, there is often an axial key of various geometries locating the disc to the spindle for overspeed protection. The keys cause discontinuity to fit and stress and are a region where cracking will often initiate.

Blade root/rotor interfaces. Figure 6.2.3 shows a portion of a rotating blade row, which was corroded when some form of compound gained access to the unit. The blade roots are of the tangential entry straddle type. In a situation such as this, it is inevitable that corrodents will have found, or will find their way into the load transfer area of the fastening. Here a number of high-stress areas initiate corrosive action when the environmental conditions support such activity. This is known to be the cause of a number of failures. In the worst circumstances, the wheel rim will crack, causing expensive corrections. In the most severe situation, it will call for removal of the blades from the stage and possibly the replacement of the rotor.

The most commonly occurring damage in the steam path and on the blades is corrosion pitting, stress corrosion cracking, and corrosion fatigue. It is necessary to consider the forms in which they manifest themselves.

FORMS OF THE CORROSION PROCESS

Corrosion is defined as damage or deterioration of a metal that occurs when it reacts with its environment. Corrosion can be "wet" or "dry," depending upon the environment and its moisture content. The rate of corrosive reaction is also influenced considerably by the environmental temperature—the rate of metal loss being faster at the higher temperatures.

The mechanism of corrosion can involve either an electrochemical reaction, or can be the result of direct chemical reaction between a corrodent and a metallic surface. In the steam turbine, corrosion is almost always "wet" and the result of direct reaction between a corrosive compound and the metallic surface with which it is in contact.

Corrosion will always initiate at a surface exposed to the corrosive medium. It can occur over an entire surface or locally. Local corrosion will cause small pits to develop on an otherwise unaffected surface. In the case of localized corrosion, the surface is normally in contact with the corrodent in a concentrated form.

Figure 6.8.1—The horizontal joint of a unit which has not been used for a number of years.

It is common when units are "stood down" for an extended period without the advantage of protective gas blanketing, they will suffer surface corrosion on susceptible surfaces that have even the most dilute concentration present at the time of shut down. As stated, in these instances it is not necessary for strong corrodent concentration to be present. The action of condensed moisture will be sufficient to cause the corroding process to begin.

Figure 6.8.1 shows a portion of the horizontal joint of a unit that has stood for a number of years without being used (the horizontal joint bolts were "finger tight"). Similarly, Figure 6.8.2 shows a valve stem from this same unit. This form of corrosion does not occur during operation and will not be considered further. However, whenever a unit is shut down, and unlikely to be restarted for some period, it is prudent to consider the consequences of this period, and the advantages of a nitrogen or similar inert gas blanketing system.

For wet corrosion—in addition to being dependent upon the temperature of the environment surrounding the surface—the rate of reaction is also dependent upon the concentration of the corrodent. This is an important consideration in the steam turbine, because in many situations—as when a unit is shutdown for a relatively short time but cools—pools of condensation will form on the various steam path surfaces. As the water evaporates, these pools will concentrate the corrosive ions in this water and in hideouts throughout the steam path. As the water is driven off or evaporated, the corrosive concentration becomes greater, exposing the metallic surfaces to a higher concentration of the corrodent, and therefore to higher rates of material damage.

Figure 6.8.3 shows the rotating blades in a unit that have been allowed to stand and cool. Possibly the local temperature was higher than the normal operating temperature, as steam continued to leak past valves. Here corrosion pits can be seen, and it is possible from their pattern that corrodents have concentrated at the edge of puddles causing the very characteristic circular pits.

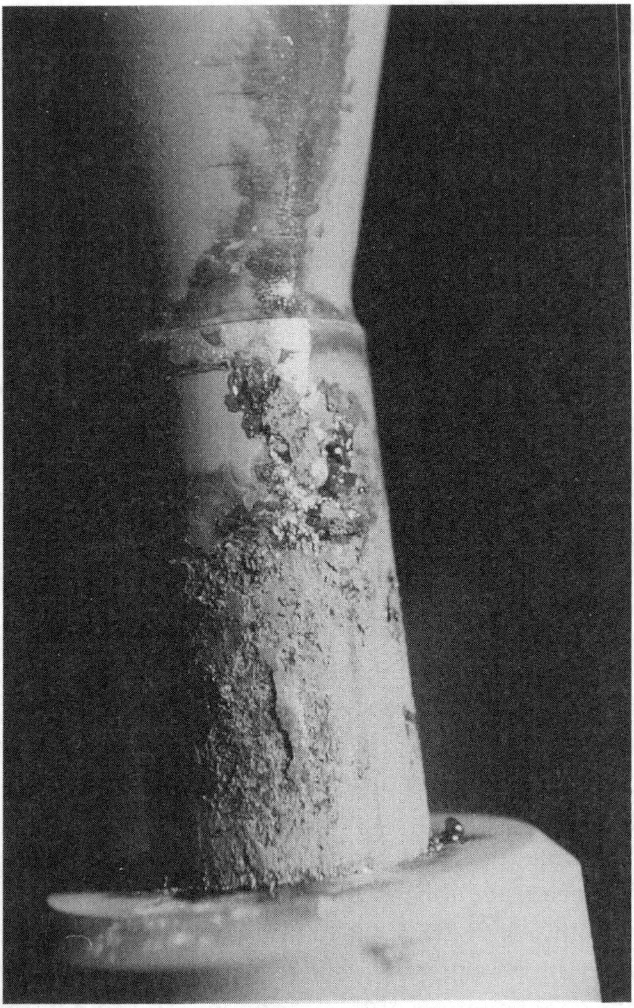

Figure 6.8.2—A control valve stem from the same unit seen in figure 6.8.1.

Another operation that can assist the corrosive process is placing the unit on turning gear for extended periods, when the internals will be hot, and any corrodent present will become active as water deposited from the steam is evaporated.

Figure 6.8.3—Rotating blade elements with corrosion pits shown on its suction face.

Impurity effect on blade failures

Figure 6.8.4 is the Mollier Diagram for steam, indicating the expansion line for the low-pressure section of a typical unit, and identifying the regions in which impurity concentrations occur. This diagram indicates the reason why a significant number of failures occur in the region where moisture formation and corrodent deposition and concentration exist.

Possibly most significant in utility-type turbines is the L-1 stage, where repeated failures continue to force units from service. The total blade failure rate in fossil-fueled utility units (shown as Fig. 6.8.5) is where the preponderance of L-1 stage failure can be seen. While not all these failures are due to corrosive action, a large proportion of them can be attributed to the various forms of corrosive action in the blades' vanes and roots.

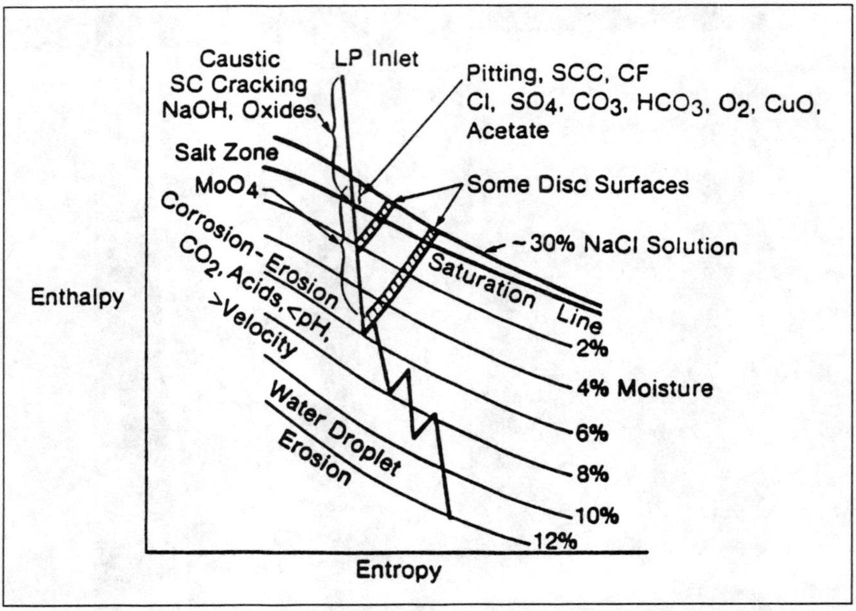

Figure 6.8.4—The Mollier diagram showing the regions, or conditions, where corrosive ions come out of solution in from the gaseous steam.

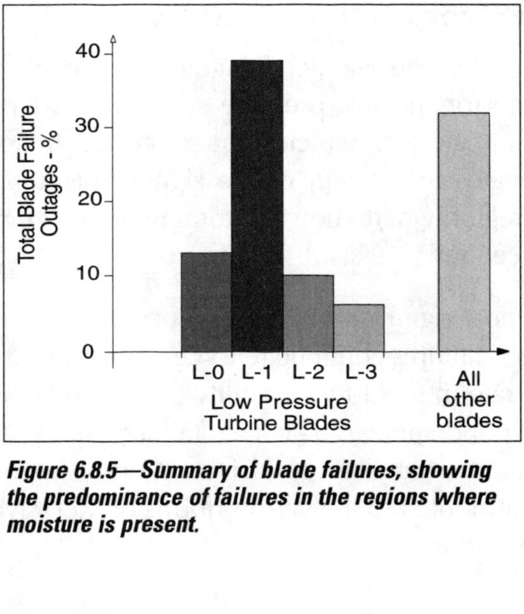

Figure 6.8.5—Summary of blade failures, showing the predominance of failures in the regions where moisture is present.

Within the steam turbine there are a number of corrosive mechanisms that can occur. These damage and degrade the quality of the unit either ultimately forcing it from service, requiring expensive replacement parts, or extending planned maintenance outages. The most common corrosive mechanisms follow:

Crevice corrosion. Crevice corrosion initiates in small crevices or hideouts into which aggressive ions can migrate and collect. In the turbine steam path—in a complex arrangement of assembled components—there are many such hideouts. There is a continued risk that such hideouts will act as centers within which corrosion will initiate when environmental conditions are favorable. The damage occurring and the pits forming can then act as a stress concentration center for further damage.

Within the steam path, many of these hideouts are in areas that move from a steam condition (where their environments are alternately "wet" or "dry"). They are also subject to changes in temperature as a consequence of changing load, or even changing conditions in the steam delivered from the boiler superheater or reheater.

Corrosive conditions within a crevice or hideout can become chemically very aggressive, as the concentration of corrosive ions increase. This concentration can occur locally, because steam conditions within any stage are not constant, but vary with radial position, the possible effects of steam secondary flow and leakage, and a number of other circumstances that occur within a very complex set of changing parameters. Under these conditions, a corrosive reaction can occur when remaining portions of the steam path are in a corrosively stable condition.

Possibly one of the most damaging conditions within the hideouts occurs when a unit is shut down. At this time, a temperature increase throughout the low-pressure sections of the unit can turn a relatively inert collection of ions into an aggressive solution.

Pitting corrosion. Components located above the moisture region are particularly susceptible to corrosion pitting. This phenomenon occurs on the surface of various components, most often the stationary and rotating blade vanes. However, this can often be noted on the wheel or disc faces of many units, and in the castellations at shaft seal points (Fig. 6.3.4). Pitting is usually indicative of the presence of chlorides.

Pitting corrosion is a surface phenomenon, and results from the chemical reaction of an aggressive corrodent on the metal surface. The pits forming on an otherwise unaffected surface with this type of corrosion are basically semicircular, with pit depth being about half the width of the pit. Pit growth is normally slow, and is to a degree self-propagating, with aggressive ions retained in the pit and corrosion continuing, using these same free ions. Pitting is similar to crevice corrosion, as there can be a level of accumulation and concentration sufficient to cause damage.

Like crevice corrosion, pitting can become active when the unit is shut down and environmental conditions surrounding the surface change (see Fig. 6.8.3). Figure 6.8.6 shows a blade row that has been subjected to heavy corrosive attack. Both the blade vanes and coverbands show evidence of severe surface degradation.

Maintenance engineers are often faced with the difficult assignment of finding pitting corrosion of the type shown in Figure 6.8.7. In this situation, pitting corrosion appears on the wheel face, but little or no evidence of material loss shows from the blade roots. It is clear the wheel material is more susceptible than the blades, but what is not known is the condition of the wheel rim. It will have sustained an ingress of corrosive ions through the axial gap between blade root blocks and into the wheel root load transfer surfaces.

Figure 6.8.8(a) shows the base of a root groove in a drum-type rotor after blade removal. Erosion at the base of the groove indicates ingress of corrosive ions and the corrosive pitting is quite clear. This

Figure 6.8.6—A rotating blade row showing pitting corrosion on the blade vanes and cover band.

Figure 6.8.7—Pitting corrosion of the wheel of a stage, possibly indicating corrodents have gained access to the root portion of the row.

form of corrosion is not considered too dangerous in this location. This pitting was also evident on the serrations (load bearing surfaces) shown in Figure 6.8.8(b). In these regions of high tensile stress, other forms of corrosive damage can occur. Figure 6.8.9 shows the same form of damage on the stage of a rotor rim. Corrosive pits are clearly evident. In this region is the blade root access window, a highly stressed region of the wheel, and one with the potential for failure to occur. This is the form of root shown in Figure 6.2.1.

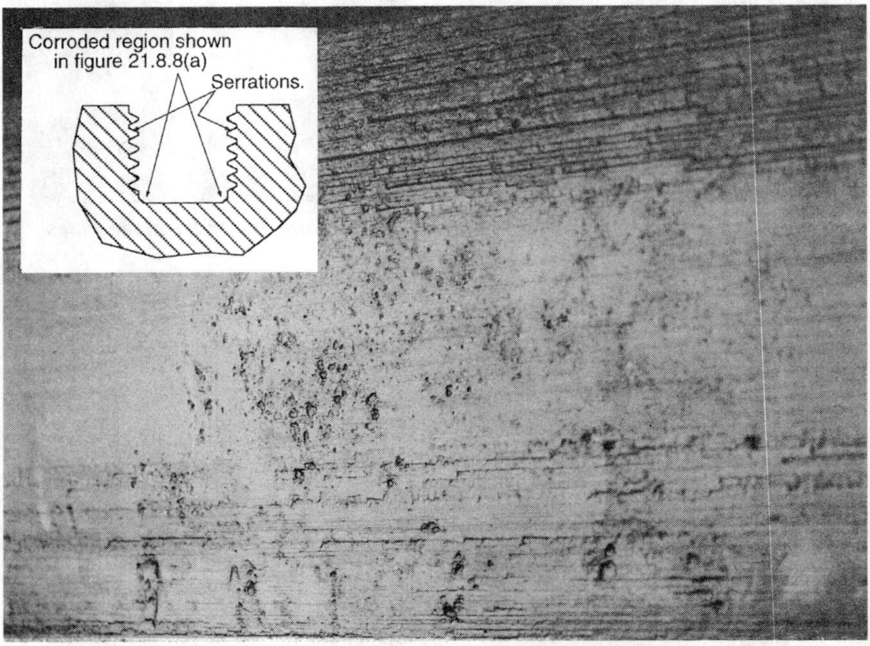

Figure 6.8.8—Showing corrosion in the root groove on a drum rotor in (a). In (b) is shown the form of the root groove.

Another region of the unit where pitting corrosion is not common—but, if present, has the potential to cause damage—is the shaft-end positions (Fig. 6.8.10). While not a common occurrence, it reduces the effect of the seals in this region when it does, and makes the shaft more susceptible to cracking damage, particularly if there is also a change of cross section in this region.

Figure 6.8.9—Corrosive pits on the wheel rim where corrodents have gained access. This is the form of root shown in chapter 5, figure 5.7.1.

The variation of steam conditions through a shaft gland is close to a throttling operation (see chapter 10). The change in thermal conditions shown in Figure 6.3.3 demonstrates how steam expands past the shaft end seals, and how it can move rapidly from a wet to a dry condition, becoming superheated. Therefore, corrodents in the leaking steam, deposited on the shaft, will corrode, possibly causing a severe material loss. While low-pressure section glands are "packed," there will be "ingress" of air that will promote corrosion due to oxygen content.

Another factor to be considered is localized rubbing between the rotor and seal knife-edges. This causes localized heating. Steam

Figure 6.8.10—Corrosive pits at the shaft end sealing position of a rotor. Corrosive pits can be seen on the rotor body and castellations where seal strips have located.

quenching will produce hardening, and in consequence, a greater susceptibility to corrosive attacks adjacent to the hardened surface.

It is also possible if the generator is not correctly grounded, there can be a static discharge across the gland regions, again causing a form of spark corrosion.

Stress corrosion cracking (SCC). SCC is the phenomenon by which cracks initiating at a metallic surface propagate through the material, breaking down the crystalline structure of the material. SCC is a relatively common condition found within the turbine steam path, most often affecting the unit's rotating components. This mechanism has been the damaging mode that has caused many expensive outages—forced and extended. It has also caused catastrophic failures.

For SCC to occur, components must be subjected to a combination of tensile stress and a corrosive environment. Component materials subjected to these corrosive ions must also be susceptible to the particular corrodent present. Corrosion-induced cracks are brittle in nature, usually branching, and may be either trans-granular or inter-granular depending upon the material composition, mechanical properties, the local stress levels, and environment.

Stress corrosion cracking can lead to a brittle failure at stresses substantially lower than those necessary to cause failure in a non-corrosive environment. The rate at which corrosion-induced or -assisted cracks initiate and then propagate depends upon stress and the temperature present to support the chemical reaction, influenced by the environment. It may take extensive periods of exposure to a corrosive environment to initiate a crack, but once formed it can propagate to failure within a relatively short period. There are many components in the steam path where residual tensile stress exists, both by design and incidentally, as a consequence of manufacture. At these locations SCC can readily occur.

Unfortunately, many of the alloy steels used to construct the turbine are susceptible to the corrodents formed or gain access to the steam path. Stainless steel is particularly susceptible to SCC in chloride and caustic environments. This means that SCC can occur on rotor steels. It will also occur on 12% chrome alloy steels such as AISI403 and AISI410 if these steels have been overheated during manufacture (*e.g.*, during the attachment of an erosion shield or weld connection of tie wire bosses). SCC can also occur in the HAZ of welds, where a BHN exists in excess of about 260. Such damage can also occur in an AISI 422 steel if it operates in the moisture region.

When molten metal cools, solidification begins at a number of randomly spaced nuclei, which grow in a regular atomic array to form grains of the alloy. These grains have an identical atomic construction. Because they are not aligned on the same axis, gaps exist between them—and this gap has a different composition—are known as the *grain boundary*.

There are two forms of SCC phenomena: Trans-granular and inter-granular. Results depend upon the variables involved. However, each form of cracking will grow at a rate determined by the environment, the stress levels involved, and local temperature. These two forms of cracking are next described.

Inter-granular corrosion. Under normal circumstances, the possibility of material inconsistencies and deficiencies at the grain boundaries are anticipated and allowed for in defining mechanical properties of the material. Similarly, dimensions of the individual components in terms of the stress levels they will experience in operation are selected so they may operate safely when produced from these materials. However, when the material from which these components are produced is subjected to a corrosive environment, the grain boundaries can become reactive, and will corrode preferentially to the purer crystals of the alloy, *i.e.*, the grain boundaries could disintegrate at a rate faster than the alloy crystals. This allows a crack to initiate and grow between them. Under these circum-

STI

Figure 6.8.11—Stress corrosion cracks on a section of a wheel in the region of the load bearing surfaces.

stances, pure alloy crystals will remain unaffected but could eventually break loose from the solid material structure.

Corrosion that occurs and preferentially attacks the grain boundary constituents is called "inter-granular corrosion." Such corrosion results in alloy disintegration at the grain boundaries, possibly leaving small crystals of the base alloy, detached from the solid. Inter-granular cracking is a common mechanism for driving cracks in the steam turbine. Protective films at the base of the crack are continually ruptured by the local stress, with grains unaffected by the corrosive ions present in the crack void.

Cracks can be either branching or run directly through the material. Figure 6.8.11 is a cut-away section of a turbine wheel, which had carried blades with a tangential entry straddle roots. This wheel portion has suffered inter-granular cracks initiating at the corner fillet radii. Figure 6.8.12 is a micrograph of such a cross section with branching cracks.

Inter-granular corrosion will often occur in material regions that have been subjected to some form of heat treatment. Other heat treatment processes such as a stress relieving cycle can often reduce the propensity of a material to this form of damage.

Trans-granular corrosion. Trans-granular corrosion is somewhat less common than inter-granular, but occurs under certain circumstances within the materials of the steam path. Austenitic steels—as used in some steam path components—are more susceptible to this form of damage.

Unlike inter-granular cracking, trans-granular cracks initiate at the surface of the alloy crystals. These cracks propagate in a series of discrete steps—as there is a cleavage of the individual grains, the cracks run across the crystal plane. The growth of the fracture is interrupted each time a grain boundary is reached, and at the boundary the crack must again be initiated. These cleavage cracks of the crystals represent a brittle fracture across the individual alloy crystals.

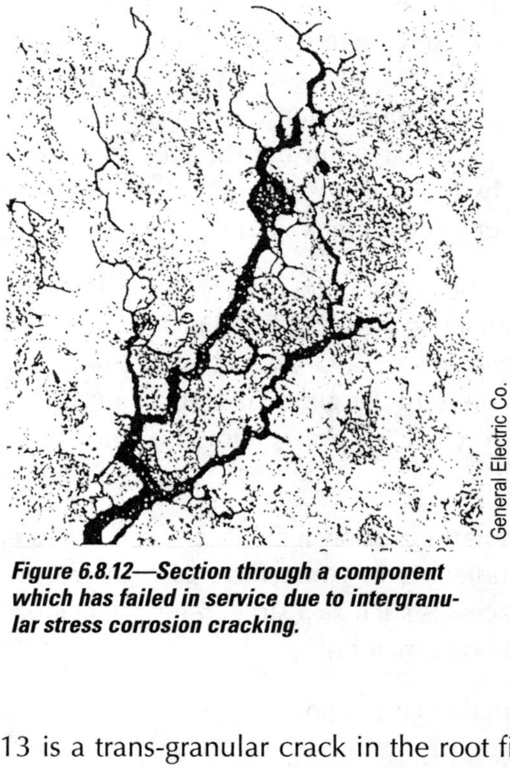

Figure 6.8.12—Section through a component which has failed in service due to intergranular stress corrosion cracking.

Figure 6.8.13 is a trans-granular crack in the root fillet radius of a rotating blade. Figure 6.8.14 shows the axial entry root with the crack clearly visible. An examination of the surface deposits of this root indicated a high sulfur content present, and the corrosive agent was hydrogen sulfide (H_2S).

By making the grains of the metal smaller, the probability of trans-granular cracks propagating at a significant rate is reduced. Unfortunately the exact mechanism of trans-granular cracking is not fully understood.

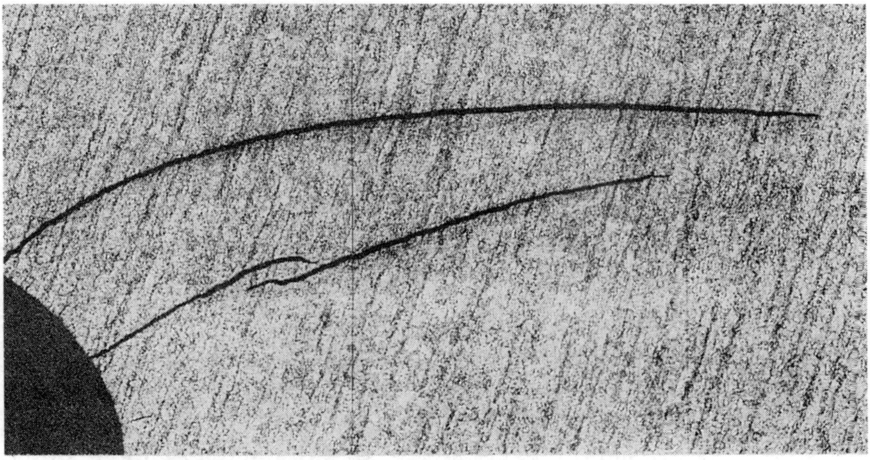

Figure 6.8.13—A transgranular crack in a blade root.

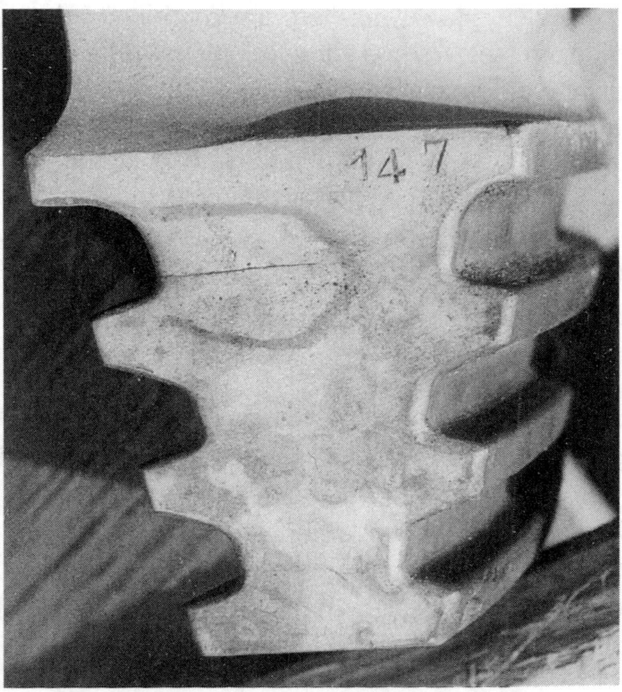

Figure 6.8.14—The blade root showing the crack initiating one side of the outer load bearing surface.

Those turbine stages most likely to be affected by stress corrosion cracking are those in the low-pressure section where the moisture of condensation begins to form (Fig. 6.8.4). The regions of the steam path most susceptible to this problem are those subject to high tensile stress, compounded by stress concentration. This typically exists at the tie-wire holes, coverbands and rivets, and at the root or fastening of both the blade and rotor attachment portion of the fastening. In these regions the direct tensile stresses can be high, and the problem possibly compounded by the effects of alternating loads. These areas are therefore considered to be at risk. Such failures are normally inter-granular.

The corrosive action that occurs at shut down (when there are no centrifugal stresses present) would primarily be the pitting type. Pits could produce a stress concentration and cracking when the unit is returned to service. However, there are portions of the steam path where tensile stresses are present even at shut down. These regions of the unit include the shrink-fits, which will exist where any component—including discs—are shrunk onto a central spindle. Also included in this group of components are those where stresses may be present, such as in a tie wire and vane deformed by poor alignment of the hole/wire interface and forced assembly is introduced (Fig. 6.7.3). Coverbands are attached by rivets that are often "cold worked" to form an acceptable head that can also contain tensile stresses in relatively complex patterns (Fig. 6.7.2).

The large number of blade root forms that are common can introduce tensile stresses in various locations in many designs. This includes the closing blocks (or blades) in some rows, which are attached in a number of configurations. The rolling of side grips also introduces tensile stresses. However, if all these design parameters are controlled at their design specified values, the induced stress should be sufficiently low, and damage will not result.

Within the shrink-fit—between the rotor spindle surface and the wheel bore—is a region of the rotors where stress corrosion cracking can be a major concern. A common location for crack initiation is at the keyway, which is required to locate the wheel on the central spindle at emergency overspeed when the shrink-fit may be lost. The keyway represents a major discontinuity and is often where stress cracks will initiate. Figure 6.8.15 shows corrosion found in the spindle/wheel region. It is indicative of the presence of corrodents in this critical region. One can also see corrosive pits on the key material. Figure 6.8.16 shows a crack originating at a keyway, running in a radial direction. Figure 6.8.17 is a cross section of a keyway intergranular crack.

Figure 6.8.15—Corrosion pitting around a shrink fit, showing corrosion on the wheel, the central spindle and the key.

Many key designs are now used to minimize the effects of stress concentration. These designs help eliminate discontinuities in the disc or wheel. Also, key designs are now available that locate them in areas of low stress. These changes represent a design improvement.

However, while keyways are known for cracks, they are also known for initiating them at other tangential positions away from the keyway. Such cracks can be seen at a number of tangential positions in Figure 6.8.18, which is a portion of a failed disc.

Figure 6.8.16—Evidence of an SCC crack emanating from a keyway.

In general, if design parameters are followed in producing the various assemblies required within the total steam path, the resulting stresses should be below threshold values required to initiate failure. Therefore, when failures do occur, they are nearly always the result of some abnormal condition that is difficult to control.

Corrosion fatigue. A material can fail due to the phenomenon of high-cycle fatigue (HCF) if subjected to a cyclic load of sufficient magnitude. Also, failure will occur at a significantly accelerated rate if:

Figure 6.8.17—A crack due to SCC originating in the keyway of a disc.

- a component is subjected to a cyclic load of sufficient magnitude that HCF would ultimately cause failure in a steam environment
- it is simultaneously subjected to a corrosive environment

This phenomenon—failure under the influence of a corrosive environment—is called "corrosion fatigue." Corrosion fatigue can initiate on a clear, non-pitted surface, *i.e.*, there is no requirement for a mechanical discontinuity as a pit or crack for this form of corrosion to initiate.

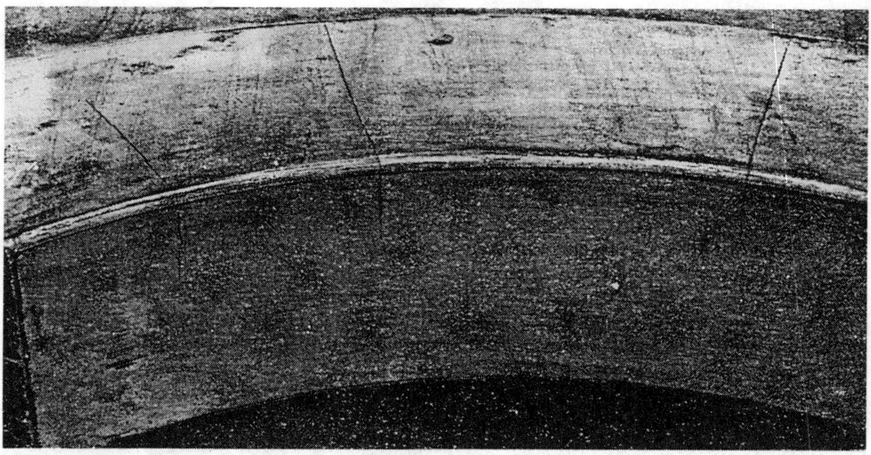

Figure 6.8.18—Portion of a disc which has ruptured. The cracks can be seen running radially from the center bore.

Corrosion fatigue is, in reality, a special case of stress corrosion cracking, in which a cyclic force drives the crack. In this type of failure, a large failure surface is normally covered in the products of corrosion, and the final failure surface—reasonably clean—occurs by mechanical rupture. This form of failure is distinct from SCC in that the applied stress is cyclic and the failure is a fatigue-type mechanism, exhibiting the conchoidal marks associated with HCF. Such failures have a tendency to initiate at those points where corrosive products can concentrate, and a level of stress concentration causes high local tensile effects. Such cracks can be either trans-granular or inter-granular. However, these corrosion fatigue failures are most often trans-granular in nature.

The higher-stressed regions of a blade, close to the root section, and at the higher load-carrying portions of the root appear to be locations most susceptible within the steam path. This is because both direct and alternating stresses are at high levels in these regions. The low-pressure section of the unit is relatively easy for corrodents to collect and concentrate. However, there are many significant failures that initiate at tie-wire holes and coverbands, where similar conditions can exist.

The total conditions surrounding a corrosion fatigue situation are complicated because the stress levels can vary from those of a low-level direct stress with a high level alternating stress imposed on it; or, it can be a high-level direct stress with a low-amplitude alternating stress applied. It can also be any combination of magnitude of these forms of load.

Other variables affecting the rate at which damage will occur include the concentration of the corrodent applied to the component that induces failure, the local environmental temperature, and possibly the number of cycles through which the component experiences a "wet" and "dry" environment. Unfortunately, it is impossible to predict the occurrence of corrosion fatigue, and these types of failure have been known to force a unit from service for extensive periods.

A typical low-pressure L-1 failure at a blade root is shown in Figure 6.8.19. The general appearance of the wheel face with surface corrosion is seen, with one blade cracked at the outer load bearing surface fillet radius. Figure 6.8.20 is a closer view of this same stage with the blade vane removed. Figure 6.8.21 is a photo-spectrograph of the surface deposits from the failed blade vane and root platform.

Figure 6.8.19—A blade with one element having a fracture near at the upper load bearing surface.

Factors affecting crack growth rate

Once a crack has initiated and grown to a size sufficient that it will continue to propagate, there are certain conditions promoting this growth, and their presence at some level will affect the rate at which the crack propagates. Important among these are listed below:

Stress levels. SCC requires a tensile stress to initiate and propagate cracks. These cracks grow mainly in a direction perpendicular

Figure 6.8.20—The blade shown in figure 6.8.19 with the blade above the fracture surface removed.

to the direction of the principal stress, and can under certain conditions become branched. Each branch grows in essentially this same perpendicular direction. The actual mechanism by which the crack propagates is one of continuing to grow from the base of the crack by rupturing the oxide or other film existing there. The higher the stress level, the greater the ease with which the film can be ruptured, and therefore the faster the crack will grow.

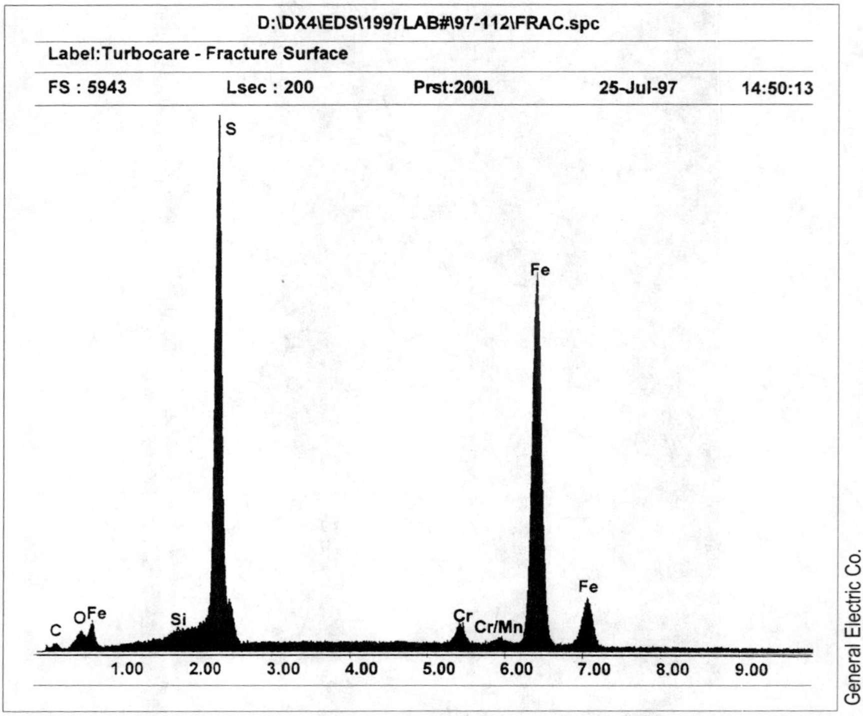

Figure 6.8.21—A photospectragraph of the surface deposits from the failed blade root.

For cracks to initiate, a threshold of stress is required. The value of this stress is sensitive to the material, its composition, its mechanical properties (yield strength, hardness, and impact strength), and any heat treatment to which it has been subjected. Type of corrodent also influences it.

Corrodent concentration. The more concentrated the active corrodent, the faster the crack will propagate. This is obvious and any environmental conditions acting to concentrate or dilute the concentration will influence the rate of crack growth.

Environmental temperature. The temperature of the piece suffering the corrosive attack is influenced directly by the temperature of the environment surrounding it. Temperature will influence crack growth by concentrating the corrodent—making it more aggressive—and, as a secondary consideration, modifying the mechanical properties of the material by a small amount. For this reason, as temperatures change in the steam path it can be expected that crack growth rate will change.

Tensile stress and high temperatures are always present in the steam turbine during operation, and often are present to some degree during shut down when the unit remains hot. Therefore, the potential for corrosive action is an ever-present possibility, and its prevention—or minimization—is achieved by the control of water quality, since normally the majority of corrodents are carried into the steam path with the working fluid.

REFERENCES

1. *Sawers Turbomachinery Maintenance Handbook, Vol. 2, Chapter 3,* "Geothermal Steam Turbines"

2. *Boiler Water Limits and Steam Purity Recommendations for Watertube Boilers,* American Boiler Manufacturers Association, 3rd Edition, 1982

3. Jonas, O., and L.G. Bias. "Analyzing Steam Turbine Deposits," *Power,* December 1976

4. Cotton, K.C., and J. Angelo. *Observed Effects of Deposits on Steam Turbine Efficiency,* ASME Paper 57-A-116

5. Forster, V.T. "Performance Loss of Modern Steam Turbines due to Surface Roughness," *Proceedings,* I. Mech. E. Vol. 181, Pt. I, No. 17

   Schofield, P. "Maintaining Optimum Steam Turbine-Generator Performance," Presented at EPRI Fossil Plant Heat Rate Improvement Workshop, Charlotte, NC, August 1981

6. *Surface Texture,* Published by the American Society of Mechanical Engineers, United Engineering Center, New York, New York

7. Traupel, W. *Thermische Turbomashinen,* Published Springer Verlag, Berlin, 1966

8. *Cycle Chemistry Corrosion and Deposition: Correction, Prevention. and Control,* EPRI report TR-103038, Prepared by GPS Technologies, San Diego, California, December 1993

9. Fontana, M.G. *Corrosion Engineering, 3rd Edition,* Published by McGraw-Hill Book Company, Inc., 1989

McCord, T.G., B.W. Bussart, R.M. Curran, and G.C. Gould. *Stress Corrosion Cracking of Steam Turbine Materials,* International Corrosion Forum, Toronto, April 1975

10. Jonas, O. *Determination of Steam Purity Limits for Industrial Turbines,* 49th International Water Conference, 1988

    Jonas, O. *Kirk-Othmer Encyclopedia of Chemical Technology, Vol. 21, 3rd Edition,* John Wiley and Sons, 1983

11. *EPRI Journal,* April 1980

12. Samuels, L.E. *Metals Engineering, A Technical Guide,* Published by ASM 1988

    Jonas, O. "Steam Turbine Corrosion," *Materials Engineering,* Vol. 24, No. 2, February 1985, pp 9-18

13. Hodge, J.M, and I.L. Mogford. "UK Experience of Stress Corrosion Cracking in Steam Turbine Discs," *I. Mech. E. Proceedings, Vol. 193, No. 11,* 1979

INDEX

A

Abrasion, 41-46, 66, 74-77, 448-449
Accept as-is, xxii-xxiii, 23-24
Acronyms, ix
Actual steam force diagram, 539-541
Aging units, xiv-xv, 11, 15, 585-586
Air pressure (blasting), 641
Alignment/adjustment settings, 173-174, 192-228
Alignment methods (casing-to-casing), 173-174, 192-228
  laser optical, 174
  piano wire, 173-174
  settings, 192-228
Alternating stress, 522-525, 569-570, 578
  magnitude, 578
Alternative power, 11-12
Aluminum oxide, 640
Amplification factor (Ad), 575-576
Applied load frequency, 578
Area change effects (erosion), 460-461
  sidewall material loss, 461
  vane material loss, 460-461
Area-pressure increase, 461
Arithmetic average, 656
Arrangement/sizing (component), xiii
Assembly/installation quality, 31-32
Associated stage hardware, 455-459
  coverbands, 455-456
  material loss, 455-459
  radial seal strips, 457-459
  tenons, 456-457
Attemperating spray water, 611
Axial deformation, 186

Axial gap, 100-103, 212-213, 438-439
adjustment, 100-103
Axial lean (vane), 254-255
Axial placement, 212, 241-243, 249-258
error (vane), 249-258
inlet/discharge edge, 243
rotating blade elements, 243
stationary blade elements, 243
Axial placement error (vane), 249-258
axial lean, 254-255
coverband distortion, 256-257
position on root platform, 249-252
root form position, 252-254
method of location, 257-258
untwist, 255-256
Axial positioning (diaphragm), 212-217
Axial pressure deflection, 94
Axial space, 225
Axial stiffeners, 515-516
Axial/tangential planes, 433-435

B

Backpressure (condenser), 28
Beach marks/striations, 38, 515-516
Bearing loading, 105
Bearing oil rise, 103
Bearing spherical seating, 105
Bearing supports flexibility, 105
Bearing temperature effect, 104-105
Bending stress, 567-568
Beta factor, 542-546
Binding, 179
Blade annulus thrust, 138-139
Blade damage, 137-138
Blade root-rotor interfaces, 676-677
Blade tip deformation, 401
Blade tip leakage, 595

Blade trailing-edge erosion, 352-369
- damage mechanisms, 359-361
- erosion penetration and time, 367
- material loss model, 361-366
- monitoring damage, 368-369
- proud blades, 368
- root section recirculation, 356-358
- water sprays (exhaust hood), 359

Blade untwist, 272-275
Blade vane, 77-78, 90, 229-248, 292-298
- cascade geometry, 229-248
- crack indications, 77-78
- moisture, 292-298

Blade vane cascade geometry, 229-248
- axial placement, 241-243
- cascade, 243-248
- tangential placement/pitch, 243
- vane form requirements, 229-241

Blade vane crack indications, 77-78
Blade vane moisture, 292-298
Blast cleaning, 639-643
- air pressure, 641
- blasting media, 640-641
- nozzle distance, 641
- nozzle size/shape, 641
- spray jet inclination angle, 641

Blasting media, 640-641
Boiler cleaning/operation changes, 459
Boiler excursions, 590
Boilers, 459, 590
- cleaning, 459
- excursions, 590
- operation changes, 459

Bowing, 176-180
Braze-attached resistant material, 331-333
Brittle fracture *(SEE Chipping mechanism)*

C

Campbell diagram, 550-551, 553-561
    frequency distribution/tuning, 556-558
    frequency tolerances, 554-556
Campbell equation, 553
Carbon dioxide pellets, 641
Cascade blade arrangement, 243-248
    discharge area, 247
    expansion passage form, 244-245
    pitch, 246
    swallowing capacity, 248
    throat, 246
    throat-pitch ratio, 246-247
    vane placement, 244
    vane twist, 245
Casing-blade carriers, 518-520
Casing boring, 182-183
Casing-casing alignment, 173-174
    laser optical, 174
    piano wire, 173-174
Casing erosion, 386-389
Casing humping (hogging), 180, 475
Casing ovality (bowing), 176-180
    horizontal joint diameter increase, 177-180
    horizontal joint diameter reduction, 177-178
Casing temperature distribution, 592-593
Casings, 173-174, 176-180, 182-183, 268, 386-389, 393-394, 474-477, 518-520, 592-593
    alignment, 173-174
    blade carriers, 518-520
    boring, 182-183
    erosion, 386-389
    humping/hogging, 180, 475
    inner surface, 474-477
    ovality/bowing, 176-180
    temperature distribution, 592-593
Categorization (susceptibility), 55-56

Causes/mechanisms (failure), 25-52
- design, 25-29
- failure examination, 34-36
- installation quality, 31-32
- maintenance procedures, 32-34
- manufacture quality, 29-30
- mechanisms, 35-52
- operating practices, 32-33

Caving, 452-454
Cavitation losses, 361
Center of gravity position, 156
Centerline average, 656-657
Centerline support (casings), 105
Centrifugal-Coriolis forces (interaction), 361
Centrifugal loading, 41-42, 361
Checking with leads, 175
Chemical contamination/deposits, 45-52, 60, 65, 138, 637-638
- removal, 637-638

Chemical deposit removal, 637-638
Chemical reaction, 45-52
Chipping mechanism, 427, 448
Chord/thickness (vane), 232-233
Classification (deposit), 627-632
Classification (impact damage), 409-426
- deformed vane inlet edges, 416
- elements trapped between rows, 425
- large surface craters, 417-420
- massive craters/damage, 420-422
- material ruptures, 421-423
- peening, 411-416
- vane damage/damage assessment, 424-426

Cleaning material at outages, 612
Cleaning methods, 639-644
- blast cleaning, 639-643
- hand cleaning, 643
- on-line water cleaning, 643-644

Clearance (seal), 79
Cold alignment, 196
Cold stationary clearances, 119, 514
Collection/drainage (water), 305

Combined factors (radial growth), 147-149
Combined stress (vibration), 532-537
  example problem, 534-535
  solution, 535-537
Compliance (design requirements), 85-90
Component adjustment, 170-173
  means, 170-172
  process, 172-173
Component adjustment means, 170-172
  diaphragm construction, 170-171
  inserted blade design, 171
  other stationary components, 172
Component alignment/stage spatial requirements, 85-282
  blade vane/cascade geometry, 229-248
  diaphragm deflection (pressure/temperature), 159-165
  differential expansion, 108-145
  factors affecting design clearance, 93-94, 175-180
  field alignment methods, 167-175
  radial expansion, 145-158
  rotor vertical deflection, 95-107
  stage operating definition, 190-192
  stage terminology, 90-93
  steam path area requirements, 188-190
  steam path arrangement, 192-228
  unit lateral alignment, 166
  vane placement errors, 248-281
Component arrangement (steam path), xiii, 192-228
  diaphragm/stationary blade assembly, 212-217
  discharge annulus form, 206-211
  rotor machining/assembly (axial), 197-203
  rotor machining/assembly (radial), 204-206
  rotating portion (circumferential), 211-213
  seal configuration, 222, 224-228
  seal geometry/arrangement, 220-224
  seal production/configuration, 218-219
  stage details, 218
Component frequency variation, 575

Component operation (high temperature), 600-603
material properties, 601
number of cycles, 602-603
temperature change rate, 601-602
Component rubbing/misalignment, 32, 394, 462-477
Components operation (high temperature), 600-603
Component susceptibility (deterioration), 52-56
categorization, 55-56
high, 53
intermediate, 54
low, 54
Component/unit speed variation, 575-576
Compressive stress, 568-569
Condensation, 284-290
Condenser pressure, 28, 351-352
backpressure, 28
Consequences (radial growth), 154-158
lifting coverband, 154-155
lifting coverband segment, 155-157
slanted outer sidewalls, 156, 158
Contaminant deposition (steam path), 607-705
cleaning methods, 639-644
contaminant/impurity sources, 610-619
composition of deposits, 619-637
corrosion, 671-677
corrosion forms, 677-703
deposition patterns, 644-658
efficiency deterioration, 659-670
removal of chemical deposits, 637-638
Contaminant/impurity sources, 45-52, 60, 65, 138, 610-619
attemperating spray water, 611
cleaning material at outages, 612
dissolved from system equipment, 613
exhaust hood spray water, 611-612
external sources, 611
geothermal steam, 614-618
internal sources, 611
manufacturing/repair lubricants, 612-613

shipping/storage preservatives, 613-614
sodium sulfide (oxygen scavenger), 612
water quality, 617-619
Contamination (chemical), 45-52, 60, 65, 138
Control stage nozzles, 443-444
Cooling stages, 462
Copper oxide, 628
Corrective actions, 2-4, 23, 57-59, 67-69
Corrective options (maintenance), xii, 19-24
accept as-is, 23-24
nonconformance (field-found), 19
repair, 21-22
rework/refurbish, 22-23
scrap/replace, 21
Corrodent concentration, 702
Corrosion concentration, 45, 80, 674-677
blade root-rotor interfaces, 676-677
coverband tenon holes, 674-675
shrink-fits, 675-676
tie wire holes, 674-675
Corrosion damage, 79-80
Corrosion effects, 45-48, 50-51
corrosion fatigue, 47-48
corrosive pitting, 47
fretting corrosion, 50-51
stress corrosion cracking, 46-48
Corrosion fatigue, 47-48, 696, 698-702
Corrosion forms, 677-703
corrosion mechanisms, 683-702
crack growth rate factors, 700-703
impurity effect (blade failures), 681-683
Corrosion hideouts, 674-677
blade root-rotor interfaces, 676-677
coverband tenon holes, 674-675
shrink-fits, 675-676
tie wire holes, 674-675
Corrosion mechanisms, 683-702
corrosion fatigue, 696, 698-702
crevice corrosion, 683
inter-granular corrosion, 690-692

pitting corrosion, 684-688
stress corrosion cracking, 688-690
trans-granular corrosion, 691-698
Corrosion (steam path), 671-703
forms, 677-703
hideouts, 674-677
Corrosive agents, 671-677
Corrosive attack, 638
Corrosive environment modification, 632-633
Corrosive pitting, 47
Coverband deterioration, 76
Coverband distortion, 256-257
tenon hole misplaced, 257
tenon position incorrect, 257
vane pitching, 257
Coverband effect, 267
Coverband erosion, 341-342
Coverband lifting, 154-157
Coverbands, 76, 154-157, 256-257, 267, 341-342, 455-456, 469-472, 481-482, 674-675
deterioration, 76
distortion, 256-257
effect, 267
erosion, 341-342
lifting, 154-157
segment interface, 481-482
tenon holes, 674-675
Coverband segment interface, 481-482
Coverband tenon holes, 674-675
Crack form, 577-578
Crack growth, 77-78, 576-578, 700-703
crack form, 577-578
growth rate factors, 700-703
measurement, 77-78
prediction, 578
Crack growth measurement, 77-78
Crack growth prediction, 578
alternating stress magnitude, 578
applied load frequency, 578

direct stress magnitude, 578
material properties, 578
material temperature, 578
Crack growth rate factors, 700-703
corrodent concentration, 702
environmental temperature, 703
stress levels, 700-702
Cracking (fatigue), 39-41, 497-501
Cracking (stress corrosion), 46-48
Crack tip shape, 577
Creep deflection, 163-164, 511-518
Creep deformation, 41-42, 94, 162-165, 488-496
characteristic curve families, 495-496
deflection, 163-164
primary phase, 494
secondary phase, 494
tertiary phase, 494-495
Creep fatigue, 511, 581-582
Creep (high temperature), 490-493
Creep mechanism, 497-504
creep progression, 503-504
grain/boundary layer relative strength, 501-502
slip planes, 502-503
Creep progression, 503-504
Creep rate, 504-506
prediction, 506
Creep (steam path components), 507-520
casings/blade carriers, 518-520
diaphragms, 511-518
rotating blades, 508-509
rotors, 510-511
threaded components, 520
Creep strength, 165
Crevice corrosion, 683
Curved suction face, 260-261
Cyclic stress, 48

D

Damage classification, 409-426
Damage/deterioration mechanisms, 35-52, 79-81
  corrosion effects, 45-48, 79-80
  creep deformation, 41-42
  fretting corrosion, 50-51, 80-81
  high-cycle fatigue, 37-39
  low-cycle fatigue, 39-40
  moisture effects, 48-50
  solid-particle erosion, 41-46
  transient operating conditions, 50
Damage/deterioration monitoring, 66-82, 368-369
  mechanisms, 69
  techniques, 70-82
Damage/deterioration signals, 7-11
  enthalpy drop test, 10-11
  extraction pressure changes, 11
  output increase, 10
  output vs. input, 9-10
  noise, 8-9
  vibration, 8
Damage evaluation process, 35-37
Damage/failure, 57-59
  parts, 57
  power, 57-58
  time, 58-59
Damage forms (water ingestion), 391-394
  erosive damage, 393
  impact-type damage, 391-392
  quenching, 393-394
  steam path rubs, 394
Damage mechanisms, 283-705
  contaminant deposition, 607-705
  material property degradation, 397-605
  operational events, 397-486
  water-induced, 283-396
Damage penetration, 349

Damping, 566-567
dry friction, 567
ferromagnetic action, 567
fluid action, 567
internal friction, 566-567
Debris disintegration, 402, 408
Debris from boiler, 402, 405
Debris from external source, 402, 405
Debris impact damage, 398-401
Debris size, 398-399, 403-406
Debris transport, 402, 413-414
Decarburization, 566
Defining surface finish, 652-654
Deflection form (rotor), 94-107
Deformation (creep), 41-42, 94, 162-165
deflection, 163-164
Degraded blades operation, 410
Demineralization, 628
Deposit composition, 619-637
classification, 627-632
steam-corrosive environment modification, 632-633
steam path condition changes, 633-636
unit shutdown, 633, 636-637
Deposition patterns (contaminant), 644-658
defining surface finish, 652-654
metal surface finish specifications, 654-658
Deposits (on components), 65, 138
Design clearance factors, 93-94,175-188
axial deformation (stationary blades), 186
casing boring after adjustment, 182-183
casing humping (hogging), 180
casing ovality (bowing), 176-180
diaphragm concentricity, 186-188
diaphragm dishing, 183-185
horizontal joint leakage, 180-182
rotor bends, 188
Design error, 25-29
Design specifications, xiii, 25-29, 85-90
error, 25-29,
requirements, 85-90

Deterioration factors, 59-66
carry-over considerations, 66
chemical contamination, 65
material property, 64-65
operating environment, 60-61
stress level, 63
Deterioration susceptibility, 52-56
categorization, 55-56
high, 53
intermediate, 54
low, 54
Detachment of components, 413, 419
Diagnostic tools/techniques, xii
Diaphragm concentricity, 186-188
Diaphragm construction, 170-171
Diaphragm deflection, 159-165, 511-518
creep deformation, 162-165
high pressure diaphragms, 159-161
high temperature diaphragms, 159-161
stationary blades, 161-162
Diaphragm dishing, 183-185, 473
Diaphragm erosion, 383-386
Diaphragm horizontal joints, 374-375
Diaphragm inner webs, 473-474
Diaphragm leakage, 595
Diaphragms, 79, 91, 159-165, 170-171, 183-188, 196, 212-217, 225-226, 374-375, 383-386, 471-474, 511-518, 595
concentricity, 186-188
construction, 170-171
definition, 91
deflection, 159-165, 511-518
dishing, 183-185, 473
erosion, 383-386
horizontal joints, 374-375
inner webs, 473-474
leakage, 595
stationary blade assembly, 212-217, 225-226
vane cracks, 79
vanes, 79, 471-473

Diaphragm/stationary blade assembly, 212-217, 225-226
Diaphragm vanes, 79, 471-473
    cracks, 79
Differential axial movement, 94
Differential expansion, 108-145, 223-225
    pressure valance pistons, 140-143
    thrust bearing failure, 143-144
    thrust reversal effects, 128-140
    transient operating conditions, 119-128
    with flexible couplings, 144-145
Differential heat rates, 14
Dimensional audit, 268-271
Direct/alternating stress combination, 569-570
    Gerber diagram, 570
    Goodman diagram, 570
Direct stress, 569-570, 578
    magnitude, 578
Discharge angle, 240-241
Discharge annulus form, 206-211
Discharge area, 247, 277
    angle, 277
Discharge edge, 453-454
Discharge tail, 234-236, 294-298
Discs/wheels temperature, 594-596
Dishing (diaphragm), 183-185, 473
Dissolved contaminants, 613
Distribution voltage/power factors adjustment, 12
Drag force, 287
Driving mechanism(s), 35-37
Drop-ins (debris), 409
Dry friction, 567

E

Economic evaluation, 5
Efficiency deterioration/changes, 608, 635-636, 659-670
    mean effective surface finish, 660-661
    rotating blade row (impulse), 663

stationary and rotating rows (reaction), 663-665
stationary blade row (impulse), 661-662
surface deterioration audit, 668-670
surface finish requirements, 665-667
Elliptical diaphragm, 473
Emergency, 3-4, 57-59, 124-125, 143, 152, 475, 590-591
conditions, 124-125, 152, 590-591
overspeed, 125, 152, 475, 590-591
stop valves, 143
Emergency conditions, 124-125, 152, 590-591
Emergency overspeed, 125, 152, 475, 590-591
Emergency stop valves, 143
Emissivity, 601-602
Endurance limit (S-N curve), 563-565
Enthalpy drop test, 10-11
Environmental temperature, 572-575, 703
change, 572-575
Erosion, 41-46, 49-52, 70-77, 81-82, 325-334, 344-349, 367, 393, 431-439
classification, 347-349
mechanisms, 325-330, 431-439
moisture-impact, 49-50, 70-73
penetration, 71-73, 344-347, 367
resistant inlet edge, 331-334
solid-particle, 41-46, 74-77
time-dependent, 344-347, 367
trailing-edge, 49, 51, 73-74
washing, 49, 51, 81-82
wire drawing, 49, 52
worming, 49, 52
Erosion classification, 347-349
Erosion influence (stage performance), 460-462
area change effects, 460-461
cooling stages, 462
increased area/local pressure, 461
steam discharge angles, 461
Erosion/material loss mechanisms, 325-330, 431-439
operational events, 431-439
rotating blade row, 432-439
stationary blade row, 431-432

Erosion penetration, 71-73, 344-347, 367
time dependent, 367
Erosion protective measures, 459-460
boiler cleaning/operating changes, 459
protective coatings, 460
stage geometry changes, 459
turbine start-up procedures, 460
Erosion-resistant inlet edge, 331-334
braze-attached resistant material, 331-333
laser hardening, 334
thermal hardening, 333-334
weld-attached resistant material, 333
Erosion-time relationship, 344-347, 367
Eutectic form, 489
Excessive overspeed, 475
Excessive windage, 590
Exfoliation (scale), 53, 65, 74-77
Exhaust hood spray water, 611-612
Exhaust steam conditions, 317-318
Expansion passage form, 244-245
External debris, 405, 409
drop-ins, 409
External removal (water), 311-312
External sources (contamination), 611
Extraction locations, 143
Extraction pressure changes, 11

F

Factors affecting design clearance, 93-94, 175-188
Failure causes/mechanisms, 25-52
causes, 25-34
failure examination, 34-36
mechanisms involved, 35, 37
mechanisms of damage/deterioration, 37-52
Failure/damage, xvi, 25-52, 57-59, 68-69
causes/mechanisms, 25-52
definition, xvi

examination, 34-36
parts, 57
power, 57-58
time, 58-59
Failure probability, 410
Fatigue, 37-40, 47-48, 564-576
corrosion, 47-48
high-cycle, 37-39
low-cycle, 39-40
strength factors, 564-569
stresses, 569-576
Fatigue strength factors, 564-569
damping, 566-567
decarburization, 566
heat treatment, 566
residual stress, 567-569
shot peening, 568
stress concentration, 565
surface marks, 565
residual compressive stress, 568-569
temperature, 565-566
Fatigue stresses, 569-576
combined direct/alternating stresses, 569-570
operational modification effects, 570-576
Feed water heaters, 28
Ferromagnetic action, 567
Field alignment methods, 167-175
alignment (casing-to-casing), 173-174
checking with leads, 175
component adjustment, 170-173
stationary component, 167-169
Finite element analysis, 569-570
Flexible couplings, 144-145
Fluid action (damping), 567
Forced outage, 11-12, 18, 25
Force magnitude, 577
Foreign object impact damage, 398-426
classification of damage, 409-426
sources of objects, 401-409

Fox-holed tenon, 343
Fracture rings, 329-330
Frequency bands unacceptable, 558-561
Frequency distribution/tuning, 556-558
Frequency tolerances (manufacturer's), 554-556
Fretting corrosion, 50-51, 477-485
  coverband segment interface, 481-482
  root interfaces, 482, 485
  root load-bearing surfaces, 482-484
  tie wire hole interface, 478-480
Fretting damage, 50-51, 80-81, 477-485
  corrosion, 50-51, 477-485
Friction, 566-567, 652
  surface loss, 652
Fuel costs, 14

G

Geothermal cycle, 283, 317-318, 615-617
Geothermal steam, 614-618
Gerber's parabola/diagram, 532-534, 570, 572-573
Gland ring, 379
Glass bead, 640
Goodman diagram, 570-576
Goodman linear relationship (modified), 533-534, 536
Gouging, 352-356, 427, 448
Gradual deterioration, 59-66, 68-69
  carry-over considerations, 66
  chemical contamination, 65
  material property, 64-65
  operating environment, 60-61
  stress level, 63
Grain boundary, 501-502, 689
  layer (relative strength), 501-502
Grain/boundary layer, 501-502
Grid system security, 410, 415
Grit blasting, 640
Gullying angle, 364-366

H

Hand cleaning, 643
Hardening, 333-334
Heat excess, 50-51
Heat exchanger, 616-617
Heat rate, 585
Heat transfer, 285-286
Heat treatment, 566
Heat-affected zone, 566
Hi-lo configuration, 221-224
    simple, 221-223
Hideouts (corrosion), 80, 674-677
High-cycle fatigue, 37-39, 477-478, 484, 488, 520-525, 576, 579-581, 583-584
    failure, 579-581
    machining/assembly marks, 521
    material deficiencies, 521
    phenomena, 522-525
    section changes, 521
High-cycle fatigue failure, 579-581
High-cycle phenomena, 522-525
High pressure diaphragms, 159-161
High pressure two-phase flow, 370
High susceptibility, 53
High temperature, 159-161, 487-488, 490-497, 582-583, 600-603
    components, 600-603
    creep, 490-493
    diaphragms, 159-161
High temperature components, 600-603
High temperature creep, 490-493
    operating stress level, 492
    operating temperature, 492
High temperature diaphragms, 159-161
Hogging, 180, 475
Horizontal joint diameter, 177-180
Horizontal joint leakage, 180-182

Horizontal joints, 177-182, 373-375
    diameter, 177-180
    leakage, 180-182
Hot running clearance, 514
Hot well water, 104
Humping, 180, 475
Hydrogen sulfide, 627
Hysteresis loop, 596

I

Impact damage classification, 409-426
    deformed vane inlet edges, 416
    elements trapped between rows, 425
    large surface craters, 417-420
    massive craters/damage, 420-422
    material ruptures, 421-423
    peening, 411-416
    vane damage/assessment, 424-426
Impact damage (steam path), 326, 391-392, 398-426
    erosion, 326
    classification, 409-426
    object sources, 401-409
    pit/crater, 399, 417, 420-421
Impact erosion, 326
Impacting object sources, 401-409
    debris from boiler, 402, 405
    debris from external source, 402, 405
    drop-ins, 409
    external to steam path, 405
    formed in boiler, 405
    formed internally, 403-405
    mechanical failure result, 401-402
    parts left during outage, 403
Impact pit/crater, 399, 417, 420-421
Impurity effect on blade failures, 681-683
Impurity/inclusion (metal), 489-490, 499

Ingestion of water. *SEE Water ingestion.*
Initiating mechanism(s), 35-37
Inlet angle, 239-240
Inlet noise, 233-235
Inlet/discharge edge, 243
Inner web, 91
Inserted blade design, 171
Inspection/maintenance outage, 17-18
Instantaneous damage/failure, 57-59, 68-69
    parts, 57
    power, 57-58
    time, 58-59
Integral coverband form, 155-156
Inter-granular corrosion, 690-692
Intermediate susceptibility, 54
Internal collection/control (water), 305-311
Internal debris, 403-405
    boiler, 405
Internal friction, 566-567
Internal sources (contamination), 611
Iron oxide, 628
Irregular surface, 380, 400, 654-657, 669-670

K

Kinetic energy, 283
Known problems (unit), 15

L

Labor costs, 14
Labyrinth seal, 220-221
Laser hardening, 334
Laser optical alignment method, 174
Laser probe, 301-304

Lateral alignment, 166
Leakage location potential, 373-379
    between blade root blocks, 376
    diaphragm gland ring locating slots, 379
    diaphragm horizontal joints, 374-375
    diaphragm-inner casing steam seal faces, 377-379
    split blades, 374-375
    turbine casing horizontal joints, 373-374
Leakage path, 370-379, 595
Life cycle basis, 14-15
Life expenditure indices, 61
Lifting coverband, 154-157
    segment, 155-157
Lifting coverband segment, 155-157
    center of gravity position, 156
    integral coverband form, 155-156
    machining thickness, 156
Light-scattering technique, 302
Load changes, 588-589
Load factor, 14
Load frequency, 577-578
Load rejection, 151-152, 589-591
    testing, 589-590
Load/stress, 487-488, 492, 497-501
Locator key positions/adjustments, 170-172
Lodging of debris, 402, 424
Logic review process, 20, 35-36
    diagram, 35-36
Long/hot rotor, 119-124
Low-cycle fatigue, 39-40, 488, 583-586
    thermal, 583-586
Low-cycle thermal fatigue, 583-586
Low susceptibility, 54
Lubricants, 612-613

M

Machining thickness, 156
Machining/assembly marks, 521
Magnetite scale, 428-429
Maintenance actions, xxi-xxiv
  nonconformance, xxi-xxii
Maintenance outage, 13-18
  equivalent operating hours, 16-17
  inspection, 17-18
  intervals, 13-17
Maintenance outage intervals, 13-17
  equivalent operating hours, 16-17
Maintenance period anticipated, 14
Maintenance policy, 2-3
Maintenance strategy, 1-84
  component susceptibility, 52-56
  considerations, 2-5
  corrective options, 19-24
  failure causes/mechanisms, 25-52
  gradual deterioration, 59-66
  inspection/maintenance outage, 17-18
  instantaneous damage/failure, 57-59
  maintenance outage intervals, 13-17
  monitoring damage/deterioration, 66-82
  outage scheduling, 12-13
  replacement parts, 82-83
  turbine outage, 6-7
  turbine shutdown, 7-12
Maintenance/repair options, xx-xxi, 19-24
  accept as-is, 23-24
  nonconformance (field-found), 19
  repair, 21-22
  rework/refurbish, 22-23
  scrap/replace, 21
Manner of operation, 15
Manufacturer's frequency tolerances, 554-556
Manufacturer's recommendations, 15

Manufacturing error, 29-30
Manufacturing method, 268
Manufacturing/repair lubricants, 612-613
Martensitic steel, 448-449
Material data availability, 562-563
Material deficiencies, 521
Material degradation, 487-605
Material loss, 41-46, 81-82, 208-210, 361-366, 439-459
 model, 361-366
 patterns, 439-459
Material loss model, 361-366
Material loss patterns, 439-459
 associated stage hardware, 455-459
 control stage nozzles, 443-444
 rotating blade rows, 450-454
 stationary blade rows, 444-450
 turbine valve system, 441-443
Material properties, 60, 64-65, 561-569, 578, 601
 factors affecting fatigue strength, 564-569
 material data availability, 562-563
 S-N curve (endurance limit), 563-564
 thermal, 601
Material ruptures, 421-423
Material structure considerations, 489-490
Material temperature, 576, 578
Mean effective surface finish, 660-661
Mechanical failure, 401-402
Mechanisms/causes (failure), 25-52
 design, 25-29
 failure examination, 34-36
 installation quality, 32-33
 maintenance procedures, 32-34
 manufacture quality, 29-30
 mechanisms involved, 35-52
 operating practices, 32-33
Mechanisms (damage/deterioration), 35-52
 corrosion effects, 45-48
 creep deformation, 41-42
 fretting corrosion, 50-51
 high-cycle fatigue, 37-39

low-cycle fatigue, 39-40
moisture effects, 48-50
solid-particle erosion, 41-46
transient operating conditions, 50
Melt crystallization, 64-65, 489-490, 689
metallurgy, 64-65
Metal surface finish specifications, 654-658
arithmetic average, 656
centerline average, 656-657
root mean square value, 656
surface specification comparisons, 658
Metallurgy, 64-65, 489-490, 654-658, 689
Moisture collection/drainage, 305-311, 331
Moisture content effects, 48-50, 70-73, 320-322
moisture-impact erosion, 49-50, 70-73
trailing-edge erosion, 49, 51
washing erosion, 49, 51
water ingestion, 49
wire drawing erosion, 49, 52
worming, 49, 52
Moisture deposition, 292-301
on blade vanes, 292-298
on sidewalls, 298-301
water flow along casing, 301
Moisture-impact damage, 322-323, 361
Moisture-impact erosion, 49-50, 70-73, 322-352, 361
condenser pressure, 351-352
erosion/material loss mechanism, 325-330
erosion penetration and time, 344-347
erosion severity classification, 347-349
off-shield erosion, 336-344
last stage blade erosion, 351-352
monitoring damage, 350
protection methods, 331-334
secondary/concentrated erosion, 335-336
Moisture-induced damage, 322-323, 361
moisture-impact erosion, 322
trailing-edge erosion, 322
water washing erosion, 323
wire-drawing erosion, 323

Moisture measurement, 301-304
Moisture removal, 318-320, 331
effects, 318-320
Monitoring (damage/deterioration), 55-56, 66-82, 368
mechanisms, 69
techniques, 70-82
Monitoring techniques, 70-82
crack indications (blade vane), 77-78
corrosive damage, 79-80
diaphragm vane cracks, 79
fretting damage, 80-81
moisture-impact erosion, 70-73
seal system wear, 79
solid-particle erosion, 74-77
trailing-edge erosion, 73-74
water washing erosion, 81-82
Moving in (deformation), 518-519

N

Noise, 8-9
Nonconformance/nonconforming situation, xiii, xxi-xxii, 2, 18-19, 191
accept as-is, xxii-xxiii
field-found, 19
repair, xxii-xxiii
rework, xxii-xxiii
scrap/replace, xxii-xxiii
Nondestructive testing/examination, 17, 639, 652
Normal operating hours, 17
Normal overspeed, 152
Nose deformation, 400-401
Nozzle distance from surface, 641
Nozzle impulse effects, 527-528
Nozzle passing effect (frequency), 528-532
Nozzle size/shape, 641
Nuclear cycle, 283, 381-382
Nucleation, 288

O

Off-design conditions, 28
Off-shield erosion, 336-344
    below-shield, 338-339
    between shield segments, 339-340
    beyond-shield, 337-338
    coverband, 341-342
    pressure surface, 341
    squealer tips, 343-344
    tenon, 342-343
On-line water cleaning, 643-644
Opening time, 13-16
Operating condition transients, 1-2, 50, 60-61, 151-154
    effects, 151-154
Operating environment, xiv-xv, xviii, 1, 50, 60-61
    transient conditions, 50
Operating stress level, 492
Operating temperature, 492
Operating transients effects, 151-154
    load rejection, 151-152
    overspeed transients, 152-153
    temperature up-ramps, 153
    temperature down-ramps, 153
    water injection, 154
Operational events damage, 397-486
    component rubbing, 462-477
    erosion influence (stage performance), 460-462
    erosion mechanisms, 431-439
    foreign object impact damage, 398-426
    fretting corrosion, 477-485
    impact damage classification, 409-426
    impacting object sources, 401-409
    material loss patterns, 439-459
    protective measures against erosion, 459-460
    scale formation, 428-430
    solid-particle erosion (abrasion), 426-462

Operational modifications (effects), 570-576
component frequency variation, 575
component/unit speed variation, 575-576
environmental temperature change, 572-575
fatigue stress, 570-576
residual stress, 571-572
stress concentration, 570-571
Outage intervals (maintenance), 13-17
equivalent operating hours, 16-17
Outage rate (unit), 14
Outage scheduling, 12-13
Outage (turbine), 6-7, 12-17
extended, 7
forced, 7
intervals for maintenance, 13-17
maintenance, 6
outage rate (unit), 14
planned, 6
scheduling, 12-13
Outer ring, 91
Output increase, 10
Output vs. input, 9-10
Ovality (casing), 176-180
Overspeed testing, 28
Overspeed transients, 152-153
Oxide scale, 428-429
Oxidizing/oxidation damage, 478
Oxygen access, 610
Oxygen scavenger, 612

P

Partial admission effect, 537-538
Parts damage/failure, 57
Parts left during outage, 403
Peening, 407, 411-416, 568
Performance potential (unit), 59-60
Piano wire alignment method, 173-174

Pilot valve, 429-430
Piston diameter, 142
Piston thrust, 129-135
Pitch, 211-212, 243, 246, 257-268
error, 258-268
error vs. height, 267
Pitch error, 258-268
casting, 268
coverband effect, 267
curved suction face, 260-261
manufacturing method, 268
pitch error vs. height, 267
root block thickness (radial entry roots), 262
root block thickness (tangential entry roots), 261
root block wedge angle, 262-265
setting angle, 265
straight suction face, 259-260
vane lean, 266-267
vane position (root platform), 265
vane untwisting, 268
welding, 268
Pitting corrosion, 47, 626, 684-688
Placement error (vane), 249-258
Planned outage, 11, 25
delay, 11
Plastic deformation *(SEE Creep deformation)*
Positive/negative deviation (vane placement), 241-242
Power damage/failure, 57-58
Power factors/distribution voltage adjustment, 12
Preservatives, 613-614
Pressure balance pistons, 140-143
Pressure changes (extraction), 11
Pressure differential, 370
Pressure point connection, 142
Pressure/suction faces, 237-238
Pressure surface erosion, 341
Pressure/temperature deflection, 159-165
Pressure valance holes, 143
Pressure valance pistons, 143
Profile (airfoil), 90

Profile shape (vane), 229-231
Protection (moisture-impact erosion), 331-334
  erosion-resistant inlet edge, 331-334
  moisture removal, 331
Protective coatings, 460
Protective measures (erosion), 459-460
  coatings, 460
Proud blades, 368
Purchasing power, 11-12

Q

Quality control, 29-30
Quenching, 393-394
  casings, 393-394
  rotors, 393

R

Radial creep deformation, 516-517
Radial distribution (moisture), 289-292
Radial expansion (steam path parts), 145-158
  combined factors (radial growth), 147-149
  operating transients effects, 151-154
  radial growth consequences, 154-158
  radial stress effects, 146-147
  stationary blade locating method, 150-151
  thermal effects, 145-147
Radial flow, 298-301, 306
Radial growth consequences, 154-158
  lifting coverband, 154-155
  lifting coverband segment, 155-157
  slanted outer sidewalls, 156, 158
Radial growth/expansion, 94, 145-158
  consequences, 154-158
  factors, 147-149

Radial growth factors, 147-149, 154-158
consequences, 154-158
Radial plane, 435-438
Radial seals, 218-219, 457-459
radial seal strips, 457-459
Radial seal strips, 457-459
Radial stress effects, 146-147
Rankine cycle, x
Rate of startup/load/shutdown, 28
Refurbish/rework options, xxii-xxiii, 4-5, 22-23, 56-59
cost, 23
Regions of tolerances, 196
Reheat temperature changes, 635
Reliability factors, xiv-xv
Removal (chemical deposits), 637-638
corrosive attack considerations, 638
Removal (parts), 5
Repair cost, 23, 410
Repair options, xxii-xxiii, 4-5, 21-23, 57-59, 410
cost, 23, 410
Replacement availability, 24
Replacement cost, 14, 23-24, 410
Replacement parts (strategy/supply), 82-83
Replace options, xxii-xxiii, 4, 14, 21, 23-24, 82-83, 410
availability, 24
cost, 14, 23-24, 410
parts strategy/supply, 82-83
Reserve capacity, 14
Residual stress, 517, 567-569, 571-572
Reversing flow, 143
Rework/refurbish options, xxii-xxiii, 4-5, 22-23, 56-59
cost, 23
Reynolds number, 608-609
Risk/risk analysis, 14, 18
Root block thickness, 261-262
Root block twist, 272
Root block wedge angle, 262-265

Root blocks, 261-265, 272, 376
thickness, 261-262
twist, 272
wedge angle, 262-265
Root/fastening, 90
Root form position, 252-254
Root interfaces, 482, 485
Root load-bearing surfaces, 482-484
Root mean square value, 656
Root platform, 90, 249-254
Root section recirculation, 356-358
Rotating blade row, 380-382, 388, 432-439, 450-454, 663-665
axial gap, 438-439
axial/tangential planes, 433-435
caving, 452-453
discharge edge, 453-454
impulse, 663-665
radial plane, 435-438
vane inlet noise, 450-452
Rotating blades, 2, 41-46, 53, 75-76, 90, 226-228, 243, 298, 306, 326-328, 380-382, 388, 415-416, 423-424, 426, 432-439, 450-454, 467-468, 508-509, 525-561, 595-596, 663-665
blade row, 380-382, 388, 432-439, 450-454, 663-665
tips, 226-228
vibration, 525-561
Rotating blade tips, 226-228
Rotating blade vibration/vibratory stresses, 525-561
actual steam force diagram, 539-541
Campbell diagram, 553-561
combined stress, 532-537
frequency bands unacceptable, 558-561
nozzle impulse effects, 527-528
nozzle passing effect (frequency), 528-532
partial admission effect, 537-538
rotational stiffening effect, 541-552
Rotating portion (circumferential arrangement), 211-213
axial gap, 212-213
vane axial placement, 212
vane pitch, 212
vane setting angle, 212

Rotational stiffening effect, 541-552
  beta factor, 542-546
  tolerance band, 547-552
Rotor bends, 188
Rotor diameter, 143
Rotor grabbing, 180
Rotor machining/assembly, 196-206
Rotor mass (loss), 465
Rotors, 94-107, 143, 180, 188, 196-206, 393, 465-467, 510-511, 593-594
  bends, 188
  diameter, 143
  grabbing, 180
  machining/assembly, 196-206
  rotor mass loss, 465
  temperature distribution, 593-594
  vertical deflection, 94-107
Rotor temperature distribution, 593-594
Rotor vertical deflection, 94-107
  sideways shift, 106-107
  vertical movement factors, 103-106
Rubs/rubbing (component misalignment), 32, 394, 462-477
  casing inner surface, 474-477
  coverbands, 469-472
  rotating blades, 467-468
  rotors, 465-467
  diaphragm inner webs, 473-474
  diaphragm vanes, 471-473

S

Sacrificial erosion, 368
Safe frequency range, 547-552
Salt content, 611, 618
Saturated steam cycles, 315-317
Scale deposit, 53, 65, 74-77
Scale formation/erosion, 426-459

Scheduled outage, 11-13
    delay, 11
    scheduling of, 12-13
Scrap options, xxii-xxiii, 4, 14, 21, 23-24, 82-83
Seal castellation, 47
Seal clearance, 79
Seal configuration selection, 222, 224-228
    diaphragms/stationary blade rows, 225-226
    rotating blade tips, 226-228
    shaft-end seal positions, 222, 225
Seal geometry/arrangement, 220-224
    hi-lo configuration, 221-224
    straight through, 220-221
Seal monitoring, 55-56
Seal production/adjustment, 196, 218-219
Seal production/configuration, 218-219, 222, 224-228
Seal production cost, 225
Seals, 47, 55-56, 79, 196, 218-228
    castellation, 47
    clearance, 79
    cost, 225
    geometry/arrangement, 220-224
    monitoring, 55-56
    production/adjustment, 196, 218-219
    production/configuration, 218-219, 222, 224-228
    system wear, 79
Seal system wear, 79
Secondary erosion, 335-336
Section changes, 521
Setting angle, 236-237, 265, 271-276
    error, 271-276
    vane, 236-237
Setting angle error, 271-276
    blade untwist, 272-275
    root block twist, 272
    vane lean, 275-276
Shaft-end seal positions, 222, 225
Shear stress, 567-568
Shield attachment, 62
Shipping/storage preservatives, 613-614

Short/cold rotor, 119, 124-128
Shot peening, 568
Shrink-fits, 675-676
Shutdown option, 6-7, 12-13
Shutdown thermal transients, 587-588
Shutdown (turbine), 6-13, 587-588, 633
    damage/deterioration signals, 7-11
    option, 6-7, 12-13
    system requirements, 11-12
    thermal transients, 587-588
Sidewall material loss, 461
Sidewall moisture, 298-301
Sidewall slant, 156, 158
Sideways shift, 106-107
Signals (damage/deterioration), 7-11
    enthalpy drop test, 10-11
    extraction pressure changes, 11
    noise, 8-9
    output increase, 10
    output vs. input, 9-10
    vibration, 8
Silicate, 628-629
Silicon carbide, 640
Simple hi-lo configuration, 221-223
Situation evaluation, xix-xx
    maintenance/repair disposition, xx
Slanted outer sidewalls, 156, 158
Slip planes, 502-503
Slippage/shear, 502-503
S-N curve (endurance limit), 563-564
Sodium chloride, 620-621
Sodium hydroxide, 620-622, 628
Sodium sulfate, 628
Sodium sulfide, 612, 627
Solid-particle erosion (abrasion), 41-46, 66, 74-77, 426-462
    erosion mechanisms, 431-439
    influence on stage performance, 460-462
    material loss patterns, 439-459
    protective measures, 459-460
    scale formation, 428-430

Solids control, 66
Solvent cleaning, 643
Spatial requirements, 93
Spinning reserve/stand-by, 11
Split blades, 374-375
Spray jet inclination angle, 641
Spray water, 285, 359, 611-612
Squealer tips erosion, 343-344
Stage details, 218
Stage geometry changes, 459
Stage mechanical damage, 321-322
Stage operating definition, 190-192
Stage performance, 460-462
Stage spatial requirements/component alignment, 85-282
  blade vane/cascade geometry, 229-248
  diaphragm deflection (pressure/temperature), 159-165
  differential expansion, 108-145
  effect of vane placement errors, 248-281
  factors affecting design clearance, 93-94, 175-180
  field alignment methods, 167-175
  radial expansion (steam path parts), 145-158
  rotor vertical deflection, 95-107
  stage operating definition, 190-192
  steam path area requirements, 188-190
  steam path component arrangement, 192-228
  terminology, 90-93
  unit lateral alignment, 166
Stage terminology, 90-93
Standstill condition, 554, 557
Start-up thermal transients, 587
Stationary blade rows, 383-387, 431-432, 444-450, 661-665
  impulse, 661-665
  vane discharge edges, 447-449
  vane inlet nose, 446-447
  vane sidewalls, 449-450
Stationary blades, 41-46, 53, 75-76, 91, 150-151, 159-162, 186, 196, 225-226, 243, 306, 383-387, 421-426, 431-432, 444-450, 661-665
  locating method, 150-151
  rows, 383-387, 431-432, 444-450, 661-665

Stationary components, 91, 167-169, 172
Stationary vanes, 91
Steam bending stress, 281
Steam conditions control, 305, 312-318, 633-636
  exhaust steam conditions, 317-318
  saturated steam cycles, 315-317
  superheated steam cycles, 312-315
Steam-corrosive environment modification, 632-633
Steam discharge angle, 280-281, 461
Steam discharge velocity, 279-280
Steam momentum thrust, 139-140
Steam path area requirements, 188-190
Steam path component arrangement, 188-190, 192-228
  area requirements, 188-190
  diaphragm/stationary blade assembly, 212-217
  discharge annulus form, 206-211
  rotating portion arrangement, 211-213
  rotor machining/assembly (axial), 197-203
  rotor machining/assembly (radial), 204-206
  seal configuration selection, 222, 224-228
  seal geometry/arrangement, 220-224
  seal production/configuration, 218-219
  stage details, 218
Steam path condition changes, 633-636
  efficiency changes, 635-636
  initial steam condition, 634-635
  reheat temperature changes, 635
  unit load changes, 634
Steam path damage/deterioration (contaminant deposition), 607-705
  cleaning methods, 639-644
  composition of deposits, 619-637
  corrosion, 671-677
  corrosion forms, 677-703
  deposition patterns, 644-658
  efficiency deterioration, 659-670
  removal of chemical deposits, 637-638
  sources of contaminant/impurity, 610-619
Steam path damage/deterioration (material degradation), 487-605
  components operation (high temperature), 600-603
  crack growth, 576-578

creep deformation, 493-496
creep fatigue, 581-582
creep in components, 507-520
creep mechanism, 497-504
creep rate, 504-506
fatigue stresses, 569-576
high-cycle fatigue, 520-521
high-cycle fatigue failure, 579-581
high-cycle phenomena, 522-525
high-temperature creep, 490-493
low-cycle thermal fatigue, 583-586
material properties, 561-569
material structure considerations, 489-490
rotating blade vibration, 525-561
temper embrittlement, 582-583
thermal stresses, 596-600
thermal transients, 586-596
Steam path damage/deterioration (contaminant deposition), 607-705
cleaning methods, 639-644
composition of deposits, 619-637
corrosion, 671-677
corrosion forms, 677-703
deposition patterns, 644-658
efficiency deterioration, 659-670
removal of chemical deposits, 637-638
sources of contaminant/impurity, 610-619
Steam path damage/deterioration (material degradation), 487-605
components operation (high temperature), 600-603
crack growth, 576-578
creep deformation, 493-496
creep fatigue, 581-582
creep in components, 507-520
creep mechanism, 497-504
creep rate, 504-506
fatigue stresses, 569-576
high-cycle fatigue, 520-521
high-cycle fatigue failure, 579-581
high-cycle phenomena, 522-525
high-temperature creep, 490-493
low-cycle thermal fatigue, 583-586

material properties, 561-569
material structure considerations, 489-490
rotating blade vibration, 525-561
temper embrittlement, 582-583
thermal stresses, 596-600
thermal transients, 586-596
Steam path damage (operational events), 397-486
component rubbing, 462-477
erosion mechanisms, 431-439
erosion influence (stage performance), 460-462
foreign object impact, 398-426
fretting corrosion, 477-485
impact damage classification, 409-426
impacting objects sources, 401-409
material loss patterns, 439-459
protective measures against erosion, 459-460
scale formation, 428-430
solid-particle erosion (abrasion), 426-462
Steam path damage (water-induced), 283-396
moisture deposition, 292-301
moisture distribution/content (measuring), 301-304
moisture-impact erosion, 323-352
moisture-induced damage, 322-323
radial distribution, 289-292
trailing-edge erosion, 352-369
water condensation, 285-290
water ingestion, 389-394
water removal, 304-322
water washing erosion, 379-389
wire drawing erosion, 370-379
Steam path dimensional audit, 268-271
Steam path parts expansion, 145-158
Steam path rubs, 32, 394, 462-477
Steam pressure rate, 27
Steam purity, 617-618
Steam reheating, 317
Steam seal faces, 377-379

Steam seals, 47, 55-56, 79, 377-379
  castellation, 47
  faces, 377-379
  monitoring, 55-56
  wear, 79
Steam temperature, 28, 574-575, 586-596
  increase, 574-575, 586-596
Steam turbine performance, 92-92
Steam velocity, 284
Stiffening effect, 541-542
Straight suction face, 259-260
Straight through arrangement, 220-221
Strain range, 596-600
Strategy (maintenance), 1-84
  component susceptibility, 52-56
  considerations, 2-5
  corrective options, 19-24
  failure causes/mechanisms, 25-52
  gradual deterioration, 59-66
  inspection/maintenance outage, 17-18
  instantaneous damage/failure, 57-59
  maintenance outage intervals, 13-17
  monitoring damage/deterioration, 66-82
  outage scheduling, 12-13
  replacement parts, 82-83
  turbine outage, 6-7
  turbine shutdown, 7-12
Strategy (replacement parts), 82-83
Streamlines, 293-298, 356-357
Stress amplification factor (Ad), 576
Stress concentration, 1, 565, 570-571, 652
Stress corrosion cracking, 46-48, 688-690
Stress level, 63, 487-488, 492, 497-501, 577-578, 700-702
  crack growth, 700-702
Stress/load, 63, 487-488, 492, 497-501, 577-578, 700-702
Striations/beach marks, 38, 515-516
Structural deterioration, 2
Structural integrity, 608
Superheated steam, 283, 312-315
  cycles, 312-315

Superheated steam cycles, 312-315
Super-saturation, 288-289
Supply (replacement parts), 82-83
Surface craters, 417-422
Surface deterioration audit, 668-670
Surface finish, 652-658, 665-667, 669-670
 establishing requirements, 665-667
 losses, 669-670
Surface finish losses, 669-670
Surface irregularity/roughness, 380, 400, 654-657, 669-670
Surface marks, 565
Surface specification comparisons, 658
Susceptibility (deterioration), 52-56
 categorization, 55-56
 high, 53
 intermediate, 54
 low, 54
Swallowing capacity, 248, 277-281, 516
 steam bending stress, 281
 steam discharge relative angle, 280-281
 steam discharge velocity, 279-280

T

Tangential placement (pitch), 243
Temper embrittlement, 582-583
Temperature change, 153, 574-575, 586-596, 601-603
 change rate, 601-602
 down-ramps, 153
 factors, 565-566
 steam temperature increase, 574-575, 586-596
 up-ramps, 153
Temperature change rate, 601-602
Temperature cycles, 602-603
Temperature distribution, 591-596
 casings, 592-593
 discs/wheels, 594-596
 rotor, 593-594

Temperature down-ramps, 153
Temperature factors, 565-566
Temperature up-ramps, 153
Tenon erosion, 76, 342-343
Tenon hole misplaced, 257
Tenon position incorrect, 257
Tenons, 76, 90, 257, 342-343, 456-457
  erosion, 76, 342-343
  hole misplaced, 257
  position incorrect, 257
Tensile stress, 567-568
Tensile stress/load, 46-47, 497-501, 567-568
Thermal cycling, 40
Thermal effects, 145-147
Thermal fatigue, 583-586
Thermal hardening, 333-334
Thermal stresses (determination), 596-600
Thermal transients, 586-596
  causes, 587-591
  severity, 591
  temperature distribution, 591-596
Thermal transients (causes), 587-591
  boiler excursions, 590
  excessive windage, 590
  load changes, 588-589
  load rejections, 590-591
  unit shutdown, 587-588
  unit start-up, 587
  valve/load rejection testing, 589-590
Threaded components, 179, 520
  binding, 179
Throat, 246-247
  ratio to pitch, 246-247
Throat-pitch ratio, 246-247
Thrust bearing failure, 143-144
Thrust reversal effects, 128-140
  blade annulus thrust, 138-139
  piston thrust, 129-135
  steam momentum thrust, 139-140
  wheel face thrust, 135-138

Tie wire hole, 47, 77-78, 90, 478-480, 674-675
    interface, 478-480
Tie wire hole interface, 478-480
Time damage/failure, 58-59
Tolerance band, 547-552
Tolerance regions, xiii, 196
Trailing-edge erosion, 49, 51, 73-74, 322, 352-369
    damage mechanisms, 359-361
    erosion penetration and time, 367
    material loss model, 361-366
    monitoring damage, 368-369
    proud blades, 368
    root section recirculation, 356-358
    water sprays (exhaust hood), 359
Trans-granular corrosion, 691-698
Transient operating conditions, 50, 60-62, 119-128
    long/hot rotor, 120-124, 126
    short/cold rotor, 124-128
Transport water removal, 305
Trapped elements (between rows), 425
Turbine casing horizontal joints, 373-374
Turbine outage, 6-7
    extended, 7
    forced, 6-7
    maintenance, 6
    planned, 6
Turbine shutdown, 7-12
    damage/deterioration signals, 7-11
    system requirements, 11-12
Turbine start-up procedures, 460
Turbine valve system, 441-443
Turning angle, 241

U

Unbalance/vibration, 466, 474-475
Under-cooling, 289, 291
Undercutting, 41-46, 452-454

Unit heat rate, 14
Unit load changes, 634
Unit shutdown, 587-588, 633, 636-637
Unit speed/swing limitations, 28
Unit start-up, 587

V

Vacuum deflection, 103-104
Valve/load rejection testing, 589-590
Valve testing, 589-590
Vane axial placement, 212
Vane cracks, 77-79
    blade vane, 77-78
    diaphragm vane, 79
Vane damage assessment, 424-426
    rotating blades, 426
    stationary blades, 425-426
Vane discharge edges, 447-449
Vane form requirements, 229-241
    chord/thickness, 232-233
    discharge angle, 240-241
    discharge tail, 234-236
    inlet angle, 239-240
    inlet noise, 233-235
    pressure/suction faces, 237-238
    profile shape, 229-231
    setting angle, 236-237
    turning angle, 241
    vortex form/profile change, 238-240
Vane inlet edge deformation, 416
Vane inlet nose, 446-447, 450-452
Vane lean, 266-267, 275-276
    axial, 276
    tangential, 276
Vane location, 257-258
Vane material loss, 460-461
Vane pitch, 212, 257

Vane placement, 212, 241-244, 248-281
    errors, 248-281
    position, 249-252, 265
Vane placement errors, 248-281
    axial, 249-258
    discharge area/angle, 277
    pitch, 258-268
    setting angle, 271-276
    steam path dimensional audit, 268-271
    swallowing capacity, 277-281
Vane position, 249-252, 265
    root platform, 265
Vane setting angle, 212
Vane sidewalls, 449-450
Vane twist, 245
Vane untwist, 255-256, 268
Vertical movement factors, 103-106
    bearing loading, 105
    bearing oil rise, 103
    bearing spherical seating, 105
    bearing temperature effect, 104-105
    centerline support of casings, 105
    flexibility of bearing supports, 105
    hot well water, 104
    vacuum deflection, 103-104
Vibration/combined stress, 532-537
    example, 534-535
Vibration (rotating blades), 525-561
Vibration signature, 67-68, 545-546
    analysis, 67-68
Vibrations/unbalance, 8, 67-68, 466, 474-475, 525-561
    combined stress, 532-537
    levels, 8, 67-68
    rotating blades, 525-561
    signature/analysis, 67-68, 545-546
Visual inspection, 17
Void formation, 500
Vortex effect, 595
Vortex form/profile change, 238-240

W

Walnut shell, 640
Warranty inspection, 55
Washing/impact material loss, 81-82
Water accumulation, 66
Water catcher belt, 307-309
Water condensation (expanding steam), 285-290
  water formation, 287-289
Water content measurement, 301-304
Water film corrosive ions, 380
Water film gases, 380
Water flow along casing, 301
Water formation, 287-289
  condensation on particles, 287
  condensation on surfaces, 287-288
  spontaneous nucleation, 288
Water-induced damage, 283-396
  blade trailing-edge erosion, 352-369
  moisture deposition, 292-301
  measuring moisture distribution/content, 301-304
  moisture-impact erosion, 323-352
  moisture-induced damage, 322-323
  radial distribution, 289-292
  water condensation, 285-290
  water ingestion, 389-394
  water removal, 304-322
  water-washing erosion, 379-389
  wire-drawing erosion, 370-379
Water ingestion, 49, 154, 389-394, 465
  forms of damage, 391-394
Water injection, 389-394
Water quality, 617-619
Water removal, 304-322
  collection/drainage, 305
  during transport, 305
  external removal, 311-312
  internal collection/control, 305-311

internal moisture removal effects, 318-320
moisture content effects (efficiency), 320-322
steam conditions control, 305, 312-318
Water sprays (exhaust hood), 359
Water-washing erosion, 49, 51, 81-82, 323, 379-389
diaphragms, 383-386
material loss, 81-82
on casings, 386-389
rotating blade row, 380-382, 388
stationary blade rows, 383-387
Wear (seal system), 79
Weld bead debris, 411-413
Weld-attached resistant material, 333
Welding, 268
Wetness probe, 301-304
Wheel face thrust, 135-138
Wire drawing erosion, 49, 52, 323, 370-379
leakage location potential, 373-379
Work hardening, 505
Worming, 49, 52

Y

Young's modulus, 565-566

Z

Zirconium oxide, 641